Communications
in Computer and Information Science　　1713

More information about this series at https://link.springer.com/bookseries/7899

Wenhui Fan · Lin Zhang · Ni Li ·
Xiao Song (Eds.)

Methods and Applications for Modeling and Simulation of Complex Systems

21st Asia Simulation Conference, AsiaSim 2022
Changsha, China, December 9–11, 2022
Proceedings, Part II

 Springer

Editors
Wenhui Fan
Tsinghua University
Beijing, China

Lin Zhang
Beihang University
Beijing, China

Ni Li
Beihang University
Beijing, China

Xiao Song
Beihang University
Beijing, China

ISSN 1865-0929   ISSN 1865-0937 (electronic)
Communications in Computer and Information Science
ISBN 978-981-19-9194-3   ISBN 978-981-19-9195-0 (eBook)
https://doi.org/10.1007/978-981-19-9195-0

This Springer imprint is published by the registered company Springer Nature Singapore Pte Ltd.
The registered company address is: 152 Beach Road, #21-01/04 Gateway East, Singapore 189721, Singapore

Preface

These two volumes contain the papers from the 21st Asia Simulation Conference (AsiaSim 2022), which is an annual simulation conference organized by the ASIASIM societies: CSF (China Simulation Federation), JSST (Japan Society for Simulation Technology), KSS (Korea Society for Simulation), SSAGsg (Society for Simulation and Gaming of Singapore), and MSS (Malaysian Simulation Society). The conference started in the 1980s and is held each year in a different Asian country. This conference provides a forum for scientists, academicians, and professionals from around the world. The purpose of the AsiaSim conference is to provide a forum in Asia for the regional and national simulation societies to promote modelling and simulation in industry, research, and development.

This year AsiaSim was held in Changsha China, together with the 34th China Simulation Conference. Research results on various topics, from modeling and simulation theory to manufacturing, defense, transportation, and general engineering fields, which combine simulation with computer graphics simulations, were shared at AsiaSim 2022. Three reviewers evaluated each contribution. A total of over 200 submissions were received and only 96 papers were accepted and presented in the online and offline oral sessions. The selected papers were finally accepted for this CCIS volume.

We thank the keynote speakers for giving great insights to the attendees. Furthermore, we wish to thank the external reviewers for their time, effort, and timely responses. Also, we thank the Program Committee and Organizing Committee members who made the conference successful. Finally, we thank the participants who participated remotely despite the difficult circumstances.

Due to the Covid pandemic AsiaSim 2022 has been postponed to January 2023

October 2022

Wenhui Fan
Lin Zhang
Ni Li
Xiao Song

Organization

Honorary Charis

Bo Hu Li   Beihang University, China
Axel Lehmann   Universität der Bundeswehr, München, Germany

General Chair

Jianguo Cao   China Simulation Federation, China

General Co-chairs

Wenhui Fan   Qinghua University, China
Lin Zhang   Beihang University, China
Satoshi Tanaka   University of Tokyo, Japan
Yahaya Md Sam   UTM, Malaysia
Gary Tan   National University of Singapore, Singapore

Organizing Committee Chair

Ni Li   Beihang University, China

Publication Committee Chair

Xiao Song   Beihang University, China

International Program Committee

Lin Zhang (Chair)   Beihang University, China
Kyung-Min Seo   Korea University of Technology Education,
                          South Korea
Jangwon Bae   Korea University of Technology Education,
                          South Korea
Kyoungchan Won   Center for Army Analysis & Simulation,
                          South Korea
Gyu M. Lee   Pusan National University, South Korea
Bohu Li   Beijing University of Aeronautics and
                          Astronautics, China
Liang Li   Ritsumeikan University, Japan

Contents – Part II

Application of Modeling/Simulation in Energy Saving/Emission Reduction, Public Safety, Disaster Prevention/Mitigation

Modeling/Simulation Applications in the Military Field

Modeling/Simulation Applications in Education and Training

Modeling/Simulation Applications in Entertainment and Sports

Contents – Part I

Complex Systems and Open, Complex and Giant Systems Modeling and Simulation

Integrated Natural Environment and Virtual Reality Environment Modeling and Simulation

Networked Modeling and Simulation

**Flight Simulation, Simulator, Simulation Support Environment,
Simulation Standard and Simulation System Construction**

**High Performance Computing, Parallel Computing, Pervasive
Computing, Embedded Computing and Simulation**

CAD/CAE/CAM/CIMS/VP/VM/VR/SBA

Big Data Challenges and Requirements for Simulation and Knowledge Services of Big Data Ecosystem

Artificial Intelligence for Simulation

Application of Modeling/Simulation in Science/Engineering/Society/Economy/Management/Energy/Transportation/Life/Biology/Medicine etc.

Flux Modelling of Membrane Bioreactor Process Plant Using Optimized-BPNN

Liu Yin, Fatimah Sham Ismail[✉], and Norhaliza Abdul Wahab

Faculty of Electrical Engineering, Universiti Teknologi Malaysia, 81310 Skudai, Johor Bahru, Johor, Malaysia

liuyin@graduate.utm.my, {fatimahs,norhaliza}@utm.my

Abstract. Membrane bioreactor (MBR) is one of the most popular sewage treatment technologies. However, membrane fouling, a complicated process, has a negative effect on the membrane service life and effluent quality. A model with high accuracy, stability, generalization ability was needed to overcome this problem. Artificial neural network (ANN) stands out from numerous machine learning modeling methods with self-learning and sufficient capacity to capture the nonlinear complexity processes. In this paper, back-propagation neural network models (BPNN) with different hyper parameters were proposed using back-propagation algorithm. To improve the efficiency of learning process, batch module was introduced into training dataset. 4000 samples experimental data have been collected with the MBR pilot plant, 60% was used for training, 20% was used for validation, the rest for testing. With the simulation result, in theory a three-layer ANN have the ability to fit any mapping problem was proved with an average of 98% for R^2 performance. However, with the comparison of models with different hyper parameters, two hidden layer models have a better performance with appropriate neurons, within an acceptable computational load. Over-fitting phenomenon occurs when the number of nodes is too large, resulting in larger MAE.

Keywords: ANN · MBR · Back-propagation · Modelling · Simulation

1 Introduction

Membrane bioreactor (MBR) technology has become one of the most popular membrane filtration processes, such as good effluent quality, small floor space, high efficiency and, reliability and easy automatic control, while compared with conventional activated sludge process.

Membrane bioreactor process is technology that can intercept particles of different sizes, but large molecules and soluble solids such as inorganic salts are allowed to pass the membrane pore. Hence, the sewage was cleaned. According to the research data, membrane bioreactor has been applied in large-scale applications and the stations with a capacity exceed 107 m^3/d are countless in the worldwide treatment [1].

Membrane fouling has a positive effect on permeate flux reduction, pressure increment, productivity reduction [2], and plays a negative role in the wider application of ultra-filtration membrane in waste water treatment industry. Therefore, how to solve the fouling problem is very important for sewage treatment process and prolong the membrane service life.

Concentration and viscosity of mixture are the major influence factors, deposition and adsorption of colloidal particles, solutes and inorganic solutes have great impact on the membrane contamination (internal fouling) and external fouling [3]. However, external fouling consists by the deposition of large particles, organic mixture, inorganic solutes on the membrane surface, forms a cake layer. The cake layer on the membrane surface forms quickly, once formed it grows rapidly in a short time, that will cause permeate pressure difference increasing rapidly and cause decrease of permeate flux, effluent quality decrease and energy cost increment [4]. Unfortunately, peristaltic pump and membrane module may be damaged in an extremely abnormal working condition.

The complexity and diversity of membrane fouling process makes a contribution on getting a model with high accuracy of the filtration process and makes it becomes a harder job.

A reliable model with high accuracy is a foundation for the control system to improve the performance of MBR plant. Mathematical deterministic model is one of the most often used method for control system, especially for fully understood systems. Actually, mathematical model is a reflection on the filtration mechanism by different equations, but the complexity and diversity of fouling impact on permeate process makes it harder for prediction purpose. Another popular model is machine learning data-driven model, such as artificial neural network (ANN) has a widely use in the waste water treatment industry. ANNs are quite popular for its sufficient capacity to capture the nonlinear, complexity and hard to model by mechanism processes within an affordable computational cost [5]. [6] used multilayer artificial neural networks to predict the aerobic granular SBR process with high accuracy predictions, the correlation R2 > 99% and RMSE < 5%. In [7], adaptive neuro-fuzzy inference system (ANFIS) and support vector regression (SVR) for data-driven modeling was proposed for sewage treatment process.

An Bandelet neural network prediction model was used in [8] for MBR system. By the work of Dornier et al. [9] a BPNN hydraulic resistance prediction model for cross flow MBR was established. BPNN model was used for analysis the effects of operational parameters on effluent quality of a submerged membrane bioreactor (SMBR) in [10]. Wavelet neural network, feedforward neural network (FFNN) model and PSO-ANN model were also used in [11–13] for filtration modeling in MBR and SMBR.

According to the Kolmogorov theorem [14], the feedforward neural network with three layers has the ability to approximate any complex, nonlinear mapping in a closed dataset. In this paper, a multilayer back-propagation neural network (BPNN) structure modeling consists of an input layer, one or more hidden layers, and an output layer was used to predict the flux. The accuracy of the established by comparing different hidden layers, activation function, hyper parameters of the ANN model to achieve an accurate prediction of dynamic behavior.

This paper presents a back-propagation ANN based model from laboratory scale MBR plant. Experimental data have been collected for 4000 data samples, peristalsis pump voltage as input, permeate flu as output. To validate the performance of models with different hyper parameters, the dataset was divided into two parts, 60% (2400 samples) was used for training process, 20% (800 samples) was used for validation and 20% (800 samples) was used for testing.

2 Artificial Neural Network Principle

Artificial neural network (ANN), a data-driven black-box modeling method, is one of the most famous techniques as a branch of machine learning technique, based on a bionics approach to modeling by simulating the structure and function of biological neural network structures can learn the mapping relationship of input and output data through iterative training. Feedforward network, cyclic network, dynamic network and competitive network are the most widely used structures for different research contents [15]. In this paper, all research are based on the feedforward neural network. M-P neuron model, shown in Fig. 1, is the basic unit in neural network, which have the ability to receive input data from other neurons and external data. A weight is needed for every input of the neuron to measure the importance of the data.

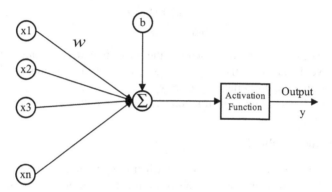

Fig. 1. M-P neuron model.

For every neuron a bias is also introduced as the input of the activation function, the equation can be expressed as follows:

$$y = f\left(\sum_{i=1}^{n} w_i x_i + b\right) \tag{1}$$

The main use of activation function is transmitting the data of the current neuron to the neuron in the next layer after nonlinear processing, to improve the nonlinear fitting ability of the model. Therefore, nearly all the equations of the activation function are nonlinear. In this paper tan-sigmoid function was selected as the activation function of the hidden layers.

$$f(x) = \tanh(x) = \frac{e^x - e^{-x}}{e^x + e^{-x}} \tag{2}$$

The earliest neural network is a two-layer neural network with one input layer and single output layer. It was also called as single layer perceptron with the only calculation layer. In the later research, the majority of researchers' attention has focused on network with three or more layers, which including at least one hidden layer, referred to multi-layer perceptron. Each layer of the model has a specific number of neurons, differently, input layer and output layer neuron number were determined by the training data. As a result, the learning ability of nonlinear function, robustness and generalization ability are improved significantly.

2.1 Data Preprocessing

Data normalization is the first step of ANN modelling, the task of normalization is scaling the data set into appropriate range. Data normalization has a positive contribution to search for the best optimal solution, that speeds up the network convergence process. In this work, Eq. (3) was used for simplify the difficulties during the train and prediction process, normalized data was scaled into the range $[-1, 1]$.

$$x_c = \frac{(x_{c\,max} - x_{c\,min}) * (x - x_{min})}{x_{max} - x_{min}} + x_{c\,min} \tag{3}$$

x_c denotes the x after scaled, x_{cmax} denotes the upper boundary of range, while x_{cmin} denotes the lower boundary of the rang. X_{max}, x_{min} denotes the maximum and minimum value of the original input-output respectively.

In this study, "mapminmax" instruction was used for data normalization.

2.2 Performance Evaluation

Performance is the determination factor for when to stop the neural network training iteration. So, the performance evaluation is a critical operation. The most commonly used indicators to evaluate the performance are Mean Absolute Error (MAE), Mean Square Error (MSE), Root Mean Square Error (RMSE) and also the Integral Absolute Error (IAE). Correlation Coefficient (R^2), given by the Eq. (4), is a relative parameter, the size of sample data is crucial to the validity of the result. In general, larger value indicates a higher degree of correlation between two data.

$$R^2 = (\frac{\sum (y_{out} - \overline{y_{out}})(y_{pre} - \overline{y_{pre}})}{\sqrt{(\sum (y_{out} - \overline{y_{out}})^2) * (\sum (y_{pre} - \overline{y_{pre}})^2)}})^2 \tag{4}$$

where y_{pre}, y_{out} depicts the prediction sample data and output of the model, respectively.

2.3 Back-Propagation Neural Network

Back-propagation neural network is a feed forward network structure, belongs to super-visor learning algorithm. For the BPNN model, the tan-sigmoid activation function was employed for the hidden layer neurons and purelin function for output layer neurons.

Hyper parameters determination is the most crucial determination factors for effective learning and performance of the network after structure selection [16]. In this work, number of hidden layers varies from 1 to 2, hidden neurons number varied from 5 to 20 and the performance was estimated by the standard MAE. To make a deeper comparison different batch-size was employed. With the limitation of the period cyclic filtration process, batch-size was selected between 1 and 5. Number of iteration and learning rate was assigned with *net.trainParam.epochs* = 1000 and *net.trainParam.lr* = 0.03 shown. Development of BPNN is described by flow chart shown in Fig. 2.

When dealing with non-linear least squares problems, Levenberg-Marquardt (LM) method is one of the most popular optimization algorithm used in the literature due to its convergence speed and performance [17–19]. The conventional optimization of weights and bias, the implementation of BP algorithm, is shown in Fig. 3, where P depicts the sample size, Q depicts the max iteration time and E is the cost function as shown in Eq. (5). y_r is the real output and y_p is the predict output. Weights and bias can be calculated with Eqs. (6) and (7).

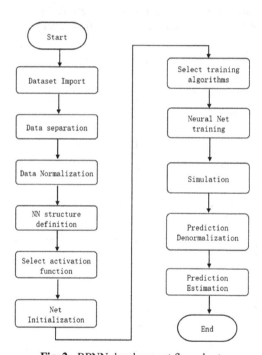

Fig. 2. BPNN development flow chart.

$$E = \frac{1}{2}(y_r - y_p)^2 \tag{5}$$

$$\Delta w = \frac{\partial E}{\partial w} \tag{6}$$

$$\Delta b = \frac{\partial E}{\partial b} \tag{7}$$

3 ANN Based MBR Model Development

The main purpose of this work is establishing an ANN model with high accuracy after training with the experimental data from the MBR filtration data sets, which contains 4000 data samples that were utilized to obtain the model. Generally, building an ANN model contain two phases, training process and prediction process, so the data set was separated into two parts, 60% (2400 samples) for training, 20% (800 samples) for validation and 20% (800 samples) for testing.

The experiments of MBR pilot plant were set to switch between relaxation and permeate state. Such changes can be described by a step function shown in Eq. (8). In order to improve computing efficiency and reduce hardware resources consumption, batch module was adopted in training process. Take the experiment period into consideration, the batch size was set to 5. As a result, the cost function was optimized as Eq. (9).

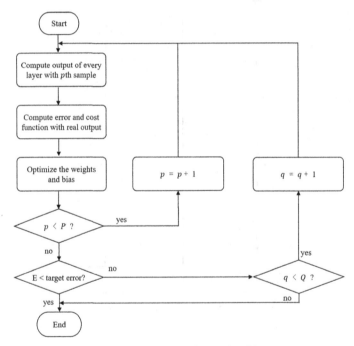

Fig. 3. Conventional BPNN algorithm.

$$workstate = \begin{cases} OFF & t = t_{relaxation} \\ ON & t = t_{permeate} \end{cases} \tag{8}$$

$$E=\frac{1}{2}\sum (y_r - y_p)^2 \tag{9}$$

4   Results and Discussion

This section presents the modelling simulation results for the MBR pilot plant. In this work, tan-sigmoid function and purelin function were employed for hidden layers and output layer respectively (Fig. 4).

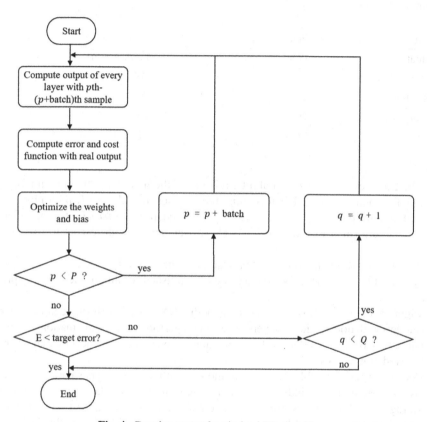

Fig. 4. Development of optimized BP algorithm.

For ANN models, a model with very simple structure has a faster calculation speed with a short training time. However, with a poor learning effect, it is difficult to get a high-precision model with underfitting problem. Meanwhile, with a higher complexity than the actual problem, the model performs well on the training dataset, but cannot be applied on the validation dataset, showing poor generalization ability. In order to realize a desired model, multiple models with different hyper parameters were built, neuron numbers, training algorithm, hidden layer numbers comparison of different models were available from Table 1.

Table 1. Table captions should be placed above the tables.

Parameter	Model I (single hidden layer)	Model II (multi hidden layer)
Architecture	BPNN	BPNN
Optimization algorithm	Levenberg Marquad	Levenberg Marquad
Activation function	Tan-sigmoid (hidden layer) Purelin (output layer)	Tan-sigmoid (hidden layer) Purelin (output layer)
Hidden layer	1	2
Number of neurons	5–20	5–20 5–15
Input	Pump voltage	Pump voltage
Output	Permeate flux	Permeate flux
Training data	60%	60%
Validation data	20%	20%
Testing data	20%	20%
Estimation	MAE, R^2	MAE, R^2

The accuracy, generalization ability, reliable of the model are the most important aspects for control system. For modeling, the data set from the MBR pilot plant was divided into two parts, 60% for training, 20% for validation and 20% for testing. Meanwhile, the performance of different models was compared with respect to MAE, R^2 criteria.

The training result and predict of validation are plotted in Fig. 5 and Fig. 6, respectively. Two different type BPNN models were established for flux of the filtration pilot plant.

Figure 5 and Fig. 6 show the results that both ANN models with single hidden layer and models with double hidden layers have a good performance for fluctuation of actual output data. However, models with double hidden layers are more accurate and stable in the validation process.

As can be seen in Table 2, the ANN models with different hyper parameters performs vary. Both single hidden layer models and multi hidden layer models showed an average of 98% for R^2 performance. The MAE for different models was shown in Table 2, with the increase of hidden layer neuron number the MAE is on the decline as expected. Nevertheless, with the neurons increment of multi hidden layer models MAE was respectively 0.7865, 0.7857 and 0.7997, decreases first and then increases, R^2 shows the exact opposite trend, due to the overfitting problem.

With these comparison, Model 22 was shown to have a best performance on the balance in generalization ability and precision requirement.

For the case at hand, theoretically, three-layer neural network can fit any nonlinear mapping, but multi-layer networks perform better within an acceptable calculate load.

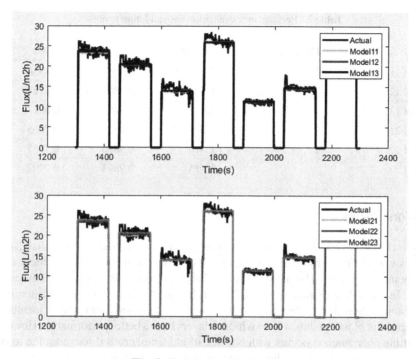

Fig. 5. Training models for flux.

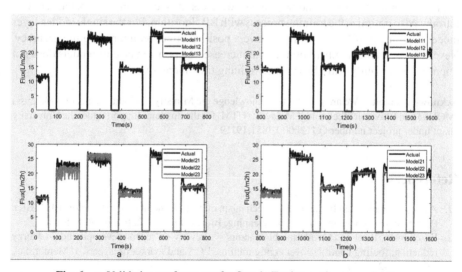

Fig. 6. a. Validation performance for flux; b. Testing performance for flux.

Table 2. Performance evaluation for validation results.

Models for flux	Hidden layer	Neurons	R2	MAE
Model 11	1	5	0.9863	0.8954
Model 12	1	15	0.9865	0.8113
Model 13	1	20	0.9867	0.8017
Model 21	2	10-5	0.9895	0.7865
Model 22	2	15-10	0.9896	0.7857
Model 23	2	20-15	0.9893	0.7997

5 Conclusion

This work has presented the basic theory of artificial neural model, and establish BPNN models with different hyper parameters for flux model of wastewater treatment using membrane technology. The performance of different models was shown by figures and tables. In general, after training with BP algorithm, all ANN models show the nonlinear fitting ability. Based on comparative analysis with the collected data from the membrane filtration pilot plant, models with two hidden layers have a better performance. However, overfitting phenomenon occurs with big size of hidden neurons throughout the testing process. Although the model has a better performance with training data, but cannot be fitted for other work conditions that show a poor generalization ability. In order to improve training efficiency and reduce hardware resource consumption, batch size was introduced to optimize the training process with BP algorithm. For the single hidden layer model, increment of hidden neurons have a positive effect on improving the accuracy. Nevertheless, as the number continues to increase, the performance of models does not improve significantly with the obvious training time increases.

Acknowledgement. We are grateful to acknowledge the Ministry of Higher Education Malaysia (MOHE) and Universiti Technologi Malaysia (UTM) for the financial support under the University Grant under project number Q.J130000.3851.19J19.

References

1. Zheng, Y., et al.: Membrane fouling mechanism of biofilm-membrane bioreactor (BF-MBR): Pore blocking model and membrane cleaning. Biores. Technol. **250**, 398–405 (2018)
2. Krzeminski, P., et al.: Membrane bioreactors – a review on recent developments in energy reduction, fouling control, novel configurations, LCA and market prospects. J. Membr. Sci. **V527**, 207–227 (2017)
3. Wu, M., et al.: Membrane fouling caused by biological foams in a submerged membrane bioreactor: mechanism insights. Water Res. **181**, 115932 (2020)
4. Du, X., et al.: A review on the mechanism, impacts and control methods of membrane fouling in MBR system. Membranes **10**, 10020024 (2020)
5. Abdul Wahab, N., et al.: Permeate flux control in SMBR system by using neural network internal model control. Processes **8**, 1672 (2020)

6. Zaghloul, M.S., et al.: Performance prediction of an aerobic granular SBR using modular multilayer artificial neural networks. Sci. Total Environ. **645**, 449–459 (2018)
7. Zaghloul, M.S., et al.: Comparison of adaptive neuro-fuzzy inference systems (ANFIS) and support vector regression (SVR) for data-driven modelling of aerobic granular sludge reactors. J. Environ. Chem. Eng. **8**, 103742 (2020)
8. Zhao, B.: Cleaning decision model of MBR membrane based on Bandelet neural network optimized by improved Bat algorithm. Appl. Soft Comput. J. **91**, 106211 (2020)
9. Yusuf, Z., Wahab, N.A., et al.: Soft computing techniques in modelling of membrane filtration system: a review. Desalin. Water Treat. **161**, 144–155 (2019)
10. Ren, N., Chen, Z., Wang, X., Hu, D., Wang, A.: Optimized operational parameters of a pilot scale membrane bioreactor for high-strength organic wastewater treatment. Int. Biodeterior. Biodegrad. **56**, 216–223 (2005)
11. Wei, A.L., Zeng, G.M., Huang, G.H., Liang, J., Li, X.D.: Modeling of a permeate flux of cross-flow membrane filtration of colloidal suspensions: a wavelet network approach. Int. J. Environ. Sci. Tchnol. **6**, 395–406 (2009)
12. Lee, Y.G., et al.: Artificial neural network model for optimizing operation of a seawater reverse osmosis desalination plant. Desalination **247**, 180–189 (2009)
13. Yusuf, Z., Wahab, N.A., Sahlan, S.: Modeling of filtration process using PSO-neural network. J. Telecommun. Electron. Comput. Eng. **9**, 15–19 (2017)
14. Schmidt-Hieber, J.: The Kolmogorov-Arnold representation theorem revisited. Neural Netw. **137**, 119–126 (2021)
15. Zhang, Z.: Research on Modeling and Predictive Control of Heavy Duty Gas Turbine Based on Neural Network. D Beijing China, pp. 10–15 (2020)
16. Xu, H., Jagannathan, S.: Stochastic optimal controller design for uncertain nonlinear networked control system via neuro dynamic programming. IEEE Trans. Neural Netw. Lean. Syst. **24**(5), 471–484 (2013)
17. Barello, M., Manca, D., Patel, R., Mujtaba, I.M.: Neural network based correlation for estimating water permeability constant in RO desalination process under fouling. Desalination **345**, 101–111 (2014)
18. More, J.J.: The Levenberg-Marquardt Algorithm: Implementation and Theory (1978). https://doi.org/10.1007/bfb0067700
19. Madaeni, S.S., Shiri, M., Kurdian, A.R.: Modeling, optimization, and control of reverse osmosis water treatment in kazeroon power plant using neural network. Chem. Eng. Commun. **202**, 6–14 (2015)

An Efficient Method of Calculating Stress Intensity Factor for Surface Cracks in Holes Under Uni-variant Stressing

Han Jing[✉], Chen Jian, and Liu Liu

AECC CAE, 3998 Lianhua South Road, Shanghai, China
jean_9412@outlook.com

Abstract. An efficient method of calculating stress intensity factors for surface cracks in holes under univariant stressing is proposed in this paper. General weight function (GWF) is utilized for its simple use and efficiency. Some deduction is made in this method to convert the integral operation of weight function to the simple polynomial operation, in which way calculation efficiency can be improved. Also, to construct the matrix of reference solutions in GWF, the sensitivity of each critical geometric parameters is investigated, based on which some simplification can be made, which saves much time in constructing the matrix of reference solutions.

Keywords: Uni-variant stress · Stress intensity factor · Surface crack · Hole surface

1 Introduction

Surface anomalies in holes introduced during manufacturing may result in uncontained events in turbine engines [1], which makes it important to assess the damage tolerance capability in hole surface properly. A probabilistic risk assessment process is then required to address this rare surface anomalies according to airworthiness requirements, where accurate stress intensity factors (SIF) calculation is necessary [2]. Also, as turbine engines work in complicated conditions with pressure load, thermal load and so on, the local stress fields in the vicinity of the uncracked hole can exhibit stress gradients [3]. In this way, published SIF solutions expressed in terms of remote loads are not applicable.

Although finite element method can calculate SIFs properly, it is time-consuming to construct singular mesh which generates $r^{-1/2}$ singularity on complicated structures of engine components and thus unrealistic for engineering use considering time efficiency. Stress intensity factor manuals [4–6] lacks SIF solutions for complicate stressing and structures of engine components. The weight function method proposed by Bueckner [7] and Rice [8] removes restrictions on the stress distributions by separating load and geometric parts in its equations and is applied quite extensively. Wu and Carlsson proposed the weight function based on an edge crack surface displacement expression [4, 9]. Shen and Glinka proposed general weight function (GWF) for semi-elliptical surface

© The Author(s), under exclusive license to Springer Nature Singapore Pte Ltd. 2022
W. Fan et al. (Eds.): AsiaSim 2022, CCIS 1713, pp. 14–25, 2022.
https://doi.org/10.1007/978-981-19-9195-0_2

cracks in finite thickness plates [10, 11]. The general weight function contains only three unknown parameters, which is considered to be simple-to-use and efficiency. To solve the three parameters in GWF, two reference solutions and a geometric condition are needed in total. The reference solutions in weight functions can be obtained by finite element methods, boundary elements methods and so on [12–14]. Usually, the reference solutions adopt the simplest loading condition, such as uniform tension and linear tapered stress distribution. The loads are applied directly to the crack surface instead of remote boundary.

As is known to all, SIF is affected by load, geometry and crack length. To construct a weight function method for complicated structures, feature dimensions which may affect SIFs need to be recognized first. For example, for surface cracks (SC) in the plate of finite width, feature dimensions include plate width W, plate thickness t, crack length a in the thickness direction, crack length c in the width direction assuming the shape of a surface crack is half-elliptical. Combining the critical dimensions together, then three parameters which may affect the final SIF calculation are selected, which are a/t, c/W and a/c. Each parameter has a rational range, such as [0, 1] for a/t as crack length cannot be smaller than zero nor larger than the cracked body boundary. In order to solve SIFs under arbitrary geometric dimensions in this case, a three dimensional matrix of reference solutions need to be constructed in advance [15–17]. Interpolation is executed between the adjacent matrix points when inputting certain combination of feature parameters (eg. a/c = 1, a/t = 0.5, c/W = 0.5) to calculate SIFs. The range of the parameters defines the solving scope.

Similarly, in purpose of constructing a weight function method for surface cracks in hole, feature parameters are considered including plate width W, plate thickness t, hole diameter D, hole off center distance B, crack length a in thickness direction, crack length c in width direction, crack off center distance T, as is seen in Fig. 1. To construct the matrix of reference solutions, the influence of totally seven parameters, namely, D/t, B/W, T/t, D/B, a/T, c/(B-D/2), a/c need to be engaged. Assuming three data points are selected in the range of each parameter, then $2187(3^7)$ reference solutions need to be calculated for one loading scheme, which is rather heavy workload. Also, interpolation in a seven-dimension matrix is also time-consuming in ensuing calculations of SIFs. According to the above, the reduction of the size of matrix of reference solutions is vital for the construction and application of the GWF method.

In this paper, an efficient method of calculating SIFs using GWF for surface crack in hole for uni-variant stressing case is proposed. Some deduction is made in GWF to convert the integral operation of weight function to the simple polynomial operation, in which way calculation efficiency can be improved. Also, to construct the matrix of reference solutions in moderate sizes, the sensitivity of each critical parameter is investigated, based on which some simplification can be made. Finally, the whole procedure is tested with a group of cases and results are compared with commercial software NASSGRO.

2 General Weight Function Method

For crack planes bearing uni-variant stress which varies along the width direction of the cracked body, stress intensity factor can be obtained from the weight function formulation

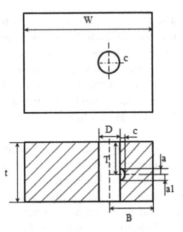

Fig. 1. Geometric dimensions for surface crack in hole

given by Eq. (1) proposed by Glinka and Shen [8].

$$K_I = \int_0^a \sigma_I(x) m(x, a) dx \qquad (1)$$

where $\sigma_I(x)$ is the stress distribution on the crack plane, and $m(x, a)$ is the weight function.

For a-tip (in the thickness direction):

$$m_a = \frac{2}{\sqrt{\pi x}} \left[1 + M_{1a}\sqrt{\frac{x}{c}} + M_{2a}\frac{x}{c} + M_{3a}\left(\frac{x}{c}\right)^{3/2} \right] \qquad (2)$$

For c-tip (in the width direction):

$$m_c = \frac{2}{\sqrt{2\pi(c-x)}} \left[1 + M_{1c}\sqrt{\frac{c-x}{c}} + M_{2c}\frac{c-x}{c} + M_{3c}\left(\frac{c-x}{c}\right)^{3/2} \right] \qquad (3)$$

When knowing two reference solutions and a geometric condition, then parameters $M_{ia,c}$ in Eqs. (2)–(3) can be figured out. As different geometric configuration corresponds to different reference solutions, coefficients $M_{ia,c}$ can be expressed as a function of multiple geometric parameters. Assume the arbitrary stress distribution along the width direction in the polynomial form:

$$\sigma(X) = \sum C_i X^i \qquad (4)$$

where X denotes the normalized distance, namely $X = \frac{x}{W}$.

Substitute Eq. (4) and corresponding reference solutions into Eq. (1), the stress intensity factor under arbitrary stress distribution with certain geometric configuration can be solved. To improve the efficiency of this process, the integral operation in Eq. (1) can be converted to polynomial operation. The deduction process is followed.

Let t = $\frac{c-x}{c}$ and substitute it into Eq. (1), general form m_i can be obtained.

$$m_i = \sqrt{\frac{2c}{\pi}} \int_0^1 (t^{i-0.5} + M_{1c}t^i + M_{2c}t^{i+0.5} + M_{3c}t^{i+1})dt \qquad (5)$$

Then stress intensity factor solution can be expressed as:

$$K_{ci} = c^i \sum_{p=0}^i (-1)^p C_i^p m_p \qquad (6)$$

$$K_c = \sum_{i=0}^n C_i K_{ci}/width^i \qquad (7)$$

where K_{ci} denotes stress intensity factor corresponds to stress component X^i, K_c denotes the final solution of c-tip. The SIF solution of a-tip can be deducted in the same way.

3   Construction of the Matrix of Reference Solutions

3.1   Sensitivity of Parameters

As is referred above, totally seven parameters D/t, B/W, T/t, D/B, a/T, c/(B-D/2), a/c may influence the value of SIF for surface crack in holes, though six of them can decide all the dimensions of the cracked body. In order to investigate the sensitivity of each parameter, data points from broad range are selected for each parameter, which are listed in Table 1. For parameters with higher influence, more data points are arranged, which is based on former experience. The total number of the combinations of parameters is 1280 (5*4*2*4*4*2). Furthermore, bending and tapered loading condition are both included as complicated stress distribution can be decomposed into these two components.

Table 1. Range of affecting parameters

Parameters	Iteration order	Selected data point	Data num
a/T	1	0.1, 0.2, 0.5, 0.8, 0.9	5
a/c	2	0.5, 1, 2, 4	4
T/t	3	0.1, 0.5	2
c/(B-D/2)	4	0.1, 0.2, 0.5, 0.8	4
D/t	5	0.25, 0.5, 1, 1.5	4
B/W	6	0.1, 0.5	2

The influence of all the six parameters listed in Table 1 can be shown in Figs. 2, 3, 4, 5, 6 and 7. Figure 2 displays the SIF values of c-tip under uniform tension and tapered loading respectively under different a/T ratios (B/W = 0.5). The value of X axis is the sequence number of testing cases, while the value of Y axis is the SIF result assuming crack size a = 1.As can be seen from Fig. 2, when the value of X is identical, the values of SIFs under different a/T ratios differ greatly, which are represented in different colors. This indicates when keeping other parameters the same, the parameter a/T have comparatively large influence on SIFs, although to different degrees at different cases (at different X values). Same conclusions can be derived from Fig. 3 and Fig. 5 that parameters a/c and c/(B-D/2) both have comparatively large influence. In contrast, parameters T/t, D/t and B/W have relatively small influence, which is illustrated in Fig. 4, Fig. 6 and Fig. 7.

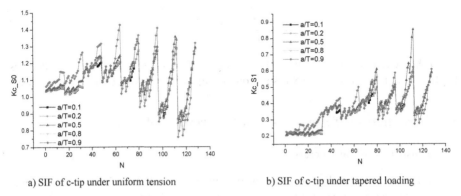

a) SIF of c-tip under uniform tension   b) SIF of c-tip under tapered loading

Fig. 2. SIF of c-tip under uniform tension and tapered loading under different a/T ratios (B/W = 0.5)

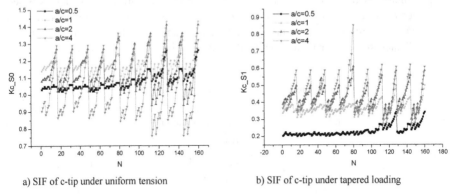

a) SIF of c-tip under uniform tension   b) SIF of c-tip under tapered loading

Fig. 3. SIF of c-tip under uniform tension and tapered loading under different a/c ratios (B/W = 0.5)

3.2 Influence of Parameters of Hole Shape

Based on the sensitivity research above, hole shape related parameters B/W and D/t seems to have relatively smaller influence on the SIF calculation. To learn more about the

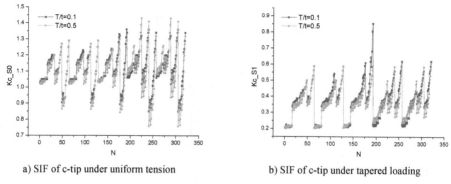

a) SIF of c-tip under uniform tension

b) SIF of c-tip under tapered loading

Fig. 4. SIF of c-tip under uniform tension and tapered loading under different T/t ratios (B/W = 0.5)

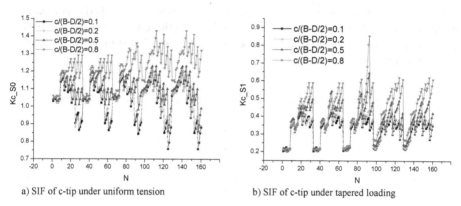

a) SIF of c-tip under uniform tension

b) SIF of c-tip under tapered loading

Fig. 5. SIF of c-tip under uniform tension and tapered loading under different c/(B-D/2) ratios (B/W = 0.5)

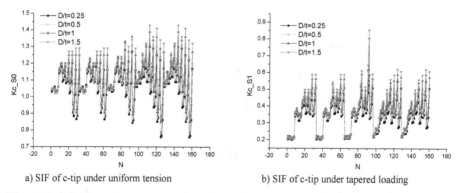

a) SIF of c-tip under uniform tension

b) SIF of c-tip under tapered loading

Fig. 6. SIF of c-tip under uniform tension and tapered loading under different D/t ratios (B/W = 0.5)

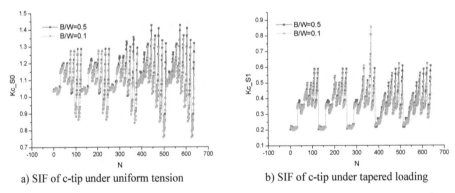

a) SIF of c-tip under uniform tension　　　　b) SIF of c-tip under tapered loading

Fig. 7. SIF of c-tip under uniform tension and tapered loading under different B/W ratios

influence of hole shape on SIFs, surface cracks in plate is also researched in comparison to the hole case. To keep parameters in accordance in these two cases, the equivalent conversion of parameters from hole crack to plate crack is listed in Table 2, which is also intuitively demonstrated in Fig. 8.

Table 2. Equivalent conversion from hole crack to plate crack

Hole crack	Plate crack
t	W
B-D/2	t
c	a
a	c

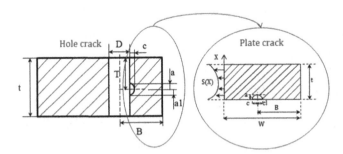

Fig. 8. Conversion from hole crack to plate crack

The results of comparison can be seen in Fig. 9 and Fig. 10. The value of X axis is the sequence number of testing cases, while the value of Y axis is the relative error of plate crack to hole crack, which is calculated in the following equation.

$$\text{Error} = (SIF_{hole} - SIF_{plate})/SIF_{hole} \tag{8}$$

As is seen in Fig. 9 and Fig. 10, SIFs calculated under plate case are generally larger than those calculated under hole case. Rare exceptions happen when crack size is relatively large comparing to the cracked body. In addition, when the crack is relatively small to the cracked body, the influence of the hole shape is relatively small. As crack becomes significantly large, namely the value of a/T increases, the influence of hole shape becomes more remarkable. Based on this observation, surface cracks in holes can also be treated as surface cracks in plate for quick calculation, in which case conservative results will be obtained. This also gives some light to the idea that for complicated shape where hole diameters change along the thickness direction, it is hard to decide which diameter size to use in the calculation, simplify this condition to plate surface crack may help to get a conservative result.

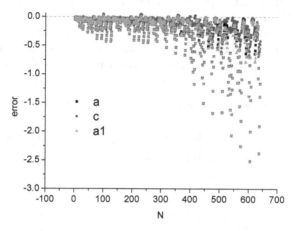

Fig. 9. Comparison of results between hole crack and plate crack for uniform tension

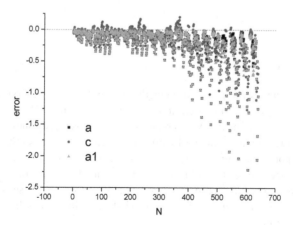

Fig. 10. Comparison of results between hole crack and plate crack for tapered loading

3.3 Conclusions of Sensitivity Analysis

According to the analysis in 3.1 and 3.2, two conclusions can be drawn. The first is, for all the six parameters a/c, D/t, B/W, T/t, a/T, c/(B-D/2), parameters of a/T, a/c, c/(B-D/2) have larger influence on SIF than parameters of B/W, D/t and T/t. The second is, plate surface crack can get more conservative SIF results than the hole crack when keeping all the parameters the same. Combining these two conclusions, we can get the idea that parameter B/W has limited influence on SIF result and ignoring it will get relatively larger SIF value. As is mentioned above, the size of the matrix of reference solutions will both affect the construction and the application (mainly interpolation efficiency) of GWF methods. If we ignore the influence of parameter of B/W, large calculation amount of FEM analysis to construct the reference solutions can be reduced. To make it more clearer, if consider all six parameters (D/t, T/t, a/T, c/(B-D/2), a/c, B/W), a total of 2400(6*4*2*5*2*5) reference solutions need to be solved, as can be seen in Table 3. If neglecting the influence of B/W, 1200(6*4*2*5*5) reference solutions can be omitted, which means half of the workload can be relieved. In addition, the calculating speed can be improved further in a five-dimensional matrix than in a six-dimensional matrix.

Table 3. Construction of reference solution matrix

Factor	Selected data point	Data num
a/T	0.01, 0.1, 0.2, 0.5, 0.8, 0.9	6
a/c	0.5, 1, 2, 4	4
T/t	0.1, 0.5	2
c/(B-D/2)	0.01, 0.1, 0.2, 0.5, 0.8	5
B/W	0.1, 0.5	2
D/t	0.25, 0.5, 1, 1.5, 2	5

4 Application

TO validate the accuracy of the proposed method, including the conversion from integral operation to simple polynomial operation, as well as omitting B/W in constructing the matrix of reference solutions in GWF, a bunch of testing cases are conducted. The results of the proposed method are compared with commercial software NASGRO.

Firstly, the matrix of reference solutions is constructed with factors (D/t, T/t, a/T, c/(B-D/2), a/c), neglecting the influence of B/W, namely, assuming 2B/W = 1. The selection of data points for each parameter is mainly based on the sensitivity analysis conducted in 3.1. For parameters a/T, a/c and c/(B-D/2), which are of higher importance, more points are arranged.

Secondly, in order to consider the stress concentration effects in the proposed method, stress distribution table is established before the calculation of SIFs. The stress distribution table is created with factors (D/t, B/W, D/B) in reference to stress concentration factor Kt. Interpolation will be executed in calculation to get a proper stress distribution when inputting different combination of (D/t, B/W, D/B). This stress distribution is then treated as the weight function input exerted on the crack plane.

Thirdly, a group of testing cases are designed as is seen in Table 4. Remote tension stress is applied to the cracked body in all testing cases. Results obtained from the proposed method and NASGRO under the same calculating settings are compared.

Table 4. Testing matrix

Factor	Selected data point	Data num
a/T	0.01, 0.1, 0.9	3
a/c	0.5, 1, 4	3
T/t	0.1, 0.5	2
B/W	0.1, 0.5	2
D/B	0.1, 1, 1.8	3
D/t	0.25, 1, 2	3

Figure 11 demonstrates the calculating error of the proposed method corresponds to NASGRO. The value of axis X is the sequence number of testing case, while the value of axis Y is the corresponding error calculated in Eq. 9. In Fig. 11, error_Ka denotes the error of a-tip (thickness direction) and error_Kc denotes the error of c-tip(width direction).As some parameters exceed the calculation boundary of NASGRO, the results of these combinations are not recorded. As is seen in Fig. 11, the errors correspond to commercial software is in the 10% range, most in a 5% range, which indicates a good agreement. This means the accuracy does not degrade when neglecting B/W in constructing the matrix of reference solutions.

$$\text{Error} = (SIF_{test} - SIF_{NASGRO})/SIF_{NASGRO} \qquad (9)$$

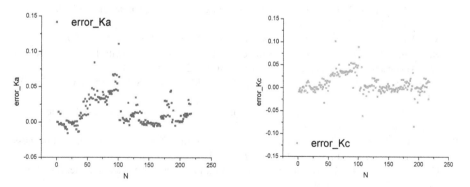

Fig. 11. The errors of test cases corresponding to NASGRO

5 Conclusion

IN this article, General weight function is utilized to calculate the stress intensity factors of surface cracks in holes. Some deduction is made to convert the integral operation of weight function to the simple polynomial operation. Also, to construct the matrix of reference solutions, the sensitivity of each critical parameter is investigated in advance, based on which the influence of parameter B/W is ignored. Actually, the selection of feature parameters is a tradeoff between conservatism and time efficiency. If all shape parameters (D/t, B/W, D/B) are neglected, hole crack can be converted to plate crack, the results of which may be too conservative. However, if all parameters are considered when constructing matrix of reference solutions, the calculation amount of FEM analysis can be huge, sometimes unrealistic.

The whole procedure is verified comparing to the results of commercial software NASGRO and good agreement is achieved, which means accuracy does not decrease much due to the simplification of B/W. Also, this procedure gives some light to construct the database of reference solutions of more complicated structures, which involves the identification and screening of feature size parameters in consideration of accuracy, conservatism and time-efficiency. All of this will be investigated in future research.

References

1. Corran, R., Gorelik, M., Lehmann, D., Mosset, S.: The development of anomaly distributions for machined holes in aircraft engine rotors. ASME Turbo Expo 2006: Power for Land, Sea and Air, GT2006-90843 (2006)
2. McCInng, R.C., Lee, Y.D., Cardinal, J.W., Guo, Y.: The pursuit of K: Reflections on the current state-of-the-art in stress intensity factor solutions for practical aerospace applications. In: 27th ICAF, pp. 1–18 (2013)
3. McClung, R.C., Enright, M.P., Lee, Y.-D., Huyse, L.J.: Efficient fracture design for complex turbine engine components. In: Proceedings of ASME Turbo Expo 2004 Power for Land, Sea and Air, GT2004-53323 (2004)
4. Wu, X.R., Carlsson, A.J.: Weight Functions and Stress Intensity Factor Solutions. Pergamon Press, New York (1991)
5. Tada, H., Paris, P.C., Irwin, G.R.: The stress analysis of cracks handbook. Paris Production Incorporated, St. Louis (1985)
6. Murakami, Y.: Stress Intensity Factors Handbook. Pergamon Press, Oxford (1992)
7. Buckner, H.F.: A novel principle for the computation of stress intensity factors. Zeitschrift fuer Angewandte Mathematik und Mechanik 50(9), 529–546 (1970)
8. Rice, J.: Some remarks on elastic crack-tip stress field [J]. Int. J. Solids Struct. 8(6), 751–758 (1972)
9. Wu, X.R.: Analytical wide-range weight functions for various finite cracked bodies. Eng. Anal. Bound. Elem. 9, 307–322 (1992)
10. Shen, G., Glinka, G.: Weight function for a surface semi-elliptical crack in a finite thickness plate. Theoret. Appl. Fract. Mech. 15, 237–245 (1991)
11. Glinka, G., Shen, G.: Universal features of weight functions for cracks in mode. Eng. Fract. Mech. 40, 1135–1146 (1991)
12. Ojdrovic, R.P., Petroski, H.J.: Weight functions from multiple reference states and crack profile derivatives. Eng. Fract. Mech. 39(1), 105–111 (1991)

13. Beghini, M., Bertini, L., Vitale, E.: A numerical approach for determining weight functions in facture mechanics. Int. J. Numer. Meth. Eng. **32**, 595–607 (1991)
14. Lorenzo, J.M., Cartwright, D.J., Aliabadi, N.H.: Boundary-element weight function analysis for crack-surface displacements and strip-yield cracks. Eng. Anal. Bound. Elem. **13**, 283–289 (1994)
15. Wang, X., Lambert, S.B.: Stress intensity factors for low aspect ratio semi-elliptical surface cracks in finite-thickness plates subjected to nonuniform stress. Eng. Fract. Mech. **51**, 517–532 (1995)
16. NASGRO Reference Manual, Version9.1, NASA Johnson Space Center and Southwest Research Institute (2019)
17. Southwest Research Institute. DARWIN theory. Southwest Research Institute, Design Assessment of Reliability with Inspection, San Antonio, US (2008)

Modeling and Simulation Based on Concurrent FC-AE-1553 Network

Gao Chi[1], Dai Zhen[1(✉)], and Kong Xuan[2]

[1] Chengdu Aircraft Design and Research Institute, Chengdu 610091, China
zhendai91@163.com
[2] Xi'an Aeronautics Computing Technique Research Institute, Xi'an 710000, China

Abstract. In order to meet the demand of avionics system for high reliability, high determinacy and high bandwidth utilization of airborne network, this paper proposes a concurrent FC-AE-1553 network model based on the command/response communication scheme of FC-AE-1553 network protocol. By executing different transmission strategies for frames with different service levels, the problem of competition among different communication tasks is effectively solved, the communication delay of the whole network is reduced, and the utilization efficiency of network bandwidth is improved. The feasibility of the network model is analyzed through the simulation experiment. In the experiment of the four-node simulation test platform, compared with the traditional serial FC-AE-1553 network model, the average bandwidth of the concurrent FC-AE-1553 network model is 5.8 times of the latter, and the maximum bandwidth is 4.6 times of the latter.

Keywords: FC-AE-1553 · Concurrent network · Modeling and simulation

1 Introduction

As an important part of aviation aircraft, avionics system is becoming more and more important in the face of increasingly complex aviation environment. The development of avionics system has become an important reason to promote the continuous improvement of aircraft performances [1].

In recent years, Fiber Channel (FC), as a high-speed serial communication protocol, has attracted more and more attention from scholars and institutions around the world, and proposed Fiber Channel Avionics Environment (Fiber Channel Avionics Environment, FC-AE) protocol cluster, in which the FC-AE-1553 bus protocol is a method proposed by referring to the mature MIL-STD-1553B bus protocol and fully considering the characteristics of strong real-time and high determinism in the aviation environment, a Fiber Channel communication protocol for aviation environment [2]. It has the characteristics of high

Supported by CADI Innovation Foundation.

W. Fan et al. (Eds.): AsiaSim 2022, CCIS 1713, pp. 26–38, 2022.
https://doi.org/10.1007/978-981-19-9195-0_3

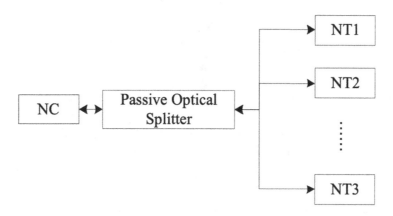

Fig. 1. Bus topology model of FC-AE-1553.

reliability, high bandwidth, strong real-time performance and strong compatibility [3], and has broad application prospects in avionics systems. However, the protocol does not provide a description of the specific implementation method of the network [4]. Therefore, it is necessary to study the specific model of the aviation bus network based on the FC-AE-1553 protocol, and to conduct specific simulation experiments to verify the feasible and advance of the model.

This paper studies the application of FC-AE-1553 protocol in avionics system, proposes a network model based on concurrent FC-AE-1553 network, and designs simulation experiments according to the proposed network model. Through simulation, the feasibility and advanced nature of the network model based on concurrent FC-AE-1553 are verified.

2   Topology Modeling of FC-AE-1553 Networks

The Fiber Channel network under the FC-AE-1553 protocol supports basic network topologies such as bus type and switching type. In specific applications, a reasonable choice is made in combination with the characteristics and actual needs of each topology network. The following briefly introduces the specific situations of various network topology schemes.

2.1   Bus Topology Model

The bus-type network model is the FC-AE-1553 protocol network using passive optical network technology [5,6]. The first purpose of this technology is to solve the bandwidth bottleneck problem of the access network. The technology replaces active multiplexers or switches with passive optical splitters, simplifying the design of fiber optic networks and improving the overall reliability of fiber optic networks.

As shown in Fig. 1, the upstream and downstream processes of the bus-type FC-AE-1553 network are shown. In the uplink, the passive optical splitter of the

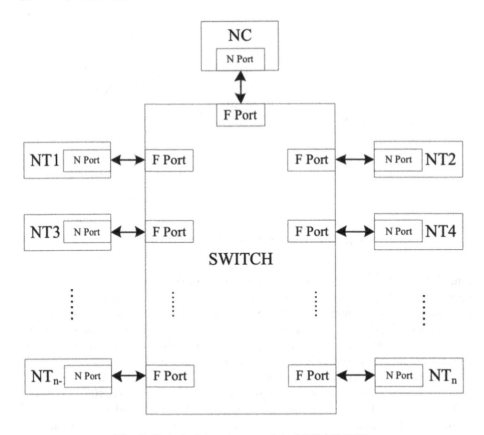

Fig. 2. Switched topology model of FC-AE-1553.

optical fiber is used as a combiner. Due to the limitation of the characteristics of the passive optical splitter, the data packets sent by any NT can only be transmitted to the NC end. Therefore, all NTs share the uplink data channel according to the principle of time division multiplexing. Because of this, when NT1 needs to send data to NT2, it needs to send the data to NC first, and then NC sends it to NT2 instead. In the downlink, after the data packets sent by the NC pass through the passive splitter, the same data are sent to each NT device respectively, and then each NT device decides whether the received data needs to be received or not.

It can be seen that the bus-type FC-AE-1553 network has a relatively simple topology, and each node is connected through a passive optical splitter. Active devices are not used in the entire optical fiber network, and there are no complex operations such as addressing, so it has the advantage of fast transmission speed. But it is also limited by the special topology structure, it is difficult to expand the network, and it is impossible to realize the direct communication from NT to NT point-to-point, which increases the network communication delay. At the same time, because each NT needs time-division multiplexing for data communication, these restrict the overall communication efficiency and application scenarios of the network.

2.2   Switched Topology Model

The switched FC-AE-1553 network is different from the bus network in that it does not use passive optical splitters for fiber line expansion, but uses switches to connect NC and multiple NTs [7,8]. Figure 2 is a schematic diagram of the switched FC-AE-1553 network topology. Each NC and NT node is connected to the switch through optical fibers, and the switch determines the transmission direction of data information, that is, when the NC needs to send data packets with NT1, only NT1 will receive packets from NC. At the same time, since the FC-AE-1553 protocol defines command/response deterministic communication that is centrally controlled by the NC, during the communication process, the NC sends command frames to control the entire network communication behavior. The following takes the NT to NT data transmission process as an example to illustrate the data transmission process under the FC-AE-1553 network [9,10].

1. The NC sends a command frame to the source NT through the switch, and the control source NT sends data to the destination NT.
2. After receiving the command frame, the source NT returns a status frame to the NC to confirm that the command frame has been received.
3. The source NT sends a command frame to receive data to the destination NT, and then sends multiple data frames to the destination NT.
4. After the destination NT finishes receiving the data, it sends a status frame to the NC for information confirmation.

In the above process, all frames are transmitted through the switch and only pass through once, so that each data transmission process is relatively deterministic, ensuring the certainty of the entire network transmission delays. At the same time, after receiving the command frame, the NT confirms the information to the NC through the status frame, which ensures the reliability of network communication.

It can be seen from this that the switched FC-AE-1553 network has a reliable topology structure, which is convenient for the expansion and connection of the network in the future. At the same time, it also has the advantages of high delay determinism and high reliability of data packet transmission [11,12], and has high use value.

Comprehensive analysis of the above two FC-AE-1553 network topology design methods, considering the strong scalability, high concurrency, and high network bandwidth utilization requirements of the avionics system, this paper chooses the switched FC-AE-1553 network topology for research. The following statements are based on the network model of the concurrent FC-AE-1553 network.

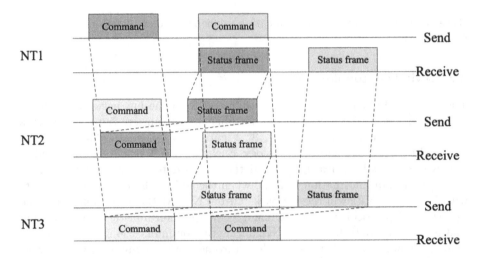

Fig. 3. Message model with status frame reply.

3   Message Modeling of Concurrent FC-AE-1553 Network

Considering that the traditional switched FC-AE-1553 network is uniformly scheduled for communication by the NC, all switching sequences must be initiated by the NC, resulting in the network bandwidth being limited by the maximum bandwidth of the NC node, the insufficient utilization of the communication bandwidth of the NT node, and the inability to fully utilize the network. At the same time, because the control function of the NC node is crucial to the entire network, once the NC node fails, it will cause the network to fail and reduce the reliability of the network. Therefore, combined with the characteristics of distributed network, this paper proposes a simplified FC-AE-1553 network communication process, which cancels the restriction that the exchange sequence must be initiated by the NC, so as to retain the high reliability of communication between FC-AE-1553 network nodes. The overall reliability of the network is improved, the concurrency capability of the network and the utilization efficiency of the network bandwidth are improved.

Considering that the main data transmission requirements in modern avionics systems are sensing parameters and status parameters, certain requirements are put forward for the bandwidth and concurrent performance of communication. At the same time, for event and mission-critical data, it is necessary to ensure high reliability and determinism of communication. Therefore, according to the characteristics of the above usage scenarios, this paper proposes three communication models based on concurrency, status frame reply, suppressed status frame reply, and stream message.

1. Message model with status frame reply. For command frames, the receiving terminal responds to the status frame, as shown in Fig. 3.

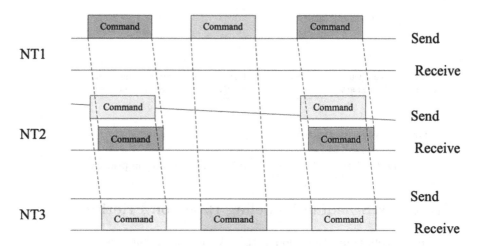

Fig. 4. Message model with suppressed status frame reply.

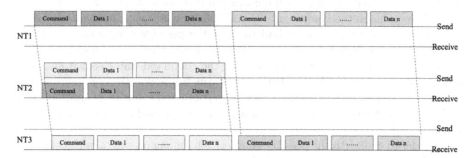

Fig. 5. Stream message model.

2. Message model with suppressed status frame reply. For a command frame, the receiving end does not need to respond to the status frame, as shown in Fig. 4.

3. Stream message model, the length of flow message data is long, one data packet cannot complete data sending, so it needs to be unpacked for sending, the command frame is followed by multiple data frames, and the flow data message model does not need to respond to state frames, as shown in Fig. 5.

As shown in Fig. 3, 4 and 5, it shows the schematic diagram of the three types of message communication mechanism based on concurrency, which the network model removes the restriction on the exchange sequence initiated by the NC.

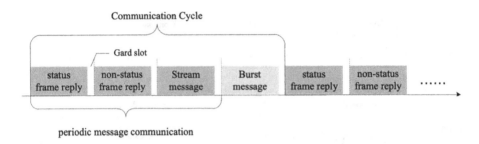

Fig. 6. Network periodic scheduling model based on priority.

4 Network Schedule Modeling Based on Priority

In the above communication model, the NC no longer initiates each exchange sequence, but each node initiates communication freely. Therefore, the method of defining and controlling all communication processes of the network through the NC bus table in the traditional FC-AE-1553 network is no longer applicable. It is necessary to propose a method to solve the possible competition problem between different exchange sequences. For example, when the sender is waiting for the reply of the status frame, it receives the command frame sent from other nodes at the same time. At this time, the status frame and the command frame are in competition. Or when a node receives a command frame from another node in the process of receiving a stream message, the command frame competes with a data frame in the stream message.

In order to deal with the possible competition among messages in the network and ensure the normal operation of the network, a network periodic scheduling model based on priority is proposed in this paper.

As shown in Fig. 6, in the scheduling optimization mechanism proposed in this paper, the bandwidth resources of the network are first allocated according to the communication cycle based on time. In each bandwidth allocation period, periodic message communication is performed first, and according to the message type priority, status frame reply communication is performed first, non-status frame reply communication is followed, and stream message communication is finally performed. For burst messages generated within each cycle time, a fixed time is reserved for communication in each bandwidth cycle. Since the clocks between the devices are not synchronized, a guard time slot is reserved during each bandwidth allocation cycle to prevent collisions between different cycles.

4.1 Periodic Scheduling Optimization Mechanism

For communication tasks that exist periodically in the network, as shown in Fig. 7, the network bandwidth is allocated by a periodic static bandwidth allocation method. At the beginning of each bandwidth allocation period, the communication of periodic messages is performed first, and the communication is performed in sequence in the order of status frame reply messages, non-status frame

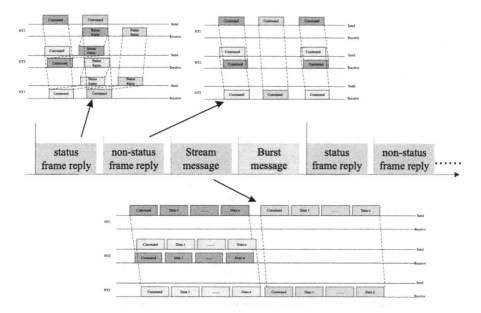

Fig. 7. Periodic scheduling optimization mechanism.

reply messages, and stream messages. Since the restriction that the sequence is initiated by the NC is canceled, the same type of periodic messages of different devices can be concurrently implemented during the periodic message sending process.

By executing the status frame reply message, the non-status frame reply message and the stream message separately in the periodic communication task, the competition between messages of different priorities can be avoided, and the certainty and reliability of the communication can be guaranteed to the greatest extent. At the same time, the network scheduling mechanism is simplified, making the network easy to implement.

4.2   Burst Scheduling Optimization Mechanism

As for the burst communication tasks generated in the bandwidth allocation period, this network model stipulates that they should be carried out after the periodic message communication is completed. As shown in Fig. 8, the frames that may exist in the network are divided into 4 levels according to the priority order: status frame > command frame with status frame reply > command frame with suppressed status frame reply > data frame. The nodes preferentially process frames with high priority in the sending and receiving buffers. At the same time, for different frames that need to be sent to the same port at the same time, the switches in the network preferentially send frames with higher priorities. Figure 8 is a schematic diagram of priority-based burst communication.

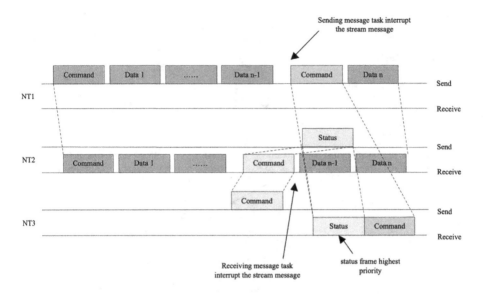

Fig. 8. Burst message communication scheduling model.

For the communication task that adopts the status frame to reply to the command frame, it is necessary to ensure the high reliability and certainty of the communication. Therefore, it is stipulated that the status frame has the highest priority. Any time the node receives the status frame reply command frame, it can confirm the message through the status frame at the fastest speed, and pass through the switch and receive processing at the fastest speed. Status frame reception times out due to network congestion to ensure communication reliability. At the same time, it is stipulated that the command frame replied by the status frame also has a higher priority, so that this type of exchange sequence can be executed preferentially in the competition with other types of exchange sequences, and the certainty of the communication delay is guaranteed.

For communication tasks that reply to command frames and stream messages using suppressed status frames, they are usually not sensitive to the reliability and determinism of the network, but require the characteristics of high network concurrency and high bandwidth. Therefore, it is specified that the suppressed status frame replying to the command frame and the data frame has a lower priority and can be executed when the network is relatively idle. Due to the large amount of stream message data, it may take a lot of time, so the priority of the data frame is specified to be the lowest, so that the stream message task can be interrupted at any time during the sending and receiving process, without affecting other communication tasks, so that the total time of all tasks is reduced. In the following, the network model proposed in this paper is verified with specific simulation experiments.

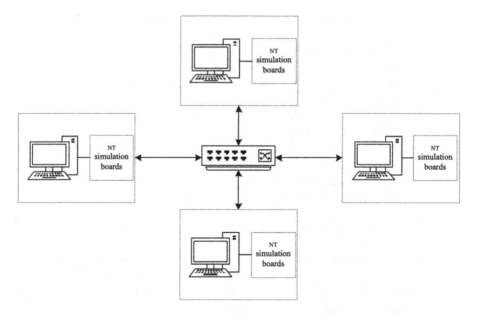

Fig. 9. Test platform topology diagram.

5   Simulation and Analysis

As shown in Fig. 9, a concurrency-based FC-AE-1553 network simulation verification test platform is built using 1 switch, 4 computers, and 4 node simulation boards. Each node emulation board is connected with the computer through the PCIe channel, and the computer controls the communication tasks in the network.

In order to meet the requirements of the avionics system and ensure the determinism of network communication, the network model in this paper retransmits frames that receive timeouts and errors. Since the network model in this paper supports the status frame reply confirmation function, it is only necessary to perform the judgment of the retransmission requirement and the execution of the retransmission task at the FC layer, while other network models do not support the status frame reply confirmation function, so this function needs to be implemented at the application layer. As shown in Table 1, the FC-AE-1553 network model in this paper compares the delay time of the FC-AE-1553 network model with other non-status frame networks when a frame data error occurs or a receiving timeout occurs and retransmission occurs.

As shown in Table 1, by using the status frame reply function, the network model in this paper has a significantly shorter communication delay than other networks when the frame error occurs or the reception times out and the message needs to be retransmitted. The network model proposed in this paper realizes the characteristics of high reliability and determinism of FC-AE-1553 network.

Table 1. Network communication delay.

No.	Communication mode	Number of experiments	Average delay
1	Retransmission with status frame reply (frame error)	10000	18 μs
2	Retransmission with status frame reply (frame loss)	10000	31 μs
3	Retransmission without status frame reply (frame error)	10000	3.273 ms
4	Retransmission without status frame reply (frame loss)	10000	4.521 ms

Table 2. Bandwidth of communication.

No.	Communication mode	Average bandwidth (Gbps)	Max bandwidth (Gbps)
1	Switched (no stream)	1.69	1.73
2	Bus (no stream)	0.29	0.38
3	Switched (with stream)	1.72	1.87
4	Bus (with stream)	0.33	0.46
5	Switched (single node failure)	1.71	1.88
6	Bus (NC failure)	–	–

As shown in Table 2, it is the comparison result of the available bandwidth in the communication between the network model in this paper and the bus network model. The experimental conditions of each group are as follows:

1. In the no-stream message experiment, the average bandwidth is the network band-width measured after the status frame reply communication and the suppressed status frame reply communication which are performed at 1:1. The maximum band-width is the network bandwidth measured after only performing the suppressed status frame reply communication task.
2. In the stream message experiment, the average bandwidth is the network bandwidth measured after the status frame reply communication, the suppressed status frame reply communication and the stream message communication, which are performed at 1:1:1. The maximum bandwidth is the network bandwidth measured after only performing stream message communication tasks.
3. In the node failure experiment, the average bandwidth is the network band-width measured after the status frame reply communication, the suppressed status frame reply communication and the stream message communication, which are performed at 1:1:1. The maximum bandwidth is the network band-width measured after only performing stream message communication tasks.

Through the first two sets of experiments, it can be seen that because the method in this paper realizes the concurrency of multiple communication tasks under the switched network, and only one communication task can be performed at the same time under the bus network, the network model in this paper is improved compared with the bus network. The average bandwidth is 5.8 times that of the bus network, and the maximum bandwidth is 4.6 times that of the bus network under the condition of the 4-node simulation verification test platform, which reflects the support of the network model in this paper for high concurrency and high bandwidth characteristics. At the same time, in the third set of experiments, the bus network cannot continue to communicate after the NC node fails, while the switched network using the network model in this paper can continue to communicate after random node failures, and the bandwidth remains basically unchanged. It shows that the switched network model in this paper has higher reliability than the bus network model.

6   Conclusion

This paper reviews the development history of aviation airborne network, looks forward to the broad application prospect of FC-AE-1553 network in the field of avionics system, and analyzes the characteristics of two topological structures of FC-AE-1553 network. Considering the reliability, determinism and utilization efficiency of network bandwidth comprehensively, a network model based on concurrent FC-AE-1553 network is proposed, and the specific communication optimization mechanism of the network model is described, combined with the specific experimental simulation, the advantages of the network model in feasibility, reliability and network utilization efficiency are verified.

References

1. Liu, P.L.: Design and Implementation of Flight Control System Simulation Demonstration System Based on FC-AE-1553. School of Communication and Information Engineering (2017)
2. Guo, S.H., Zhang, X.L., Zhang, H.M.: Optimization design on complex switching topology of FC-AE-1553 protocol. J. Guilin Univ. Electron. Technol. **39**(161(02)), 10–15 (2019)
3. Chen, X.X.: Design and Implementation of Command and Response Network Emulator Based on Fibre Channel. School of Communication and Information Engineering (2019)
4. Fu, P., Zhang, J.N., Zhao, X.Y., et al.: Research on the structure of FC-AE-1553 verification system. Electron. Measur. Technol. **36**(01), 124–128 (2013)
5. Cao, S.Z., Fang, L., Wu, S.J., Zhang, S.: A bus network topology structure of FC-AE-1553 and real-time analysis. Semicond. Optoelectron. **35**(05), 858–861 (2014)
6. Kou, X.X.: Software Design and Implementation of FC-AE-1553 Node Based on PON Topology. School of Communication and Information Engineering (2020)

7. Jiao, J.: Design of FC-AE-1553 Simulation and Verification System. Harbin Institute of Technology (2012)
8. Wu, S.J., Zhao, G.H., Wang, L.Q., et al.: Dynamic bandwidth allocation mechanism with parallel and switching for FC-AE-1553 network. J. Beijing Univ. Aeronaut. Astronaut. **42**(12), 2579–2586 (2016)
9. Lan, X.: The Simulation Modeling of Deterministic Fiber Channel Network. School of Communication and Information Engineering (2019)
10. Tang, J.: Software Design of FC-AE Node Card Based on Real-Time Operating System Support Time Triggered Communication. School of Communication and Information Engineering (2018)
11. Wang, W.H., Wu, Y., Zhang, J.D.: Fibre channel bus transport model based on the SPN. Fire Control Command Control **37**(04), 185–187 (2012)
12. Chen, Y., Lin, B.J., Zhang, S.C.: Research on FC-AE-1553B network performance with different topology structure. Comput. Eng. **37**(022), 79–81 (2011)

Research on Modeling and Simulation Method of Laser System for Multi-domain Unified Model

Weijian Huang[1], Dong Li[1], Tao Ma[1], Siqiang Yi[2], and Baoran An[1(✉)]

[1] Institute of Computer Application, China Academy of Engineering Physics, Mianyang, China
anbaoran@qq.com
[2] Suzhou Tongyuan Software & Control Technology Co., Ltd., Suzhou, China

Abstract. The optical axis accuracy and beam efficiency of laser equipment are the key indicators that affect the performance of the equipment. The modeling and simulation of its system dynamics model is an important means for its analysis and evaluation. The existing system dynamics modeling methods do not support optical system modeling, which leads to the fragmentation of the multi-domain system model of laser equipment and the inefficiency of frequency domain analysis. Focusing on the co-simulation analysis requirements of the opto-mechanical system of laser equipment, a laser system modeling and simulation method for multi-domain unified models is proposed in this paper. Based on the lumped parameter method and geometric optics, the energy flow model of the laser system is established, and the unified theoretical modeling of the opto-mechanical coupling system is achieved. The research results of this paper will be applied to the unified modeling and co-simulation of the opto-mechanical coupling system, providing a theoretical and technical basis for the high-precision and efficient design verification of laser equipment.

Keywords: Laser equipment · Multi-domain unified model · Lumped parameter method · Geometric optics

1 Introduction

Laser equipment is a set of opto-mechanical equipment with precise structure, complex composition and advanced technology. It needs to irradiate the target continuously and directionally during work, which puts forward extremely high requirements for its tracking accuracy and irradiation power [1]. In the design and verification process of laser equipment, the modeling and simulation [2, 3] of the multi-domain coupling system is an important means for the analysis and evaluation of laser equipment's ergonomic performance, and also a necessary basis for the design of servo control algorithm of tracking and pointing system [4]. Therefore, the research on unified modeling and co-simulation technology for laser equipment has high theoretical significance and application value for improving the quality and efficiency of the design verification process.

The modeling and simulation of laser equipment is a typical multi-domain coupling system modeling and simulation problem. The modeling and simulation methods

W. Fan et al. (Eds.): AsiaSim 2022, CCIS 1713, pp. 39–49, 2022.
https://doi.org/10.1007/978-981-19-9195-0_4

of multi-domain systems are mainly divided into two categories: co-simulation and unified modeling. Co-simulation is the integration of simulation software in a single domain, so as to achieve multi-domain system simulation, which can be divided into model conversion and data communication from the working mode. The advantage of co-simulation is that it can make full use of the modeling and simulation capabilities of single-domain simulation software, while the disadvantage is that it must rely on corresponding software interfaces. With the increase of the number of software participating in co-simulation, the interface complexity increases sharply. For this reason, it is hoped that interface development can be made easier through standardization. At present, typical standardized interfaces for co-simulation include S-Function [5] proposed by MathWorks, HLA [6, 7] proposed by the Office of Modeling and Simulation of the U.S. Department of Defense, and FMI [8–10] led by Daimler Group. S-function is based on model transformation, HLA is based on runtime data communication, and FMI supports two co-simulation modes.

Interface standardization significantly reduces the interface complexity in multi-software co-simulation, but it does not reduce the difficulty of model transformation, nor does it solve the accuracy or stability problems caused by communication step size during simulation. Therefore, system simulation scholars try to achieve unified modeling and simulation of multi-domain systems in a single software environment. Elmqvist [11] proposed Dymola, a unified modeling language for multi-domain continuous systems, in 1978, which established the multi-domain coupled system as a single model and generated a unique simulation program. Subsequently, a large number of physical modeling languages emerged, such as gPROMS, Omola, NMF, Smile and so on. In order to achieve the standardization of modeling languages, The European Simulation Society summarized the advantages of various physical modeling languages and proposed the multi-domain unified modeling language Modelica [12] in 1997. Since then, the system simulation technology based on Modelica language has been developed rapidly, and a large number of industry application model libraries have been accumulated. Liping Chen [13, 14] made an in-depth study of the multi-domain unified simulation technology based on Modelica and established the system modeling and simulation platform of Modelica, including a complete compilation system, model solution and model library.

In summary, multi-domain unified modeling avoids the problems related to software interfaces in co-simulation, and its efficiency may be lower than that of dedicated software when only simulating a single domain. However, there is a lack of laser system and corresponding component library in the current multi-domain unified modeling theory, and a unified system dynamics model cannot be established in the modeling and simulation of laser equipment. Frequency domain analysis of laser beam optical axis and efficiency relies on multi-software co-simulation based on data communication [15]. It increases the difficulty and cost of co-simulation analysis of optical, mechanical, electronic, thermal and control and other multi-domain systems in laser equipment, and ultimately leads to disjointing modeling and simulation methods in the process of design verification of laser equipment, and low precision and low efficiency in frequency domain analysis.

Aiming at the above bottlenecks, a modeling method of laser system based on the lumped parameter method is proposed in this paper. Based on the unified theory, the system dynamics model of the opto-mechanical coupling system including mechanical components, optical components and laser beam is established. The co-simulation of the opto-mechanical coupled system is carried out by using the simulation method of the mechanical system dynamics model. Consequently, it will support the modeling and simulation of the multi-domain coupling system model of laser equipment, the mature and reliable dynamic analysis method could be used to evaluate the comprehensive performance of the opto-mechanical system, which provides the theoretical and technical basis for the high quality and efficient design verification.

2 Modeling and Simulation Method of Laser System for Multi Domain Unified Model

2.1 Modeling Method of Laser System Based on Lumped Parameter Method

Complex system modeling adopts a bottom-up approach to establish a corresponding system model according to the physical topology of the modeling object. Therefore, the basis of laser system modeling is the various optical component models it contains. The lumped parameter law simplifies the model of each component and considers that the state quantities of it are completely uniformly distributed, which greatly reduces the amount of simulation calculation and facilitates high-efficiency simulation of multi-domain coupled systems.

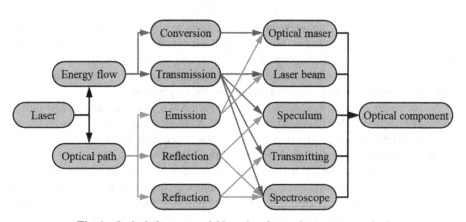

Fig. 1. Optical element model based on lumped parameter method

Based on the lumped parameter method, the laser beam can be regarded as a uniform energy flow transmitted in a three-dimensional optical path from a macroscopic perspective, and the spot power of it is uniformly distributed without optical phase deviation. The 3D optical path model of the laser beam completely follows geometric optics and is divided into emission, reflection and refraction. The energy flow model follows the principle of energy conservation and is divided into conversion and transmission. Therefore, the optical component models in various laser systems can be regarded as an arrangement and combination of the three-dimensional optical path and energy flow, as shown in Fig. 1.

The core basic components of laser system modeling are speculums, transmitting mirrors and spectroscope. For a speculum, according to the reflection law of light, the component model of it can be expressed as:

$$\mathbf{P}_{r,out} = \mathbf{P}_{r,in} \tag{1}$$

$$\vec{v}_{r,in} \times \vec{v}_m = -\vec{v}_{r,out} \times \vec{v}_m \tag{2}$$

$$p_{r,out} = k_r \cdot p_{r,in} \tag{3}$$

where $\mathbf{P}_{r,out}$ and $\mathbf{P}_{r,in}$ are the intersection of the reflected light and mirror, the incident light and mirror. $\vec{v}_{r,out}$, $\vec{v}_{r,in}$ and $\vec{v}_m$ are the optical axis vector of the reflected light and the incident light, and the normal vector of mirror. $P_{r,out}$, $P_{r,in}$, and k_r are the reflected light power, incident light power and reflectivity, respectively.

For a transmitting mirror, according to the law of refraction of light, the component model of it can be expressed as:

$$\mathbf{P}_{t,out} = \mathbf{P}_{t,in} \tag{4}$$

$$n_{in} \cdot \vec{v}_{t,in} \times \vec{v}_m = n_{out} \cdot \vec{v}_{t,out} \times \vec{v}_m \tag{5}$$

$$p_{t,out} = k_t \cdot p_{t,in} \tag{6}$$

where $\mathbf{P}_{t,out}$ and $\mathbf{P}_{t,in}$ are the intersection of the refracted light and mirror, the incident light and mirror. $\vec{v}_{t,out}$, $\vec{v}_{t,in}$ and $\vec{v}_m$ are the optical axis vector of the refracted light and the incident light, and the normal vector of mirror. n_{out} and n_{in} are the refractive indices of the transmission medium and the incident medium. $P_{t,out}$, $P_{t,in}$, and k_t are the transmitted light power, incident light power and transmissivity, respectively.

A spectroscope can be equivalent to a combination of a speculum and a transmitting mirror, and has both reflectivity and transmissivity. Without considering the optical phase deviation and the uneven distribution of the spot power density, other optical components can be equivalent to the finite combinations of these three basic mirrors.

The constraint relationship between optical elements and optical axes in geometric optics is an algebraic equation, which is consistent with the basic mathematical model of joint constraints in mechanical dynamics. Meanwhile, the laser beam energy flow model based on the lumped parameter method only contains algebraic equations and ordinary differential equations. Therefore, the established laser system model can be jointly simulated with the lumped parameter model of the mechanical, electronic, thermal and control system of the laser equipment under the unified modeling standard.

2.2 Co-simulation Method of Opto-Mechanical System Based on Multi-domain Unified Model

The multi-domain coupled system model of laser equipment established based on the lumped parameter method includes subsystem models in optical, mechanical, electromagnetic, thermal, control and other fields. The model equation of the mechanical part is the differential-algebraic equation set of index 3 (high index), while the set in other fields is usually index 1 or 0 (low index). Therefore, the numerical solution method of the mechanical system model is generally backward compatible with the solution simulation of the system model in other fields. The opto-mechanical coupling system model of the laser equipment is shown in Fig. 2.

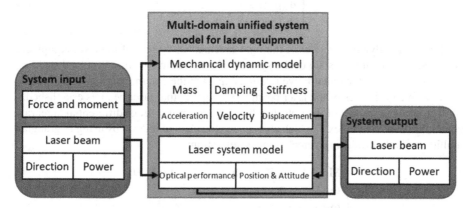

Fig. 2. Multi-domain unified system model for laser equipment

The dynamic model of the mechanical system is to establish a mathematical model between each mechanical component and the generalized displacement (translation and rotation) according to the assembly constraint relationship of the mechanical structure:

$$\begin{cases} \mathbf{M}\ddot{q} + \mathbf{C}\dot{q} + \mathbf{K}q = \mathbf{F}(t) \\ \Phi(q, t) = 0 \end{cases} \tag{7}$$

where $\mathbf{M}$, $\mathbf{C}$ and $\mathbf{K}$ are the generalized mass, damping and stiffness matrices of the system, $\mathbf{F}(t)$ are the external input forces of the system over time, q are the generalized displacements of the system, and Φ are the system constraints.

On the basis of the dynamic model of the mechanical system, the dynamic model of the laser system coupled with it can be established. The system dynamics model of optical components based on geometrical optics is the geometric constraints of the position and attitude of optical components on the optical axis of the laser beam. The mathematical model between each optical component and the optical axis of the laser beam is established according to the constraint relationship between the optical component and the optical axis:

$$\Phi_l(q_l, t) = 0 \tag{8}$$

where q_l is the macroscopic property of the laser beam and Φ_l is the optical component constraint on the optical axis of the laser beam.

Combined with the installation position of the optical component on the mechanical structure, the geometric position relationship between the generalized displacement of the mechanical component and the position and attitude of the optical component can be established, and the dynamic model of the opto-mechanical coupling system can be established by simultaneous equations. It can be considered that the optical path propagation of the laser beam has no influence on the dynamic model of the mechanical system without considering the influence of the light pressure effect:

$$\begin{cases} \mathbf{M}\ddot{q} + \mathbf{C}\dot{q} + \mathbf{K}q = \mathbf{F}(t) \\ \Phi(q, t) = 0 \\ \Phi_l(q_l, q, t) = 0 \end{cases} \tag{9}$$

Finally, the dynamic model of the opto-mechanical system of the laser equipment is solved to simulate the state of the system. The dynamic equation is solved by kinematic analysis, and the quantitative relationship from the external driving force and the macroscopic properties of the laser beam to the final output laser beam is established.

The dynamic model of the opto-mechanical system is a set of differential algebraic equations (DAEs) of index 3, which are reduced to DAEs of index 1 by using Lagrangian multipliers and introducing velocity and acceleration:

$$\begin{cases} \mathbf{M}a + \mathbf{C}v + \mathbf{K}q + \Phi_q^T(q, t)\lambda - \mathbf{F}(t) = 0 \\ \Phi(q, t) = 0 \\ \Phi_l(q_l, q, t) = 0 \\ v - \dot{q} = 0 \\ a - \dot{v} = 0 \\ \Phi(q, v, t) = 0 \\ \Phi(q, v, a, t) = 0 \end{cases} \tag{10}$$

where λ is the Lagrange multiplier, v is the velocity, and a is the acceleration.

The position, velocity, acceleration and Lagrangian multipliers are simultaneously used as the generalized coordinates of the system, and the simultaneous equations are overdetermined DAEs of index 1. Then two unknown parameters are introduced according to the compatibility of the system, and the overdetermination is eliminated to obtain the final reduced model for solving and simulation.

3 Modeling and Simulation Experiment of Laser System

According to the theory mentioned above, the basic laser emission, propagation, and reflection components can be established for modeling and simulation testing of optical systems based on geometric optics. This test is based on the Modelica 3.2.3 language standard and MWorks.Sysplorer 5.1 of MWORKS platform.

3.1 Optical Component Model Based on Modelica

At present, the Modelica standard library does not contain an optical component library, so it is necessary to establish three basic optical components: optical maser, macro laser beam, and speculum, so as to establish an optical system model for simulation testing.

In order to build a Modelica-based optical component model, it is first necessary to establish a Modelica-based optical system interface. For the optical path, in addition to the energy flow information, additional 3D optical path geometry information needs to be included.

Therefore, the Modelica-based optical system interface needs to include variables: optical path existence, optical axis intersection, optical axis direction, spot radius, and power density. For connected optical system interfaces, the corresponding variables are equal (Table 1).

Table 1. Variables for the optical system interface

Name	Value type	Unit
Optical path existence	Boolean	
Optical axis intersection	{Real, Real, Real}	mm
Optical axis direction	{Real, Real, Real}	
Spot radius	Real	mm
Power density	Real	W/mm^2

The laser beam is first generated by a optical maser, and currently only a pure ideal continuous laser beam is considered, that is, a completely parallel beam with a stable energy output. Macroscopically, it can be considered that the laser beam is a space ray, and the power is uniformly distributed along the spot and remains stable, and the energy source and conversion rate of the laser are not considered.

Therefore, the laser beam needs to include the following parameters: light exit point, emission direction, spot radius, and power density. This model has only the output optical system interface, which represents the emitted laser, and assigns its own parameters to the corresponding variables of the output interface.

In the system model, the macroscopic model of the laser beam is established instead of the microscopic model, and the connection line of the optical system represents the preset optical path, that is, the modeler believes that the optical components connected together may generate the optical path. If the detection optical path does not contact the connected optical element in the actual calculation, the actual optical path does not intersect, and some optical paths may not exist.

Therefore, the macroscopic laser beam model needs to include the following variables: optical path connection, incoming point, beam direction, beam length, outgoing point, spot radius and power density. The macroscopic laser beam model requires two optical system interfaces, one input and one output corresponding to incoming light and outgoing light respectively (Table 2).

The laser beam model needs to contain the equation:

$$\mathbf{P}_{out} = \mathbf{P}_{in} + \vec{v}_L \cdot l_L \tag{11}$$

where $\mathbf{P}_{out}$, $\mathbf{P}_{in}$, $\vec{v}_L$ and l_L are the outgoing point, incoming point, beam direction and beam length of the macroscopic laser beam model respectively.

For the speculum in the system model, it is not only must follow the specular reflection principle in geometric optics, but also need to determine whether the preset incident light

Table 2. Variables for the macroscopic laser beam model

Name	Value type	Unit
Optical path connection	Boolean	
Incoming point	{Real, Real, Real}	mm
Beam direction	{Real, Real, Real}	
Beam length	Real	mm
Outgoing point	{Real, Real, Real}	mm
Spot radius	Real	mm
Power density	Real	W/mm^2

is in contact with the mirror surface. If it does not actually contact, there is no reflected light. In order to simplify the model, the mirror is selected to be circular (Table 3).

Table 3. Parameters and variables for the speculum model

Name	Type	Value type	Unit
Speculum center	Parameter	{Real, Real, Real}	mm
Speculum normal vector	Parameter	{Real, Real, Real}	
Speculum radius	Parameter	Real	mm
Optical path connection	Variable	Boolean	

The speculum model needs to include the followings parameters: speculum center, speculum normal vector, and speculum radius. And the variable is optical path connection. The speculum model requires two optical system interfaces, one input and one output correspond to the incoming light and the outgoing light respectively.

According to the variables of the input optical system interface and the parameters of the speculum model, it is judged whether the preset optical path truly intersects with the speculum surface. The intersection of the incoming light and the plane where the speculum is located could be calculate via Eq. (12):

$$\mathbf{P}_{isec} = \mathbf{P}_{in} + \frac{(\mathbf{P}_m - \mathbf{P}_{in}) \cdot \vec{v}_m}{\vec{v}_{in} \cdot \vec{v}_m} \cdot \vec{v}_{in} \tag{12}$$

where $\mathbf{P}_{isec}$ is the intersection of the incoming light and the plane where the speculum is located, $\mathbf{P}_m$ is the speculum center, $\vec{v}_m$ is the speculum normal vector and $\vec{v}_{in}$ is the direction of the incoming light respectively.

According to the intersection, it can be judged whether the optical path truly intersects the speculum:

$$|\mathbf{P}_{isec} - \mathbf{P}_m| < r_m \tag{13}$$

where r_m is the speculum radius.

When the in Eq. (13) is established, it means that the optical axis and the speculum surface really intersect, and the mirror will work normally at this time.

3.2 Laser System Simulation in Sysplorer

The optical component model described in the previous section is used to build a simple reflection light path model for the optical system simulation test. The test model is composed of 1 ideal optical maser model, 3 macroscopic laser beam models, and 2 speculum models in series. The initial geometric properties of them are set so that the laser beam emitted from the optical maser is reflected by the speculumin and finally emitted. The pose parameters of the mirror in the simulation are adjusted to test the simulation effect of the three-dimensional optical path. The 6 optical component models are connected in the sequence shown in Fig. 3.

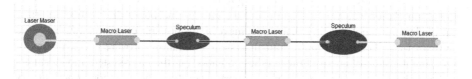

Fig. 3. Model of laser reflection system in Sysplorer

In the simulation, the parameters of the components are set as the value shown in Table 4.

Table 4. Parameters of the components

Component	Parameters	Value
Optical Maser	Positon	$\{0, 0, 0\}$
	Attitude	$\{1, 0, 0\}$
Speculum 1	Positon	$\{400, 0, 0\}$
	Attitude	$\{[-\cos 10°\sim-\cos 30°], [\sin 10°\sim\sin 30°], 0\}$
	Radius	25
Speculum 2	Positon	$\{100, [100\sim400], 0\}$
	Attitude	$\{\cos 10°, -\sin 10°, 0\}$
	Radius	25

Figure 4(a), (b) are the normal vector of the mirror, the optical axis direction of the incident light and the reflected light of the two speculums, respectively, which fully conforms to the reflection law in geometric optics. When the laser beam is not contacting the Speculum 2, the reflected light of it no longer exists. Therefore, the reflection simulation

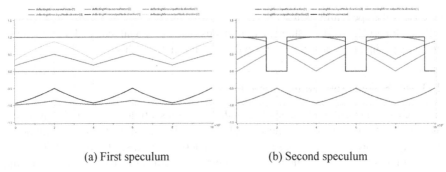

(a) First speculum   (b) Second speculum

Fig. 4. Simulation of laser reflection system in Sysplorer

results of the laser system conform to the optical theory, and can be used for modeling and simulation analysis of the opto-mechanical system.

At present, in the simulation of laser equipment, it is convenient to use professional optical simulation software for analyzing the energy and phase distribution information of the light spot, but it is hard to achieve multi-domain co-simulation based on multi-software. The use of multi-domain system simulation software is convenient for analyzing the state of energy flow in other domains, but the laser system model cannot be established without the optical model library. Therefore, the laser component library based on Modelica established in this paper will support the unified modeling of the multi-domain system of laser equipment, and realize the high-precision and efficient analysis and evaluation of the performance of the laser equipment system.

4 Conclusion

Focusing on the unified modeling, simulation, performance analysis and evaluation requirements of multi-domain coupled systems in the design and verification process of laser equipment, aiming at the problem of fragmentation of existing system modeling and simulation methods, a unified modeling and simulation method for laser system dynamics based on the lumped parameter method is proposed in this paper. Combined with the energy flow model of the laser system and the three-dimensional space constraint of the optical element on the laser beam optical axis in geometric optics, the system dynamics model of the laser system based on the lumped parameter method is established. According to this theory, a model library of laser system components is established based on Modelica, and the validity of the modeling theory and model library is verified by the simulation analysis of the laser system in Sysplorer. The modeling and simulation method of laser system proposed in this paper provides the analysis and evaluation foundation under the unified theory for the multi-domain coupling system of laser equipment. The established component model library will support the high-precision and efficient analysis and evaluation of laser equipment system performance, and improve the quality and efficiency of laser equipment design and verification.

Acknowledgements. This work was supported by the National Natural Science Foundation of China (61903348). The authors would like to thank Baokun Zhang and Yinqi Tang for their assistance with the experimental work.

References

1. Zhang, H., Song, N., Dai, Z., et al.: Multi- ple physical field system simulation for high energy laser weapon target attacking. J. Appl. Opt. **38**(4), 526–532 (2017)
2. Wang, P., Li, Y., Tian, J.: Simulation system and analysis of airborne laser weapon. Infrared Laser Eng. **40**(7), 1238–1242 (2011)
3. Fu, J., Maré, J.C., Fu, Y.: Modelling and simulation of flight control electromechanical actu- ators with special focus on model architecting, multidisciplinary effects and power flows. Chinese J. Aeronaut. **30**(1), 47–65 (2017)
4. Zhang, Y.: A Study on Key Technology for Improve Tracking Accuracy of Ship-based Opto- electronic Equipment. Changchun Institute of Optics, Fine Mechanics and Physics Chinese Academy of Sciences (2013)
5. Ozana, S., Machácek, Z.: Implementation of the mathematical model of a generating block in matlab and simulink using s-functions. In: 2009 Second International Conference on Computer and Electrical Engineering, pp. 431–435. IEEE (2009)
6. Kuhl, F., Weatherly, R., Dahmann, J.: Creating computer simulation systems: an introduction to the high level architecture. Prentice Hall PTR, USA (1999)
7. Zhang, Z., Li, B., Chai, X., et al.: HP-HLA/RTI prototype oriented on shared memory environment. J. Syst. Simul. **26**(2), 315–322 (2014)
8. Blochwitz, T.: The functional mockup interface for tool independent exchange of simulation models. In: 8th International Modelica Conference. DLR (2011)
9. Tian, H., Shan, L., Li, C.: Functional mock-up interface (FMI) standard. In: 2011 International Conference on Information, pp. 1401–1405. Services and Management Engineering (2011)
10. Vanfretti, L., Bogodorova, T., Baudette, M.: Power system model identification exploiting the Modelica language and FMI technologies. In: 2014 IEEE International Conference on Intelligent Energy and Power Systems (IEPS), pp. 127–132. IEEE (2014)
11. Elmqvist, H.: A Structured Model Language for Large Continuous Systems. Sweden: Lund Institute of Technology (1978)
12. Elmqvist, H., Mattsson, S.E., Otter, M.: Modelica: the new object-oriented modeling language. In: 12th European Simulation Multiconference, Manchester, UK, pp. 127–131 (1998)
13. Zhao, J., Ding, J., Zhou, F., Chen, L.: Modelica and its mechanism of multi-domain unified modeling and simulation. J. Syst. Simul. **18**(2), 570–573 (2006)
14. Zhou, F., Chen, L., Wu, Y., et al.: MWorks: a modern IDE for modeling and simulation of multidomain physical systems based on Modelica. In: Proceedings of the 5th International Modelica Conference, pp. 725–731 (2006)
15. Zhao, Y.: Research on Parallelization of Simulation and Computing of Complex Mechatronics Systems. Huangzhong University of Science & Technology (2017)

Fast Electromagnetic Scattering Modeling of Complex Scenes Based on Multi-scale Coherent Computation

Gong Ningbo[1](✉), Diao Guijie[2](✉), Chen Hui[3], Ni Hong[1], Du Xin[1], and Liu Zhe[1]

[1] Beijing Electro-Mechanical Engineering Institute, Beijing 100074, China
ningbogong15@163.com
[2] Science and Technology on Complex System Control and Intelligent Agent Cooperation Laboratory, Beijing 100074, China
dgj1002@163.com
[3] South-East University, Nanjing 210096, China

Abstract. Modeling of scattering characteristics between ground/offing targets and environment is one of the necessary conditions to realize the internal field reconstruction of complex electromagnetic scenes. According to the high-frequency approximation assumption of electromagnetic scattering localization, the coupling scattering electromagnetic modeling method of composite targets and ground/offing scenes is studied in the framework of the bouncing ray method. Based on the curvature adaptive bin separation algorithm and the principle of multi-scale coherent superposition, an integrated scattering modeling method for multi-scale targets and complex environments is developed. The scattering of typical ship target, background and the coupling scattering between ground/offing artificial targets and environment are calculated respectively. Theoretical analysis and simulation result show the correctness and effectiveness of the algorithm.

Keywords: Scattering signature · Shooting bounce ray (SBR) · Multi-scale integrated scattering modeling

1 Introduction

The ability to detect, track and identify targets in the complex clutter environment is one of the important links in evaluating the effectiveness of radar detector [1, 2]. Fortunately, the various complex electromagnetic environment faced by the radar in the process of detecting targets could be reconstructed by the array radio frequency (RF) simulation system. It becomes an effective means to verity the performance of radar detector. How to accurately and quickly obtain the scattering characteristics between ground/offing targets and environment is an important technical basis for the internal field reconstruction of complex scenes.

The spatial distribution of land surface is random, which includes large-scale topographic relief and small-scale surface random micro relief. Similarly, the state of the ocean surface is random and time-varying, which includes large-scale swell and small-scale short wave, capillary wave and other fine structures. Small-scale wave is dynamically changed with time by the hydrodynamic modulation of large-scale wave. Small-scale waves are distributed in the range of centimeter to tens of centimeter. Bragg resonance occurs with centimeter or millimeter wave, which is considered to be the main contribution mechanism of sea surface scattering [3].

The high frequency approximation method based on ray tracing technique is an effective engineering method for electromagnetic scattering computation of complex structural targets [4–8]. In this method, the shape of the complex structure should be expressed determinedly in the form of triangulation. When the target surface has random fine structures such as roughness, all the fine structures should be expressed in the form of determinate subdivision. For the sea surface, it need to reflect the scattering mechanism of the sea surface capillary wave. So, the fluctuation of the sea surface should be digitally rasterized according to the centimeter accuracy. Similarly, it is also necessary to refine the digital geometric subdivision for slightly rough land surface, vegetation branches and leaves. In this way, a huge number of partition elements will be introduced. It is difficult to deal with even for high-frequency methods with high computational efficiency.

Based on the high-frequency approximation assumption of electromagnetic scattering localization, this paper studies the coupling scattering electromagnetic modeling method of complex targets and ground sea surface scenes under the framework of the bouncing ray method. An integrated scattering modeling method of multi-scale targets and complex environment is discussed to establish electromagnetic scattering model more suitable for the composite scene of ground and sea targets. It could lay a foundation for further realization of infield reconstruction in complex scenes.

2 Integrated Scattering Modeling Principle of Multi-scale Target and Complex Environment

The ground background is divided into large-scale terrain contour and small-scale landform according to its electrical scale. The large-scale structure of the surface is characterized by an accurate digital CAD model. While the modulation effect of the micro scale random rough fluctuation on the electromagnetic wave is accounted for by the statistical random surface scattering model. According to the high-frequency approximation assumption of electromagnetic scattering localization, the coherent scattering contribution of each surface triangle element can be calculated by the high-frequency method of roughness factor correction. The incoherent contribution caused by rough fluctuation is given by the random scattering mechanism model. And the total contribution is composed of two parts of contribution superposition.

The beam tracing technique is used as the main part of the integrated scattering modeling of target and complex environment. And the reflection coefficient model is used to connect the terrain specular reflection and geomorphic diffuse reflection. The coherent scattering contribution of the geomorphic surface is taken into account by the physical optics method. The incoherent scattering contribution is with the aid of the rough surface scattering theory. The coupling between the surfaces (including the target and the surface) is included by the ray tracing method. Finally, the coherent and incoherent components are combined to obtain the total field amplitude and phase distribution of the target (Fig. 1).

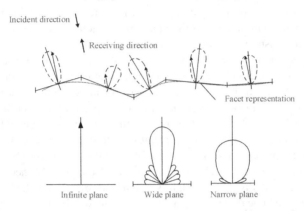

Fig. 1. Small bin representation and reradiation diagram of large-scale surface

3   Bouncing Ray Method Based on Curvature Adaptive Subdivision

The shooting bouncing ray (SBR) method based on curvature adaptive subdivision is composed of geometric optics (GO) method and physical optics method (PO). PO is used to calculate the scattering field of the target. The influence of multiple reflection field is included in the calculation of the scattering characteristics of the target. It consists of three parts: ray tracing, field intensity tracing and far-field integration, as shown in Fig. 2.

3.1   Beam Tracking Based on Curvature Adaptive Subdivision

In order to improve the efficiency of ray tracing, an adaptive segmentation algorithm of targets and background curvature is adopted. The flat areas are described by a large patch. The beam is rarely or almost not split on the large patch. So, correct calculation results could be obtained, which greatly reduces the number of ray tubes and improves the calculation speed. The large curvature area is described by more patches. It not only reduces the number of patches to be divided and the division error, but also ensures that the beam does not split too much in the area with dense patches. The ray tube with a small projection area has little contribution to the field, but it will pay a great price to track them. The combination of the two greatly improves the efficiency of ray tracing and accelerates the calculation speed of the algorithm (Fig. 3).

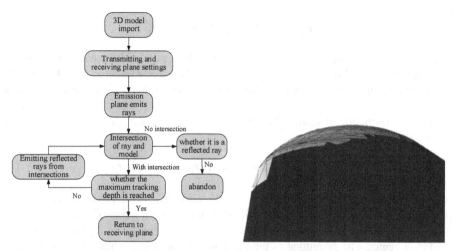

Fig. 2. Ray tracing flow chart

Fig. 3. X-ray tube splitting at the apex of NASA amygdala

3.2 Field Strength Tracking

Field strength tracking includes field strength amplitude tracking and field strength phase tracking (Figs. 4 and 5).

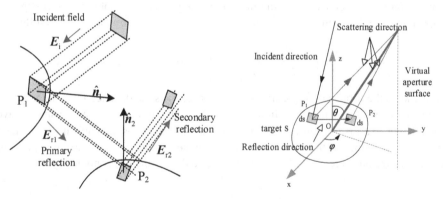

Fig. 4. Field intensity amplitude tracking

Fig. 5. Field intensity phase tracking

The incident electric field is:

$$\mathbf{E}^i = E_{//}\hat{\mathbf{e}}^i_{//} + E_\perp\hat{\mathbf{e}}^i_\perp \tag{1}$$

where $\hat{\mathbf{e}}^i_{//}$ is the unit vector parallel to the incident plane, $\hat{\mathbf{e}}^i_\perp$ is the unit vector perpendicular to the incident plane, $\mathbf{E}^i$ is the incident electric field, $E_{//}$ and $E_\perp$ are the components along and respectively.

Combined with the reflection law and electromagnetic field boundary conditions, the reflected electric field can be expressed as:

$$\mathbf{E}^\mathbf{r}_\mathbf{vv} = -2R^r_{//}(\hat{\mathbf{n}} \cdot \mathbf{E}^\mathbf{i})\hat{\mathbf{n}} - R^r_\perp\mathbf{E}^\mathbf{i} \tag{2}$$

$$\mathbf{E}^{r}_{HH} = -2R^{r}_{//}(\hat{\mathbf{n}} \cdot \mathbf{E}^{i})\hat{\mathbf{n}} + R^{r}_{//}\mathbf{E}^{i} \tag{3}$$

where, HH and VV represent horizontal polarization and vertical polarization respectively. $\hat{\mathbf{n}}$ is the normal vectors of intersections. $\mathbf{E}^{r}$ is the reflected electric field. $R^{r}_{//}$ and $R^{r}_{\perp}$ are the reflection coefficients of transverse magnetic and transverse electric components respectively.

Phase information of scattered electric field of all rays at radar receiving antenna is expressed as follows,

$$P = P_1 + P_2 + P_3 = k_o = (\hat{i} \cdot \mathbf{r}_{p1} + d + (\hat{S} \cdot \mathbf{r}_{p2})) \tag{4}$$

where $\hat{i}$ is the unit vector of incident wave direction, $\mathbf{r}_{p1}$ is the position vector of P_1 point relative to the target center, d is the distance between points P_1 and P_2, $\hat{s}$ is the unit vector of scattering direction, and $\mathbf{r}_{p2}$ is the position vector of P_2 point relative to the target center. The amplitude of the incident electric field at each reflection point is multiplied by the phase information to obtain the incident electric field at the reflection point.

3.3 Far Field Integral

The surface induced current at the emitting position of each ray tube is solved by using the equivalent current principle [9]. The far-field scattering field generated by the equivalent current on the object surface at the radar receiving antenna is obtained by coherently superimposing the contributions of all ray tubes. Kirchhoff formula [10] is as follows:

$$\sqrt{\sigma_s(\mathbf{r})} = \frac{jk_0}{4\pi} \cdot \frac{e^{-jk_0 R_r}}{R_r} \cdot \int_{S_c} \hat{s} \times (M_s(r') + Z_0\hat{s} \times \mathbf{J}_s(r'))e^{(jk_0\hat{\mathbf{r}}' \cdot \hat{s})} ds_c \tag{5}$$

where k_0 is the propagation constant in free space, R_r is the distance from the target to the antenna, $\hat{s}$ is the unit vector in the observation direction, Z_0 is the free space wave impedance, $\mathbf{J}(r')$ is the equivalent current in the bin, $\mathbf{M}_s(r')$ is the equivalent magnetic current in the bin, r is the radar antenna coordinate vector, $\mathbf{r}'$ is the current equivalent current coordinate vector on the target surface, and S_c is the target boundary.

4 Scattering Correction of Rough Surface Based on Multi-scale Coherent Superposition

According to the bin idea described by Kouali et al. [11], the phase term contributed by the bin is preliminarily assumed to be composed of fixed random initial phase and relative path delay phase. The total scattering field of the target can be obtained by superposing the scattering coherence of all small facets with random fine mechanisms.

Assuming that the partition number of a target triangular panel is M and the area of the panel is ΔS, the total field can be expressed as follows,

$$E^S = (\hat{k}_i, \hat{k}_s) \sum_{m=1}^{M} \left\{ E_{PO}^{Coherent}(\hat{k}_i, \hat{k}_s) + \frac{ikR_0}{R_0} \sqrt{\Delta S \sigma_m^{in\ coherent}(\hat{k}_i, \hat{k}_s)/4\pi} \exp(-i\varphi_m^{Add}) \right\}$$

(6)

where R_0 is the distance from the observation point to the center of the target coordinate system and the scattering contribution $\sigma_m^{Incoherent}$ is caused by the random microstructure of each bin, which can be obtained from the statistical model. $E_{PO}^{Coherent}$ is the coherent scattering contribution calculated by considering the bin as geometrically smooth and introducing the roughness correction reflection coefficient. It can be calculated according to the physical optics high frequency method. It will not be repeated here. The Fresnel reflection coefficient needs to consider the following correction.

$$\Gamma^{rough} = e^{-4\delta^2 k_n^2} \Gamma_0$$

(7)

where k_n is the projection of the incident wave vector on the panel normal vector. In addition, the phase deflection term φ_{ij} of each bin due to the influence of random microstructure can be assumed as

$$\varphi_m^{add} = \Delta\varphi_m^{max}\xi + \Delta\rho_m$$

(8)

where ξ is a uniformly distributed random number, and $\xi \in [-1/2, 1/2]$. $\Delta\varphi_m^{max} = \left(\hat{k}_s - \hat{k}_i\right) \cdot \Delta\vec{r}_m$, represents the variation range of the maximum phase difference in the bin. $\Delta\vec{r}_m$ is the length vector intercepted along the projection direction of the vector on each bin; $\Delta\rho_m = \left(\hat{k}_s - \hat{k}_i\right) \cdot \vec{r}_m$, is the phase delay caused by the relative position of each bin, and $\vec{r}_m$ is the position vector of each bin center relative to the origin of the target coordinate system.

5 Simulation Results and Analysis

5.1 Typical Target Scattering Calculation

The scattering characteristics of typical ship target are calculated by using SBR based on curvature adaptive subdivision. Its electromagnetic calculation parameters are set as follows: center frequency F_0, HH polarization, bi-static, Azimuth $0°$ to $360°$. The results

56   G. Ningbo et al.

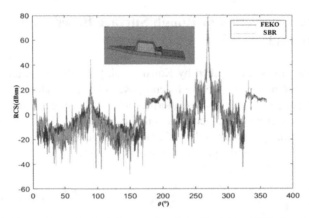

Fig. 6. Comparison of calculation results of double station RCS at typical ship

are compared with those of the electromagnetic calculation commercial software FEKO to verify the correctness of the algorithm (Fig. 6).

Obviously, the calculation results of SBR based on curvature adaptive subdivision are highly consistent with those of FEKO, which improves the calculation efficiency while ensuring the calculation accuracy.

5.2 Scattering Calculation of Sea Surface Environment

In order to verify the validity of the calculation results by the environmental scattering model, the comparison were carried out with the field measured data. The following figure shows the variation of back scattering coefficient with incident angle under different wind speeds and polarization conditions.

It is easy to draw the following conclusions through comparison: 1) from the overall trend, the predicted value under the same polarization condition is in good agreement with the measured data, while the predicted value of cross polarization is generally smaller than the measured value; 2) The difference with the measured data is mainly reflected in the near vertical incidence region and the near grazing incidence region. For the medium incidence condition, the predicted value is in good agreement with the measured data (Fig. 7).

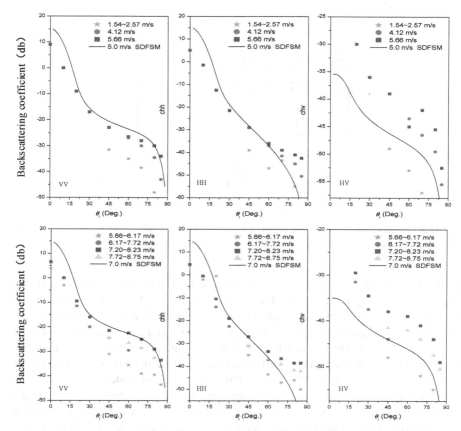

Fig. 7. Comparison between the calculated results and JOSS measured data

5.3 Application Verification of Coupling Electromagnetic Scattering in Complex Scene

Based on the calculation results of the electromagnetic scattering between ground/sea targets and environment, dynamic complex scenes are constructed via RF hardware in the loop simulation system. The imaging radar detector is used to process the scene echo modulated by electromagnetic scattering information to realize the effective reconstruction of the internal field in complex scenes. See Fig. 8. According to the flight path, the coupling scattering characteristics between ground/sea targets and environment are modeled to provide input for the radar echo simulator. During the simulation, the radar echo is simulated in real time according to the flight attitude and track of the detector, and the simulation test environment is built in the darkroom environment. After receiving the signal, the detector carries out imaging and matching recognition.

The imaging matching scene shown in Fig. 9(a) is used as the detector image matching scene, which contains typical geomorphic features such as asphalt, ground and cement buildings. Through this method, the electromagnetic scattering of the composite scene are modeled, and the broadband scattering characteristics model is constructed.

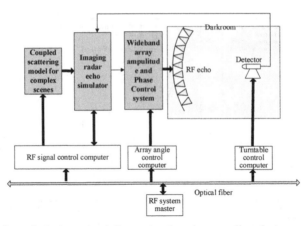

Fig. 8. RF Hardware in the loop simulation system based on coupling electromagnetic scattering in complex scenes

During the simulation, the real-time altitude is simulated by the detector mounted on the turntable. According to the relative line of sight relationship between the detector and the ground scene in flight, the broadband scattering feature model is retrieved in real time. The echo signal of the detector is simulated by the imaging radar echo simulator in combination with the relative relationship between the detector and the scene. After receiving the signal, the detector performs imaging. The imaging result is shown in Fig. 8(b). It can be seen that with this algorithm, the calculated scattering characteristics are well focused after the SAR detector imaging processing, which fully reflects the details of asphalt, grass, cement buildings and so on. Among them, the vehicles with smooth surfaces are mainly specular reflection, so the vehicle images in the SAR images show strong bright spots. And the detector successfully matched the typical buildings on the ground. The imaging processing and matching recognition performance of the detector in flight have been effectively verified.

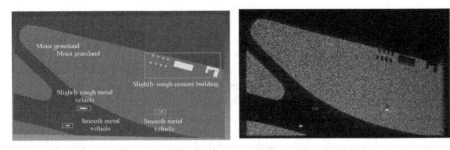

(a)modeling scenarios   (b)detector imaging results

Fig. 9. Application verification example of coupling electromagnetic scattering in complex scene

6 Summary

Based on the assumption of high frequency approximation of electromagnetic scattering localization, an integrated scattering modeling method of multi-scale target and complex environment is discussed in the framework of bouncing ray method. The correctness and effectiveness of the method are verified by comparing and analyzing the azimuth dependent scattering field of typical ship targets and the electromagnetic scattering characteristics of the sea environment. In addition, the electromagnetic scattering data of complex scene are applied to the RF hardware in the loop simulation system to complete the verification of the imaging processing and matching recognition performance of the detector in flight.

References

1. Blair, W.D., Brandt-Pearce, M.: Monopulse DOA estimation of two unresolved Rayleight targets.Trans. Aerospace Electr. Syst. **37**(2), 452–469 (2001)
2. Xin, Z., Willett, P., Bar-Shalom, Y.: Mono-pulse radar detection and localization of multiple targets via joint multiple-bin processing. In: 2003 IEEE Radar multiple-bin Conference. Huntsville Alabama, pp. 232–237 (2003)
3. Ning. W.J.: Modeling and Simulation of broadband electromagnetic scattering characteristics from time-varying sea surface. Beijing University of Aeronautics and Astronautics, Beijing (2015)
4. Johnson, J.T., Toporkov, J.V., Brown, G.S.: A numerical study of backscattering from time-evolving sea surface: comparison of hydrodynamic models. IEEE Trans. Geo Sci. Remote Sens. **39**(11), 2411–2420 (2001)
5. Zheng, H., Ye, H., Xu, F.: Extended bi-directional analytical ray tracing algorithm for lossy dielectric object. Chin. J. Radio Sci. **30**(5), 896–902 (2015)
6. Bhalla, R., Lin, L., Andersh, D.: A fast algorithm for 3D SAR simulation of target and terrain using Xpatch. In: IEEE International Radar Conference, pp. :377–382 (2006)
7. He, H.J.,Xin, G.L.: A multihybrid FE-BI-KA technique for 3-D electromagnetic scattering from a coated object above a conductive rough surface. IEEE Geosci. Remote Sens. Lett. PP(99): 1–5 (2016)
8. Bhalla, R., Ling, H.: Image-domain ray-tube integration formula for the shooting and bouncing ray technique. Radio. Sci. **30**, 1435–1446 (1995)
9. Guiru, Y.: Advanced Electromagnetic Theory. Higher Education Press, Beijing (2004)
10. Gordon, W.B.: Far-field approximations to the Kirchhoff-helmholtz representations of scattered fields. IEEE Trans. Antennas Propag. **23**(5), 590–592 (1975)
11. Kouali, M., Bourlier, C., Kubicke, G.: Scattering from an object above a rough surface using the extended Pile method hybridized with PO approximation. In: Antennas and Propagation Society International Symposium, 2012: 1–2.Author, F.: Article title. Journal **2**(5), 99–110 (2016)

Pixelated Image Abstraction via Power Diagram

Tao Li, Yuyou Yao, Wenming Wu, and Liping Zheng$^{(\boxtimes)}$

School of Computer Science and Information Engineering,
Hefei University of Technology, Hefei, China
zhenglp@hfut.edu.cn

Abstract. Image abstraction is increasingly used in the field of computer graphics and computer vision. Existing methods divide the input images into disjoint regions and generate the abstract images with a reduced color palette. However, the generated images are not satisfactory due to the presence of blurred and unsmooth boundaries, etc. To this end, a pixelated image abstraction method via power diagrams is introduced in this paper to generate high-quality abstract images. Firstly, a pre-processing step is used to capture the significant information in the input image, which is divided into core and non-core parts according to the importance of each pixel. Then, a density map is calculated on basis of the division and grayscale value of each pixel. Furthermore, the primal image is partitioned into polygonal subregions with equal area constraint based on power diagrams. Based on the average RGB color of the 3 × 3 grids centered at its centroid, the pixels in each polygonal subregion are colored to produce the output abstract image. Experimental results demonstrate that the proposed method could generate high-quality images with abstraction controllable and smooth boundaries.

Keywords: Image abstraction · Segmentation · Pixelated · Power diagram

1 Introduction

Pixel images are widely used in our daily life, i.e., the smartphones and computer monitors utilize pixel art to convey information to users [1]. With the development of software and hardware in the photograph technology, an increasing number of high-quality images has emerged, which usually contain colorful pixels. Despite these images with colorful pixels bring good experience to users, the problem of data storage and image processing also arises [2]. Hence, it is important to simplify images while preserving the significant pixels.

Image abstraction [3] is an efficient technique for simplifying images by generating a simpler image with less memory consumption and fewer colorful pixels from an input image. Existing image abstraction methods simplify images by considering two aspects: the structure and the color of an image. The former uses less structure to represent an image, while the latter uses less color. On the

W. Fan et al. (Eds.): AsiaSim 2022, CCIS 1713, pp. 60–74, 2022.
https://doi.org/10.1007/978-981-19-9195-0_6

basis of this, existing image abstraction methods could be broadly classified into two categories: segmentation based image abstraction method [4–7] and filter based image abstraction method [8–11].

Segmentation based image abstraction method divides the primal image into disjoint subregions, each of which is further filled with the same color [4]. DeCarlo et al. [12] introduce a visual perception based non-uniform image abstraction algorithm, which converts an input image into a line-drawing style consisting of multi-regions with same color. High-attention regions are preserved to remain more detailed information, while these regions with low-attention are simplify filled. However, the input image is segmented on basis of the visual perception, which requires an expensive eye tracker and is inefficient. Ren et al. [13] propose the concept of superpixels to divide the input image into irregular blocks, each of which is composed of these pixels with similar features. The superpixels greatly reduce the complex of images, and generate satisfactory abstract images. Several researches [4] have studies the superpixels and present some algorithms for image abstraction [5,6]. Nevertheless, the image abstraction method based on superpixels may generate low-quality abstract images with unsmooth boundaries (the boundaries between significant and non-significant pixels in the following).

Different from the segmentation based method, the filter based image abstraction method use the Mean-Shift filter to solve the image abstraction problem in video [8]. Specifically, Wang et al. [8] develop an anisotropic Mean-Shift filter to segment video data into continuous data, thus solving the abstraction problem of offline video to achieve good spatio-temporal correlation [14]. However, the smoothness and accuracy of the segmentation are still the challenges, and the large amount of time consumption is also a problem to be solved with the filter based image abstraction method.

Therefore, we focus on the limitations of previous work, namely the unsmooth boundaries and the uncontrollability of generated abstract images. In this paper, we proposed a novel method for pixelated image abstraction with power diagrams. A pre-processing step is required to capture these pixels with important information in the input image, which is divided into core (with significant pixels) and non-core (without significant pixels) parts. Then, a density map is initialized based on the division and grayscale value of each pixel in the input image. Additionally, the primal image is partitioned into polygonal subregions with the same area constraint based on the power diagram, each of which is filled with the average RGB color of the 3×3 grids centered on its centroid. Benefiting from the properties of polygonal subregions, our method could generate high-quality abstract images with smooth boundaries. Besides, the number of sites in the power diagram based image segmentation and the initialized density map give our method with the ability of generating controllable abstract images. The main contribution of our work is as follows:

- We introduce a pixelated image abstraction method via power diagrams to generate high-quality abstract images with a controllable abstraction degree.
- An image segmentation method based on the power diagram to produce polygonal subregions (hexagon is optimal), which makes the generated abstract images with smooth boundaries.

The remainder of this paper is organized as follows. Section 2 briefly reviews the concepts of power diagrams and the image abstraction methods. Our method is introduced in Sect. 3. Some experimental results are provided in Sect. 4, and some conclusions are given in Sect. 5.

2   Related Work

In this section, we introduce the concepts of the power diagram and the centroidal capacity constraint power diagram (CCCPD). Besides, a briefly review of image abstraction methods is provided.

2.1   Power Diagram

Given the domain $\Omega \subset E^d$ and a set of sites $\mathbf{S} = \{\mathbf{s}_i\}_{i=1}^n$ (also called "points" or "generators"), the Voronoi diagram defines a partition, dividing the domain Ω into n disjoint subregions $V(\mathbf{S}) = \{V(\mathbf{s}_i)\}_{i=1}^n$. Each subregion (also called "Voronoi cell") $V(\mathbf{s}_i)$ of the site $\mathbf{s}_i$ is defined:

$$V(\mathbf{s}_i) = \{\mathbf{s} | \|\mathbf{s} - \mathbf{s}_i\| \le \|\mathbf{s} - \mathbf{s}_j\|, \forall j \ne i\} \tag{1}$$

where $\| \cdot \|$ is the Euclidean distance.

As an extension of the Voronoi diagram, the power diagram introduces the "weight" characteristic to each site, that is, each site $\mathbf{s}_i$ is associated with a value w_i. Similar to the Voronoi diagram, the power diagram divides the domain Ω into n subregions $P(\mathbf{S}) = \{P(\mathbf{s}_i)\}_{i=1}^n$, and each subregion (also called "power cell") $P(\mathbf{s}_i)$ of the site $\mathbf{s}_i$ is redefined:

$$P(\mathbf{s}_i) = \{\mathbf{s} | \|\mathbf{s} - \mathbf{s}_i\|^2 - w_i \le \|\mathbf{s} - \mathbf{s}_j\|^2 - w_j, \forall j \ne i\} \tag{2}$$

Notably, the power diagram degenerates to a Voronoi diagram when all weights of sites are equal [15].

2.2   CCCPD

By imposing the centroidal constraint and capacity constraint to the ordinary power diagram, a CCCPD can be obtained. In other words, a CCCPD is a special power diagram, where 1) each site $\mathbf{s}_i$ is located in the mass center of its power cell $P(\mathbf{s}_i)$; and 2) the area/capacity of each $P(\mathbf{s}_i)$ is equal to the preset capacity constraint c_i of the site $\mathbf{s}_i$. The centroid and capacity of each power cell $P(\mathbf{s}_i)$ could be calculated:

$$\begin{cases} \mathbf{s}_i^* = \dfrac{\int_{P(\mathbf{s}_i)} \mathbf{s}\rho(\mathbf{s})ds}{\int_{P(\mathbf{s}_i)} \rho(\mathbf{s})ds} \\ m_i = \int_{P(\mathbf{s}_i)} \rho(\mathbf{s})ds \end{cases} \tag{3}$$

where $\rho(\mathbf{s})$ is the density of any point $\mathbf{s}$ located in the domain Ω. Aurenhammer et al. [16] point out that a CCCPD could be generated by minimizing the following term:

$$F(\mathbf{S}, \mathbf{W}) = \sum_{i=1}^n \int_{P(\mathbf{s}_i)} \rho(\mathbf{s})\|\mathbf{s} - \mathbf{s}_i\|^2 ds - \sum_{i=1}^n w_i(m_i - c_i) \tag{4}$$

Various studies have presented different algorithms to compute the power diagram. Balzer et al. [17] use the false-position method and the Lloyd's method to generate the CCCPD, which is time-consuming. de Goes et al. [18] propose the Newton's method to optimize the weights, and use the adaptive step-size gradient descent method to optimize the sites, which greatly improves the CCCPD computation efficiency. Xin et al. [19] further present the L-BFGS method with super-linear convergence to compute the CCCPD, but the construction of power diagram is still time-consuming. Recently, Zheng et al. [20] provide a GPU-accelerated method to improve the efficiency of power diagram construction rather than CGAL based method. Furthermore, they extend the capacity constraint in CCCPD to the general case, and introduce a hybrid capacity constrained centroidal power diagram (HCCCPD) [21]. In this paper, the power diagram computation framework in [18] is utilized to compute the image segmentation results.

2.3   Image Abstraction

Existing methods for image abstraction mainly involves the segmentation based image abstraction method and the filter based image abstraction method. The former divides the input image into several subregions with similar features, each of which is filled with the same color to generate the abstract image. The Mean-Shift filter is used to separate the input image, and the abstract image is generated accordingly.

DeCarlo et al. [12] propose an interaction technique based on a visual perceptual model. The image abstraction relies on a hierarchical representation of visual forms in the input image, and the abstract images are formed by segmenting on different scales. However, an expensive eye tracker is required in this method, and the image abstraction is inefficient. Ren et al. [13] introduce the concept of superpixels, which are irregular blocks, consisting of pixels with similar features. By dividing the input image into a small number of superpixels rather than a large number of pixels, the abstract image could be generated, which greatly reduces the complexity of subsequent image processing. Achanta et al. [4] propose a Simple Linear Interative Clustering (SLIC) method for segmenting the input image into superpixels. The SLIC method iteratively computes the clusters with the smallest distance, and these pixels with small distances are clustered in the same superpixel, thereby generating the abstract image. Rose et al. [22] propose a fuzzy clustering algorithm, where each pixel is assigned to clusters on basis of the probability of the distance between each pixel to clusters. According to idea in [13,22], Gerstner et al. [5,6] present an improved version of SLIC. A color palette is used to record the color of each superpixel and the abstract image is generated by simplifying the primal image colors. However, the segmentation based image abstraction method may generate the low-quality abstract image with unsmooth boundaries. In this paper, the image segmentation based on the power diagram is used to obtain the segmentation results, which produces the high-quality abstract image with smooth boundaries.

Wang et al. [8] first introduce the Mean-Shift filter to segment video data into continuous data, so as to achieve better image abstraction. Kang et al. [23] propose a stream-based variant of the bilateral filter to address the problems of abstract drawing and region smoothing. However, this method may lead to over-abstraction or under-abstraction regions. Winnemoller et al. [14] present an automatic, real-time image abstraction framework that uses a non-linear diffusion filter to smooth regions with low color contrast. Meanwhile, a Gaussian difference operator is utilized to process regions of high color contrast to achieve image abstraction. Kyprianidis et al. [9] present a framework for automatic non-realistic image processing techniques using anisotropic filtering methods. Furthermore, they introduce a Knwahara filtering algorithm with different anisotropy, but the obtained results are not satisfactory. Other researchers [24] provide some filter based image abstraction methods [11,25], which may produce low-quality abstract images with artifacts and blurring. Besides, the degree of abstraction is difficult to control. In this paper, the number of sites and the density map provide our method with the ability of producing controllable abstract images.

3   Image Abstraction Based on Power Diagram

In this section, we will introduce the pixelated image abstraction method via power diagram to generate high-quality abstract images.

3.1   Overall Method

Given that the input image M is composed by a set of pixels $\{p_i\}_{i=1}^N$, where $N = w_M \times h_M$ is the number of pixels. Note that w_M is the width and h_M is the height of the input image. Each pixel p_i is associated with a color value $rgb(p_i)$ (RGB color is used in our work). The purpose of image abstraction is to produce an output image M' with a reduced color palette from the input image M, where the output image is of equal size to the input image. In this paper, we propose a novel pixelated image abstraction method via power diagram to generate high-quality abstract images with smooth boundaries and controllable abstraction degree.

The power diagram based pixelated image abstraction method in this paper mainly involves the following three modules:

- pre-processing for capturing the significant pixels and then calculating a density map;
- partitioning the input image into accurate polygonal subregions based on power diagram;
- coloring each polygonal subregion on basis of the average RGB color of the 3×3 grids centered at its centroid.

Consequently, the abstract images could be produced by merging color-filled polygonal subregions. The pseudo code of the pixelated image abstraction method via power diagram is provided in Algorithm 1, and the pipeline of the computational process is shown in Fig. 1.

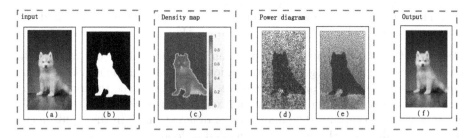

Fig. 1. The flow of the pixelated image abstraction method via power diagrams. (a) The input image; (b) image of the pre-processing; (c) the initialized density map; (d) the initialized power diagram with density adaptive; (e) the optimized power diagram, where each polygonal subregion with the same area constraint; and (f) the output image.

Algorithm 1: Pixelated image abstraction based on power diagram

---

Input: Origin image M, containing a set of pixels $\mathbf{p}_{in} = \{p_i\}_{i=1}^{w_M \times h_M}$, the number of sites n

Output: Abstract image M' (a set of pixels $\mathbf{p}_{out} = \{p_o\}_{i=1}^{w_M \times h_M}$)

1   *Pre-processing:* dividing the input image M into core and non-core parts
2   Calculating the density map $\rho(\mathbf{s})$
3   Generating n distinct sites $\mathbf{S} = \{\mathbf{s}_i\}_{i=1}^{n}$ on a domain of equal size to M, and setting each $w_i = 0$
4   Setting the area/capacity constraint of each site $\mathbf{s}_i$ with average mass
5   Optimizing the sites and weights to produce polygonal subregions with area constraint
6   Computing the average color r_i of each site $\mathbf{s}_i$
7   Filling each polygonal subregion $P(\mathbf{s}_i)$ with the color r_i

---

3.2   Pre-processing and Density Initialization

Density map initialization is an essential step in our method, which affects the optimized polygonal subregions in the second module. Before initializing the density map, we should use a pre-processing step to capture these pixels with significant information in the input image, which greatly improves the quality of the output abstract image.

Pre-processing. Typically, a simple image consists of a foreground layer and a background layer. The former shows some significant information, such as the cat in Fig. 1, while the latter is the non-significant information, such as the table and wall in Fig. 1. In the image abstraction process, the pixels with significant information in the image should be preserved while these with non-significant information are ignored. With this regard, the pre-processing is utilized to divide the primal image into core and non-core parts with different levels of pixel importance. The pre-processing in our work could be directly implemented by existing segmentation techniques, such as OpenCV [26] and deep learning based method

[27]. The core part with high level importance is represented by M_α, and the remaining is denoted by M_β, where $M = M_\alpha \cup M_\beta$.

Density Initialization. Density map plays an important role in our method. In what follows, we provide a general formulation for calculating density map of the input image based on the pre-processing results. We grayscale the input image, and then the grayscale value $g(p_i)$ of each pixel p_i is normalized to $col(p_i) \in [0, 1]$, where $p_i \in M$. To capture the importance of different pixels in the input image, the density value of each pixel p_i is calculated as follows:

$$\rho(p_i) = \begin{cases} a \cdot col(p_i) + b, & p_i \in M_\alpha \\ c \cdot col(p_i), & p_i \in M_\beta \end{cases} \tag{5}$$

where a, b and $c \in [0, 1]$ are the parameters for controlling the density map. Consequently, the density values of pixels with high-level importance (core, M_α) is larger than these with low-level importance (non-core, M_β) on basis of the input image.

3.3 Image Segmentation Based on the Power Diagram

Existing image segmentation based methods [4, 6] utilize "superpixels" to segment the input image into subregions, each of which is filled with the color of the corresponding superpixel. However, these methods may produce low-quality abstract images with unsmooth boundaries, such as these results in Fig. 8. Different from previous work, we use the power diagram to segment the input image into polygonal subregions, which are more suitable for capturing boundaries. Benefiting from this, our method could generate high-quality abstract images with smooth boundaries, which receives better visual effects to users. The polygonal division based on power diagram is introduced in the following.

Let Ω represent the domain of equal size to the input image M. $\mathbf{S} = \{\mathbf{s}_i\}_{i=1}^n$ denotes n distinct sites located in Ω, where each site $\mathbf{s}_i$ is associated with a weight w_i. The power diagram defines a partition of the domain Ω into n disjoint subregions $P(\mathbf{S}) = \{P(\mathbf{s}_i)\}_{i=1}^n$, as shown in Fig. 2(a). However, the subregions of the ordinary power diagram may be irregular, which contains many skinny power cells. Due to the possibility of producing many jagged power cells, the ordinary power diagram is useless for image abstraction.

In this paper, we use the centroidal power diagram with capacity constraint to produce the image segmentation results rather than the ordinary power diagram, as shown in Fig. 2(b). Specifically, we assign a capacity constraint c_i to each site $\mathbf{s}_i$, that is,

$$c_i = \frac{\int_\Omega \rho(\mathbf{s}) d\mathbf{s}}{n} \tag{6}$$

The area m_i of each polygonal subregion $P(\mathbf{s}_i)$ should satisfy: $m_i = c_i$. Meanwhile, the location of each site $\mathbf{s}_i$ should located in the centroid of the corresponding polygonal subregion $P(\mathbf{s}_i)$. To compute the CCCPD, researchers have provided various algorithms, including CPU-based computing algorithm [18, 19],

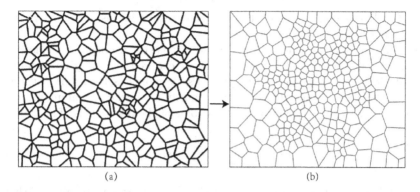

Fig. 2. Ordinary power diagram and centroidal power diagram with capacity constraints (CCCPD). (a) power diagram; and (b) CCCPD.

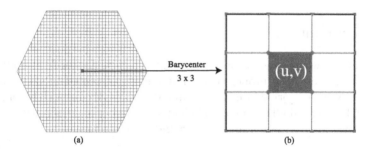

Fig. 3. The color filling of a power cell based on the average RGB color of the 3×3 grids centered at its barycenter. (a) optimized power cell; and (b) 3×3 grids centered at the barycenter of the power cell.

GPU-based computing algorithm [20], etc. In our work, we refer to previous work [18] and the CCCPD computing framework is used to compute the image segmentation based on power diagram.

Therefore, the image segmentation based on power diagram produces polygonal subregions with better characteristics that can capture the boundaries more accurately. Meanwhile, according to the initialized density map in previous text, the size of each polygonal subregion can be controlled. Specifically, the size of polygonal subregion is larger in low-level significant pixels while smaller in high-level significant pixels. Consequently, more important information (core) is preserved while less important information (non-core) is ignored during the image abstraction process of our method.

3.4   Color Filling

After obtaining the optimized polygonal subregions from the image segmentation, each polygonal subregion should be filled with the corresponding RGB color. A straightforward way is to fill each polygonal subregion with the RGB color of its barycenter. However, this way may results in unsatisfactory abstract results due to the existing noisy pixels.

To eliminate the influence of noisy pixels, we use the average pixel color to fill each polygonal subregion $P(\mathbf{s}_i)$. First of all, we choose the 3×3 grids centered at the centroid $\mathbf{s}_i^* = (x_i, y_i)$ of each polygonal subregion $P(\mathbf{s}_i)$, as shown in Fig. 3. Let $pixel_{ij} = (u, v)$ denote the exact pixel coordinate in the primal image, where $u = \lfloor x_i \rfloor$ and $v = \lfloor y_i \rfloor$. The average RGB color of the 3×3 grids of $\mathbf{s}_i^*$ could be calculated as:

$$r_i = \frac{1}{9} \sum_{k=u-1}^{u+1} \sum_{t=v-1}^{v+1} rgb(k, t) \tag{7}$$

Accordingly, the pixels in each polygonal subregion $P(s_i)$ are filled with the average RGB color r_i, and the abstract image could be directly generated by filling all polygonal subregions.

4   Experiment

In this section, we present some experimental results to verify the effectiveness of our method. The proposed method was implemented in C++, and all experiments were performed on a computer with 3.6 GHz Intel (R) Core (TM) i7-9700K CPU and 16 GB memory. The default number of sites N is $5.0K$, and the default value of these parameters a, b and c are set to 0.5.

4.1   Results Analysis

To verify the effectiveness of our method for image abstraction, we conduct an experiment on several different images. In this experiment, the number of sites n is set to $5.0K$, and the parameters for density map initialization is set to the default values. The results of pixelated image abstraction with power diagram are presented in Fig. 4. Benefiting from the density map initialization and the power diagram based image segmentation, our method could generate high-quality abstract images with smoother boundaries.

4.2   Abstract Controlling

An fundamental characteristics of our method is the controllability of the abstraction degree. Both the number of sites and the density map could be taken into consideration to control the abstraction of images. Therefore, we perform two groups of experiments to demonstrate the image abstraction controllability of our method.

Fig. 4. Results of pixelated image abstraction with power diagram. From top to bottom: input images, pre-processing results, density maps, and output abstract images.

    (a)   (b)   (c)   (d)   (e)   (f)   (g)

Fig. 5. The image abstraction results of our method with different number of sites. (a) Input images; (b) segmentation images; (c) density maps; (d)–(g) abstract images with $1.0K$, $2.0K$, $5.0K$ and $10.0K$ sites.

Number of Sites Controlling. Considering the influence of the number of sites on the generated results, we design the first group of experiments on three images, and the number of sites is selected from $1.0K$ to $10.0K$. Note that the

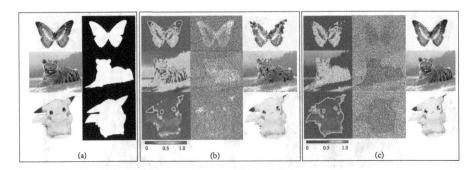

Fig. 6. The image abstraction results of density ablation experiments with our method. (a) Input images & pre-processing results; (b) results without density *Pair*1-*Exp*1; and (c) results with density *Pair*1-*Exp*2.

Fig. 7. The image abstraction results of our method with different densities. (a) Input images & pre-processing results; (b) results with density *Pair*2-*Exp*1; and (c) results with density *Pair*2-*Exp*2.

parameters used to initialize the density map are set to the default values. The results of abstract images with different number of sites are provided in Fig. 5.

According to the results in Fig. 5, we can observe that the quality of abstract image improves with the gradually increasing of the number of sites. Specifically, when the number of sites is small (as shown in the fourth column in Fig. 5), our method produces the polygonal subregions with large sizes, which may cause unsmooth boundaries of the abstract images. As the number of sites increases, the sizes of polygonal subregions become smaller and the quality of generated abstract images improves.

Density Map Controlling. To further show the generated results with different densities, the second group of experiments is designed, where the number of sites is set to the default value, that is, $n = 5.0K$.

Density Controlling 1. Firstly, the densities are selected as follows: 1) the densities are initialized without the consideration of pixels importance; and 2) the densities are initialized based on the pixels importance. Specifically, the density of each pixel in this experiment is calculated as follow:

$$Pair1\text{-}Exp1 : \rho(p_i) = col(p_i), \quad p_i \in M$$

$$Pair1\text{-}Exp2 : \rho(p_i) = \begin{cases} 0.5 \cdot col(p_i) + 0.5, & p_i \in M_\alpha \\ 0.5 \cdot col(p_i), & p_i \in M_\beta \end{cases} \tag{8}$$

The experimental results of the first pair experiments are provided in Fig. 6.

Density Controlling 2. Then, a new pair of experiments is performed, and the importance of pixels is taken into account. To be specific, the densities of pixels with high-level importance are larger than those with low-level importance. However, the difference lies in the density variability between core and non-core parts, and the density map calculation is as follow:

$$Pair2\text{-}Exp1 : \rho(p_i) = \begin{cases} 0.5 \cdot col(p_i) + 0.5, & p_i \in M_\alpha \\ 0.5 \cdot col(p_i), & p_i \in M_\beta \end{cases}$$

$$Pair2\text{-}Exp2 : \rho(p_i) = \begin{cases} 0.3 \cdot col(p_i) + 0.7, & p_i \in M_\alpha \\ 0.3 \cdot col(p_i), & p_i \in M_\beta \end{cases} \tag{9}$$

The abstract image results are presented in Fig. 7.

According to the results in Fig. 6 and Fig. 7, we can observe that our method achieves controllable abstract images with different densities. Specifically, when the importance of pixels is not considered (as shown in Fig. 6(b)), many detailed features in the abstract images are lost, such as the textures on the tiger. On the contrary, the results in Fig. 6(c) demonstrate the validity of the density map in our method, which takes into account the importance of pixels. Thus, the generated results retain more features of the primal images. Considering the density variations of core and non-core parts, the results in Fig. 7 show that higher densities could retain more detailed information in the core part, while lower densities could ignore the non-important information in the primal image. More generally, the density map initialization in Eq. 5 could be extended to more general cases, such as Gaussian or non-linear densities.

4.3   Comparison

To further illustrate the effectiveness of our method with power diagram, we compare against the improved SLIC method of Gerstner et al. [6] as a representative of state-of-the-art image abstraction methods based on image segmentation. The improved SLIC method [6] divides the input image into superpixels and uses a simple linear iterative clustering method to optimize the superpixels. In this experiment, we select six different images as inputs of our method and the improved SLIC method, and the generated abstract images of different methods

Fig. 8. Comparison results of different image abstraction methods. From top to bottom: input images, results of the improved SLIC method [4], and results of our method.

are presented in Fig. 8. To maintain the consisty, the number of superpixels in the improved SLIC method is set to 4800, that is same to the number of sites in our method. Notably, the parameters used for density map initialization in our method are set to the default values.

From the results in Fig. 8, we can observe that the method proposed in this paper outperforms the improved SLIC method. Specifically, compared with the improved SLIC method [6], the abstract images generated by our method retains more features, i.e., the eyes of the rabbit in the second column and the ears of the cats in the last column in Fig. 8. Besides, in terms of the boundaries of the generated abstract images, our method produces higher quality results with smooth boundaries, while the improved SLIC method obtains results with unsmooth boundaries.

5   Conclusion

We propose a novel image abstraction method based on power diagram to generate high-quality abstract images with smooth boundaries and controllable abstraction degree. Based on the power diagram, the domain with equal size to the input image is partitioned into several polygonal power cells, which is then optimized to meet the equal capacity constraints. The abstract image is generated by coloring each pixel in the power cell with the average color of 3×3 grids centered at the barycenter of its power cell. Compared with other segmentation based image abstraction method, the proposed method use the power diagram to partition the primal domain into polygonal cells, which is more suitable for capturing the boundaries. Besides, both the density and the number of sites could dominate the degree of image abstraction.

Limitations and Future Work. Despite the density map of equal size to the input image is built based on the saliency segmentation image, the generated image is not satisfactory for complex images. In our future work, we will extend the current work to make it more applicable for complex image abstraction.

Acknowledgements. This work was supported in part by a grant from the National Natural Science of Foundation of China (No. 61972128).

References

1. Kumar, M.P.P., Poornima, B., Nagendraswamy, H.S., Manjunath, C.: A comprehensive survey on non-photorealistic rendering and benchmark developments for image abstraction and stylization. Iran J. Comput. Sci. **2**(3), 131–165 (2019). https://doi.org/10.1007/s42044-019-00034-1
2. Dickinson, S.: Challenge of image abstraction. In: Object Categorization: Computer and Human Vision Perspectives, vol. 1 (2009)
3. Pavan Kumar, M., et al.: Image abstraction framework as a pre-processing technique for accurate classification of archaeological monuments using machine learning approaches. SN Comput. Sci. **3**(1), 1–30 (2022). https://doi.org/10.1007/s42979-021-00935-8
4. Achanta, R., Shaji, A., Smith, K., Lucchi, A., Fua, P., Süsstrunk, S.: SLIC superpixels. Technical report (2010)
5. Gerstner, T., DeCarlo, D., Alexa, M., Finkelstein, A., Gingold, Y.I., Nealen, A.: Pixelated image abstraction. In: NPAR@ Expressive, pp. 29–36 (2012)
6. Gerstner, T., DeCarlo, D., Alexa, M., Finkelstein, A., Gingold, Y., Nealen, A.: Pixelated image abstraction with integrated user constraints. Comput. Graph. **37**(5), 333–347 (2013)
7. Kumar, M., Poornima, B., Nagendraswamy, H., Manjunath, C.: Structure-preserving NPR framework for image abstraction and stylization. J. Supercomput. **77**(8), 8445–8513 (2021). https://doi.org/10.1007/s11227-020-03547-w
8. Wang, J., Xu, Y., Shum, H.Y., Cohen, M.F.: Video tooning. In: ACM SIGGRAPH 2004 Papers, pp. 574–583 (2004)
9. Kyprianidis, J.E., Döllner, J.: Image abstraction by structure adaptive filtering. In: TPCG, pp. 51–58 (2008)
10. Kyprianidis, J.E.: Image and video abstraction by multi-scale anisotropic Kuwahara filtering. In: Proceedings of the ACM SIGGRAPH/Eurographics Symposium on Non-Photorealistic Animation and Rendering, pp. 55–64 (2011)
11. Sadreazami, H., Asif, A., Mohammadi, A.: Iterative graph-based filtering for image abstraction and stylization. IEEE Trans. Circ. Syst. II Express Briefs **65**(2), 251–255 (2017)
12. DeCarlo, D., Santella, A.: Stylization and abstraction of photographs. ACM Trans. Graph. (TOG) **21**(3), 769–776 (2002)
13. Ren, X., Malik, J.: Learning a classification model for segmentation. In: IEEE International Conference on Computer Vision, vol. 2, p. 10. IEEE Computer Society (2003)
14. Winnemöller, H., Olsen, S.C., Gooch, B.: Real-time video abstraction. ACM Trans. Graph. (TOG) **25**(3), 1221–1226 (2006)
15. Aurenhammer, F.: Power diagrams: properties, algorithms and applications. SIAM J. Comput. **16**(1), 78–96 (1987)

16. Aurenhammer, F., Hoffmann, F., Aronov, B.: Minkowski-type theorems and least-squares clustering. Algorithmica **20**(1), 61–76 (1998). https://doi.org/10.1007/PL00009187

17. Balzer, M.: Capacity-constrained Voronoi diagrams in continuous spaces. In: 2009 Sixth International Symposium on Voronoi Diagrams, pp. 79–88. IEEE (2009)

18. De Goes, F., Breeden, K., Ostromoukhov, V., Desbrun, M.: Blue noise through optimal transport. ACM Trans. Graph. (TOG) **31**(6), 1–11 (2012)

19. Xin, S.Q., et al.: Centroidal power diagrams with capacity constraints: computation, applications, and extension. ACM Trans. Graph. (TOG) **35**(6), 1–12 (2016)

20. Zheng, L., Gui, Z., Cai, R., Fei, Y., Zhang, G., Xu, B.: GPU-based efficient computation of power diagram. Comput. Graph. **80**, 29–36 (2019)

21. Zheng, L., Yao, Y., Wu, W., Xu, B., Zhang, G.: A novel computation method of hybrid capacity constrained centroidal power diagram. Comput. Graph. **97**, 108–116 (2021)

22. Rose, K.: Deterministic annealing for clustering, compression, classification, regression, and related optimization problems. Proc. IEEE **86**(11), 2210–2239 (1998)

23. Kang, H., Lee, S., Chui, C.K.: Flow-based image abstraction. IEEE Trans. Vis. Comput. Graph. **15**(1), 62–76 (2008)

24. Zhao, H., Jin, X., Shen, J., Mao, X., Feng, J.: Real-time feature-aware video abstraction. Vis. Comput. **24**(7), 727–734 (2008). https://doi.org/10.1007/s00371-008-0254-8

25. Koga, T., Suetake, N.: Structural-context-preserving image abstraction by using space-filling curve based on minimum spanning tree. In: 2011 18th IEEE International Conference on Image Processing, pp. 1465–1468. IEEE (2011)

26. Bradski, G.: The openCV library. Dr. Dobb's J. Softw. Tools Prof. Program. **25**(11), 120–123 (2000)

27. Minaee, S., Boykov, Y.Y., Porikli, F., Plaza, A.J., Kehtarnavaz, N., Terzopoulos, D.: Image segmentation using deep learning: a survey. IEEE Trans. Pattern Anal. Mach. Intell. **44**(7), 3523–3542 (2022)

SMT Component Defection Reassessment Based on Siamese Network

Chengkai Yu[1,2], Yunbo Zhao[2,3,4(✉)], and Zhenyi Xu[2(✉)]

[1] AHU-IAI AI Joint Laboratory, Anhui University, Hefei 230601, China
[2] Institute of Artificial Intelligence, Hefei Comprehensive National Science Center,
Hefei 230088, China
xuzhenyi@mail.ustc.edu.cn
[3] Department of Automation, University of Science and Technology of China,
Hefei 230026, China
[4] Institute of Advanced Technology, University of Science and Technology of China,
Hefei 230088, China
ybzhao@ustc.edu.cn

Abstract. In the SMT process, after component placement, checking the quality of component placement on the PCB board is a basic requirement for quality control of the motherboard. In this paper, we propose a deep learning-based classification method to identify the quality of component placement. This is a comparison method and the novelty is that the siamese network is trained to extract the features of the standard placement component map and the placement component map to be inspected and output the probability of similarity between the two to determine the goodness of the image to be inspected. Compared to traditional hand-crafted features, features extracted using convolutional neural networks are more abstract and robust. In addition, during training, the concatenated network pairs the sample images to expand the amount of training data, increasing the robustness of the network and reducing the risk of overfitting. The experimental results show that this method has better results than the general model for the classification of placement component images.

Keywords: Siamese network · Component placement · Binary classification

1 Introduction

In the production process of laptop motherboard, Surface Mount Device (SMD) needs to be mounted on the surface of PCB board, this process is called Surface Mounted Technology (SMT). The link of SMT mainly has several steps

This work was supported in part by the National Natural Science Foundation of China (62103124), Major Special Science and Technology Project of Anhui, China (202104a05020064), China Postdoctoral Science Foundation (2021M703119).

such as solder paste printing, component placement, reflow oven soldering, AOI inspection, etc., The purpose of SMT is to firmly solder SMD such as resistors and capacitors to the PCB board through solder paste [1]. In order to make the SMD firmly welded to the PCB board, the PCB board is put into the reflow oven after the SMD placement to melt the solder paste by high temperature, and then the SMD can be firmly welded to the PCB board after waiting for cooling. Then it is also necessary to detect whether the soldered components have poor welding, such as standing monument, displacement, empty solder and other defects. Detecting these defects is a very important step in the SMT production line, because the proper mounting of electronic components has a critical impact on the motherboard, and even a missing part can affect the use of the entire motherboard.

In recent years, automatic optical inspection (AOI) technology has been used to detect defects in the patch, which is more efficient than manual inspection and reduces inspection costs, but its false detection rate and leakage rate is still high [3,5]. The most widely used method in AOI inspection is the reference inspection method. In this defect detection method, the exact alignment of the reference image and the detection image is required to calculate the correlation between the two window sections of the detection image and the reference image. This detection method is susceptible to image angle changes, illumination, etc., resulting in a high false detection rate and the subsequent need to arrange manual reassessment of the location of the detected defects, which increases production costs.

The rapid development of deep learning has injected new strength into many industries. Deep learning methods have shown good performance in automatic feature extraction and end-to-end control. However, because deep learning networks generally require a large amount of data for training, and the defects of component placement are complex and variable, it is difficult to collect a large number of samples to meet the needs of deep learning methods, and some classical AI vision detection methods cannot show good results.

This paper proposes a new solution to the problems that exist in the process of AOI inspection of patch quality in SMT production lines and summarizes the main contributions of this paper as follows:

(1) We propose an intelligent comparison method. Unlike the traditional matching method of template matching, our proposed intelligent matching method maps both the image to be detected and the template image into the feature space, and then compares them through feature learning and feature matching. This approach is not only less susceptible to interference from factors such as machine vibration, but also significantly improves inspection efficiency.

(2) We use a model of a siamese network [2]. We train a feature extraction module for extracting the features of the image to be detected and a feature correlation module for calculating the similarity between the image to be detected and the standard image, so as to judge the quality of the placement components to be detected.

(3) We obtained images of placement components in actual SMT production lines and generated training data by sample pairing, which greatly increases the amount of data, improves the robustness of the network, and enhances the network detection effect. At the end of this paper, we verify the effectiveness of our proposed method by comparing it with various models through experiments and analysis.

2 Proposed Method

2.1 Siamese Network

In [13], to compare the correlation between pairs of images, three types of CNN models are presented: siamese, pseudosiamese, and 2-channe. The siamese and pseudosiamese networks consists of two structurally identical convolutional neural network branches. While general convolutional neural networks use a single image data as input to extract features, the input of siamese and pseudosiamese networks is a pair of image data. A pair of data is fed into each of the two convolutional neural network branches, and the image data are mapped to a new feature space by convolutional operations, forming a representation of the input in the new space, i.e., features. The above are their similarities, while they differ in that both branches of the siamese network share the same set of weights for both training and prediction, while the two branches of the pseudosiamese network have their own weights. In the 2-channe network, the approach taken is to merge two images into a single two-channel image that is fed directly into the network. All three methods can be used to perform the similarity calculation of the two images. In fact, all three methods can be used to perform the similarity calculation of two images.

In the approach proposed in this paper, we use the network framework of Siamese Network. Figure 1 shows the structure of the siamese network we designed. A pair of sample images is input to two network branches separately, and the features of the images are obtained after a series of convolution and pooling layer operations. The features are then fed into the feature correlation module to calculate the distance between two features, and the similarity probability of the input image pair is output through the fully connected layer. When the similarity probability is greater than the threshold value, judge that the two samples are the same, and output 0; otherwise, output 1.

Siamese networks have a wide range of applications and we can see them in areas such as face recognition/verification [10,11], image retrieval [7], Deep Metric Learning (DML) [6], and object tracking [4,12]. In this paper, we also adopt a conjoined network structure mainly due to the following considerations. First, the two convolutional neural network branches of the siamese network not only have the same structure, but also share parameters, which greatly reduces the storage space required for the network and parameters, and speeds up the training of the network. Second, this design enables both convolutional neural networks to use the same method to extract features from images, ensuring consistent predictions. Since the pair of input images we use for detection are acquired for the same component sites under equal lighting, distance, and sensor conditions, the images have similar features, and this weight-sharing approach ensures that two similar samples are not mapped to different parts of the feature space [14].

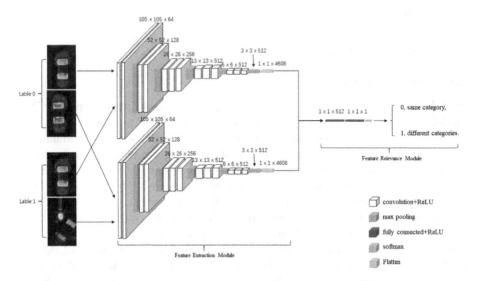

Fig. 1. A pair of sample images is extracted by the feature extraction module to get the corresponding features, and the two features are then input into the feature correlation module to calculate the similarity, and when the similarity probability is greater than the threshold value, the two samples are judged to be the same and output 0, and vice versa, output 1.

2.2   Loss Function

When training the designed network, we need an appropriate objective function whose optimization can yield good performance in feature extraction and classification.

Let $\{p_1, p_2\}$ denote the input image sample pair, $f(p_1)$, $f(p_2)$ denote the features generated by the image through the convolutional neural network, and $D(p_1, p_2)$ denote the Euclidean distance function of the feature vector, which can reflect the correlation between the two images. That is

$$D(p_1, p_2) = \|f(p_1), f(p_2)\|_2 \tag{1}$$

Let $Y = 0$ indicate that both input images belong to the same category and $Y = 1$ indicates a different category. Since we know whether each input image pair belongs to the same class or not, we can perform the following analysis.

When $Y = 0$, the loss is

$$L_0 = (1 - Y) L_S \left(D^i\right) \tag{2}$$

when $Y = 1$, the loss is

$$L_1 = Y L_D \max \left\{0, (M - D)^i\right\} \tag{3}$$

We adopt such a loss function due to the fact that when $Y = 1$, we expect the distance between p_1 and p_2 to be as large as possible. So we can set a maximum distance M. When $D(p_1, p_2) > M$, it means that the distance between these two samples is large enough, just when the loss at this point is 0, to simplify the operation.

Then, the loss function in its general form is

$$\ell = \sum_{k=1}^{p} L\left((Y, p_1, p_2)^k\right) = \sum_{k=1}^{p} (1 - Y) L_S \left(D^i\right) + Y L_D \max \left\{0, (M - D)^i\right\} \tag{4}$$

where $(Y, p_1, p_2)^k$ is the kth labeled training sample pair and P is the number of training sample pairs. L_S, L_D are constants, and the default is 0.5. The default $i = 2$ is the same as the commonly used contrastive loss, which is the square of the Euclidean distance.

During the training process, the minimization loss function is able to bring the Euclidean distance between intra-class samples close to zero and the Euclidean distance between inter-class samples close to M, which is a large value relative to zero. Ultimately, the network is able to produce features that show lower intra-class variation and higher inter-class variation through learning. This results in the ability to have better image classification performance after feature extraction, as will be demonstrated subsequently in experiments.

2.3 Detailed Detection Scheme

After a pair of test images are extracted features by two neural network branches respectively, the distance vector between two feature vectors is then calculated, and the final prediction result, i.e., the probability of similarity between two images, is obtained using the fully connected layer and the sigmoid activation function as a classifier. In the design and training of the network, we have adopted some techniques to obtain more efficient algorithms.

In the convolutional network branch, we use multiple 3×3 convolutional kernels in series, which not only has less number of parameters but also has more nonlinear transformations compared to using a larger convolutional kernel alone. Moreover, we apply the RELU activation on all the linear layers except the output label layer, apply batch normalization after every layer to reduce internal covariance-shift, and apply a dropout of 0.1 to prevent overfitting.

Data enhancement is very important to improve the robustness of the target detection algorithm, we can make the image more diverse by changing the brightness, image flipping, etc.

3   Experiments

In this section, we perform an evaluation to demonstrate the performance of our proposed method and compare it with other methods.

3.1   Experimental Data

To train the proposed network and evaluate the method, we acquired image data of actual circuit components on the production line of Lianbao (Hefei) Electronic Technology Co. This data contains 899 images of normal placement components, and 1015 images of defective placement components. The normal component diagram is shown in Fig. 2, and the defective component diagram is shown in Fig. 3. These data are intercepted from the PCB board after component placement on a real SMT production line, and the label of the component is obtained by manual determination by the enterprise experts.

In deep learning, a larger number of samples can reduce the risk of overfitting in the training process of the model, enhance the generalization ability of the model, and the better the effect of the trained model. Therefore, we adopted the data augmentation method to expand the number of samples. Finally, we obtained 2728 images of normal and defective component placement by data enhancement. We divide the above data into a training set and a test set in the ratio of 9:1 to obtain a training set containing 2462 images of placement components and a test set of 266 images of placement components. The number of samples is shown in Table 1.

The input of the network is fixed, we first convert the image to a size of 105×105. Since our proposed detection method compares features through a siamese network, it is also necessary to pair the samples. Randomly pair images of the same category with two, set $label = 0$, as a positive sample. The images of the current category are randomly paired with images of another category, and set $label = 1$ as a negative sample. The above operations are handled separately for the training and validation sets. The ratio of positive and negative samples in this design is 1:1.

Table 1. Sample size of placement components

No.	Class	Original	Data enhancement	Training set	Test set
0	ok	899	2728	2462	266
1	ng	1015	2728	2462	266
	Total	1914	5456	4924	532

3.2   Optimization

We trained the proposed network with 500 epochs, the batch size was set to 32, and this was done in a single NVIDIA GTX 3060 GPU with 24G memory. The training took about half an hour.

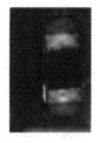

Fig. 2. Normal placement components

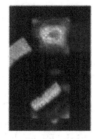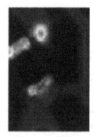

Fig. 3. Defective placement components.

Table 2. Precision, recall and F1-score of each model on the training set

Model	Siamese network			MobileNetV2			GoogLeNet			ResNet50			ResNet101		
	p	r	F1	p	r	F1	p	r	F1	p	r	F1	p	r	F1
0	0.969	0.950	0.959	0.888	0.935	0.911	0.883	0.935	0.908	0.914	0.950	0.932	0.921	0.957	0.938
1	0.951	0.970	0.960	0.932	0.882	0.906	0.931	0.876	0.903	0.948	0.911	0.929	0.955	0.918	0.936
Average accuracy	0.960			0.909			0.906			0.931			0.946		

Table 3. Precision, recall and F1-score of each model on the test set

Model	Siamese network			MobileNetV2			GoogLeNet			ResNet50			ResNet101		
	p	r	F1	p	r	F1	p	r	F1	p	r	F1	p	r	F1
0	0.966	0.952	0.959	0.889	0.872	0.880	0.902	0.895	0.898	0.914	0.921	0.918	0.926	0.940	0.933
1	0.951	0.966	0.958	0.875	0.891	0.883	0.896	0.902	0.899	0.921	0.914	0.918	0.939	0.925	0.932
Average accuracy	0.959			0.882			0.899			0.918			0.933		

We used the Adam optimizer [9] for training. During training, the initial learning rate is 0.001 and we decay 0.92 times in each epoch. The decay factor for each epoch is 0.92. We apply the RELU activation function on all linear layers except the output label layer, and apply batch normalization [8] after each layer to reduce the internal covariance bias. We also apply a dropout of 0.1 to prevent overfitting.

3.3   Results and Evaluation

To demonstrate the classification power of the features extracted by the proposed siamese Network, we compared several traditional feature extraction methods, such as GoogLeNet, MobileNetV2 and ResNet50. First, some metrics need to be defined to evaluate the classification performance of the algorithm. Model training evaluation metrics:

$$Precision = \frac{TP}{TP + FP} \tag{5}$$

$$Recall = \frac{TP}{TP + FN} \tag{6}$$

$$F1 = \frac{2TP}{2TP + FP + FN} \tag{7}$$

The concept of F1 value is proposed based on Precision and Recall to evaluate Precision and Recall as a whole. F1 is the harmonic mean of precision rate and recall rate. When both values are high, F1 is also high. Therefore, F1 can reflect precision rate and recall rate.

The evaluation results of each model in the training set are shown in Table 2, and in the test set are shown in Table 3 (p:precision, r:recall). As shown in Fig. 4, 5, 6, 7, and 8, the classification confusion matrix of each model is clear.

We can see that the highest F1 value of the siamese network reaches 0.96 on training set, while ResNet101, which has the highest F1 value among the other networks, only reaches 0.938. In addition, our siamese network achieves up to 96% average accuracy in the quality inspection of placement components. The results of the experiments on the test set lead us to the same conclusion. The average accuracy of the siamese network reached 95.9%, with an F1 score of 0.959. The experimental results show that the proposed siamese network-based classification model in this paper has good results in classifying good and bad placement components.

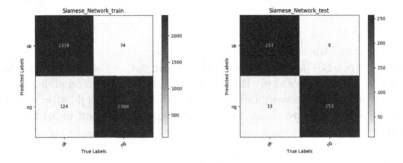

Fig. 4. Confusion matrix for siamese network

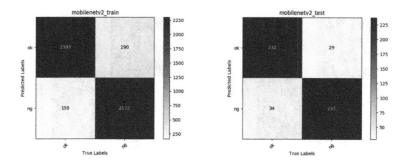

Fig. 5. Confusion matrix for mobilenetv2

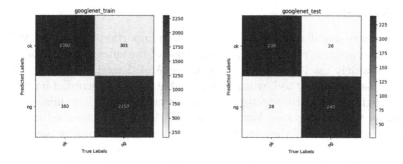

Fig. 6. Confusion matrix for googlenet

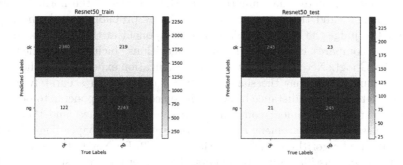

Fig. 7. Confusion matrix for resnet50

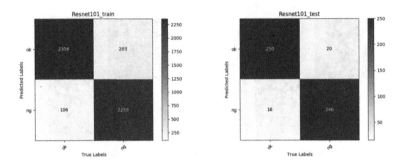

Fig. 8. Confusion matrix for resnet101

3.4 Conclusion

The component installation defect detection is an important link in the SMT production line, which is related to the normal use of the subsequent motherboard. The most widely used method of the existing AOI detection equipment is the reference inspection method, which is vulnerable to the influence of image angle changes, lighting and other factors, leading to problems such as missed inspection, many false inspections, low detection efficiency, and the need to arrange manual re evaluation of the detected defect location, which increases the production cost.

In this paper, the method of deep learning is used to solve the problem of quality inspection of placement components. In this paper, the features of the placement component image to be detected are extracted based on a siamese network, and calculates the similarity with the features of the normal mounting component, and then compares the similarity probability with the preset threshold to judge the quality of placement component at this location. Finally, this paper compares the commonly used algorithms, including MobileNetV2, ResNet50, GoogLeNet and ResNet101, for classification experiments. The experimental results show that the siamese network method can accurately extract features, determine the distance between the placement components to be tested and the standard placement components in the feature space, and then judge the goodness of the placement components. Finally, this paper compares the commonly used algorithms, including MobileNetV2, ResNet50, GoogLeNet and ResNet101, for classification experiments. The experimental results show that the siamese network method can accurately extract features, determine the distance between the placement components to be tested and the standard placement components in the feature space, and then judge the goodness of the placement components.

References

1. Akhtar, M.B.: The use of a convolutional neural network in detecting soldering faults from a printed circuit board assembly. HighTech Innov. J. **3**(1), 1–14 (2022)
2. Shah, R., Säckinger, E., et al.: Signature verification using a "siamese" time delay neural network. Int. J. Pattern Recogn. Artif. Intell. **07**(4), 669 (1993)
3. Chin, R.T., Harlow, C.A., et al.: Automated visual inspection: a survey. IEEE Trans. Pattern Anal. Mach. Intell. **PAMI-4**(6), 557–573 (1982)
4. Bertinetto, L., Valmadre, J., Henriques, J.F., Vedaldi, A., Torr, P.H.S.: Fully-convolutional siamese networks for object tracking. In: Hua, G., Jégou, H. (eds.) ECCV 2016. LNCS, vol. 9914, pp. 850–865. Springer, Cham (2016). https://doi.org/10.1007/978-3-319-48881-3_56
5. Chang, Y.M., Wei, C.C., Chen, J., et al.: Classification of solder joints via automatic mistake reduction system for improvement of AOI inspection. In: 2018 13th International Microsystems, Packaging, Assembly and Circuits Technology Conference (IMPACT), pp. 150–153. IEEE (2018)
6. Duan, Y., Zheng, W., Lin, X., et al.: Deep adversarial metric learning. In: Proceedings of the IEEE Conference on Computer Vision and Pattern Recognition, pp. 2780–2789 (2018)
7. Gordo, A., Almazan, J., Revaud, J., et al.: End-to-end learning of deep visual representations for image retrieval. Int. J. Comput. Vis. **124**(2), 237–254 (2017). https://doi.org/10.1007/s11263-017-1016-8
8. Ioffe, S., Szegedy, C.: Batch normalization: accelerating deep network training by reducing internal covariate shift. In: International Conference on Machine Learning, pp. 448–456. PMLR (2015)
9. Kingma, D.P., Ba, J.: Adam: a method for stochastic optimization. arXiv preprint arXiv:1412.6980 (2014)
10. Rana, S., Kisku, D.R.: Face recognition using siamese network. In: Bhattacharjee, D., Kole, D.K., Dey, N., Basu, S., Plewczynski, D. (eds.) Proceedings of International Conference on Frontiers in Computing and Systems. AISC, vol. 1255, pp. 369–376. Springer, Singapore (2021). https://doi.org/10.1007/978-981-15-7834-2_35
11. Schroff, F., Kalenichenko, D., Philbin, J.: FaceNet: a unified embedding for face recognition and clustering. In: Proceedings of the IEEE Conference on Computer Vision and Pattern Recognition, pp. 815–823 (2015)
12. Yao, S., Han, X., Zhang, H., et al.: Learning deep Lucas-Kanade Siamese network for visual tracking. IEEE Trans. Image Process. **30**, 4814–4827 (2021)
13. Zagoruyko, S., Komodakis, N.: Learning to compare image patches via convolutional neural networks. In: Proceedings of the IEEE Conference on Computer Vision and Pattern Recognition, pp. 4353–4361 (2015)
14. Zhan, Y., Fu, K., Yan, M., et al.: Change detection based on deep siamese convolutional network for optical aerial images. IEEE Geosci. Remote Sens. Lett. **14**(10), 1845–1849 (2017)

Real Time Traffic Sign Recognition Algorithm Based on SG-YOLO

Qianfan Wang[1], Xin Sun[1(✉)], Kaixiang Yi[1], and Tianhong Feng[2]

[1] School of Mechatronic Engineering and Automation, Shanghai University, Shanghai 200444, China
xsun@staff.shu.edu.cn
[2] Shanghai Pinghe School, Shanghai 201206, China

Abstract. Precise and rapid recognition of road traffic signs can enhance the environment perception of autonomous vehicles, which is an essential component for the safe driving. Aiming at addressing the problem that the recognition of traffic signs performs poorly in real-time and generally with accuracy, this paper proposes a real-time traffic sign recognition algorithm based on SG-YOLO. Combining the light weight of Ghost convolution and the perceptual capability of SE attention, a new SG-Bottleneck module is proposed and introduced into the backbone network of YOLOv5, which decreases the model parameters of the algorithm while speeding up the detection. Based on the characteristics of the target shape of the traffic signs, this paper modify the loss function to Distance Intersection over Union to obtain higher accuracy. The experimental results show that the mean average precision can reach 74.95% and detection speed can reach 41.7 FPS. When compared with other traffic sign recognition methods, SG-YOLO demonstrates better real-time performance and guarantees accuracy.

Keywords: YOLO · GhostNet · Attention mechanism · Object recognition

1 Introduction

As automotive industry grows at a rapid pace, autonomous driving technology also keeps pushing its boundaries. For autonomous driving, the recognition of traffic signs plays a decisive role in aspects such as path planning and navigation, as well as providing drivers with functions that include speeding and safety alerts.

The way to detect and recognize traffic signs more timely and accurately has been a hotspot and challenge of research. Real time traffic sign detection is of a great difficulty owing to the susceptibility to light, weather, and photographic angles during vehicle movement. In addition, due to the sheer quantities and small sizes of traffic signs, it also results in inaccurate traffic sign detection.

Recently, target detection algorithms, such as You Only Look Once (YOLO) [1], ResNet [2], and Single Shot MultiBox Detector (SSD) [3], have received wide attention for excellent performance. Although these algorithm networks have enormous numbers of parameters and considerable depth, which implicate that more features can be

W. Fan et al. (Eds.): AsiaSim 2022, CCIS 1713, pp. 86–99, 2022.
https://doi.org/10.1007/978-981-19-9195-0_8

extracted, they suffer from the problems of excessive computation and are unfavorable for end-device deployment.

Therefore, tiny target detection networks are highly worthy of attention. For the sake of improving the detection speed as well as reduce the computational cost of traffic sign recognition network, this paper proposes a tiny traffic sign recognition network SG-YOLO.

2 Related Work

Early research on the detection and recognition of traffic signs preferred traditional methods, which primarily used obvious features such as colors and shapes of traffic signs to detect. With the advancement of deep learning, target detection algorithms continue to emerge and have achieved remarkable outcomes in traffic sign recognition field.

2.1 Traditional Detection Methods

As most traffic signs are of bright colors and fixed shapes, these features allow them to be easily separated from the real road background. Therefore, based on the characteristics described above, traditional detection methods for traffic signs can be grouped into color-based and shape-based.

For the color-based algorithms, images taken by cameras are mainly stored in RGB three-channels format. Processing images directly based on RGB color space has certain speed advantage over converting to other color spaces. Huang et al. [4] set the corresponding color thresholds by the different R, G, and B components to separate the desired color regions from the images and carried out target localization based on the color regions. To address the issue that the traffic sign images are susceptible to external factors, Wang et al. [5] proposed a method by combining the achromatic model with the normalized RGB space and the results showed that the detections were robust for the images with severe lighting variations. Alternatively, sacrificing speed and converting the original images to other color spaces can improve the accuracy of recognition. El Baz et al. [6] combined the HSV and RGB segmentation by the logic "AND" operation to remove the noise generated during segmentations and achieved a high-level accuracy in the traffic signs with red color.

The shape-based algorithms mainly use the geometry of the features to extract candidate areas of the images, and use a classifier to distinguish the extracted candidate areas to complete the detection. Bae et al. [7] used the Hough transformation to detect traffic signs illuminated by LEDs and the detection ratio reached 85.37%. Tang et al. [8] also introduced the Hough transformation to shape detection after the recognition in YUV color space. Via Tang's method, excellent detection accuracy could be achieved, but false detections still occurred for small targets and in cases of interference by other objects. Although the shape-based algorithms can detect traffic signs with obvious shapes such as circles, triangles, and rectangles, when the images are distorted and skewed, it will increase the probability of missing detection.

2.2 Deep-Learning Based Detection Methods

Traditional detection algorithms are limited in their ability to extract features, while deeper feature information is required to improve detection and recognition performance. In deep-learning based methods, the convolutional neural networks can be trained to extract feature to compensate for the shortcomings of traditional methods. Target detection algorithms are mainly classified into two categories.

The two-stage algorithms fall into the first category, represented by Faster R-CNN [9], R-FCN [10], and Cascade R-CNN [11], which are based on region proposals to find bounding boxes where target objects may exist. Li et al. [2], based on Faster R-CNN, used ResNet50-D as feature extractor along with AutoAugment and the algorithm could detect traffic signs precisely and reduce the risk of overfitting. This type of networks has been developed over several years for traffic sign detection and recognition, with relatively satisfactory detection accuracy, but the speed of the networks falls short.

In addition, the other category is the one-stage detection algorithms represented by YOLO [12–14], FCOS [15], etc., which generate prediction boxes directly when classifying and regressing. Wang et al. [1] detected traffic signs by a modified YOLOv3, which introduced the Fusion Information module to fuse the three scale feature maps, and improved the mean Average Precision (mAP) by 11.1% compared with the original YOLOv3. Zhang et al. [16] used YOLOv4 network model for traffic sign detection and employed a cross layer connection to improve YOLOv4, satisfying the accuracy while meeting the real-time detection requirements. You et al. [3] reduced the computation of SSD network by streamlining and removing some convolutional layers, improving the accuracy of traffic sign recognition by 3% compared with the baseline SSD network.

The deep-learning based methods are more sensitive to adverse factors such as light changes, distortions and occlusions, gradually becoming the mainstream of traffic sign detection and recognition.

3   Principles of Algorithm

The YOLO series algorithm has evolved into several versions since its inception in 2016. Bochkovskiy et al. [17] proposed YOLOv4 in April 2020, who combines Cross Stage Partial (CSP) with Darknet-53 to establish CSP-Darknet53 as the backbone network. The feature extractor of YOLOv4 is formed using Spatial Pyramid Pooling (SPP). In the composition of YOLOv4, CSP [18] can reduce inference computation and enable more diverse gradient combinations, while SPP [19] can increase the perceptual field and separate contextual features more effectively.

Immediately afterwards, YOLOv5, proposed by Ultralytics in May 2020, continues the implementation of the CSP structure and introduces it to both the backbone network and the neck for stronger feature fusion capability. YOLOv5 accelerates the inference speed and reduces the model size by a large margin, which makes it more suitable for the application of traffic sign recognition than previous versions of YOLO. The structure of YOLOv5 is shown in Fig. 1. Where, the composition of CBL, SPP and CSP is shown in Fig. 2.

The YOLOv5 network model is mainly divided into four sections: the input block, the backbone block, the neck block, and the head block. For the purpose of increasing

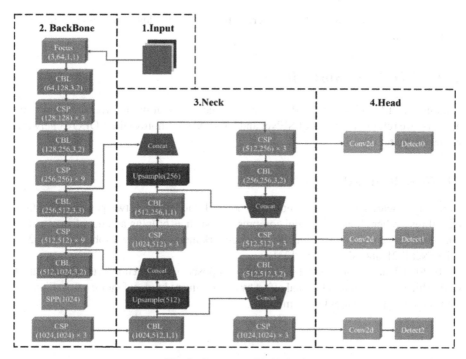

Fig. 1. Structure of YOLOv5

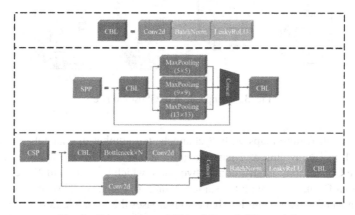

Fig. 2. Composition of CBL, SPP and CSP module

the discrimination and accuracy of detection, YOLOv5 incorporates adaptive image padding, anchor frame calculation, and Mosaic enhancement to process the input data. The backbone part mainly uses the CBL modules and the CSP modules, where the CSP module effectively prevents the gradient disappearance caused when the network is deepened by residual structures. The neck block adopts the structure of Feature Pyramid Network (FPN) and Path Aggregation Network (PAN). The FPN employs up-sampling to improve semantic transmission, whereas the PAN uses down-sampling to improve feature localization.

4 SG-YOLO Network Model

Although YOLOv5 has many advantages in target recognition, in order to lighten the network, we refer to the ideas of SENet and GhostNet to improve the YOLOv5 network model for better detection.

4.1 Ghost-Bottleneck

Currently, researchers mostly lightweight network models from two perspectives. One part of researchers focus on model compression, such as pruning and quantization. The other part concentrate on building delicate network models, such as MobileNet [20, 21], MicroNet [22], and MobileDets [23].

In 2020, Huawei Noah's Ark Laboratory proposed a novel neural network GhostNet [24], which reduces computational effort by decreasing redundant feature maps. Figure 3 shows the composition of Ghost module.

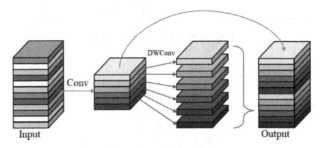

Fig. 3. Composition of Ghost module

For the input feature maps with N channels, the number of channels is condensed to $N/2$ by 1×1 convolution. Then, generate another $N/2$ feature maps by DWConv (Depthwise Convlution). Lastly, concat two parts, which both contain $N/2$ channels, to get the output feature layer with the same channels of the input. This structure not only reduces the amount of convolution, but also increases the receptive field of the network via DWConv. Ghost-Bottleneck is similar to the residual structure proposed by ResNet [25]. The compositon of Ghost-BottleNeck is illustrated in Fig. 4.

As is shown in Fig. 4, both Ghost-Bottlenecks with stride one and stride two are composed of two Ghost modules. The input of Ghost-Bottleneck connects to the output of the second Ghost module via shortcut. When the width and the height of feature layers need to be compressed, we set the stride of Ghost-Bottleneck to two and increase an extra DWConv with between two Ghost modules.

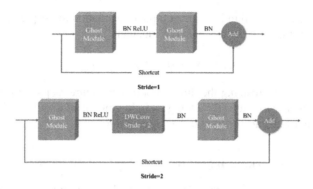

Fig. 4. Composition of Ghost-Bottleneck

4.2 Channel Attention Mechanism

Hu et al. [26] researched the relationship between channels and introduced Squeeze and Excitation (SE) module. SE module allows the system learning global information to emphasize dominant features by adjusting the weights of channels. Figure 5 shows the structure of SE module.

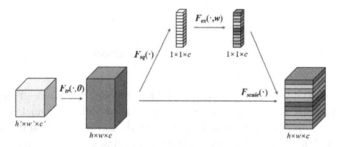

Fig. 5. Structure of SE module

SE module consists of squeeze and excitation. First of all, the feature map $X \in R^{h' \times w' \times c'}$ is transformed into $U \in R^{h \times w \times c}$ by a standard convolution operation, where $U = [u_1, u_2, ..., u_c]$. Subsequently, a global average pooling of U is applied to obtain a feature map Z_c with size of $1 \times 1 \times c$, which is named squeeze operation. The calculation of Z_c is shown in Eq. 1.

$$Z_c = Fsq(u) = \frac{1}{h \times w} \sum_{i=1}^{h} \sum_{j=1}^{w} u(i,j) \tag{1}$$

With the aim of obtaining the correlations between channels further, the generated feature maps are passed through two fully connected layers, then the corresponding weights between channels can be obtained using Sigmoid function, as shown in Eq. 2.

$$S = Fex(Zc, W) = sigmoid(W_2 ReLU(W_1 Z_c))$$
$$W1 \in R^{\frac{c}{r} \times c}, W2 \in R^{\frac{c}{r} \times c} \tag{2}$$

Finally, as demonstrated in Eq. 3, the output of excitation to the previous features is reweighted by multiplying them channel by channel.

$$\tilde{x}_c = Fscale(uc, sc) = sc \cdot uc \tag{3}$$

SE module permits resizing the dimension of the feature maps without modifying the original network structure. By adding a slight amount of additional compution, it can greatly enhance the accuracy of target recognition.

4.3 SG-YOLO Network Architecture

For the purpose of decreasing the complexity of the YOLOv5 model and requiring the hardware configuration in autonomous driving system. We propose SG-Bottleneck module to lighten the YOLOv5 network model. The component of the SG-Bottleneck module shows in Fig. 6.

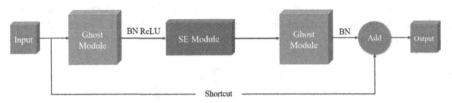

Fig. 6. Structure of SG-Bottleneck

An SE module is incorporated into the Ghost-Bottleneck with stride of one, between two Ghost modules, to enhance the sensitivity of the model to channel features. The output of the second Ghost module is connected via shortcut to form the whole SG-Bottleneck module. The SG-Bottleneck module integrates the advantages of both the SE and Ghost module, and not only achieves the reduction of redundant feature maps in the network, but also the extracted feature maps reflect the significance of the feature channels. We replace the bottleneck of original network with the proposed SG-Bottleneck module, and the improved YOLOv5 network is named SG-YOLO.

4.4 DIoU Loss Function

Intersection over Union (IoU), which is an value describes the overlap ratio between prediction and ground truth boxes, is a criterion to evaluate the prediction accuracy.

Despite the excellent performance of the IoU, there are still some shortcomings. On one hand, when prediction boxes fails to intersect with ground truth boxes, it can be obtained from the definition of IoU that the value is zero. At this time, IoU loss function fails to reflect the distance, which lowers the learning efficiency. On the other hand, if ground truth boxes and prediction boxes have the same intersection area with different distances, the IoU results will also be consistent. Thus, it is difficult to provide an accurate description of the overlap ratio.

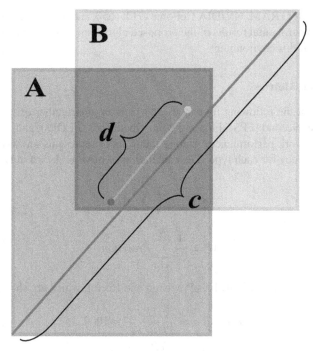

Fig. 7. Principle of DIoU

To deal with the above problems, this paper adopts DIoU as loss function. Figure 7 gives the principle of DIoU.

Equation 4 defines the calculation formula of DIoU loss function.

$$L_{DIoU} = 1 - IoU + \frac{\rho^2(b^{pd}, b^{gt})}{c^2} = 1 - IoU + \frac{d^2}{c^2} \tag{4}$$

where ρ indicates the Euclidean Metric between the center of the prediction box b^{pd} and the ground truth box b^{gt}, and c represents the diagonal distance between b^{pd} and b^{gt}. Equation 5 shows the experssion of IoU, where A and B represent the prediction and the ground truth box respectively.

$$IoU = \frac{A \cap B}{A \cup B} \tag{5}$$

DIoU loss have better perception of distance and high accuracy. In addition, it can be used in the calculation of Non-Maximum Suppression (NMS) to accelerate the convergence of model.

5 Experiments and Analysis

5.1 Experimental Environment

The experimental environment is configured as follows: software conditions are Windows 10, Python3.8 with PyTorch framework, and the hardware conditions are Intel

Core i5-9300H, 16G RAM, NVIDIA GeForce GTX1650, and the training is accelerated by GPU. The experimental results of the proposed algorithm and its comparison models are all derived in this environment.

5.2 Evaluation Metrics

We mainly adopt the following four metrics, which are the number of model parameters, Frames Per Second (FPS), Floating Point Operations (FLOPs), and mAP, to better evaluate the network performance. Among them, mAP represents an evaluation value for average accuracy for each type. The calculation of mAP is shown in Eq. 6.

$$mAP = \frac{1}{N} \sum_{1}^{N} AP$$

$$AP = \frac{1}{N} \sum_{1}^{N} Precision$$

(6)

The calculation of mAP involves Precision and Recall, which are shown in Eq. 7.

$$Precision = \frac{TP}{TP + FP} \times 100\%$$

$$Recall = \frac{TP}{TP + FN} \times 100\%$$

(7)

where, in prediction process, the true positive samples are marked as TP, the false positive samples are marked as FP, and the false negative samples are marked as FN.

5.3 Dataset and Processing

The experimental dataset is obtained from CSUST Chinese Traffic Sign Detection Benchmark (CCTSDB), established by Zhang [27], which contains a total of 15724 images with original, stretched, and adjusted brightness. The dataset classifies traffic signs into three categories: mandatory, warning, and prohibitory.

Table 1. Composition of CCTSDB

Category	Picture	Bounding box
Warning	3478	3962
Prohibitory	7255	10475
Mandatory	5118	6996
Total	15851	21433

There are 15851 original labeled images in the CCTSDB dataset, and the total number of labeled bounding boxes is 21133, as shown in Table 1.

In this experiment, the training set is made up of 2726 photographs from the CCTSDB dataset which have low similarity and high variability in different scenes, and tested on 681 images.

To enrich the local features in the dataset, we use Mosaic data enhancement. Mosaic data enhancement is equivalent to an augmentation of the samples selected for training.

Fig. 8. Mosaic data enhancement

As is shown in Fig. 8. Mosaic data enhancement, Mosaic data enhancement is applied on four traffic sign images selected from the CCTSDB dataset, and stitched together to form a new image. In this way, Mosaic data enhancement fully exploits all features of images by stitching.

5.4 Results and Analysis

Quantitative Results Analysis. To verify the performance of SG-YOLO, we conduct comparison experiments with SSD, YOLOv3, YOLOv4, and YOLOv5. The results for each algorithm are given in Table 2.

Table 2. Comparison of Params and FLOPs

Algorithm	Params(M)	FLOPs(G)	mAP/%	FPS/f·s^{-1}
SSD	26.79	31.4	73.49	13.7
YOLOv3	61.54	154.9	71.17	15.4
YOLOv4	63.99	29.91	74.51	22.4
YOLOv5	7.07	16.4	**75.32**	26.9
SG-YOLO	**4.82**	**9.3**	74.95	**41.7**

The proposed SG-YOLO requires a total of 4.82M parameters. Compared with SSD, YOLOv3, YOLOv4 and YOLOv5, SG-YOLO reduces the number of parameters by

82%, 92.2%, 92.5% and 31.8% respectively. The number of FLOPs in SG-YOLO is 9.3G, which is less than the other four algorithms by 70.4%, 94.0%, 68.9% and 43.3%. It can be concluded that SG-YOLO has the lowest number of parameters and FLOPs among the five algorithms, which gives a huge advantage in terms of model size.

According to the training results, it can be obtained that SG-YOLO achieves 74.95% mAP and 41.7(f·s^{-1}) FPS. Compared with the remaining four algorithms, SG-YOLO achieves a faster detection speed at the expense of a little accuracy.

Qualitative Results Analysis. To validate the performance of SG-YOLO more visually, three videos with different road conditions and weather are selected for identification.

Fig. 9. Traffic sign recognition in different weathers

Figure 9 gives the recognition results of SG-YOLO. The recognition results for road traffic signs are 64%, 84%, 95%, and 96% for different weather and road conditions respectively. As can be seen from the left two pictures, the influence of light leads to poor recognition results at night and on rainy days. However, the traffic signs can still be recognized accurately, and there is no leakage or false detection.

On the meanwhile, SG-YOLO also presents outstanding recognition ability for three distinct classes of traffic signs on the road, as shown in Fig. 10. The confidence levels of the prohibitory, mandatory, and warning traffic sign are 96%, 92%, and 91%, respectively. In the case of complex traffic intersections, SG-YOLO can detect the traffic signs without any false detection.

From the analysis of the above experimental results, it can be concluded that SG-YOLO algorithm has both ubiquity and real-time performance, which provides accurate guidance for subsequent applications in autonomous driving.

Fig. 10. Traffic sign recognition for three types

6 Conclusions

In this paper, for the purpose of solving the problem of poor real-time and low accuracy of traffic sign recognition, we propose a real-time recognition algorithm based on SG-YOLO. The main idea is that we construct a novel SG-Bottleneck module by combining SE module with Ghost- Bottleneck and use DIoU as the loss function. The experimental results show that the proposed SG-YOLO algorithm can achieve 74.95% of mAP and 41.7 FPS on CCTSDB dataset. Compared with mainstream algorithms, SG-YOLO is remarkably accelerated in terms of detection speed and lessened the quantity of model parameters by maintaining detection accuracy. Considering the speed and accuracy, SG-YOLO algorithm requires lower hardware equipment for autonomous driving systems, making it more suitable to be deployed in mobile and embedded devices.

Acknowledgments. This work is supported by Natural Science Foundation of Shanghai under Grant 22ZR1424200.

References

1. Bu, W., Yang, H.E.: Traffic sign detection based on improved YOLOv3. J. Sichuan Univ. (Nat. Sci. Ed.) **59**(1), 012004 (2022)
2. Li, X., Xie, Z., Deng, X., Wu, Y., Pi, Y.: Traffic sign detection based on improved faster R-CNN for autonomous driving. J. Supercomput. **78**(6), 7982–8002 (2021). https://doi.org/10.1007/s11227-021-04230-4
3. You, S., Bi, Q., Ji, Y., et al.: Traffic sign detection method based on improved SSD. Information **11**(10), 475 (2020)
4. Huang, H., Hou, L.-Y.: Traffic road sign detection and recognition in natural environment using RGB color model. In: Huang, D.-S., Bevilacqua, V., Premaratne, P., Gupta, P. (eds.) ICIC 2017. LNCS, vol. 10361, pp. 345–352. Springer, Cham (2017). https://doi.org/10.1007/978-3-319-63309-1_32

5. Wang, Q., Liu, X.: Traffic sign segmentation in natural scenes based on color and shape features. In: 2014 IEEE Workshop on Advanced Research and Technology in Industry Applications (WARTIA), pp. 374–377. IEEE (2014)
6. El Baz, M., Zaki, T., Douzi, H.: An improved method for red segmentation based traffic sign detection. In: 2021 IEEE 9th International Conference on Information, Communication and Networks (ICICN), pp. 490–494. IEEE (2021)
7. Bae, G.Y., Ha, J.M., Jeon, J.Y., et al.: LED traffic sign detection using rectangular hough transform. In: 2014 International Conference on Information Science & Applications (ICISA), pp. 1–4. IEEE (2014)
8. Tang, J., Su, Q., Lin, C., et al.: Traffic sign recognition based on HOG feature and SVM. In: Proceedings of the 2020 4th International Conference on Electronic Information Technology and Computer Engineering, pp. 534–538 (2020)
9. Ren, S., He, K., Girshick, R., et al.: Faster R-CNN: towards real-time object detection with region proposal networks. In: Advances in Neural Information Processing Systems, vol. 28 (2015)
10. Dai, J., Li, Y., He, K., et al.: R-FCN: object detection via region-based fully convolutional networks. In: Advances in Neural Information Processing Systems, vol. 29 (2016)
11. Cai, Z., Vasconcelos, N.: Cascade R-CNN: delving into high quality object detection. In: Proceedings of the IEEE Conference on Computer Vision and Pattern Recognition, pp. 6154–6162 (2018)
12. Redmon, J., Divvala, S., Girshick, R., et al.: You only look once: unified, real-time object detection. In: Proceedings of the IEEE Conference on Computer Vision and Pattern Recognition, pp. 779–788 (2016)
13. Redmon, J., Farhadi, A.: YOLO9000: better, faster, stronger. In: Proceedings of the IEEE Conference on Computer Vision and Pattern Recognition, pp. 7263–7271 (2017)
14. Jiang, P., Ergu, D., Liu, F., et al.: A review of yolo algorithm developments. Procedia Comput. Sci. **199**, 1066–1073 (2022)
15. Tian, Z., Shen, C., Chen, H., et al.: FCOS: fully convolutional one-stage object detection. In: Proceedings of the IEEE/CVF International Conference on Computer Vision, pp. 9627–9636 (2019)
16. Gan, Z., Wenju, L., Wanghui, C., et al.: Traffic sign recognition based on improved YOLOv4. In: 2021 6th International Conference on Intelligent Informatics and Biomedical Sciences (ICIIBMS), vol. 6, pp. 51–54. IEEE (2021)
17. Bochkovskiy, A., Wang, C.Y., Liao, H.Y.M.: Yolov4: optimal speed and accuracy of object detection. arXiv preprint arXiv:2004.10934 (2020)
18. Wang, C.Y., Liao, H.Y.M., Wu, Y.H., et al.: CSPNet: a new backbone that can enhance learning capability of CNN. In: Proceedings of the IEEE/CVF Conference on Computer Vision and Pattern Recognition Workshops, pp. 390–391 (2020)
19. He, K., Zhang, X., Ren, S., et al.: Spatial pyramid pooling in deep convolutional networks for visual recognition. IEEE Trans. Pattern Anal. Mach. Intell. **37**(9), 1904–1916 (2015)
20. Sandler, M., Howard, A., Zhu, M., et al.: Mobilenetv2: inverted residuals and linear bottlenecks. In: Proceedings of the IEEE Conference on Computer Vision and Pattern Recognition, pp. 4510–4520 (2018)
21. Howard, A., Sandler, M., Chu, G., et al.: Searching for mobilenetv3. In: Proceedings of the IEEE/CVF International Conference on Computer Vision, pp. 1314–1324 (2019)
22. Li, Y., Chen, Y., Dai, X., et al.: Micronet: improving image recognition with extremely low flops. In: Proceedings of the IEEE/CVF International Conference on Computer Vision, pp. 468–477 (2021)
23. Xiong, Y., Liu, H., Gupta, S., et al.: Mobiledets: searching for object detection architectures for mobile accelerators. In: Proceedings of the IEEE/CVF Conference on Computer Vision and Pattern Recognition, pp. 3825–3834 (2021)

24. Han, K., Wang, Y., Tian, Q., et al.: Ghostnet: more features from cheap operations. In: Proceedings of the IEEE/CVF Conference on Computer Vision and Pattern Recognition, pp. 1580–1589 (2020)
25. He, K., Zhang, X., Ren, S., et al.: Deep residual learning for image recognition. In: Proceedings of the IEEE Conference on Computer Vision and Pattern Recognition, pp. 770–778 (2016)
26. Hu, J., Shen, L., Sun, G.: Squeeze-and-excitation networks. In: Proceedings of the IEEE Conference on Computer Vision and Pattern Recognition, pp. 7132–7141 (2018)
27. Zhang, J., Wang, W., Lu, C., et al.: Lightweight deep network for traffic sign classification. Ann. Telecommun. **75**(7), 369–379 (2020)

Research on Task Oriented Reliability Simulation Evaluation and Maintenance Strategy Optimization of Degraded System

Du Haidong, Cao Junhai, and Huang Xsiying[✉]

Army Academy of Armored Forces, Beijing 100072, People's Republic of China
dkaito@163.com

Abstract. For the state changes characteristic of degradation system under imperfect maintenance, the traditional analysis method become useless, the thesis presents a simulation method based on Agent for repairable system, according to which the maintenance strategy is studied and optimized. Firstly, the reliability variation law of degraded system in task execution is analyzed; Secondly, the performance degradation process of system components is described, the system reliability modeling method of agent is proposed, and the simulation algorithm is given; Finally, combined with an example, the system reliability under different maintenance strategies is simulated and evaluated. The research results can provide a basis for the formulation and improvement of maintenance support scheme of degraded system.

Keywords: Mission · Degradation system · Reliability · Maintenance strategy · Simulation

1 Introduction

For the repairable degraded system, after a fault occurs during the task, the system function is generally restored by replacing some components. However, due to the aging of non fault components caused by wear or corrosion [1], it can not be restored to a new state, there is a certain degree of performance degradation existing in system with the accumulation of operation and the increase of maintenance times, thus showing polymorphic characteristics. Considering that the above system status is affected by the working time of the components, and the intact system status does not mean that the system status is updated, and the time when the system is put into operation again after maintenance cannot be used as the system regeneration point, at this time, the update process theory will no longer be applicable [2]. Therefore, Tan Lin et al. Described the system degradation process by using gamma function [3], Di Peng and others studied the system reliability model under n-type maintenance strategy according to Markov arrival process [4]. The change law of component state of degraded system can be described better with the increase of component scale using the above methods, a NP problem will be caused easily. Moreover, the analytical calculation model is difficult to be applied in

W. Fan et al. (Eds.): AsiaSim 2022, CCIS 1713, pp. 100–112, 2022.
https://doi.org/10.1007/978-981-19-9195-0_9

practical engineering systems. Therefore, through the analysis of the reliability change law of degraded system during the mission, this paper constructs the reliability algorithm model of multi-component degraded system, puts forward the reliability simulation evaluation method of agent system, and optimizes the selection of maintenance strategy of degraded system under multi maintenance scheme combined with an example. The relevant results can provide a basis for the reliability evaluation of degraded system in the mission stage and the formulation and improvement of maintenance scheme.

2   Introduction

2.1   Component System

For repairable degraded systems, each maintenance behavior during task execution will reduce the failure probability of the system under certain conditions, which is usually expressed as a ratio of the actual failure rate or the function of system operation process. Assuming that the system fails at time t(i), it is the failure probability of the system at the current time, and t(i + 1) is the instantaneous time after the system fails, it means that the system is initialized after maintenance. Then the probability of system failure can be recorded as:

$$\omega_{i+1}(t) = \omega(t)(1 - \rho) \tag{1}$$

In the above formula, ρ It is the maintenance effectiveness index, generally $0 \le \rho \le 1$, it can be found that when $\rho = 0$, the above model is equivalent to the system reliability model under the minimum maintenance condition, that is, the probability of system failure remains constant before and after maintenance; When $\rho = 1$, after the system is repaired, the probability of instantaneous failure becomes 0, that is, the system is repaired as new. In fact, for most equipment systems, the degradation of system performance due to the wear and aging of some components should be taken as $0 < \rho < 1$. It is used to describe the change of reliability state of degraded system. It's degradation law is shown in Fig. 1, that is, after the system function is restored at different fault times (t_1, t_2,..., t_N, $n = 1, 2, 3...$), the reliability level should be between the repaired as new and the repaired as old. If $\rho = 0.1$, it indicates that most maintenance activities are minimum maintenance, and the system will be updated or replaced every 10 maintenance times on average.

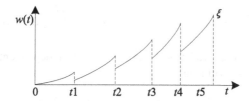

Fig. 1. Variation law of reliability state of degraded system during mission

2.2 Multi-component System

For the component degraded system, its performance continues deteriorating after multiple maintenance, the failure rate increases faster and faster, and the corresponding maintenance time continues is increasing until its function cannot be restored through maintenance. In the above degradation process, it is assumed that the distribution density function of the performance degradation x of a component in the system at time t is g $(x; t)$. For the performance degradation of the complex system is a monotonic increasing process, set the failure threshold as D and the degradation failure time as T_D. Under the degradation failure mode, the reliability function of the system is:

$$R_D(t) = P(T_D > t) = P(x < D) = \int_0^D g(x; t)dx \tag{2}$$

where, $R_D(t)$ is the reliability of degradation failure at time t, and T_D is the performance degradation failure time; x is the degradation amount and D is the degradation failure threshold. For the multi-component system, the performance degradation failure itself has multiple degradation channels. The above model is transformed into a competitive failure problem. The typical performance degradation failure process is shown in Fig. 2.

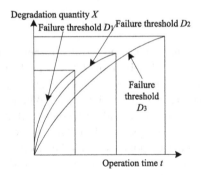

Fig. 2. The competitive degradation failure process of multi-component repairable system

At this time, it can be assumed that the system has n degraded failure modes, the probability density function for the i-th failure mode is $g_i(x; t)$, and the reliability function for the i-th failure mode is:

$$R_{Di}(t) = P(T_{Di} > t) = P(x < D_i) = \int_0^{D_i} g_i(x; t)dx \tag{3}$$

Assuming that each degradation failure process is independent with each other, Eq. (2) can be further rewritten as:

$$R_D(t) = P(T_{D_1} > t, T_{D_2} > t, \cdots, T_{D_n} > t) = P(x_1 < D_1, x_2 < D_2, \cdots, x_n < D_n)$$
$$= \int_0^{D_1} g_1(x; t)dx_1 \int_0^{D_2} g_2(x; t)dx_2 \cdots \int_0^{D_n} g_n(x; t)dx_n \tag{4}$$

The analysis of the above degradation process models is usually carried out in combination with random processes. Commonly used processes include gamma process [5], winner process [6], Markov [7], Brownian motion process [8] and multiple renewal process [9]. In practice, the appropriate model should be determined and described in combination with the model assumptions and the monitoring results of system degradation status.

3 Reliability Model of Degraded System

3.1 Multi-state Transfer Process

Definition 1: For random variables ξ, $\eta(\xi > \eta)$, for real numbers α, if $P(\xi > \alpha) \geq P(\eta > \alpha)$, then $\xi \geq_{st} \eta$ or $\eta \leq_{st} \xi$. So for random sequences $\{X_n, n = 1, 2, \cdots\}$, if $X_n \geq_{st} X_{n+1}$, the sequence will be named decreasing process; if $X_n \leq_{st} X_{n+1}$, then the sequence is increasing;

Definition 2: If $\{\alpha^{n-1}X_n, n = 1, 2, \cdots, \alpha > 0\}$ is an update system composed of random sequences $\{X_n, n = 1, 2, \cdots\}$. If $\alpha > 1$, it's an incremental process; If $0 < \alpha < 1$, which will became a decreasing process.

For repairable degraded systems, as shown in Fig. 3, it is assumed that:

(1) There are $2n + 1$ states in the system, of which: state 1 is the initial state of the system, indicating that the system state is intact; Status $2n + 1$ means the degraded system needs to be replaced after multiple maintenance; $2i$-1($1 = 1, 2, ..., n$)is the random state of the degenerate system; State $2i$ is the random fault state during the use of degraded system;

(2) With the accumulation of working time, the system is assumed to degenerate from state $2i$-1 to state $2i + 1$, and the system degradation rate is α_i, $i = 1, 2, \cdots, n$;

(3) It is assumed that the failure probability from degraded state $2i$-1 to failure state $2i$ is λ_{i-1}, there $\lambda_{i-1} < \lambda_i$, $i = 1, 2, \cdots, n$;

(4) During the operation of the system, it is restored from the maintenance state $2i$ to the degraded state $2i$-1, and the repair rate is recorded as μ_i. Among them $\mu_{i-1} > \mu_i$, $i = 1, 2, \cdots, n$;

(5) After several times of maintenance, when the system enters the state $2n + 1$, the system will be changed to make it "repaired as new";

(6) It is assumed that each fault and maintenance state of the system are independent of each other.

3.2 System Availability Model

For a multi-component system, assuming that all its components are degraded failure modes, the system performance depends on the change of the state of each component, and its reliability can be expressed by the component reliability function. Assuming that a system contains n-key components, the state function of the i-th component at time t is

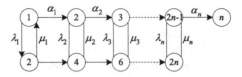

Fig. 3. System state transition diagram under maintenance conditions

$x_i(t)$, and its probability density function is $f_i(x; t)$, then the state function of the system at time t can be expressed as $X(t) = (x_1(t), x_2(t),..., x_n(t))$.

For the above multi-component system, assuming that there may be k-states, the system state output function at time t can be expressed as $Y(t) = (y_1(t), y_2(t),..., y_n(t))$, where $y_i(t) = g_i(X(t)) = g_i(x_1(t), x_2(t),..., x_n(t))$. If only component y_i works at time t, the change of system state only depends on the performance of the component. At this time, the system reliability function is expressed as:

$$R(t) = P\{y_i(\tau) \in \Omega, \forall \tau \in [0, t]\} \tag{5}$$

For multi-component systems, the reliability function is expressed as:

$$R(t) = P\{y_1(\tau) \in \Omega_1, y_2(\tau) \in \Omega_2, \cdots, y_k(\tau) \in \Omega_k, \forall \tau \in [0, t]\} \tag{6}$$

In the above two equations, Ω is the variation range of component reliability output state. Since the system components show degradation failure, assuming that there is a monotonic decreasing function $g_i(X(t))$, Eq. (6) can be expressed as:

$$R(t) = P\{y_i(\tau) \in \Omega, \forall \tau \in [0, t]\} = P\{g_i(x_1(\tau), x_2(\tau), \cdots, x_n(\tau))\} \in \Omega, \forall \tau \in [0, t]\} \tag{7}$$

If $\Omega = [y_{iL}, +\infty]$, y_{iL} is the lower limit of component performance output result, then the above formula can be written as:

$$R(t) = P\{g_i(x_1(\tau), x_2(\tau), \cdots, x_n(\tau))\} \in \Omega,$$
$$\forall \tau \in [0, t]\} = P\{g_i(x_1(t), x_2(t), \cdots, x_n(t)) \in \Omega\} = \tag{8}$$
$$P\{g_i(x_1(t), x_2(t), \cdots, x_n(t)) \geq y_{iL}\}$$

For the random variable $x_i(t)$, it is assumed that the joint probability density function at time t is written as $f(x_1, x_2, \cdots, x_n; t)$, when only component y_i is considered, the calculation expression of system reliability is:

$$R(t) = P\{g_i(x_1(t), x_2(t), \cdots, x_n(t)) \geq y_{iL}\} = \int \cdots \int \int_{E_i} f_t(x_1, x_2, \cdots, x_n) dx_1 dx_2 \cdots dx_n \tag{9}$$

The integral region E_i is expressed as $E_i = \{(x_1, x_2, \cdots, x_n) : g_i(x_1(t), x_2(t), \cdots, x_n(t)) \geq y_{iL}\}$, when all components are considered, the integral calculation formula of system reliability function is:

$$R(t) = P\{g_i(x_1(\tau), x_2(\tau), \cdots, x_n(\tau))\} \in \Omega, \forall \tau \in [0, t]\} =$$
$$P\{g_i(x_1(t), x_2(t), \cdots, x_n(t)) \in \Omega\} = P\{g_i(x_1(t), x_2(t), \cdots, x_n(t)) \geq y_{iL}\} \tag{10}$$

In the above formula, the integral region E_i is $E = E_1 \cap E_2 \cap \cdots \cap E_k$, then the system failure rate can be calculated by the following formula:

$$\int \cdots \int \int_E f_t(x_1, x_2, \cdots, x_n) dx_1 dx_2 \cdots dx_n \tag{11}$$

There $\bar{E} = \bar{E}_1 \cap \bar{E}_2 \cap \cdots \cap \bar{E}_k$. It can be found that with the increase of spatial dimension, the complexity of analytical calculation model increases exponentially, which will be difficult to be applied to engineering practice.

4 Research on Reliability Modeling of Degraded System

4.1 Multi Component System Interaction

For complex equipment system, its task execution is based on the realization of one or more subsystem level functions. Combined with the decomposition of system function structure, the interface and association relationship between lower level functions gradually appear, and its logical associate on and function flow diagram are also reflected, as shown in Fig. 4. It can be found that although the above structural model can accurately reflect the structural hierarchical relationship between the system function and its subsystems and key components, it only describes the static composition relationship of the system. In the product operation state, with the progress of the task, the execution of tasks in each stage requires the dynamic connection and combination of different subsystems, and randomly joins and exits the current task process, which strengthens the timing correlation and stage dependence of subsystems and related components, and further increases the difficulty of system reliability analysis and evaluation.

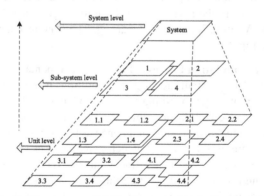

Fig. 4. The system reliability hierarchy block diagram model

4.2 Simulation Model Design

Component Agent. The component Agent is the basic element of the product. In the process of system design and operation, it may experience the following states: standby

(InStore), which is the initial state of components; Operational, component operation status; Failure, the failure state of the product during use; WaitForRepair: maintenance delay caused by waiting in line; Repair, replacement and maintenance process of the system; Spareparts, parts repaired as new. At this time, standby is the initial state of component simulation. Here, three communication mechanisms are defined, namely class M, class T and class C, which are respectively represented as message trigger, delay trigger and condition trigger. The state mechanism of this kind of agent is shown in Fig. 5.

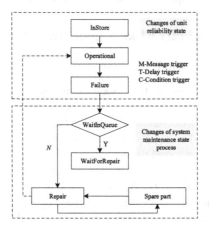

Fig. 5. Component agent state mechanism model

System Agent. For the multi-component functional module subsystem, it will experience the following states in the running state: InStore, Operational, FailureButOperational, StopWorking, WaitForRepair and Repairing. The state mechanism of this kind of agent is shown in Fig. 6.

System Degradation Simulation Process. For multi-component system, assuming that all components of the system are in good condition, it is recorded as state (*a*); With the accumulation of system working time, components will have random faults, which are recorded as state (*b*); At this time, the system will carry out the first fault maintenance. With the increase of fault maintenance times, the performance of each component of the system will deteriorate, which is recorded as system state (*c*); With the advance of the working time of the task, the number of repair of a single component gradually increases until it reaches the failure threshold, that is, the faulty component has no maintenance value, so it will be replaced for maintenance and recorded as state (*d*). For the above evolution process, we can first build the component agent, complete the construction of the system model through layer by layer packaging, determine the random fault events in combination with the random sampling of component states, and determine the system maintenance strategy by recording the maintenance times of a single component during the operation of the system, so as to accurately describe the change state of system reliability, as Fig. 7 shown.

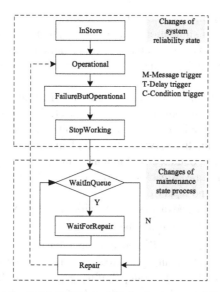

Fig. 6. State diagram of agent system model

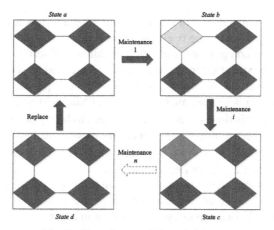

Fig. 7. State change of degraded system

4.3 Simulation Algorithm Design

During task execution, due to the existence of component degradation process, the normal operation of the system does not mean that the system is in good condition. Its polymorphism includes work, degradation, failure, maintenance, repair and replacement. Each event occurs randomly with the change of system state. At this time, the influence of the working state before the system failure on the subsequent use can not be ignored. The simulation can simulate the reliability degradation law in the process of system operation, record the system state transition and update time, and realize the statistics of system shutdown and normal operation time. According to the corresponding calculation model, the purpose of simulation evaluation of system reliability can be achieved. For the k-component equipment system, under the maintenance strategy $M = (N_1, N_2, \cdots, N_k)$, the system reliability evaluation simulation algorithm is:

Step 1: System initialization, set simulation duration is T, simulation run times is m, and the initial failure times of component i is $F_i = 0$, $i = 1, 2, \cdots, k$;

Step 2: According to the reliability and maintainability distribution function of each component of the system, the service life z_i and maintenance time μ_i, $i = 1, 2, \cdots, k$ of components are obtained by random sampling. The system operation time is W_S, and the working time of each component is recorded as w_i, $i = 1, 2, \cdots, k$ (the initial value for W_S and w_i is 0).

Step 3: The simulation clock advances and the simulation step is set to Δt, then: $W_S' = W_S + \Delta t$, $w_i' = w_i + \Delta t$, W_S' and w_i' are the working hours after system and component i status update respectively. If $W_S' > T$, then remember that the system is in good condition during the simulation operation, and return to step 2, the simulation times is $m = m + 1$; if $m \geq M$, shift to step 8, if not, go to step 2; if $w_i' \geq z_i (z_i = min\{z_1, z_2, \cdots, z_k\})$, then the component fails in the simulation operation (at this time, the system state needs to be judged according to the structural relationship of the component), then $F_i' = F_i + 1$, go to step 4;

Step 4: Judge the maintenance mode according to the component i maintenance information: if $F_i' \geq N_i$, go to step 6; else if, go to step 5;

Step 5: Assuming that component i is in series, its failure will lead to system shutdown. At this time, the component enters the maintenance state, and its state is updated after repair, let $F_i = 1$, then the reliability life value of the component after maintenance becomes $z_{F_i=1}$, the corresponding maintenance time changes to $z_{\mu_i=1}$, and go to step 7;

Step 6: If $F_i' \geq N_i$, at this time, the component i needs to be replaced, that is, "repair as new", record the replacement time as μ_{T_i}, initialize its state, and go to step 7;

Step 7: Judge the working time of the system, if $W_S' > T$, at that time, the simulation ends and returns to step 2; Otherwise, go to step 3;

Step 8: End of simulation. The simulation results are counted and the experimental report is formed. According to the above simulation steps, the system reliability simulation evaluation algorithm designed in this paper is shown in Fig. 8.

5 Case Analysis

5.1 Model Assumptions

The two component series system is considered here. Due to the degradation process in the use of the system, the component performance will degrade to a certain extent after fault repair. Under the imperfect maintenance strategy, it is assumed that the degradation coefficients after component repair are α_1 and α_2. The increasing coefficient of maintenance time is γ_1 and γ_2. Among them, the system will be repaired immediately after failure. Assuming that the resources required for maintenance are sufficient, the

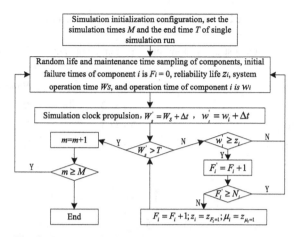

Fig. 8. System reliability simulation evaluation algorithm

impact of maintenance delay on system repair will not be considered temporarily. The system will be shut down during maintenance until all the faulty parts are repaired and the system function is restored. Set the maintenance strategy of two parts as N_1 and N_2 respectively, that is, when the maintenance times of parts reach or exceed n, the parts no longer have maintenance value and need to be replaced. The specific parameters of system components are shown in Table 1.

Table 1. Reliability and maintainability parameter data of system components

Components	λ	μ	μ'	α	γ
1	$4.58 \times 10{-}4$	$2.79 \times 10{-}3$	237	0.968	1.534
2	$3.90 \times 10{-}4$	2.79×10^{-3}	256	0.963	1.231

5.2 Result Analysis

Under the above system composition conditions, the component maintenance strategy shown in Table 2 is given.

Under the imperfect maintenance strategy, due to the existence of the system degradation process, all components in the system operation will experience the changes of degradation, failure, maintenance, repair and renewal status. Under maintenance strategy 3, $N_1 = 3$ and $N_2 = 3$, that is, system components 1 and 2 need to be replaced when the number of maintenance reaches 3. The periodic change process of reliability and maintainability parameters of the two components during the operation of the simulation system is shown in Fig. 9.

Using the simulation algorithm and simulation model designed in this paper, set the simulation running time as $100000h$ and the number of simulation iterations as 10000.

Table 2. The component maintenance strategy

Maintenance and replacement strategy of system components	System component1	System component2
Maintenance strategy 1	$N_1 = 1$	$N_2 = 1$
Maintenance strategy 2	$N_1 = 2$	$N_2 = 2$
Maintenance strategy 3	$N_1 = 3$	$N_2 = 3$
Maintenance strategy 4	$N_1 = 4$	$N_2 = 4$

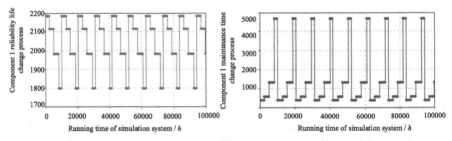

Fig. 9. The changes of reliability and maintainability parameters of system component 1 during the operation of simulation system

The system availability evaluation results are shown in Fig. 10, and the statistical results of system unavailable time in the operation of the simulation system are shown in Fig. 11.

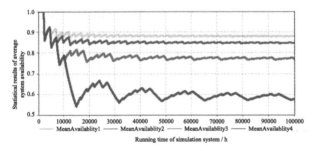

Fig. 10. System availability simulation evaluation results

Through the comparison results of Figs. 10 and 11, it can be found that maintenance strategy 1 is actually a perfect maintenance strategy, that is, when the system components fail, the replacement repair method is adopted, that is, "repair as new". At this time, the system has high availability. Because the replaced components still have a high performance level after repair, this maintenance method needs to consume a lot of maintenance resources and does not take advantage of the saving of maintenance cost; Under maintenance strategy 4, due to the obvious degradation of the performance of the system components after multiple maintenance, the system availability will decline

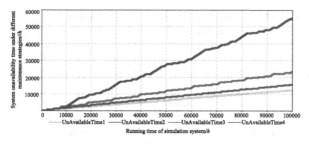

Fig. 11. Simulation statistical results of system unavailable time

significantly, and the system will be shut down for a long time, which is difficult to meet the needs of the system. If the components are still maintained at this time, the gain is not worth the loss. Comparing maintenance strategies 2 and 3, it can be found that when $N_1 = 3$ and $N_2 = 3$, that is, when the maintenance times of two parts reach 3, the system not only has a high level of availability, but also the maintenance resources can be used at a high level.

6 Summary

The thesis aiming at the wide application characteristics of imperfect maintenance strategy in current engineering maintenance, which makes the traditional renewal process and theory difficult to meet the requirements of system reliability evaluation, an Agent-based system reliability simulation evaluation method is proposed. Through the analysis of the system reliability evaluation results under different maintenance strategies, it can be found that under the imperfect maintenance strategy, different maintenance methods and timing will have a great impact on the system availability. The research results can be used to formulate and optimize the system maintenance scheme.

References

1. Gao, Y., Feng, Y., Tan, J.: Multi-principle preventive maintenance: a design-oriented scheduling study for mechanical system. J. Zhejiang Univ.-Sci. A (Appl. Phys. Eng.) **15**(11), 862–871 (2014)
2. Chen, Y., Jin, J.: A simulation algorithm for reliability index of repairable K/N(G) system. Comput. Simul. **25**(11), 115–118 (2008)
3. Peng, D., Fang, L., Tong, C.: Research for the reliability of multi-state repairable system with replacement policy N. J. Syst. Eng. Electron. **36**(3), 604–606 (2014)
4. Tan, I., Cheng, Z., Guo, B.: Availability of series repairable systems under imperfect repair. J. Natl. Univ. Defense Technol. **31**(6), 100–105 (2009)
5. Rausand, M.: System Reliability Theory: Model, Statistical Methods and Applications, 2th edn, pp. 130–147. Academic Press (2010)
6. Grall, A., Berenguer, C., Dieulle, L.: A condition based maintenance policy for deteriorating systems. Reliab. Eng. Syst. Saf. **76**, 167–180 (2002)
7. Bloclr-Mercier, S.: A preventive maintenance policy with sequential checking procedure for a Markov deteriorating system. Eur. J. Oper. Res. **147**, 548–576 (2002)

8. Van Noortwijk, J.M., Vander Weide, J.A.M., et al.: Gamma processes and peaks-over-threshold distributions for time dependent reliability. Reliab. Eng. Syst. Saf. **92**, 1651–1658 (2007)
9. Moghaddam, K.S., Usher, J.S.: A new multi-objective optimization model for preventive maintenance and replacement scheduling of multi-component systems. Eng. Optim. **43**(7), 702–710 (2011)
10. Fitouhi, M.C., Nourelfath, M.: Integrating noncyclical preventive maintenance scheduling and production planning for a single machine. Int. J. Prod. Econ. **136**, 344–351 (2012)
11. Liao, H.T., Elsayed, A., Chan, L.Y.: Maintenance of continuously monitored degrading systems. Eur. J. Oper. Res. **175**, 821–835 (2006)
12. Yang Jikun, X., Tingxue, C.L.: Availability modeling and simulation of missile weapon system based on SEBS-TOMS layered composite frame. J. Syst. Eng. Electron. **37**(2), 460–462 (2015)

A Campus Scene Navigation Scheme Based on MPCC Dynamic Obstacle Avoidance Method

Zonghai Chen, Liang Chen, Guangpu Zhao, and Jikai Wang[✉]

University of Science and Technology of China, Hefei 230026, Anhui, China
wangjk@mail.ustc.edu.cn

Abstract. The navigation system is a key module in the future application of mobile robots and is essential for the safety and robustness of mobile robot motion. Available navigation systems can already perform reasonable path planning and motion planning processes in specific scenarios. However, with the development of mobile robotics, there are higher requirements for the scenarios in which the robots operate and the response efficiency requirements. In order to solve the problem of motion planning and obstacle avoidance between path points in dynamic campus scenes, a combination of static obstacle avoidance based on voxel grid and dynamic obstacle avoidance based on MPCC is proposed on Ackermann kinematic model as well as motion control. The organic combination of static obstacle avoidance and dynamic obstacle avoidance solves the problem of quickly performing path planning and obstacle avoidance for unmanned vehicles in complex environments. The work has been experimented on simulation conditions and actual robots, and the robots have been placed in campus scenarios for validation.

Keywords: Navigation system · Campus scenarios · Ackerman model · Obstacle avoidance

1 Introduction

Mobile robots are playing an increasingly important role in the process of daily production and life. According to reliable data, robots are now widely used in medical, education, logistics, industrial production, national defense and other fields. The implementation of robots frees human hands and replaces human beings in dangerous environments to perform the corresponding work and safeguard human lives.

At the moment, the research of scientists on robots has made remarkable achievements in the field of front-end vision [1] as well as macroscopic high-level planning. Meanwhile, the path planning of mobile robots performs well in specific scenarios [2], in contrast to the mobile robots' ability to avoid obstacles [3] and the reasonable use of computational resources in highly dynamic scenarios such as campus scenarios, which still have more room for improvement.

In considering to solve the dynamic obstacle avoidance problem [4] of mobile robots in complex campus scenes [5], to improve the stability of dynamic obstacle avoidance, and to make reasonable use of computational resources, we propose a combination

W. Fan et al. (Eds.): AsiaSim 2022, CCIS 1713, pp. 113–124, 2022.
https://doi.org/10.1007/978-981-19-9195-0_10

method of static obstacle avoidance based on voxel grid and dynamic obstacle avoidance based on MPCC [6]. This method can realize the navigation work of the cart in complex campus scenes based on global or local path planning [7]. Since the static obstacle avoidance work is unified under the overall framework of the navigation system, no separate calculation is required, saving a lot of computational resources while being able to accomplish stable obstacle avoidance of dynamic obstacles in campus scenes. Section 2 details the architecture of the navigation system; Section 3 presents the static obstacle avoidance method based on voxel grid and the dynamic obstacle avoidance method based on MPCC; Section 4 partly presents a series of experimental results including simulation experiments and experiments in campus scenarios; Section 5 summarizes the work of this paper and shows the direction for future work.

2   System Overview

The overall process overview of the system is shown in Fig. 1: firstly, the local planner determines whether the obstacles obtained from the current terrain analysis are in the corresponding grid according to the collision detection correspondence between the offline path group and the voxel grid production, and if so, the corresponding path is removed from the selectable paths to get the path group after static obstacle avoidance. Then the MPCC method is applied to predict the future trajectory and avoid the dynamic obstacles after modeling to get the path after dynamic obstacle avoidance, and then the two paths are fused and selected to get the final path and handed over to the system for processing to get the control command of the chassis and complete the navigation process.

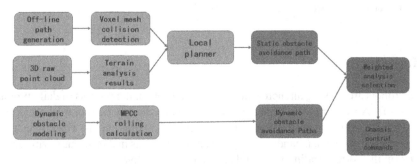

Fig. 1. System overview block diagram

2.1   System Hardware

The proposed navigation system is built on the YUHESEN FR07 Ackermann chassis platform. The sensing device uses the velodyne VLP-16 lidar, which has a measurement range of up to 100m, an error range of ± 3 cm, a vertical field of view of 30° (±15°), and a horizontal field of view of 360°. This 16-channel sensor provides a vertical angular resolution of 2°. The horizontal angular resolution varies from 0.1° to 0.4°, based on the rotation rate. The outdoor positioning uses a dynamic state carrier phase differential positioning device, HI-TARGET iNAV2 model RTK, with horizontal positioning accuracy at 2 cm + 1 ppm and elevation accuracy at 4 cm + 1 ppm. The heading angle accuracy error is within 0.1°. This device can provide high accuracy positioning information in campus scenarios, which is important for the whole navigation system. The computing device is a laptop with AMD rayzon7 5800H CPU and GPU NVIDIA RTX 3070. The above devices are the experimental equipment used in the system in the paper and is shown in Fig. 2.

Fig. 2. Mobile robot hardware platform

2.2 System Modules

Path Generation Based on Kinematic Constraints of Ackermann Model
In general, there are two widely used physical structures for the chassis of mobile robots: differential steering chassis and Ackermann structure chassis. Mobile intelligent robots based on Ackermann structures must satisfy certain kinematic constraints when performing motion control. We equate a mobile robot based on the Ackermann structure to a two-wheeled structure that has little impact on accuracy but can greatly improve the ease of our calculations.

The distance traveled by the left and right wheels of the Ackermann structure when steering is different: R is the rotation radius, θ is the rotation angle, and L is the wheelbase of the Ackermann [8].

$$\begin{cases} s_l = \theta R \\ s_r = \theta (R + L) \end{cases} \tag{1}$$

Dividing both sides of the equation by time at the same time, we further obtain

$$\begin{cases} v_l = \omega R \\ v_r = \omega (R + L) \end{cases} \tag{2}$$

Combining these two equations again, the angular velocity can be inverted as formula (3). The ackermann structure steering schematic is shown as Fig. 3

$$\omega = (v_r - v_l)/L \tag{3}$$

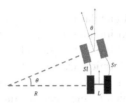

Fig. 3. Ackermann structure steering schematic. s_l indicates the arc of the left wheel turning and s_r indicates the arc of the right wheel turning, L indicates wheelbase.

The angular velocity of the Ackermann structure is the same, so the overall angular velocity of the Ackermann platform at this time is. Therefore, we can equate the Ackermann model to a two-wheeled bicycle structure. The velocities in the three dimensions of x, y and, respectively, for the case of forward motion of the mobile robot are

$$
\begin{cases}
\dot{x} = vcos \\
\dot{y} = vsin \\
\dot{\theta} = \frac{vtan(\delta)}{L}
\end{cases}
\tag{4}
$$

In order to represent it in the computer and to facilitate the calculation, we discretize the kinematic model based on Ackermann's chassis as a series of points in space.

Thus, starting from the initial position at the initial moment, the position and rotation angle at moment $t+1$ can be recursively derived from the position at the previous moment t. The equation is expressed as

$$
\begin{cases}
x_{t+1} = x_t + v_t \cos(\theta_t)d_t \\
y_{t+1} = y_t + v_t \sin(\theta_t)d_t \\
\theta_{t+1} = \theta_t + \omega_t d_t
\end{cases}
\tag{5}
$$

Accordingly, the corresponding discrete path set applicable to Ackermann kinematic constraints can be obtained according to this discrete formula.

Terrain Analysis Based Local Point Cloud
In this paper, the ground is first extracted, and the relative height of points in the point cloud relative to the ground is calculated based on the extracted ground, and this height is used as a basis for judging the actual situation of the cart (e.g., wheel height). Those higher than the ground threshold but lower than the obstacle threshold are treated as slopes and left as penalty terms for subsequent paths; those higher than the obstacle height will be judged as obstacles, and after joint detection, they are treated as impassable areas.

3 Proposed Method

We model the static obstacles detected in the environment separately from the dynamic obstacles and combine the corresponding obstacle avoidance strategies for path planning on this basis. Since static obstacles already exist and do not occupy waypoint positions during path planning, the processing of static obstacles during the movement of a mobile robot can be simplified to the process of identification and obstacle avoidance, without the need for prediction and collision range determination as for mobile obstacles. Dynamic obstacles, on the other hand, require prediction of the next moment's position based on their observation and the use of a modified boundary to ensure dynamic obstacle avoidance.

The MPCC rolling computational process is resource intensive, and in our approach, MPCC is used only for dynamic obstacle prediction, which greatly reduces the computational effort and improves the real-time performance. And the integration of the two new methods makes it applicable to the implementation of navigation and obstacle avoidance in campus scenarios.

3.1 Static Obstacle Collision Detection Based on Voxel Grid

After generating the discrete path groups, in order to improve the efficiency of collision detection for mobile robots in practical applications, this method corresponds the offline paths to the corresponding voxel grids, and stores the corresponding path points with the path IDs through the voxel grids. Based on the LIDAR detection range of VLP-16, we limit the size of the voxel grid to 5*5 m. In this way, an offline data corresponding to the voxel grid ID and path ID can be formed [9]. When an obstacle appears in the voxel grid range [10], the exercisable path schematic obtained after contour detection based on the mobile robot is shown in Fig. 4.

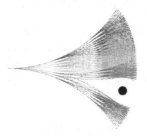

Fig. 4. After an obstacle appears on the path group, the system selects the path group that is not blocked, black dots indicate obstacles.

Specifically when detecting during usage, we let the path points within the voxel grid enclosed area, traverse the index of the voxel grid, and calculate the affected path sequence within the contour of the mobile robot and remove this sequence from the corresponding sequence to be selected. Usually, static obstacles can be captured on a grid map or scanned in real time by lidar and obtained by terrain analysis. As shown in Fig. 5, it is the lidar that detects obstacles in real time and removes the blocked path.

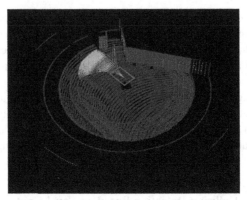

Fig. 5. The case of path selection when static obstacles such as walls are encountered. The circle indicates the drivable area on the ground.

3.2 MPCC-Based Dynamic Obstacle Avoidance Strategy

Equation of State for Ackermann Model

First, we fit the plane where the robot is located to represent $W = R^2$ and the mobile robot dynamics by the discrete-time nonlinear system [11]

$$z(t + 1) = f(z(t), u(t)) \tag{6}$$

denoted by the region occupied by the robot at the moment of state z as B(z), and in order to solve the result when MPC is pushed later, we must also obtain the state vector of the Ackerman model, which can be obtained by the following steps. The kinematic constraints on the Ackermann structure, derived from the previous Sect. 2.2, take the state quantities

$$Z = [x, y, \varphi]^T$$

$$u = [v, \delta]^T$$

where Z denotes the state of the Ackermann robot in the plane, φ is the vehicle traverse angle, u is the control quantity, and v, δ is the rear wheel center velocity and the front wheel deflection angle, respectively. To linearize it, after Taylor expansion at any point (Z_r, u_r) and neglecting the higher order terms, the linearized representation of Eq. [12].

$$\dot{\tilde{Z}} = A\tilde{Z} + B\tilde{u} \tag{7}$$

where $\tilde{Z} = Z - Z_r$, $\tilde{u} = u - u_r$, and

$$A = \begin{bmatrix} 0 & 0 & -v_r \sin\varphi_r \\ 0 & 0 & v_r \cos\varphi_r \\ 0 & 0 & 0 \end{bmatrix}$$

$$B = \begin{bmatrix} cos\varphi_r & 0 \\ sin\varphi_r & 0 \\ tan\varphi_r/L & v_r/Lcos^2\varphi_r \end{bmatrix}$$

After forward Eulerian discretization:

$$\tilde{Z}(k+1) = \tilde{A}\tilde{Z}(k) + \tilde{B}\tilde{u}(k) \tag{8}$$

At the moment, we have obtained the Ackermann model discrete state space equation. Using this discrete space state equation, the optimization problem can be built and solved.

Modeling of Dynamic Obstacles

In the fitted plane, each dynamic object i is described by an ellipse of area $Ai \subset W$, defined by its own long semi-axis a_i, short semi-axis b_i, and rotation matrix R_i. For multiple dynamic obstacles within the environment [11], consider as $i \subset \mathbb{N} : [1, ..., n]$, where the size of n can vary with time. The area occupied by all moving obstacles at any moment t is given by

$$\Omega_t^{dyn} = \cup_{i \in [1,...,n]} A_i(z_i(t)) \tag{9}$$

to describe.

It is common to select multiple waypoints on a map that form a continuous path. Between each two waypoints, we consider the intervening path segments to be defined by a cubic polynomial. θ denotes a variable that (approximately) represents the distance traveled along the reference path. When generating this continuous path on the grid map, only the motion of dynamic obstacles needs to be considered when our system is running, since the positions of static obstacles have already been determined and eliminated during path planning. The objective is now to generate collision-free motion for the mobile robot at N future steps while minimizing the cost function. Combining the equation of state of the Ackerman model, in which static obstacles have been removed by the navigation system itself and can be left out of consideration, the optimization problem can be built as

$$J^* = \min_{z_{0:N}, u_{0:N-1}, \theta_{0:N-1}} \sum_{k=0}^{N-1} J(z_k, u_k, \theta_k) + J(z_N, \theta_N)$$
$$\text{s.t. } z_{k+1} = f(z_k, u_k), \theta_{k+1} = \theta_k + v_k\tau \tag{10}$$
$$B(z_k) \cap \Omega_k^{dyn} = \emptyset$$
$$u_k \in U, z_k \in Z, z_0, \theta_0 given$$

where v_k is the forward velocity of the robot (for mobile robots it is part of the input and for cars it is part of the state), τ is the time step, and U and Z are the set of allowed states and inputs, respectively. $z1 : N$ and $u0 : N - 1$ are the sets of states and control inputs in the prediction range $T_{horizon}$ [13], respectively, divided into N prediction steps. θ_k denotes the prediction progress along the reference path at time step k. By solving the optimization problem, we obtain a locally optimal sequence of commands $\left[u_t^* \right]_{t=0}^{t=N-1}$ to guide the robot along the reference path while avoiding collisions with moving obstacles.

Dynamic Collision Avoidance

Each moving obstacle i is represented by an ellipse with its position $pi(t)$ and semiaxes

a_i and b_i and a rotation matrix $Ri(\psi)$. For each obstacle $i \in \{1, ..., n\}$ and the prediction step k, we denote the position of the robot in the plane by the circle j and force circle j not to intersect the elliptical region occupied by the obstacle. The radius of circle j is r_{disc}. Omitting i for simplicity, the inequality constraint on each disk of the robot with respect to the obstacle is

$$c_k^{dyn,j}(z_k) = \begin{bmatrix} \triangle x_k^j \\ \triangle y_k^j \end{bmatrix}^T R(\psi)^T \begin{bmatrix} \frac{1}{\alpha^2} & 0 \\ 0 & \frac{1}{\beta^2} \end{bmatrix} R(\psi) \begin{bmatrix} \triangle x_k^j \\ \triangle y_k^j \end{bmatrix} > 1 \qquad (11)$$

where the distance between the disk j and the obstacle is divided into its $\triangle x^j$ and $\triangle y^j$ components. The parameters α and β are the semiaxes of the enlarged ellipse [14], including the original ellipse and the concatenation of the circles. Although previous methods approximated the ellipse with the Minkowsky sum of the circle as an ellipse with length semi-axis $\alpha = a + r_{disc}$ and short semi-axis $\beta = b + r_{disc}$. This is shown in Fig. 6.

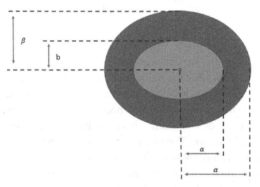

Fig. 6. Schematic diagram of the risk area of a dynamic obstacle after equating it to an ellipse. The green color indicates the obstacle body, and the purple color indicates its risk area. (Color figure online)

Consider two ellipsoids $E1 = Diag(1/a^2 1/b^2)$ and $E2 = Diag(1/(a + \delta)^2 1/(b + \delta)^{22})$. E1 is an ellipsoid with a and b being the semi-long and semi-short axes, respectively. E2 denotes the ellipsoid E1 magnified by a factor of δ on both axes. The goal is to find the smallest ellipsoid that limits the Minkowsky sum [15]. This corresponds to finding the minimum value of δ such that the minimum distance between the ellipsoids E1 and E2 is greater than r^2, the radius of the circle around the robot. This value of δ and the sum of each of the two semi-axes of the ellipsoid are sufficient to ensure that the collision space is completely bounded.

Thus, combined with the collision avoidance of dynamic obstacles, the final optimization problem is established as

$$J^* = \min_{z_{0:N}, u_{0:N-1}, \theta_{0:N-1}} \sum_{k=0}^{N-1} J(z_k, u_k, \theta_k) + J(z_N, \theta_N)$$

$$\text{s.t.} \quad z_{k+1} = f(z_k, u_k), \theta_{k+1} = \theta_k + v_k \tau$$

$$B(z_k) \cap \Omega_k^{dyn} = \emptyset$$

$$u_k \in U, z_k \in Z, z_0, \theta_0 \; given$$

$$c_k^{dyn,j}(z_k) > 1, \forall j \in \{1, \ldots, n_c\}, \forall dyn$$

where $J(z_k, u_k, \theta_k)$ is the tracking cost and velocity cost of the system, while $J(z_N, \theta_N)$ is the terminal cost of the system. The final solved control sequence value is the input to control the robot without collision.

4 Experiment

We have tested both the simulation environment and the real campus scenario in order to verify the feasibility of the method proposed in this paper. The experimental platform is the FR07 platform mentioned above. To construct a reference local path, we first define a series of waypoints and let the mobile robot travel along the established route of the waypoints. For the static obstacle experiments we compare the proposed method with the classical MPC [16] method and Dynamic Window [17] method, and the experimentally obtained running trajectory is shown in Fig. 7.

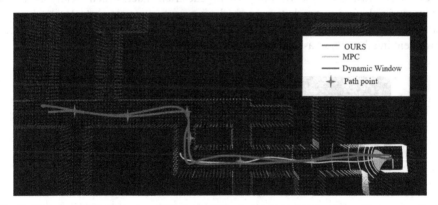

Fig. 7. Experimental trajectories of the three methods run on narrow corridors. Two corners need to be passed respectively, and the corner position is set as a static obstacle area. It can be seen that the classical MPC method fails, while the other two methods pass through all the waypoints normally.

As you can see from the figure, our method and the Dynamic method can follow the waypoint to the specified destination. The classical MPC method, on the other hand, has failed at this point because the rolling optimization process is broken after the first corner is encountered.

In purpose of comparing the dynamic obstacle avoidance effect of our method with Dynamic Window method, we choose a site in front of the campus library where there are often pedestrians walking around. The total length of the waypoint in the scenario is about 30 m, and the speed of the car is about 1 m/s. The actual scene of the site is shown in Fig. 8.

Fig. 8. Campus experiment scene.

The trajectory from Fig. 9 shows that both our method and Dynamic Window have real-time obstacle avoidance capability for dynamic obstacles, but our method simplifies the static obstacle avoidance process in the MPC calculation, so the algorithm consumes less resources, computationally, and the obstacle avoidance response is more timely. In contrast, Dynamic Window avoids pedestrians later, which greatly increases the insecurity. It can be seen that our method can react to dynamic obstacles in advance, and increase the turning angle of driving away from the direction of dynamic obstacle movement in advance, so as to avoid dynamic obstacles.

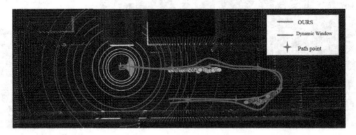

Fig. 9. Comparison chart of trajectories in the campus experiment scenario. The green and orange ellipses represent the trajectories of pedestrians. It can be seen that our method (blue line) is more responsive to dynamic obstacles and the DW method is not sensitive enough to dynamic obstacles. (Color figure online)

Meanwhile, to compare the real-time speed and efficiency of tracking the path between our method and Dynamic Window, we counted the time consumed by both methods traveling along the same path 20 times, the longest and shortest time obtained, and the average time of the 20 runs. The details are shown in Table 1.

Table 1. Table captions should be placed above the tables.

Method	Shortest time/s	Longest time/s	Average time/s	Efficiency improvement rate
OURS	36.3	42.6	39.1	9.4%
DW	39.5	46.4	43.2	0%

The analysis based on the experimental results shows that our method outperforms the traditional MPC algorithm in obstacle avoidance of dynamic obstacles. In the obstacle avoidance of dynamic scenes, not only the response to dynamic obstacles is more rapid, but also the tracking efficiency for a given route is higher, and the average time consumed to walk the same path is shorter.

5   Conclusion and Future Work

In this paper, we propose a joint obstacle avoidance method using voxels and offline path groups for static obstacle avoidance and based on model predictive contour control for dynamic obstacle avoidance. Based on this, this obstacle avoidance method is applied to a navigation system for campus scenes. It is demonstrated through experiments that the navigation system based on our proposed method can not only react quickly to the dynamic obstacles that often appear in the campus scene, but also greatly improve the operational safety of the mobile robot in the campus scene. And it can have higher execution efficiency for the path given by path planning, thanks to the rolling optimization process of stripping static obstacles out of the model predictive control in our method and fixing this process to the navigation system with offline path and grid driving.

For future work, since we have not added the corresponding a priori map for the time being, the subsequent work can be based on this, adding the global grid map and point cloud map in the campus scene [18], and can be based on the raster map to sense the dynamic obstacles in advance and do the corresponding path planning work. In this way, with the double guarantee of a priori map and local navigation obstacle avoidance, the whole system can ensure the complete obstacle avoidance navigation function in the highly dynamic campus scene.

Acknowledgments. This work was supported by the National Natural Science Found of china (Grant No. 62103393).

References

1. Bonci, A., Cen Cheng, P.D., Indri, M., Nabissi, G., Sibona, F.: Human-robot perception in industrial environments: a survey. Sensors **21**, 1571 (2021)
2. Patle, B., Pandey, A., Parhi, D., Jagadeesh, A.: A review: on path planning strategies for navigation of mobile robot. Def. Technol. **15**, 582–606 (2019)
3. Hutabarat, D., Rivai, M., Purwanto, D., Hutomo, H.: Lidar-based obstacle avoidance for the autonomous mobile robot. In: 2019 12th International Conference on Information & Communication Technology and System (ICTS), pp. 197–202. IEEE (2019)

4. Ajeil, F.H., Ibraheem, I.K., Azar, A.T., Humaidi, A.J.: Autonomous navigation and obstacle avoidance of an omnidirectional mobile robot using swarm optimization and sensors deployment. Int. J. Adv. Robot. Syst. **17** (2020). https://doi.org/10.1177/1729881420929498

5. Park, C., Kee, S.-C.: Online local path planning on the campus environment for autonomous driving considering road constraints and multiple obstacles. Appl. Sci. **11**, 3909 (2021)

6. Wang, D., Pan, Q., Hu, J., Zhao, C., Guo, Y.: MPCC-based path following control for a quadrotor with collision avoidance guaranteed in constrained environments. In: 2019 IEEE 28th International Symposium on Industrial Electronics (ISIE), pp. 581–586. IEEE (2019)

7. Chen, P., Huang, Y., Papadimitriou, E., Mou, J., van Gelder, P.: Global path planning for autonomous ship: a hybrid approach of fast marching square and velocity obstacles methods. Ocean Eng. **214**, 107793 (2020)

8. Carpio, R.F., et al.: A navigation architecture for ackermann vehicles in precision farming. IEEE Robot. Autom. Lett. **5**, 1103–1110 (2020)

9. Cao, C., et al.: Autonomous exploration development environment and the planning algorithms. arXiv preprint arXiv:2110.14573 (2021)

10. Zhang, J., Hu, C., Chadha, R.G., Singh, S.: Falco: fast likelihood-based collision avoidance with extension to human-guided navigation. J. Field Robot. **37**, 1300–1313 (2020)

11. Brito, B., Floor, B., Ferranti, L., Alonso-Mora, J.: Model predictive contouring control for collision avoidance in unstructured dynamic environments. IEEE Robot. Autom. Lett. **4**, 4459–4466 (2019)

12. Franch, J., Rodriguez-Fortun, J.M.: Control and trajectory generation of an ackerman vehicle by dynamic linearization. In: 2009 European Control Conference (ECC), pp. 4937–4942. IEEE (2009)

13. Cai, K., Chen, W., Wang, C., Song, S., Meng, M.Q.-H.: Human-aware path planning with improved virtual doppler method in highly dynamic environments. IEEE Trans. Autom. Sci. Eng. (2022)

14. Guo, B., Guo, N., Cen, Z.: Obstacle avoidance with dynamic avoidance risk region for mobile robots in dynamic environments. IEEE Robot. Autom. Lett. **7**, 5850–5857 (2022)

15. Uteshev, A.Y., Yashina, M.V.: Metric problems for quadrics in multidimensional space. J. Symb. Comput. **68**, 287–315 (2015)

16. Kang, C.M., Lee, S.-H., Chung, C.C.: On-road path generation and control for waypoints tracking. IEEE Intell. Transp. Syst. Mag. **9**, 36–45 (2017)

17. Fox, D., Burgard, W., Thrun, S.: The dynamic window approach to collision avoidance. IEEE Robot. Autom. Mag. **4**, 23–33 (1997)

18. Ajeil, F.H., Ibraheem, I.K., Azar, A.T., Humaidi, A.J.: Grid-based mobile robot path planning using aging-based ant colony optimization algorithm in static and dynamic environments. Sensors **20**, 1880 (2020)

Numerical Simulation of Ship Tank Sloshing Based on MPS Method

Yiping Zhong[1], Xiaofeng Sun[1]([✉])(iD), Feng Bian[2], Chunlei Liu[1], Jingkui Wang[3], and Yong Yin[1]

[1] Dalian Maritime University, Dalian 116026, China
xfsun@dlmu.edu.cn
[2] Liaoning Railway Vocational and Technical College, Jinzhou 121000, China
[3] Shenzhen Pilot Station, Shenzhen 518081, China

Abstract. Liquid sloshing in a 3D ship tank under the actual loading condition and the real rolling motion of the ship is numerically simulated based on moving particle semi-implicit (MPS) method. First, the validation of the present MPS method is carried out by numerically simulating the liquid sloshing in a rectangular tank. After that, numerical simulations of liquid sloshing in a 3D ship water ballast tank are carried out in which the more realistic rolling periods and the position of rolling axis are considered under ship load condition. Finally, the effect of ship rolling angle and filling ratio are investigated. Simulation results shows that MPS method can well observe the more realistic phenomenon of large deformation and nonlinear fragmentation of free surface flow in the sloshing tank. Further more, it is worth noting that both the filling ratio and rolling angle have significant effects on liquid sloshing.

Keywords: Liquid simulation · Ship tank sloshing · MPS method

1 Introduction

Under the external excitation of ship's motion, the phenomenon of liquid sloshing will inevitably take place in partially filled liquid tanks of moving ships. Although the experimental simulation can accurately studied the flow kinematics in sloshing tanks [19], it is expensive and difficult to conduct on-board experiments of liquid sloshing in ship tanks. In these cases, it is of great importance to investigate the flow kinematics in the sloshing tank by numerical simulation.

The work was supported by the National Key Research and Development Program of China (No. 2019YFE0111600), National Natural Science Foundation of China (No. 52071049, No. 61971083 and No. 51939001), LiaoNing Revitalization Talents Program (No. XLYC2002078), Dalian Science and Technology Innovation Fund (No. 2019J11CY015) and the project of intelligent ship testing and verification from the Ministry of Industry and Information Technology of the People's Republic of China (No. 2018/473).

In recent decades, many researchers shifted their attention to the computational fluid dynamics (CFD) technique to solve the liquid sloshing problem. Such as finite element method (FEM) [20], finite volume method (FVM) [5] and finite difference method (FDM) [1] in which the free surface flow is numerically simulated by volume of fluid (VOF) [4] and Level Set (L-S) methods [7]. Most of these numerical simulation discussed above are based on mesh-based methods. Resulted from the use of mesh, however, the mesh based numerical methods suffer from difficulties in dealing with the nonlinear free surface flows.

Recently, with the improvement of computing ability, meshfree methods have been widely used to simulate the violent free surface flow with large deformation and nonlinear fragmentation [2]. Smoothed particle hydrodynamics (SPH) method is one of the popular meshfree methods. SPH was developed by Gingold and Monaghan [6] and Lucy [15] in 1977. Monaghan firstly applied SPH method to address free surface flow problem in 1994 [16]. However, the kernel in the SPH method is considered as a mass distribution of each particle. The superposition of the kernels represents the physical superposition of mass. Thus, the particle is like a spherical cloud [13]. This concept may be more fitted to compressible fluids. Another popular meshfree method is the moving particle semi-implicit (MPS) method. The original MPS method was proposed by Koshizuka, S. and Oka, Y. to simulate the incompressible flow [11]. From the discrepancies of the kernel function between SPH and MPS methods, the MPS method is more suitable for simulating incompressible fluids than the SPH method.

Although the problem of liquid sloshing in ship tanks has constituted a large amount of the research efforts in recent years, most of them investigated the liquid sloshing problem in simplified tanks, not real ship tanks [3,8,10,18,24,25]. Meanwhile, the excitation period adopted in previous works came from the natural frequencies of gravity waves in an upright resting cylindrical tank [23]. However, this period is different from the rolling period of real ships in most cases. Further more, the position of rolling axis adopted in previous work is generally the gravity center of the tank not the ship's. Therefore, the liquid sloshing in ship tanks needs to be further investigated under more realistic conditions.

This paper aims to proposed a numerical simulation method based on the MPS method, to carry out numerical simulations of the free surface flow in liquid tanks under the actual loading condition and the real rolling motion of the ship. First, the MPS method used in this paper is verified via numerical simulations of liquid sloshing in a rectangular tank. After that, numerical simulations of liquid sloshing in a ship water ballast tank are carried out. Finally, the effect of ship rolling angle and filling ratio under ship load condition are investigated.

The rest of this paper is organized as follows: Sect. 2 introduces the basic theories of the MPS method. Section 3 verifies the MPS method used in this paper via liquid sloshing simulations. Section 4 gives the numerical results of the liquid sloshing in a real ship tank under different ship rolling motions. Section 5 draws conclusions.

2 MPS Method

MPS method is based on fully Lagrangian description. It uses a semi-implicit algorithm to simulate the incompressible viscous flows. In MPS method, derivatives in the governing equations are transformed to interactions among the neighboring particles [12]. The MPS method used in this paper is presented as follows.

2.1 Mathematical Formulation

Governing Equations. For an incompressible fluid flow, the governing equations which contain continuity equation and Navier-Stokes (N-S) equation can be written as follows:

$$\nabla \cdot \boldsymbol{u} = 0 \tag{1}$$

$$\frac{D\boldsymbol{u}}{Dt} = -\frac{1}{\rho}\nabla P + \nu\nabla^2\boldsymbol{u} + \mathbf{g} \tag{2}$$

where $\boldsymbol{u}$, P, ρ, ν, t, and $\mathbf{g}$ denote velocity vector of the particle, pressure, fluid density, kinematic viscosity, time, and gravity vector, respectively.

Particle Interaction Model. In MPS method, functions are approximated by a weighted average approach. In order to avoid non-physical pressure oscillation in original kernel function, a modified kernel function which was proposed by Zhang et al. [25] is employed in this paper, that is:

$$W(r) = \begin{cases} \frac{r_e}{0.85r + 0.15r_e} - 1 & 0 \le r < r_e \\ 0 & r_e \le r \end{cases} \tag{3}$$

where r_e is the radius of particle interaction and r is the distance between two particles.

Summation of the weight function is called particle number density, which is used to keep incompressibility of fluid.

$$\langle n \rangle_i = \sum_{j \neq i} W(|\boldsymbol{r}_j - \boldsymbol{r}_i|) \tag{4}$$

Gradient Model. The gradient model is mainly used to discretize pressure gradient. The gradient operator is the weighted average of a physical quantity ϕ between particle i and its neighboring particles. The model proposed by Koshizuka et al. [13] is adopted here, that is:

$$\langle \nabla\phi \rangle_i = \frac{d}{n^0} \sum_{j \neq i} \frac{\phi_j - \widehat{\phi}_i}{|\boldsymbol{r}_j - \boldsymbol{r}_i|^2}(\boldsymbol{r}_j - \boldsymbol{r}_i)W(|\boldsymbol{r}_j - \boldsymbol{r}_i|) \tag{5}$$

$$\widehat{\phi}_i = min(\phi_j, \phi_i), \{j : W(|\boldsymbol{r}_j - \boldsymbol{r}_i|) \neq 0\} \tag{6}$$

where d is number of dimension, n^0 is initial particle number density and $\boldsymbol{r}$ is position vector.

Divergence Model. The divergence model is used to discrete the velocity divergence in PPE and is similar to the gradient model. It is described as follows:

$$\langle \nabla \cdot \boldsymbol{u} \rangle_i = \frac{d}{n^0} \sum_{j \neq i} \frac{(\boldsymbol{r}_j - \boldsymbol{r}_i) \cdot (\boldsymbol{u}_j - \boldsymbol{u}_i)}{|\boldsymbol{r}_j - \boldsymbol{r}_i|^2} W(|\boldsymbol{r}_j - \boldsymbol{r}_i|) \tag{7}$$

Laplacian Model. The Laplacian model is the weighted average of the distribution of a physical quantity ϕ from particle i to its neighboring particles. This operator is used to discretize the viscosity term of PPE and is described as follows:

$$\langle \nabla^2 \phi \rangle_i = \frac{2d}{n^0 \lambda} \sum_{j \neq i} (\phi_j - \phi_i) W(|\boldsymbol{r}_j - \boldsymbol{r}_i|) \tag{8}$$

where λ is a parameter which is used to compensate for a finite range of kernel function, and can be obtained by:

$$\lambda = \frac{\sum_{j \neq i} |\boldsymbol{r}_j - \boldsymbol{r}_i|^2 W(|\boldsymbol{r}_j - \boldsymbol{r}_i|)}{\sum_{j \neq i} W(|\boldsymbol{r}_j - \boldsymbol{r}_i|)} \tag{9}$$

Pressure Poisson Equation. In MPS method, the pressure is calculated by solving the PPE, which is proposed by Lee et al. [14] as follows:

$$\langle \nabla^2 P^{k+1} \rangle_i = (\gamma - 1) \frac{\rho}{\Delta t} \nabla \cdot \boldsymbol{u}_i^* - \gamma \frac{\rho}{\Delta t^2} \frac{\langle n^* \rangle_i - n^0}{n^0} \tag{10}$$

where γ is blending parameter which range is between 0.01 to 0.05, n^* is temporal particle density, Δt is time step.

The left hand side of Eq. (10) can be discretized by applying the Laplacian model, as shown in Eq. (8). The right hand side of Eq. (10) is the mixed source term, which is gived by the divergence-free condition and the deviation of the temporal particle number density from the constant. The term which gived by the divergence-free condition can be discretized by applying divergence model, as shown in Eq. (7).

Boundary Condition

Free Surface Boundary. Since no particle can exist outside the fluid domain, the particle number density will inevitably decrease on the free surfaces. Therefore, the particles nearing the free surface can be detected by the following condition:

$$\langle n \rangle_i^* < \beta n^0 \tag{11}$$

where β is a parameter, and is less than 1.0.

Furthermore, the pressure of free surface particles are set to zero.

Solid Boundary. Two kinds of solid particles are adopted to prevent the fluid particles from penetrating the boundary. One is the wall particles, which are set along the solid boundary. The other one is the dummy particles, which are placed outside the solid wall. It's should be noted that the position of all solid particles won't update velocity and position after they gained pressure. The pressure of wall particles are involved in solving the PPE. The pressure of dummy particles are obtained by extrapolation.

2.2   Procedure of Numerical Simulation

Liquid sloshing is numerically simulated by using the MPS method. As a CFD method, the procedure of MPS method can be divided into the following three stages, as shown in Fig. 1.

Pre-processing Stage. Based on the 3D design data of ship tanks, the computational domain is firstly discretized into fluid particles. Meanwhile, two kinds of solid particles are arranged to satisfy the solid boundary condition. After that, the particle information, including position, velocity, acceleration and pressure, etc., and calculation parameters, including initial particle distance, operator support distance and time step, etc., are initialized.

Solving Stage. First, the neighbor particle list is established by using the cell-linked-list (CLL) technique [26]. After that, the semi-implicit algorithm is used to obtain particle information. In this process, the key role is to solve the PPE, as shown in Eq. (10). At each time step, PPE is firstly discretized into a large system of linear equations, whose coefficient matrix is a typical symmetric sparse matrices. Thus, the conjugate gradient (CG) method is employed to solve this linear system [2]. Then, the pressure of each particle is obtained and is used to calculate the pressure gradient, as shown in Eq. (5, 6). Finally, the pressure gradient is used to update particles' velocity, displacement and other information. These information will be recorded occasionally since the computational step is too small.

Post-processing Stage. In this stage, the output data are used for data analysis, e.g. visualization of the liquid sloshing, computation and analyzation of impact pressure acting on bulkhead, etc.

3   Validation

To verify the effectiveness of the MPS method used in this paper, numerical simulations of liquid sloshing are carried out with a rectangular tank. Simulation results are then compared with the experimental results [9].

The simulation conditions are set according to the experiments [9]. The length, width and height of this tank are 0.35 m, 0.80 m and 0.50 m, respectively

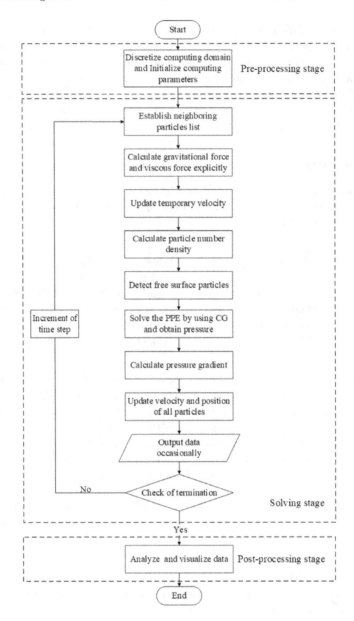

Fig. 1. Flow chart of MPS numerical simulation.

and the water depth is set to 0.15 m. This tank sways harmonically under the external excitation $x = A\sin(\omega t)$, where A is amplitude of sway ($A = 0.02\,\mathrm{m}$) and ω is excitation frequency ($\omega = 4.967$ rad/s). Meanwhile, a probe P_0 is arranged at the middle of the wall perpendicular to the excitation direction and is 0.115 m away from tank bottom. For computation parameters, the initial particle distance is set to 0.006 m, the time step is 5×10^{-4} s, and the density of fluid is 1000 kg/m^3.

Comparisons of the numerical simulation results of the free surface profiles at time 14.86 s and 15.18 s with that of the experiments [9] are illustrated in Fig. 2. From this figure, it can be seen that both the 2D and the 3D numerical simulation results make good agreement with the experimental results.

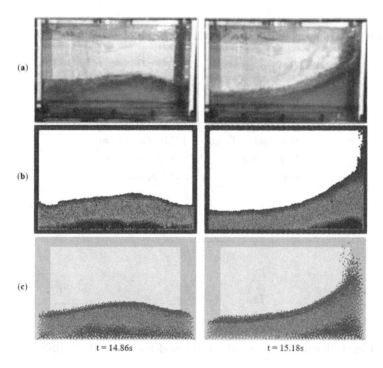

Fig. 2. Comparisons of free surface profiles results. (**a**) experimental results given by Kang et al. [9]; (**b**) 2D numerical simulation results obtained in this paper; (**c**) 3D numerical simulation results obtained in this paper.

The above numerical simulation results show that the MPS method used in this paper is effective in simulating the liquid sloshing in tanks swaying harmonically and thus can be used to numerically simulate the liquid sloshing in ship tanks.

4   Numerical Results

4.1   Description of Test Cases

In this paper, numerical simulations of liquid sloshing are performed on a 400,000 DWT Very Large Ore Carrier (VLOC) "PACIFIC VISION", which is designed by the Shanghai Merchant Ship Design and Research Institute (SDARI). Since liquid sloshing is mostly taking place in water ballast tanks, a ballast water tank of this VLOC, namely "R2.03AP", is selected to carry out the numerical simulation of liquid sloshing in ship tanks under different rolling motions.

The parameters of the water ballast tank "R2.03AP" are shown in Table 1 [17]. The 3D design models of the hull and this water ballast tank are illustrated in Fig. 3.

Table 1. Parameters of the water ballast tank "R2.03AP", from SDARI [17].

Description	Value
Compartment ident	R2.03AP
Volume	8477.14 m^3
Aft end at frame	117.00 (222.24 m)
Fore end at frame	123.00 (238.38 m)
Lowest point	0.00 m above BL
Highest point	30.97 m above BL

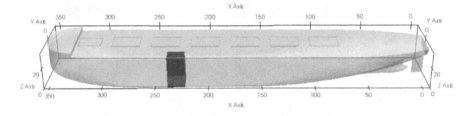

Fig. 3. Illustration of the 3D design models of "PACIFIC VISION".

It is worth noting that a scaled model of "R2.03AP" is applied to carry out the simulation due to the limitation of computing ability. The scale ratio is set to 1:50. The sizes of scale model in X, Y and Z direction are 322.80 mm, 403.60 mm and 619.48 mm respectively.

One characteristic ballast load condition, LOADA02, is used to carry out numerical simulation, in which the "T.C.G." = 0.059 m and "V.C.G." = 15.342 m indicate the center of gravity of the ship of the breadth from the centerline and the height above the base line, respectively. The position of rolling axis is set at gravity of the ship under LOADA02 condition.

The undamped natural period of the ship T_r can be determined by the equation given by wawrzynski et al. [22], as follows:

$$T_r = \frac{2cB}{\sqrt{GM}} \tag{12}$$

where $B = 65\,\mathrm{m}$ is ship's breadth moulded, $GM = 17.27\,\mathrm{m}$ is ship's metacentric height under condition LOADA02 [17]. c is the coefficient describing ship's transverse gyration radius, and is determined by:

$$c = 0.373 + 0.023\frac{B}{T} - 0.043\frac{L}{100} \tag{13}$$

where $L = 355.0\,\mathrm{m}$ and $T = 14.001\,\mathrm{m}$ are ship's length between perpendiculars and mean ship draft [17], respectively.

In order to describe ship's real motion, the dimensionless rolling period of the scaled tank and the real ship should be set to the same value. The rolling period of the scaled tank T_m can be determined by the equation given by Tan et al. [21], as follows:

$$T_m \cdot \sqrt{\frac{g}{H_m}} = T_r \cdot \sqrt{\frac{g}{H_r}} \tag{14}$$

where g is the gravitational acceleration; $H_m{:}H_r$ is the scaled ratio, which is set to 1:50 in this paper.

In this paper, the liquid sloshing in ship tank is numerically simulated in three characteristic cases, as described in Table 2, in which the rolling period is obtained by Eq. (12–14) and thus is the ship's undamped natural period under LOADA02 condition. Meanwhile, the main parameters used in the MPS method are described in Table 3.

Table 2. Description of the numerically computation cases.

ID	Filling height [mm]	Filling ratio [%]	Rolling angle [°]	Period [s]
1	100	18.8	10	1.45 s
2	300	54.5	10	1.45 s
3	100	18.8	20	1.45 s

4.2 Snapshots of Flow Field

The snapshots of flow field at several typical time steps under different filling ratio and maximum rolling angle are illustrated in Fig. 4 and 5.

Figure 4 gives the comparison of snapshots at several typical time steps under case 1 and case 2. Both of these two cases have same rolling period, 1.45 s, and maximum rolling angle, 10°, but different filling heights. The filling height in case 1 is 100 mm, while 300 mm in case 2. From this figure, it can be seen obviously that the profiles of free surface under these two cases are different, especially at time 6.53 s. This phenomenon is reasonable since the sloshing periods of the liquid in these two cases are different due to different filling height.

Table 3. Main calculation parameters used in MPS method.

Parameters	Value
Initial particle spacing, l_0	0.015 m
Operating distance of Gradient operator, r_{Grad}	$2.1\ l_0$
Operating distance of Laplacian operator, r_{Lap}	$3.1\ l_0$
Time step, Δt	0.0005 s
Density of fluid, ρ	1000 kg/m^3
Free surface coefficient, β	0.97
Blending parameter, γ	0.01

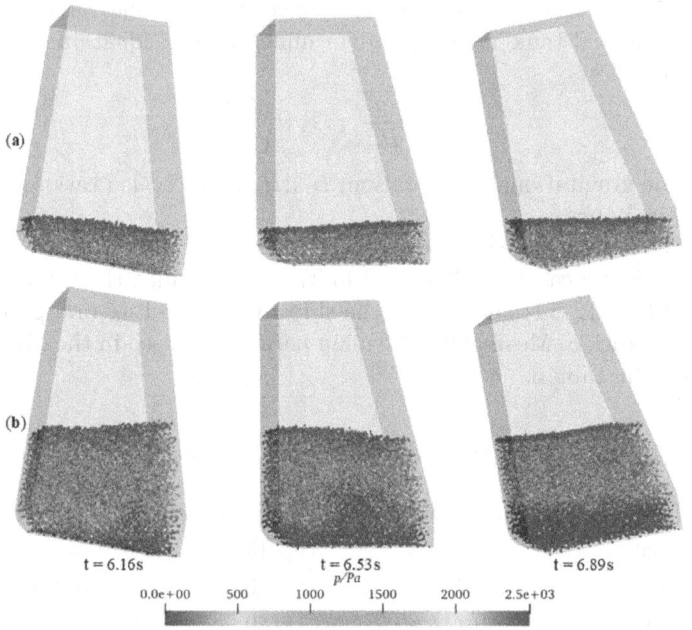

Fig. 4. Snapshots of flow field in different filling ratio. (**a**) Case 1; (**b**) Case 2.

Figure 5 gives the comparison of snapshots at several typical time steps under case 1 and case 3. Both of these two cases have same rolling period, 1.45 s, and filling height 100 mm, but different maximum rolling angles. The maximum rolling angle in case 1 is 10°, while 20° in case 3. From this figure, it can be seen that the phenomenon of fluid accumulation, bending and slamming on free surface occurs in case 3, but not in case 1. Therefore, it can be concluded that a larger rolling angle would induce a more violent liquid sloshing in the ship tanks.

From Fig. 4 and 5, it can also be seen that the free surface is hardly parallel to the sea level, even the phenomenon of fluid accumulation, bending and slamming is happening, which proves that the traditional quasi-static method can not

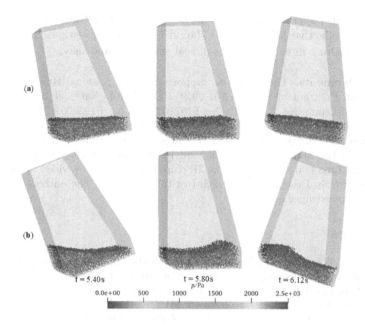

Fig. 5. Snapshots of flow field in different rolling angle. (**a**) Case 1; (**b**) Case 3.

reflect the movement of free surface realistically. Meanwhile these numerical simulation results show that the MPS method can more realistically reflect the free surface flow motion with strong nonlinear and large deformation of ship tank sloshing.

5 Conclusions

Liquid sloshing in ship tanks will cause a strong nonlinear and large deformation free surface phenomenon. This paper numerically simulates the phenomenon of liquid sloshing in a ballast tank under ship rolling motion by using the MPS method.

Conclusions can be summarized as follows:

(1) The phenomenon of large deformation and nonlinear fragmentation of free surface, e.g. accumulation, bending and slamming, etc., can be found in liquid sloshing which is numerically simulated by MPS method. In this case, meshfree methods should be used to get more realistic results.
(2) Both the filling ratio and rolling angle have significant effects on liquid sloshing, in which the different filling ratio lead to different liquid slosing period in ship tanks, and a larger rolling angle would induce a more violent liquid sloshing in partially filled liquid tanks of moving ships.

Although MPS method has its own advantages, it also has the limitation of computing ability in dealing with large-scale violent free surface flow. That's the reason why this paper uses a scaled tank model to perform the simulation. Therefore, a more effective MPS method should be further investigated to simulate the liquid sloshing in all ship tanks of real size simultaneously.

Acknowledgements. The work was supported by the National Key Research and Development Program of China (No. 2019YFE0111600), National Natural Science Foundation of China (No. 52071049, No. 61971083 and No. 51939001), LiaoNing Revitalization Talents Program (No. XLYC2002078), Dalian Science and Technology Innovation Fund (No. 2019J11CY015) and the project of intelligent ship testing and verification from the Ministry of Industry and Information Technology of the People's Republic of China (No. 2018/473). The 3D ship design data is provided by the Shanghai Merchant Ship Design and Research Institute (SDARI). The authors express the gratefully acknowledgement.

References

1. Celis, M., Wanderley, J., Neves, M.: Numerical simulation of dam breaking and the influence of sloshing on the transfer of water between compartments. Ocean Eng. **146**, 125–139 (2017). https://doi.org/10.1016/j.oceaneng.2017.09.029
2. Chen, X., Wan, D.: GPU accelerated MPS method for large-scale 3-D violent free surface flows. Ocean Eng. **171**, 677–694 (2019). https://doi.org/10.1016/j.oceaneng.2018.11.009
3. Delorme, L., Colagrossi, A., Souto-Iglesias, A., Zamora-Rodriguez, R., Botia-Vera, E.: A set of canonical problems in sloshing, Part I: pressure field in forced roll? Comparison between experimental results and SPH. Ocean Eng. **36**(2), 168–178 (2009). https://doi.org/10.1016/j.oceaneng.2008.09.014
4. Ding, S., Wang, G., Luo, Q.: Study on sloshing simulation in the independent tank for an ice-breaking LNG carrier. Int. J. Naval Archit. Ocean Eng. **12**, 667–679 (2020). https://doi.org/10.1016/j.ijnaoe.2020.03.002
5. Fu, C.Q., Jiang, H.M., Yin, H.J., Su, Y.C., Zeng, Y.M.: Finite volume method for simulation of viscoelastic flow through a expansion channel. J. Hydrodyn. Ser. B **21**(3), 360–365 (2009). https://doi.org/10.1016/S1001-6058(08)60157-2
6. Gingold, R.A., Monaghan, J.J.: Smoothed particle hydrodynamics: theory and application to non-spherical stars. Mon. Not. R. Astron. Soc. **181**(3), 375–389 (1977). https://doi.org/10.1093/mnras/181.3.375
7. He, T., Feng, D., Liu, L., Wang, X., Jiang, H.: CFD simulation and experimental study on coupled motion response of ship with tank in beam waves. J. Mar. Sci. Eng. **10**(1), 113 (2022). https://doi.org/10.3390/jmse10010113
8. Jena, D., Biswal, K.C.: A numerical study of violent sloshing problems with modified MPS method. J. Hydrodyn. **29**(4), 659–667 (2017). https://doi.org/10.1016/S1001-6058(16)60779-5
9. Kang, D., Lee, Y.: Summary report of sloshing model test for rectangular model. Daewoo Shipbuilding & Marine Engineering Co., Ltd., South Korea (001) (2019)
10. Khayyer, A., Gotoh, H., Falahaty, H., Shimizu, Y.: An enhanced ISPH-SPH coupled method for simulation of incompressible fluid-elastic structure interactions. Comput. Phys. Commun. **232**, 139–164 (2018). https://doi.org/10.1016/j.cpc.2018.05.012
11. Koshizuka, S., Oka, Y.: Moving-particle semi-implicit method for fragmentation of incompressible fluid. Nucl. Sci. Eng. **123**(3), 421–434 (1996). https://doi.org/10.13182/NSE96-A24205

12. Koshizuka, S., Oka, Y.: Moving particle semi-implicit method: fully lagrangian analysis of incompressible flows. In: Proceedings of the European Congress on Computational Methods in Applied Sciences and Engineering (ECCOMAS), Barcelona, Spain, pp. 11–14 (2000)
13. Koshizuka, S., Shibata, K., Kondo, M., Matsunaga, T.: Moving particle semi-implicit method: a meshfree particle method for fluid dynamics. Academic Press (2018). https://doi.org/10.1016/C2016-0-03952-9
14. Lee, B.H., Park, J.C., Kim, M.H., Hwang, S.C.: Step-by-step improvement of MPS method in simulating violent free-surface motions and impact-loads. Comput. Methods Appl. Mech. Eng. **200**(9–12), 1113–1125 (2011). https://doi.org/10.1016/j.cma.2010.12.001
15. Lucy, L.B.: A numerical approach to the testing of the fission hypothesis. Astron. J. **82**, 1013–1024 (1977). https://doi.org/10.1086/112164
16. Monaghan, J.J.: Simulating free surface flows with SPH. J. Comput. Phys. **110**(2), 399–406 (1994). https://doi.org/10.1006/jcph.1994.1034
17. SDARI: Final loading manual of 400,000 dwt ore carrier "pacific vision" (2018)
18. Vieira-e Silva, A.L.B., dos Santos Brito, C.J., Simões, F.P.M., Teichrieb, V.: A fluid simulation system based on the MPS method. Comput. Phys. Commun. **258**, 107572 (2021). https://doi.org/10.1016/j.cpc.2020.107572
19. Song, Y.K., Chang, K.-A., Ryu, Y., Kwon, S.H.: Experimental study on flow kinematics and impact pressure in liquid sloshing. Exp. Fluids **54**(9), 1–20 (2013). https://doi.org/10.1007/s00348-013-1592-5
20. Sriram, V., Sannasiraj, S., Sundar, V.: Numerical simulation of 2D sloshing waves due to horizontal and vertical random excitation. Appl. Ocean Res. **28**(1), 19–32 (2006). https://doi.org/10.1016/j.apor.2006.01.002
21. Tan, Q.-M.: Introduction. In: Tan, Q.-M. (ed.) Dimensional Analysis, pp. 1–6. Springer, Heidelberg (2011). https://doi.org/10.1007/978-3-642-19234-0_1
22. Wawrzyński, W., Krata, P.: Method for ship's rolling period prediction with regard to non-linearity of GZ curve. J. Theor. Appl. Mech. **54**(4), 1329–1343 (2016). https://doi.org/10.15632/jtam-pl.54.4.1329
23. aus der Wiesche, S.: Sloshing dynamics of a viscous liquid in a spinning horizontal cylindrical tank. Aerosp. Sci. Technol. **12**(6), 448–456 (2008). https://doi.org/10.1016/j.ast.2007.10.013
24. Xie, F., Zhao, W., Wan, D.: CFD simulations of three-dimensional violent sloshing flows in tanks based on MPS and GPU. J. Hydrodyn. **32**(4), 672–683 (2020). https://doi.org/10.1007/s42241-020-0039-8
25. Zhang, Y.X., Wan, D.C., Takanori, H.: Comparative study of MPS method and level-set method for sloshing flows. J. Hydrodyn. Ser. B **26**(4), 577–585 (2014). https://doi.org/10.1016/S1001-6058(14)60065-2
26. Zhu, X., Cheng, L., Lu, L., Teng, B.: Implementation of the moving particle semi-implicit method on GPU. Sci. China Phys. Mech. Astron. **54**(3), 523–532 (2011). https://doi.org/10.1007/s11433-010-4241-5

Simulation Study of Dynamic Reactive Power Optimization in Distribution Network with DG Based on Improved Lion Swarm Algorithm

Tianyang Wu[1], Qiang Li[2], Bing Fang[1], Nannan Zhang[1], Haowei Qu[1], Jiankai Fang[1], and Lidi Wang[1(✉)]

[1] Shenyang Agricultural University, Shenyang 110866, China
wanglidi@163.com

[2] State Grid Liaoyang Power Supply Company, Liaoyang 111000, China

Abstract. This paper proposes a spectral clustering method in consideration of the operational issues of distribution networks like load fluctuation, intermittent power output, reactive power flow, and daily switching frequency of reactive power compensation. We divide the daily load curve of the distribution network with distributed generation units (DG) into time periods, and set the minimum network loss and voltage offset of each time period as the objective function. Then we use this method to establish a time-divided dynamic reactive power optimization (RPO) mathematical model of DG. Since the traditional random lion swarm optimization (LSO) can hardly escape a local optimum, a random black hole mechanism is introduced to improve the LSO algorithm, and to formulate a random black hole based lion swarm optimization (RBH-LSO) algorithm. This paper takes the improved IEEE 33-node system as the sample object. The RBH-LSO algorithm, the LSO algorithm and the particle swarm optimization (PSO) algorithm are mutually used to realize the optimization of this system. After the simulation results of the optimization are analyzed, this paper demonstrates, as a summary, that the RBH-LSO algorithm has exceeding excellence in performance and proves to be an effective mechanism for dynamic RPO of distribution networks with DG.

Keywords: Distributed generation · Reactive power optimization · Lion swarm optimization · Network loss · Voltage offset

1 Introduction

The main form of voltage regulation in power systems is reactive power optimization (RPO), which can be categorized into static reactive power optimization (SRPO) and dynamic reactive power optimization (DRPO). A previous work (Work 1) [1] summarized the pros and cons of the particle swarm optimization (PSO) algorithm and the ant colony optimization algorithm, and sought to reduce network losses by applying both algorithms to the RPO of distribution networks with distributed generation units (DG).

Funding information: National Natural Science Foundation of China (61903264)

Another work (Work 2) [2] analyzed the impact of PV systems on distribution networks and used an adaptive genetic algorithm for RPO, so as to reduce the risks of voltage crossing and system losses. A third work (Work 3) [3] devised a multi-objective RPO model, in which the economy and safety of DG-integrated grids are taken as the objective function, and utilized a simulated annealing algorithm to optimize the reactive power compensation for distributed power sources. Work 1 & 2 fail to consider the reactive power provision by DG in RPO; they assumed that DG provided a constant current source and misused SRPO. Work 3 noticed the reactive power compensation by DG but still used SRPO. Considering the output fluctuation of distributed power sources such as wind power and PV, it is obvious that the traditional SRPO is no longer suitable for the status quo.

DRPO divides a day into several periods of a fixed duration [4], and performs SRPO for each period in consideration of the daily switching frequency of the compensation equipment [5] A reasonable and effective period division is the key to DRPO. A previous work [6] directly divided a day into 24 periods without paying attention to the number of the actions of the compensation equipment. Another work [7] marked out the periods by the absolute difference between the adjacent time load curves, and merged the periods that were less than an experience-based threshold set in advance; it neglected the amplitude of load fluctuations in the adjacent time periods, and failed to perform reasonable RPO for loads with large fluctuations in the adjacent time periods.

This paper proposes a DRPO model for DG-integrated distribution networks using network losses and voltage stability as the objective function. Focusing on the number of actions of the compensation equipment, this model achieves time division by a spectral clustering method, which takes in the changes of load and DG output in the adjacent time periods without a set threshold. A random black hole (RBH) mechanism is synthesized with the lion swarm optimization (LSO) algorithm to increase the rate of convergence and meanwhile make it easier to escape a local optimum. The sample object is a modified IEEE 33-node system, based on which the results reveal for us some of the characteristics of the method this paper proposes.

2   The RPO Model

2.1   The Objective Function

The existence of RPO substantially improves the power quality of distribution systems, and greatly ensures safe and economic operation of distribution networks, by which the operation of the power enterprises and the whole society is also guaranteed to a great extent [8]. For this reason, most of the traditional RPO models aim only at reducing active power losses so as to maintain economic efficiency of the networks [9]. The DRPO model of this paper takes 24 h as an optimization cycle and divides it into several time periods. The mathematical expression of the active power loss of a system is

$$\min f_{\text{loss}} = \sum_{k=1}^{N} G_{k(i,j)}\left[V_i^2 + V_j^2 - 2V_i V_j \cos(\theta_i - \theta_j)\right] \tag{1}$$

where N is the total number of branches in the system, [the G mark] is the conductance in branch k when connected to nodes i and j, V is the magnitude of voltage at each node in the system, and θ is the phase angle of voltage at each node.

The development of the economy and the society has imposed higher requirements for power quality, and power quality has already become a prerequisite factor in the development of the economy [10]. Voltage is one of the crucial indicators of power quality and has a wide impact on the operation of the whole power system. If the system operation is coupled with unstable voltage, the whole power system will suffer from equipment damage, and the smoothness and safety of the system operation may also be affected. In some serious cases, load fluctuations may cause voltage collapse of the network, the unlisted state of the grid, and widespread power outages [11]. The difference between the voltage of each node in the operating network and the desired voltage is taken as one of the objective functions for RPO, and the conditions of voltage stability are strictly ensured so that the absolute value of the difference is minimized:

$$\min f_{\Delta v} = \sum_{i=1}^{N_l} \left| V_i - V_i^{spec} \right| \tag{2}$$

where N_l is the number of load nodes, and V_i^{spec} is the expected value of node i.

A weighting method is adopted to transform the network loss and the voltage deviation into a single objective function, and the transformed objective function ensures the system network loss and the voltage deviation to be the minimum:

$$\min f = \omega_1 f_{loss} + \omega_2 f_{\Delta v} \tag{3}$$

where ω_1, ω_2 are the weighting factors that conform to $\omega_1 = \omega_2 = 1$.

2.2 Constraints

Considering the safety and economy of the actual operation of the system, we need to constrain the control and state variables [12]. In contrast to the situation for the conventional network without DG, for the DRPO of DG's connection to the grid, the reactive power provision by DG is taken into consideration; the actual supply capacity of DG to the system and that of the static reactive power compensator (SVC) to the system are taken as control variables; the node voltage is taken as the state variable [13]. In order to ensure the safety and stability of the operation of the distribution network, the state and control variables must not cross the boundaries. Based on the above listed factors, when DG is connected to the grid, the system active-reactive power balance equation is regarded as an equality constraint, and the node voltage, the DG input (of reactive power to the grid) and the SVC input are regarded as inequality constraints, as shown in order in Eqs. (4) to (8).

$$P_{Gi} + P_{DGi} - P_{li} = V_i \sum_{j=1}^{N_b} V_j \left(G_{ij} \cos \delta_{ij} + B_{ij} \sin \delta_{ij} \right) \tag{4}$$

$$Q_{Gi} + Q_{DGi} - Q_{li} = V_i \sum_{j=1}^{N_b} V_j \left(G_{ij} \sin \delta_{ij} - B_{ij} \cos \delta_{ij} \right) \tag{5}$$

$$Q_{DGg\,min} \leq Q_{DGg} \leq Q_{DGg\,max} (g = 1, 2 \ldots, N_{DG}) \tag{6}$$

$$Q_{ck\,min} \leq Q_{ck} \leq Q_{ck\,max} (k = 1, 2 \ldots, N_c) \tag{7}$$

$$V_{l\,min} \leq V_l \leq V_{l\,max} (k = 1, 2, \ldots, N_L) \tag{8}$$

where P_G and Q_G are the active and reactive power supplied by the generator to the network, respectively, P_l and Q_l are the active and reactive power delivered by the system to the load nodes, respectively, B is the electric power of the branch, Q_{DG} and Q_C are the reactive power delivered by DG and SVG to the network, respectively, N_{DG} and N_L are the number of DG connected to the network and the number of loads in the system, respectively, and *min* and *max* represent the minimum and the maximum values of each variable [14].

2.3 Reactive Power of the Distributed Power Supply

Reactive power supply to the grid can be conducted by PV power plants via inverters, and small gas turbines as well as biomass power plants via synchronous generators with regulated excitation voltage [15].

With the booming development of wind power generation technology, doubly-fed induction generators with a variable speed and a constant frequency are gradually becoming the mainstream of wind power generation. With the decoupling control of reactive power, the doubly-fed turbines can generate or absorb reactive power within its capacity according to the system dispatch.

By controlling the inverter that connects the PV to the grid, the PV can deliver reactive and active power to the grid simultaneously. The reactive power supply capacity of PV is denoted as follows

$$|Q|_{max}(t) = \sqrt{S_{max}^2 - P_{act}^2(t)} \tag{9}$$

where Q represents the reactive power delivered by the PV power plants to the system, S_{max} represents the maximum apparent power of the inverter between the PV power plants and the grid, and P_{act} represents the active power of the inverter between the PV power plants and the grid.

The operation data of the PV power plants show that most PV inverters can hardly be operated at their full load and usually output a mere 80% of their full reactive power supply. According to Eq. (9), 60% of the reactive power limit can be used in this way. The active PV output is 0 at night and the interface capacity can be used for reactive power regulation [16].

For diesel turbines, gas turbines and other generators like conventional synchronous motors, the active power output is stabilized and the reactive power output is regulated by adjusting the excitation system.

3 The Time Division Method

Theoretically, the more time periods the load curve is divided into, the more accurate the scheme for reactive power compensation will be. But the excessive division will lead to the frequent switching of the compensation equipment, which will affect its service life. Thus, the maximum number of the daily switching actions is generally 3 to 6 [17]. The reactive power output of DG is a control variable and does not participate in the division, and the load curves of active and reactive power are similar to each other. So this paper only processes the active load and the active power output of DG, where the maximum number of the daily switching actions is noted as M. The spectral clustering method is used to divide the daily load curve of the distribution network with DG, which does not require a experience-based threshold value, and generates stable results of division regardless of the fluctuations.

(1) Read the daily load curve and the daily active output curves of PV and wind power to determine M.
(2) Calculate the active load for each time period:

$$S_t = P_{l_t} - P_{w_t} - P_{r_t} \tag{10}$$

where P_{l_t} is the load for the period, P_{w_t} is the output of the wind power for the period, and P_{r_t} is the PV output for the period.
(3) Calculate the load difference between the adjacent time periods ΔS_t:

$$\Delta S_t = |S_{t+1} - S_t| \tag{11}$$

where S_{t+1} and S_t are the active loads of the adjacent periods, ΔS_t for each time period is counted in the vector D, and the smallest ΔS_t in D is selected so as to make the adjacent time periods combined into one time period.
(4) The active power of the combined load is

$$S_t = \frac{S_t i + S_{t+1} j}{i + j} \tag{12}$$

where i and j are the number of load points in periods t and t + 1, respectively.

4 An Improved LSO Algorithm

4.1 Principles of the LSO algorithm

The LSO algorithm is a group search algorithm that was proposed in 2018 and simulates lion king guarding, lioness hunting and cub followings [18].

The proportion of adult lions, i.e.the lion king and lionesses, in the pride is a random number between (0,1), and the size of the adult lions will affect the convergence speed of the algorithm.

Definitions of Parameters

α_f is the disturbance factor of lioness moving range [19], which is intended to make the global retrieval stronger when the iteration of the lionesses starts, and to make the local retrieval stronger as the number of iterations increases. The disturbance factor is defined as follows:

$$\alpha_f = \text{step} * \exp\left(-\frac{30t}{T}\right)^{10} \tag{13}$$

where

$$\text{step} = 0.1(\text{ high - low }) \tag{14}$$

denotes the step length for the lion in the search domain, *high* and *low* are the minimum mean and the maximum average of each dimension of the lion's search domain, T is the maximum number of iterations of the pride, and t is the current number of iterations.

α_c is defined as the disturbance factor of cub moving range, which aims to expand or narrow the search domain, where the cubs are allowed to explore for food, and then in a gradual linear downward trend to keep searching in small steps. The disturbance factor α_c is defined as follows:

$$\alpha_c = \text{step} * \left(\frac{T - t}{T}\right) \tag{15}$$

The Search Process

Suppose there are N lions in the pride within the D-dimensional search domain, and the number of adult lions is *nLeader*:

$$2 \leq nleader \leq \frac{N}{2} \tag{16}$$

where the number of the lion king is limited to one. The position of the i lion is:

$$x_i = (x_{i1}, x_{i2}, \ldots i_{iD}), 1 \leq i \leq N \tag{17}$$

And number of adult lions is:

$$nLeader = N\beta \tag{18}$$

The number of cubs is $N-nLeader$. Different kinds of lions move in their own ways. The lion king moves in a small area with the guarantee of the best place for food hunting:

$$x_i^{k+1} = g^k \left(1 + \gamma \left\| p_i^k - g^k \right\| \right) \tag{19}$$

The lioness needs to cooperate with another lioness in food hunting, and moves by

$$x_i^{k+1} = \frac{p_i^k + p_c^k}{2} (1 + \alpha_f \gamma) \tag{20}$$

The movement of the cubs is defined by

$$x_i^{k+1} = \begin{cases} \frac{g^k + p_i^k}{2}(1 + \alpha_c \gamma), & q \leq \frac{1}{3} \\ \frac{p_m^k + p_i^k}{2}(1 + \alpha_c \gamma), & \frac{1}{3} \leq q < \frac{2}{3} \\ \frac{g'^k + p_i^k}{2}(1 + \alpha_c \gamma), & \frac{2}{3} \leq q < 1 \end{cases} \tag{21}$$

$$g'^k = low' - high' - g^k \tag{22}$$

Among them, γ is a generated random number between $(0,1)$; P_i^k is the historical optimal position of the i-th child and the k generation; g^k is the optimal position for the k generation group; P_c^k is the historical optimal position of the lioness from the k generation; low' and $high'$ are the minimum mean and the maximum average of each dimension of the individual.

4.2 The Random Black Hole Mechanism

In the search domain, each individual in the group can be regarded as a star in the universe, the fitness value of each individual is considered as the gravitational force, and each individual is attracted to the global optimal solution during each iteration [20]. The true optimal solution in the search domain is considered a black hole which is randomly and immediately generated adjacent to the global optimal position at each iteration. This black hole can be considered the incoming likelihood of the global optimal solution, which exerts the gravitational force on each individual [21]. The RBH-LSO algorithm is thereby proposed to create a domain where a black hole is randomly generated at each iteration, the optimal position P is the center, r is the radius, and a threshold a between [0,1] is set as the attraction capacity of the global optimal position to each individual. For each individual Xi between [0,1], a random number b is randomly generated; if $b < a$, then the individual is devoured by the black hole: if the individual is not devoured, then its position is updated according to the above LSO algorithm, while the position of the devoured individual is updated according to the following formula:

$$X_i = P + 2r(c - 0.5) \tag{23}$$

$$r = \frac{P}{\sum\limits_{i=1}^{m} S_i} \tag{24}$$

where m is the number of iterations, S is the individual adaptation value of the current generation, and c is the generated random number. The flow chart of the improved lion swarm algorithm is shown in Fig. 1.

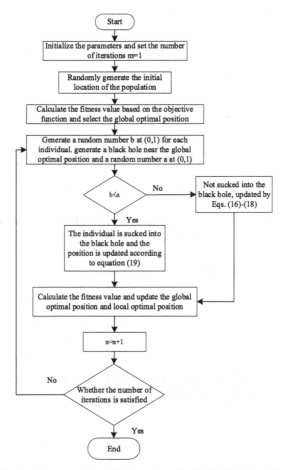

Fig. 1. Diagram of the RPO based on the RBH-LSO algorithm

5 Simulation Analysis

As shown in Fig. 2, the IEEE 33-node system is improved by connecting to PV at node 10, to wind power at node 17, and to SVCs at nodes 24 and 32, with the specific parameters shown in Fig. 2 and Table 1.

The spectral clustering method is used to divide the daily load curve of the improved IEEE 33-node system into time periods, assuming that the number of equipment switching actions in an RPO cycle is 6.

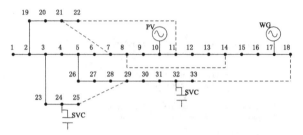

Fig. 2. The distributed network of the IEEE 33-node system

Table 1. Access location and parameters

Nodes	Access devices	Rated capacity (MWA)	Upper limit of reactive power (Mvar)	Lower limit of reactive power (Mvar)
10	PV	0.4	0.6	0
17	WG	0.4	0.6	0
24	SVC	–	0.6	0
32	SVC	–	0.6	0

As shown in Fig. 3, a 24-h load curve is divided into 6 time periods, namely, 1–4 h, 5–17 h, 18 h, 19 h, 20–22 h, and 23–24 h.

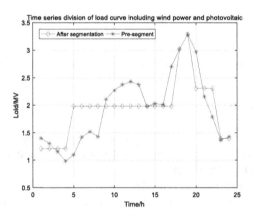

Fig. 3. Time sequence division of distribution network with distributed power

In order to reduce the active power losses of distribution systems with suitable reactive power compensation and reduce the voltage deviations, the RBH-LSO algorithm is used to obtain the reactive power output of DG and the compensation value of the reactive power compensation equipment, and to compare with other algorithms including pre-optimization, LSO and PSO. The specific parameters of the algorithm are shown in Table 2.

Table 2. Algorithm parameters

Method	Number of iterations	Group size	β	ω_{max}	ω_{min}	c_1	c_2
LSO	100	60	0.3	–	–	–	–
RBH-LSO	100	60	0.3	–	–	–	–
PSO	100	60	–	0.9	0.4	2	2

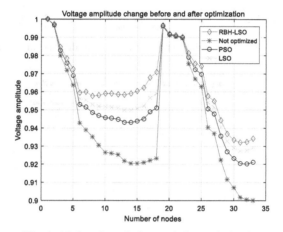

Fig. 4. Node voltage before and after optimization

As shown in Fig. 4, the comparison is made between the curves of node voltage of the system during the 6 time periods before and after the optimization by RBH-LSO, LSO, and PSO. Both the installation of DG and the reactive power compensation equipment can significantly improve the node voltage, and the node voltage after optimization by RBH-LSO is significantly better than those by LSO and PSO. Nodes, like nodes 17 and 32, where DG or the reactive power compensation equipment is installed or nearby, all have significant improvements on voltage.

Figure 5 shows the fitness function as the sum of network losses, voltage offset and active-reactive crossing penalty after the optimization by PSO, LSO, and RBH-LSO, respectively. The RBH-LSO algorithm outperforms the other two algorithms in terms of both convergence speed and algorithm accuracy.

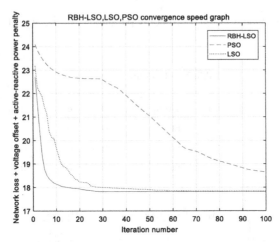

Fig. 5. Comparison of the convergence curves of different algorithms

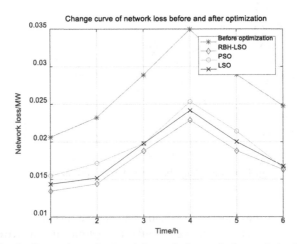

Fig. 6. Comparison of network losses before and after optimization

Figure 6 and Table 3 show the comparison of the network losses before and after optimization in each time period. The network loss of the RBH-LSO group in each time period is lower than that of the PSO group and the LSO group, and the total network loss of the RBH-LSO group after optimization is 0.057 MVA lower than that before optimization in a single optimization cycle. The voltage deviations after optimization by the RBH-LSO algorithm are smaller than those by the other algorithms.

Table 3. Comparison of network losses of the optimized systems in different periods

Time period	Before optimization (MVA)	After PSO (MVA)	After LSO (MVA)	After RBH-LSO (MVA)
1	0.0206	0.0154	0.0144	0.0134
2	0.0233	0.0172	0.0152	0.0144
3	0.0289	0.0198	0.0198	0.0188
4	0.0350	0.0253	0.0242	0.0229
5	0.0290	0.0214	0.0200	0.0188
6	0.0248	0.0165	0.0168	0.0163
Total	0.1616	0.1156	0.1104	0.1046

Table 4. Comparison of system voltage deviations before and after optimization by different algorithms

	Before optimization	PSO	LSO	RBH-LSO
Voltage deviation	1.8440	1.3185	1.1870	1.1502

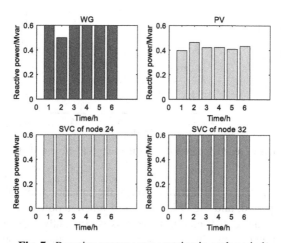

Fig. 7. Reactive power compensation in each period

Figure 7 shows the reactive power compensation provision in each time period by PV, wind power, and SVCs. The reactive power compensation output of PV basically focuses on the second time period, namely, 5–17 h, during which the sunlight intensity is relatively high. For wind power, the output is relatively low in the second time period, which makes wind power complement PV perfectly.

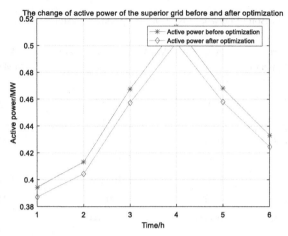

Fig. 8. The change of active power of the superior grid before and after optimization

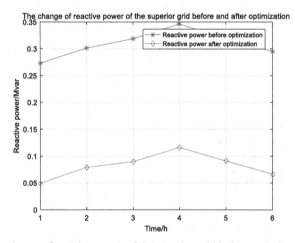

Fig. 9. The change of reactive power of the superior grid before and after optimization

Figure 8 shows the change of active power of the superior grid before and after the RPO in each time period, and Fig. 9 is for that of reactive power. The reactive power of the superior grid in the distribution network is significantly reduced after RPO by the improved LSO algorithm. The reactive power flow in the distribution network is controlled, so as to reduce the active power losses of the system, and the purchased electricity from the superior grid is therefore reduced.

6 Conclusion

In consideration of DG's regulation capability for reactive power and the daily switching frequency of the compensation equipment, this paper proposes a spectral clustering method for time-phased RPO of the daily load curve of distribution networks with DG, and an RBH-LSO algorithm in order to solve the problem that the LSO algorithm is easy to fall into a local optimum. The reactive power input from DG and SVCs to the grid is used as the control variable, the system network losses and voltage offset are used as the objective function, and the improved IEEE 33-node system is taken as a sample to simulate the time-phased DRPO of DG-integrated grids. A series of constraints are considered for the optimization by RBH-LSO, LSO and PSO, respectively, and the network losses and voltage offset after the optimization by RBH-LSO, LSO and PSO are compared. The results show that the RBH-LSO algorithm can conduct effective optimization to reduce system network losses and improve the stability of node voltage in the dynamic reactive power compensation of distribution networks.

References

1. Hailong, D., Xi Huixing, F., Hailin, L.W.: Reactive power optimization for DG-containing distribution network based on hybrid PSO-ACO algorithm J. Power syst. clean energ. **33**(01), 50–56 (2017). (in Chinese)
2. Yajun, R., Haichao, F., Liwei, S., Cong, L., Chao, L.: Reactive power optimization of distribution network with photovoltaic power supply based on adaptive genetic algorithm. Electr. Appliances Energ. Effi. Manage. Technol. **11**, 85–91 (2020). (in Chinese)
3. Hong, C.: Study on reactive power compensation optimization of distribution network with distributed power supply. Inf. Technol. **44**(11), 132–136 (2020). (in Chinese)
4. Liying, L., Qingjie, Z., Shaokun, Y.: Summary of reactive power optimization in power system. Power inf. **03**, 69–74 (2002). (in Chinese)
5. Zechun, H., Xifan, W.: Time division control strategy for reactive power optimization of distribution network. Power Syst. Autom. **06**, 45–49 (2002). (in Chinese)
6. Yanwei, M., Jianguang, Z., Jian, S., Jiaming, C., Xing, Z., Shiping, D.: Research on dynamic reactive power optimization of distribution network considering distributed photovoltaic and energy storage. Electrotechnical **10**, 56–59 (2020). (in Chinese)
7. Junjun, Y.: Research on dynamic reactive power optimization of distribution system based on pso-dp algorithm. Shandong University (2011) (in Chinese)
8. Jia, H., Yaowu, W., Suhua, L., Xingen, X.: Dynamic reactive power optimization of power system based on particle swarm optimization algorithm Power Grid Technol. **02**, 47–51+79 (2007). (in Chinese)
9. Wu, X., Liu, T., Jiang, X., Sheng, G.: Reactive power optimization of offshore wind farm based on improved genetic algorithm. Electr. Measu. Instrum. **57**(04), 108–113 (2020). (in Chinese)
10. Lei, L., Sipeng, H.: Hierarchical optimal operation strategy of active distribution network with microgrid. Electr. Measu. Instrum. **56**(20), 76–81 (2019). (in Chinese)
11. Jiguang, X.: Improved particle swarm optimization algorithm based on adaptive grid density and its application on multi-objective reactive power optimization. Adv. Power Syst. Hydroelectric Eng. **33** (02), 21–25 (2017). (in Chinese)
12. Yanjun, L., Youbo, L., Jinzhou, R., Junyong, L.: Optimal control strategy model of reactive power-voltage in power grid driven by deep learning. Electri. Measur. Instrum. **21**, 1–10 (2021). (in Chinese)
13. Liu, J., Li, D., Gao, L., Song, L.: Vector evaluation adaptive particle swarm optimization algorithm for multi-objective reactive power optimization. Chinese J. Electr. Eng. **31**, 22–28 (2008). (in Chinese)

14. Xianjun, G., Xiaodong, Y., Hao, T., Yuxiang, W., Leijiao, G.: Reactive power optimization model of distribution network integrating renewable energy generation. Electr. Measur. Instrum. 1–10 (2021). (in Chinese)
15. Braun, M.: Reactive power supply by distributed generators. In: 2008 IEEE Power and Energy Society General Meeting, Pittsburgh, USA, pp. 1–8. IEEE (2008)
16. Deyun, T., Huan, T., Xin, L., da, K.: Multi objective reactive power optimal scheduling considering multiple distributed generators connected to distribution network. Electr. Meas. Instrum. 56(13), 39–44 (2019). (in Chinese)
17. Shan,Y.: Research on dynamic reactive power optimal allocation method of distribution network. Xi'an University of science and technology (2009). (in Chinese)
18. Sheng-jian, L., Yan, Y., Yongquan, Z.: A swarm intelligence algorithm-lion swarm algorithm. Pattern Recogn. Artifi. Intell. 31(05), 431–441 (2018). (in Chinese)
19. Shengwei, D.: Research and application of multi-objective lion swarm optimization algorithm Shandong University (2021). (in Chinese)
20. Yu, X., Peng, W.: Multi objective particle swarm optimization with black hole mechanism and chaotic search. Control Eng. 26(02), 251–257 (2019). (in Chinese)
21. Jing, L., Xianjue, L.: Environmental economic power generation scheduling using multi-objective stochastic black hole particle swarm optimization algorithm. Chinese J. Electr. Eng. 30(34), 105–111 (2010). (in Chinese)

Observation Geometry Improvement of BDS by Near-Space Balloons in Regional Navigation Augmentation

Zhang Yang[1](✉), Tao Ping[2], Liu Rui[3], Xianghong Li[1], Li Ran[1], Yaping Li[1], Yang Guang[1], and Yuan Hong[1]

[1] Aerospace Information Research Institute, Chinese Academy of Sciences, Beijing, China
zhangyang101002@aircas.ac.cn
[2] Beijing Institute of Control and Electronic Technology, Beijing, China
[3] Non-commissioned Officer School of Space Engineering University, Beijing, China

Abstract. Aiming at the problem that the availability of BeiDou system (BDS) is reduced or even unavailable in urban canyon and complex mountain environment, the improvement of BDS observation geometry augmented by near-space balloons is researched. Firstly, the formula for assessing the observation configuration of BDS augmented by near-space balloons is derived, then the observation configuration variation of BDS at different elevations in typical regions is analysed, and the improvement of the GDOP of BDS augmented by different numbers of near-space balloons is compared, finally, the GDOP value only by near-space balloons in the independent position mode is calculated. The research shows the GDOP of BDS varies within 50 in Beijing and there are less than 4 visible BDS satellites in some epochs in Kashi when the minimum observation elevation is 50°. The availability and the observation configuration of BDS are improved by combining with near-space balloons. The GDOP is improved by 18.75% combining with one balloon at 35 km of the user's zenith, improved by 62.78% combining with three evenly distributed balloons, and improved by 66.67% combining with four networking balloons in Beijing. When the four networking balloons work in the independent position mode, the GDOP value is 1.73 in geometric visibility and 6.66 with the minimum observation elevation of 50°. The results provide theoretical support for the positioning research of BDS augmented by airborne or near-space platforms.

Keywords: Near-space platform · Regional navigation augmentation · Observation geometry · BDS · DOP

1 Introduction

The BDS is a satellite navigation system developed by China independently, which is constructed in the process of "three steps" strategy [1]. In June 2020, China launched the last BD-3 satellite and completed the networking of the BDS space constellation [2]. On July 31, 2020, the BDS-3 began to provide global navigation service. However, in typical regional environments such as urban canyons or complex mountains, the signal of the BDS and other GNSS is easily occluded, and the multipath effect is obvious. As a result, the number of visible satellites decreases, and the performance of navigation service of BDS is significantly affected or even unavailable [3].

In recent years, LEO satellites augmenting GNSS has been widely researched, LEO satellite navigation augmentation signals have been verified in orbit by "Luojia-1" and "Wangtong-1" [4–10]. However, due to the periodic motion of LEO satellites around the earth, the navigation augmentation effect of regional environment is significant only when there are a certain number of LEO satellites in orbit, therefore, it is not suitable for regional navigation augmentation due to the low performance-to-price ratio [11]. The near-space balloon is a typical aircraft flying at the altitude of 20 km~100 km, which has the characteristic of fixed point and little change of relative movement with the ground. It is a suitable platform for regional navigation augmentation [12]. By applying a certain number of near-space balloons with the ability of navigation signal broadcasting in urban canyons and complex mountainous environment, the observation configuration of BDS can be improved effectively and the navigation service performance of regional environment can be augmented significantly.

Therefore, the aim of this study is to analyse the improvement of BDS observation geometry by near-space balloons in urban canyons and complex mountainous environment. The remaining parts of this paper are organized as follows. The methodology of the observation configuration analysis by BDS and near-space balloons are given in Sect. 2. In Sect. 3, the observation geometry only by BDS, the observation geometry improvement combining with near-space balloons and the observation geometry only by the networking balloons are computed and compared in typical regions. Some typical conclusions are listed in Sect. 4.

2   Methodology

In the process of augmenting the positioning ability of the BDS by near-space balloons, the linearized ground user positioning equation is listed as follows [11].

$$\Delta \mathbf{z} = \mathbf{H} \Delta \mathbf{r} \tag{1}$$

$\mathbf{H}$ is the observation matrix, the form is listed as follows.

$$\mathbf{H} = \begin{bmatrix} \mathbf{H}_{bds} \\ \mathbf{H}_{zep} \end{bmatrix} = \begin{bmatrix} \mathbf{h}_{bds\_1} & 1 \\ \cdots & 1 \\ \mathbf{h}_{bds\_n} & 1 \\ \mathbf{h}_{zep\_1} & 1 \\ \cdots & 1 \\ \mathbf{h}_{zep\_m} & 1 \end{bmatrix} \tag{2}$$

where $\mathbf{H}_{bds}$ denotes the sub observation matrix of the BDS, $\mathbf{H}_{zep}$ is the sub observation matrix of the near-space balloons, and $\mathbf{h}$ is the unit observation vector of visible BDS satellites and near-space balloons relative to ground users, the calculation formula is listed in Eq. (3).

$$\mathbf{h} = \begin{bmatrix} \dfrac{-(x_{sz}-x_{user})}{\sqrt{(x_{sz}-x_{user})^2+(y_{sz}-y_{user})^2+(z_{sz}-z_{user})^2}} \\ \dfrac{-(y_{sz}-y_{user})}{\sqrt{(x_{sz}-x_{user})^2+(y_{sz}-y_{user})^2+(z_{sz}-z_{user})^2}} \\ \dfrac{-(z_{sz}-z_{user})}{\sqrt{(x_{sz}-x_{user})^2+(y_{sz}-y_{user})^2+(z_{sz}-z_{user})^2}} \end{bmatrix}^{\mathrm{T}} \tag{3}$$

$(x_{sz}\ y_{sz}\ z_{sz})$ is the position vector of visible satellites and near-space balloons in ECEF, $(x_{user}\ y_{user}\ z_{user})$ is the position vector of users in ECEF.

Based on Eqs. (2) and (3), $\mathbf{H}$ is determined by the spatial geometric distribution of visible satellites and balloons with users. Figure 1 shows the visibility analysis mode in the ground user view. The line marked in red is the geometric visible arc, and the arrow line marked in blue is the visible arc when considering the minimum observation elevation.

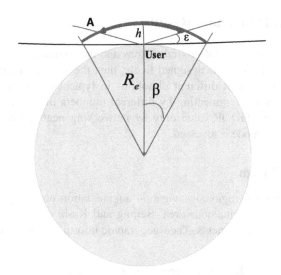

Fig. 1. Ground user visibility mathematical model (Color figure online)

The visibility calculation formula is shown in Eq. (4).

$$\beta = 2(\text{ar}\cos(\frac{R_e \cos \varepsilon}{R_e + h}) - \varepsilon) \tag{4}$$

where β is defined as geocentric angle, R_e is the radius of the earth, ε denotes the minimum observation elevation, h is the altitude of BDS satellites and the near-space balloons [13]. Assuming the observations of the BDS and the balloons have the same weight, the user state vector in Eq. (1) are calculated based on the least square method, shown in Eq. (5).

$$\Delta \mathbf{r} = \left(\mathbf{H}^{\mathrm{T}}\mathbf{H}\right)^{-1}\mathbf{H}^{\mathrm{T}}\Delta \mathbf{z} \tag{5}$$

DOP is used to assessing the observation configuration of users in regional environment, calculated by Eq. (6).

$$
\begin{aligned}
\text{GDOP} &= \sqrt{\text{tr}(\mathbf{H^T H})^{-1}} = \sqrt{\text{tr}\left(\mathbf{H_{bds}^T H_{bds}} + \mathbf{H_{zep}^T H_{zep}}\right)^{-1}} \\
\text{PDOP} &= \sqrt{(\mathbf{H^T H})_{1,1}^{-1} + (\mathbf{H^T H})_{2,2}^{-1} + (\mathbf{H^T H})_{3,3}^{-1}} \\
\text{TDOP} &= \sqrt{(\mathbf{H^T H})_{4,4}^{-1}}
\end{aligned}
\tag{6}
$$

Based on Eq. (6), the DOP value is determined by $\mathbf{H_{bds}}$ and $\mathbf{H_{zep}}$ [14–17]. Besides, The DOP value is lower compared only by the BDS, which indicates that the introduction of near-space balloons is able to improve the regional environment observation configuration. Near-space balloons augmenting the BDS will be analyzed and verified in the next section.

3   Experiment Design and Results

IN the section, the experimental ground regions and the augmenting scenarios by networking near-space balloons is designed firstly, then the visible satellite number and GDOP variation of BDS at different elevations in typical regions is analysed, and the GDOP improvement augmenting by different numbers of near-space balloons is compared, finally, the GDOP value only by networking near-space balloons in the independent position mode is assessed.

3.1   Experiment Design

Considering two typical regional navigation augmentation environments, i.e., urban canyon and complex mountainous area, Beijing and Kashi are selected respectively for the conduction of experiments. Their geographic information is shown in Table 1.

Table 1. Geographic information of users

User	Longitude(°)	Latitude(°)	Altitude (km)
Beijing	116.420	39.908	0.049
Kashi	75.929	39.505	1.255

Four networking near-space balloons are designed to improve the observation configuration. The space distribution of balloons is set as follows: one balloon is in the direction of the user's zenith at 35 km, and the other three balloons are evenly distributed at 20 km. The distribution of balloons in Beijing at the minimum observation elevation of 30°, 40° and 50° is shown in Fig. 2.

The geographic information of balloons in Beijing and Kashi at the minimum observation elevation of 50° is shown in Table 2 and Table 3.

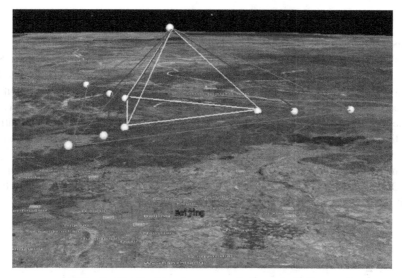

Fig. 2. Distribution of near-space balloons in Beijing

Table 2. Geographic information of balloons in Beijing

Platform	Balloon1	Balloon2	Balloon3	Balloon4
Longitude (°)	116.322	116.323	116.615	116.42
Latitude (°)	40.038	39.778	39.908	39.908
Altitude (km)	20.049	20.049	20.049	35.049

Table 3. Geographic information of balloons in Kashi

Platform	Balloon1	Balloon2	Balloon3	Balloon4
Longitude (°)	75.832	75.832	76.123	75.929
Latitude (°)	39.635	39.745	39.505	39.505
Altitude (km)	21.255	21.255	21.255	36.255

The analysis of observation configuration improvement by balloons is carried out in three experimental scenarios is carried out. The scenario is designed as follows.

Scenario 1: combining with one balloon in the direction of user's zenith at 35 km.

Scenario 2: combining with three balloons evenly distributed at 20 km.

Scenario 3: combining with all the four balloons in Table 2 and Table 3.

3.2 Experiment Design

(1) Observation Geometry analysis only by BDS

In the urban canyon and complex mountain environment, the observation of BDS is mainly affected by the minimum observation elevation. Therefore, the visible satellite number and DOP variation of BDS in typical area at different observation elevation are analysed, the results are shown in Figs. 3 and 4. It is proved that the number of visible satellites decreases significantly with the increase of the minimum observation elevation. When the minimum observation elevation angle is 50°, the number of visible satellites in some epochs in Kashi is only 3, the positioning conditions is not satisfied. Besides, when the minimum observation elevation is 50°, the GDOP in Beijing varies within 50. The GDOP in Kashi is invalid in some epochs, and greater than 50 in many epochs have GDOP, the navigation service performance is significantly reduced.

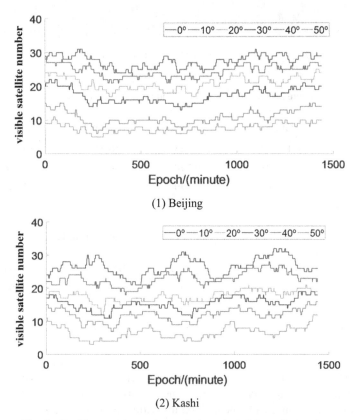

(1) Beijing

(2) Kashi

Fig. 3. Satellite visible number variation of BDS in typical regions

The average GDOP in Beijing and Kashi one day is statistics in Table 4. When the minimum observation elevation is lower than 30°, the average GDOP is less than 5. The GDOP value increases obviously at the minimum observation elevation of 40° and 50°

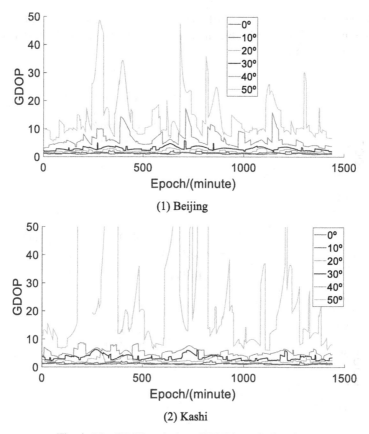

Fig. 4. The GDOP variation of BDS in typical regions

in Beijing, the average GDOP is invalid at 50°because the number of visible satellites in Kashi is less than 4 in some epochs.

Table 4. The average GDOP statistics one day

Mini-observation elevation	0°	10°	20°	30°	40°	50°
Beijing	1.10	1.44	2.11	3.03	6.26	14.4
Kashi	1.07	1.39	2.35	3.47	5.42	-

(2) Observation Geometry Improvement Combined with Near-Space Balloons

Based on the augmenting scenarios designed in Sect. 3, the observation geometry improvement by balloons in Beijing and Kashi is analysed. The GDOP variation is shown in Fig. 5 at the minimum observation elevation of 50°. It is shown that combining with one balloon in the direction of zenith has a limited improving capacity to the

GDOP. When combining with 3 or 4 balloons, the amplitude and stability of GDOP are significantly improved. Generally, the improvement performance in Beijing is slightly better than that in Kashi.

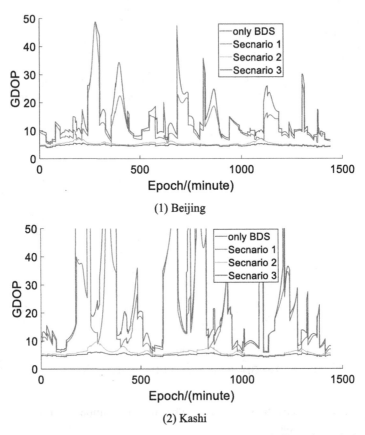

Fig. 5. The GDOP variation of BDS augmented by near-space balloons in typical regions

The average GDOP one day in different augmenting scenarios are shown is statistics in Table 5. The GDOP of all epochs in Kashi is effective due to the introduction of the balloons, and the navigation availability and continuity are improved. The improvement effect in scenario 1 is significantly lower than that in scenario 2 and scenario 3, where the GDOP value decreased to below 6. Based on Tables 4 and 5, the GDOP is improved by 18.75% when combining with one balloon in the user's zenith direction at 35 km in Beijing. When combining with three evenly distributed balloons and four networking balloons, the GDOP is improved by 62.78% and 66.67% respectively. General, when the number of balloons increases from one to three, the observation configuration can be improved significantly, but the improvement variation is not obvious when the number of balloons increases from three to four.

Table 5. The average GDOP statistics in the designed scenarios

GDOP	Scenario 1	Scenario 2	Scenario 3
Beijing	11.7	5.36	4.8
Kashi	25.5	5.94	5.02

(3) Observation Geometry Analysis Only by Networking Near-Space Balloon

In scenario 3, the networking balloons have the ability of independent navigation and position. Therefore, the DOP variation and value at different minimum observation elevation are derived, shown in Fig. 6 and Table 6. With the increase of the minimum observation elevation, the DOP is gradually getting higher, moreover, the variation of TDOP is slightly higher than that of PDOP. The GDOP is 1.73 when the minimum observation elevation is 0°, i.e., the balloons are geometrically visible to the users, and is 6.66 when the minimum observation elevation is 50°.

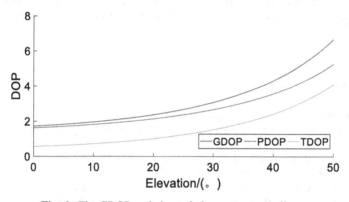

Fig. 6. The GDOP variation only by near-space balloons

Table 6. The average DOP statistics only by near-space balloons one day

Mini-observation elevation	0°	10°	20°	30°	40°	50°
GDOP	1.73	1.96	2.37	3.07	4.31	6.66
PDOP	1.63	1.82	2.14	2.67	3.57	5.25
TDOP	0.58	0.73	1.02	1.53	2.42	4.1

4 Conclusion

The improvement of the observation configuration of BDS by near-space balloons in urban canyon and complex mountain environment is researched systematically. Some typical conclusion is derived. (1) At the minimum observation elevation angle of 50°, the GDOP of BDS in Beijing varies within 50, and the average of GDOP is 14.4 one day, in some epochs the visible BeiDou satellite number is less than 4 in Kashi, the availability is reduced significantly; (2) In the analysis of BDS observation configuration augmented by near-space balloons, the observation configuration can be improved significantly with the increase of balloons from one to three, but the improvement variation is not obvious when the number of balloons increases from three to four. In Beijing, the GDOP is improved by 18.75% combining with one balloon at 35 km of the user's zenith, by 62.78% combining with three evenly distributed balloons and by 66.67% combining with four networking balloons; (3) The GDOP value is 1.73 in geometric visibility and 6.66 with the minimum observation elevation of 50° with four networking balloons in the independent position mode.

Moreover, the user position by BDS and near-space balloons will be researched in the future.

Acknowledgment. This work was partially supported by Youth Innovation Promotion Association, CAS (2022126), State Key Laboratory of Geo-Information Engineering and Key Laboratory of Surveying and Mapping Science and Geospatial Information Technology of MNR, CASM (NO. 2021-01-07).

References

1. Yang, Y., Mao, Y., Sun, B.: Basic performance and future developments of BeiDou global navigation satellite system. Sat. Navig. **1**(1), 1–8 (2020). https://doi.org/10.1186/s43020-019-0006-0
2. Ouyang, C., et al.: Evaluation of BDS-2 real-time orbit and clock corrections from four IGS analysis centers. Measurement. **168**, 1–10 (2021)
3. Liu, X., Zhan, X.-Q., Chen, A.M.-L.: Quantitative assessment of GNSS vulnerability based on D-S evidence theory. J. Aeronaut. Astronaut. Aviat. **46**(3), 1–20 (2014)
4. Murata, M., Kawano, I., Inoue, K.: Precision Onboard Navigation for LEO Satellite based on Precise Point Postioning (2020)
5. Li, X., et al.: LEO–BDS–GPS integrated precise orbit modeling using FengYun-3D, FengYun-3C onboard and ground observations. GPS Solut. **24**(2), 1–13 (2020). https://doi.org/10.1007/s10291-020-0962-8
6. Haibo Ge, et al.: Improving Low Earth Orbit (LEO) Prediction with Accelerometer Data. Remote Sensing (2020)
7. Su, M., et al.: BeiDou augmented navigation from low earth orbit satellites. Sensors. **19**(1), 1–14 (2019)
8. Li, B., et al.: LEO enhanced global navigation satellite system (LeGNSS) for real-time precise positioning services. Adv. Space Res. **63**(1), 73–93 (2019)
9. Wang, L., et al.: Initial assessment of the leo based navigation signal augmentation system from Luojia-1A Satellite. Sensors. **18**(11), 3919 (2018)
10. Li, M., et al.: Precise orbit determination of the Fengyun-3C satellite using onboard GPS and BDS observations. J. Geodesy **91**(11), 1313–1327 (2017). https://doi.org/10.1007/s00190-017-1027-9
11. Zhang, Y., et al.: Orbital design of LEO navigation constellations and assessment of their augmentation to BDS. Adv. Space Res. **66**(8), 1911–1923 (2020)

12. Chen, W., et al.: Integration of space and ground collaboration based on near space platform. In: IEEE 2017 8th International Conference on Mechanical and Aerospace Engineering (2017)
13. Yarlagadda, R., et al.: Gps gdop metric. In: IEE Proceedings - Radar, Sonar and Navigation, vol. 147, no. 5 (2000)
14. Teng, Y., Wang, J., Huang, Q.: Minimum of geometric dilution of precision (GDOP) for five satellites with dual-GNSS constellations. Adv. Space Res. **56**(2), 229–236 (2015)
15. Xue, S., Yang, Y.: Positioning configurations with the lowest GDOP and their classification. J. Geodesy **89**(1), 49–71 (2014). https://doi.org/10.1007/s00190-014-0760-6
16. Lansard, E., Frayssinhes, E., Palmade, J.-L.: Global design of satellite constellations a multi-criteria performance comparison of classical walker patterns and new design patterns (1998)
17. Massatt, P., Rudnick, K.: Geometric formulas for dilution of precision for dilution of precision calculations. J. Inst. Navig. **37**(4), 379–391 (1990)

Application of FE Simulation Method in the Field of Electrical Contact Performance Analysis

Wenbo Fan[(⊠)]

Beijing Aerospace Automatic Control Research Institute, Beijing 100854, China
13504042326@163.com

Abstract. Electrical connectors with contact finger structure are widely used in the high-power electrical interconnection occasions, which can achieve stable electrical contact and conduct large currents. At present, the most simple and accurate method to analyze and predict the electrical contact performance is to analyze and calculate it with the help of the commercial FE (finite element) simulation software. Taking the SCFS (spring contact finger structure) of the high-power electrical connectors as the research object, modeling according to its structural characteristics, the mechanical field simulation analysis and the thermoelectric coupling field simulation analysis of the SCFS model are carried out through the FE simulation software COMSOL Multiphysics, and the simulation results of the multi-physical field coupling characteristics are obtained. The changes and the influencing factors of the electrical contact performance parameters of the SCFS under three working conditions of the static connection, the mechanical insertion and the continuous power-on can lay a foundation for further guiding the design and optimization of the high-power electrical connectors.

Keywords: Spring contact finger structure · Electrical contact performance · Modeling · Finite element simulation

1 Introduction

With the development of today's electrical engineering field, the capacity and quantity of electrical equipment have increased rapidly, and the voltage and current levels have been greatly improved. Electrical connectors that can carry high power are required to transmit electrical energy in the power system [1]. The electrical contact performance parameters have positive implications for estimating system's life and reliability [2, 3].

The electrical contact performance of the contact finger structure can generally describe its insertion ability and current capacity of the electrical connectors, which can be characterized by characteristic parameters such as contact pressure, insertion force, contact area, contact resistance and contact temperature rise [4, 5]. These parameters can evaluate electrical connectors' reliability and further predict the whole system's life.

Up to now, for the contact finger structure of an electrical connector with complex structure, the most simple and accurate method for analyzing and predicting its electrical contact performance is to use the commercial FE simulation software [6]. By setting the structural parameters and the simulation conditions, the changes and the influencing factors of electrical contact performance under three working conditions of the static connection, the mechanical insertion and the continuous power-on can be studied quickly and accurately. It lays a foundation for the design and the optimization of the high-power electrical connectors [7, 8].

2   Analysis Method of Electrical Contact Performance Based on FE Simulation

2.1   Mechanical Field Simulation

In order to obtain the contact pressure and the insertion force of the SCFS by simulation method, we use the commercial FE software to simulate and analyze the mechanical insertion and extraction characteristics of the model. The material properties are selected as shown in Table 1 (Fig. 1).

Table 1. Material property table of SCFS.

Attribute	Variable	Value	Unit
Young's modulus	E	13.4×10^9	Pa
Shear modulus	G	46.8×10^9	Pa
Poisson's ratio	ν	0.35	-
Coefficient of kinetic friction	μ	0.045	-

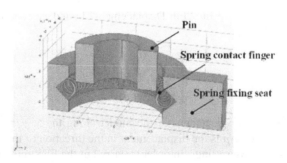

Fig. 1. The FE model imported into COMSOL Multiphysics software

The contact pressure control equation is

$$
\begin{cases}
F_p = (\lambda + G)\dfrac{\partial e}{\partial i} + G\nabla^2 u_i \\[4mm]
\lambda = \dfrac{E\nu}{(1+\nu)(1-2\nu)} \\[4mm]
e = \dfrac{\partial u_x}{\partial x} + \dfrac{\partial u_y}{\partial y} + \dfrac{\partial u_z}{\partial z}
\end{cases}
\tag{1}
$$

where λ is the lamp constant, ν is the Poisson's ratio, u_i is the deformation of the contact surface along the direction of the Cartesian coordinate system ($i = x, y, z$) component, and F_p is the force on the contact surface of the SCFS.

During the mechanical insertion and extraction process of the SCFS, there is one contact surface between the pin and the spring contact finger, and the other between the spring contact finger and the spring fixing seat. In order to accurately simulate the contact state of the SCFS during the working process, it is necessary to set the contact pairs. One SCFS has two groups of contact pairs. The first is arranged between the pin and the spring contact finger, and the second is arranged between the spring fixing seat and the spring contact finger. In the subsequent simulation process, COMSOL Multiphysics will automatically solve the electrical contact performance parameters according to the settings of each contact pair. For the FE model, we set the outside of the spring fixing seat to "Fixed Constraint", and the "Pin Displacement" at the top surface of the pin to simulate the insertion depth during the mechanical insertion and extraction process of the SCFS.

The "Fixed Constraint" boundary condition governing equations during the FE simulation is

$$
\begin{cases}
u_x = 0 \\
u_y = 0 \\
u_z = 0
\end{cases}
\tag{2}
$$

The governing equation of the "Pin Displacement" boundary condition during the FE simulation is

$$
\begin{cases}
u_x = 0 \\
u_y = 0 \\
u_z = -disp
\end{cases}
\tag{3}
$$

where u_x, u_y, u_z are the displacement of the model along the x, y, z axes in the FE simulation software, and $disp$ is the displacement in the direction of the pin insertion.

After completing above settings, in order to carry out the FE simulation analysis, it is necessary to discretize the three-dimensional geometric model of the SCFS into the mesh model, and further refine the mesh near the contact pairs, so as to make the FE simulation calculation results more accurate (Fig. 2).

Finally, COMSOL Multiphysics will complete the mechanical field simulation of the discrete FE model according to the setting conditions and obtain the results of the contact pressure and the insertion force.

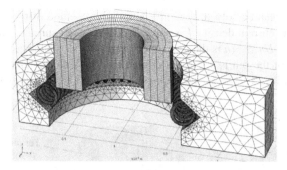

Fig. 2. The mesh model of the SCFS

2.2 Thermoelectric Coupling Field Simulation

In order to obtain the contact area and the contact resistance of the SCFS by simulation, we use COMSOL Multiphysics to carry out the thermoelectric coupling simulation of the model.

The contact resistance of the SCFS can be simulated by the current module and the heat transfer module of COMSOL Multiphysics. Material property values related to the thermoelectric coupling field should be added to each part of the model, and the material properties related to the thermoelectric coupling field are listed in Table 2.

Table 2. Material property table of the SCFS.

Conductor material	Gold plated copper
Resistivity	4.77×10^{-7} Ω/m
Thermal conductivity	400 W/(m * K)

Electrical contacts should be provided between the spring contact finger and the pin as well as the spring fixing seat respectively.

The governing equations of the current density in the FE simulation are

$$\begin{cases} \vec{J} = -\dfrac{1}{\rho}\nabla U \\ \nabla \cdot \vec{J} = 0 \end{cases} \tag{4}$$

where ρ is the resistivity, U is the potential, and J is the current density.

The governing equation of electrical insulation part is

$$\vec{n} \cdot \vec{J} = 0 \tag{5}$$

Through specially programmed thermoelectric coupling field analysis, the problems of contact resistance and temperature rise can be solved.

Firstly, Eq. (4) can be solved by the FE simulation method of the current module. In this module, the current density J is applied to the contact surface. The total current I is 500A, and $J = I/n\pi r^2$ (n is the number of contacts and r is the radius of contacts). The bottom side of the pin is set to the ground and the side boundaries are set to be electrically insulating. The power-on settings of the SCFS are shown in Fig. 3.

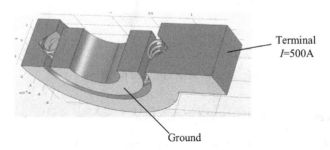

Fig. 3. The power-on settings of the model

Secondly, COMSOL Multiphysics can simulate the thermoelectric coupling of the discrete model according to the setting conditions, and obtain the results of contact resistance and contact temperature rise.

3   FE Simulation Results and Discussion

3.1   Mechanical Field

For the established SCFS model with the number of spring turns $N = 63$, the wire diameter of the spring wire $d = 0.3$ mm, and the middle diameter of the ring spring $\varphi = 13.3$ mm, the stress distribution cloud diagram is obtained as shown in Fig. 4.

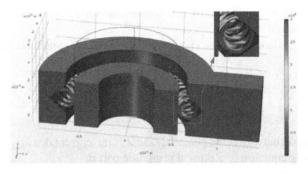

Fig. 4. The stress distribution cloud diagram of the SCFS

Through the "derived value" function of COMSOL Multiphysics, the contact pressure, the contact area and the insertion force are integrated respectively. The contact pressure between the spring contact fingers and the pin is 201.35 N, the insertion force is 9.06 N, and the contact area is 1.42 mm^2. Ignoring the machining accuracy error of each spring contact finger, the average contact pressure between the single turn spring contact finger and the pin is calculated to be about 3.20 N, the average insertion force is about 0.14 N, and the average contact area is about 2.25×10^{-2} mm^2. It can be seen from Fig. 4 that the inner root of the radial spring wire of each turn bears the largest force, which is calculated about 410.54 MPa, less than the yield strength of the spring contact finger 1000 MPa.

Changing the diameter of the spring wire, while keeping the number of turns and the center diameter unchanged, the simulation results of contact pressure, insertion force and contact area are shown in Fig. 5.

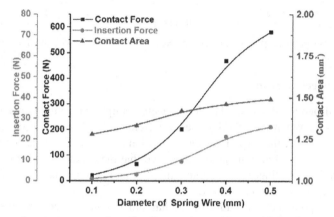

Fig. 5. Curves of contact pressure, insertion force and contact area with the diameter of the spring wire

It can be seen from Fig. 5 that when the diameter of the spring wire increases from 0.1 mm to 0.5 mm, the contact pressure between the spring contact finger and the pin increases from 21.61 N to 582.17 N, and the insertion force increases from 0.97 N to 26.18 N. The results show that when the diameter of the spring wire increases by 5 times, the corresponding contact pressure and insertion force both increase to about 26.6 time, which is approximately a quadratic relationship. However, in this process, the contact area only increases from 1.28 mm^2 to 1.49 mm^2, so it can be inferred that the structural parameter of the spring wire diameter has a very limited influence on the contact area.

Changing the number of spring turns, while keeping the diameter of the spring wire and the middle diameter of the ring spring unchanged. The contact pressure, insertion force and contact area are drawn in Fig. 6.

Figure 6 shows that in the process of increasing the number of turns from 10 to 90, the contact pressure increases from 10.17 N to 309.18 N, the insertion force increases from 0.46 N to 13.91 N, and the contact area increases from 0.18 mm^2 to 2.82 mm^2. Analysis of the data shows that they are all increase approximately linearly with the increase of the number of spring turns.

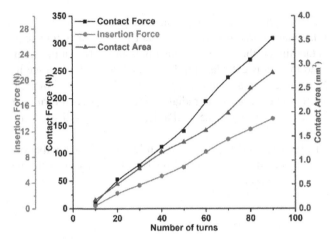

Fig. 6. Curves of contact pressure, insertion force and contact area with the number of turns

3.2 Thermoelectric Field

For the established SCFS model, the thermoelectric coupling field is simulated by using the FE software. When the SCFS passes a current of 500A, keeping the number of spring turns $N = 63$, the wire diameter $d = 0.3$ mm and the diameter of the spring contact finger $\varphi = 13.3$ mm, it can be calculated that the contact area between the pin and the spring contact finger is 1.42 mm^2, the contact resistance is 44.71 $\mu\Omega$, and the maximum temperature rise is 50.96 °C.

The surface potential distribution cloud diagram and the surface temperature distribution cloud diagram of the SCFS are shown in Figs. 7 and 8, respectively. Figure 7 shows that the potential of the SCFS gradually decreases from the terminal to the bottom of the pin. There is an obvious potential at the contact spot, which is caused by the contact resistance. Figure 8 shows that the SCFS has the highest temperature at the contact spot near the terminal, and the lowest temperature at the far terminal. The temperature of the contact spot is significantly higher than the surrounding conductors. This is due to that the contact resistance at the contact spot causes the current line to shrink, and the current density is higher than other areas. By $\varphi-\theta$ theory, the temperature of the contact spot is higher than the temperature of the outer conductor of the shrinkage zone. The surface temperature of the pin is significantly higher than that of the spring fixing seat because the spring fixing seat has a larger effective heat dissipation area.

First, we change the number of the spring turns, while keeping the diameter of the spring wire at 0.3mm, and the middle diameter of the ring spring at 13.3 mm. The simulation results of contact area, contact resistance and temperature rise changing with the number of turns are shown in Fig. 9.

It can be seen from Fig. 9 that in the process of increasing the number of spring turns from 10 to 90, the contact area of the spring contact structure increases approximately linearly, the contact resistance decreases from 112.53 $\mu\Omega$ to 38.95 $\mu\Omega$, and the contact temperature rise decreases from 137.48 °C to 44.55 °C. It shows that the relationship between the contact resistance R_j and the number of turns n can be approximately

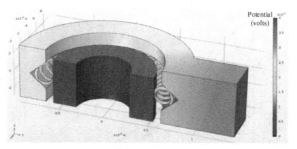

Fig. 7. The cloud map of the surface potential distribution

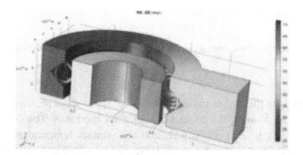

Fig. 8. The cloud map of the surface temperature distribution

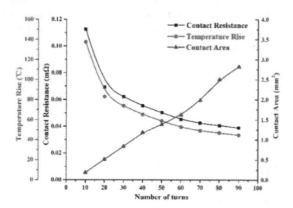

Fig. 9. Curves of contact area, contact resistance and temperature rise with the number of turns

expressed as $R_j \propto n^{-0.5}$ under the condition of continuous energization. The contact temperature rise decreases with the increase of the number of turns, and the slope is first fast and then slow.

Next, we change the diameter of the spring wire, while keeping the number of turns at 63, and the middle diameter of the ring spring at 13.3 mm. The simulation results of contact resistance, contact temperature rise and contact area changing with the diameter of the spring wire are shown in Fig. 10.

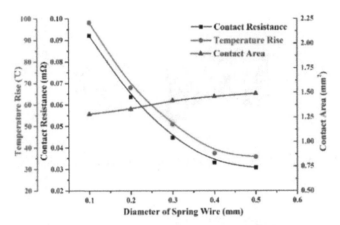

Fig. 10. Curves of contact area, contact resistance and temperature rise with the diameter of the spring wire

It can be seen from the curves in Fig. 10 that in the process of increasing the diameter of the spring wire by 5 times, the contact area slightly increases. The contact resistance decreases from 92.21 $\mu\Omega$ to 30.76 $\mu\Omega$, and the contact temperature rise decreases from 98.23 °C to 35.74 °C. The decreasing rate of each electrical contact performance parameter is first fast and then slow.

Then we change the middle diameter of the ring spring, while keeping the number of turns at 63, and the spring wire diameter at 0.3 mm. The simulation results of contact area, contact resistance and temperature rise changing with the middle diameter of the ring spring are shown in Fig. 11.

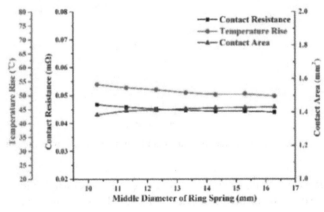

Fig. 11. Curves of contact area, contact resistance and temperature rise with the middle diameter of ring spring

It can be seen from the curves in Fig. 11 that in the process of increasing the middle diameter of the ring spring from 10.3 mm to 16.3 mm, the contact area increases from

1.385 mm² to 1.432 mm², the contact resistance decreases from 46.77 μΩ to 44.01 μΩ, and the contact temperature rise decreases from 53.96 °C to 49.73 °C. The analysis of the data shows that contact area, contact resistance and contact temperature rise of the SCFS under the condition of continuous energization are little affected by the change of the middle diameter of the ring spring.

Finally, we change the energization current of the SCFS, while keeping the number of spring turns at 63, the diameter of the spring wire at 0.3 mm, and the middle diameter of the ring spring at 13.3 mm. The change of contact temperature rise with the total current is shown in Fig. 12.

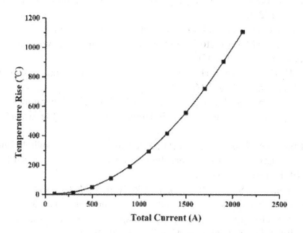

Fig. 12. Curve of temperature rise with total current

It can be seen from the curve in Fig. 12 that the contact temperature rise of the SCFS increases from 4.28 °C to 1102.33 °C during the process of the total current increases from 100 A to 2000 A. The analysis of the data shows that the change of the contact temperature rise is approximately square with the change of the energization current.

4   Conclusion

IN summary, we have demonstrated the electrical performance of the SCFS using COM-SOL Multiphysics. This paper provides a numerical simulation analysis method based on COMSOL Multiphysics, which can effectively calculate the electrical contact per-formance parameters such as contact pressure, insertion force, contact area, contact resistance and contact temperature rise of the SCFS. Meanwhile, we also simulated and analyzed the influence of some structural parameters in SCFS model, such as the diameter of spring wire, the number of spring turns and the middle diameter of ring spring, on electrical contact performance. The simulation results show that increasing the number of spring turns can significantly increase the contact area of the SCFS and reduce the temperature rise, but also lead to the increase of the contact pressure. the method of importing model parameters is convenient and effective to predict the factors affecting its electrical contact performance. This method can provide guidance for further optimization of the high-power electrical connectors.

References

1. Queffeelec, J.L., Benjemaa, N., Travers, D., Pethieu, G.: Materials and contact shape studies for automobile connector development. IEEE Trans. Compon. **14**(3), 90–94 (1991)
2. Elmanfalouti, A., Benjemaa, N., El Abdi, R.: Experimental and theoretical investigations on connector insertion phase. In: Proceeding of the 49th IEEE Holm Conference on Electrical Contacts, pp. 17–22 (2003)
3. JinSeok, K., SungHun, L., JaeChul, K.: Bus-voltage sag suppressing and fault current limiting characteristics of the SFCL due to its application location in a powerdistribution system. J. Elect. Eng. Technol. **8**(6), 1305–1309 (2013)
4. HsuYehLiang, H.-C., MingSho, H.: Shape optimal design of contacts springs of electronic connectors. J. Electron. Packag. **124**(4), 178–183 (2002)
5. Gatzsche, M., Lücke, N., Großmann, S.: Validity of the voltage-temperature relation for contact elements in high power applications. In: IEEE Holm Conference on Electrical Contacts, pp. 29–39 (2015)
6. Jianchao, Z., Lingdong, X., Yuming, Z., et al.: A new multi-gap spark switch connected with frequency dependent network for EHV overvoltage protection applications. IEEE Trans. Dielectr. Electr. Insul. **19**(4), 1369–1376 (2012)
7. Abeygunawardane, S.K., Sisira, J.R.S., Ekanayake Kumara, J.B.: A magnetic-core based fault current limiter for utility applications. J. Natn. Sci. Found. Sri Lanka. **39**(3), 227–234 (2011)
8. Srikanth, O., Sri, K., Reddy, R.: Optimal positioning of superconducting fault current limiters for smart grid. Int. J. Educ. Appl. Res. **4**(1), 77–84 (2014)

Damage Localization and Imaging of Composite Materials Based on Sparse Reconstruction Using Lamb Wave Anisotropy Propagation Model

Hui Wu, Shiwei Ma$^{(\boxtimes)}$, and Bingxu Du

School of Mechatronics Engineering and Automation, Shanghai University, Shanghai 200444, China

masw@shu.edu.cn

Abstract. In order to solve the anisotropy problem of composite materials on the impact of damage localization, a sparse reconstruction imaging method based on Lamb wave anisotropy propagation model is proposed in this paper. In this approach, the spatially distributed array of fixed PZTs sensor on the surface of the measured composite materials is employed to excite and receive Lamb wave. The received signal shows sparse under a pre-built damage dictionary with anisotropy propagation model, and the damages can be localized and imaged using a sparse reconstruction method. When constructing Lamb simulation signal waveform with Lamb wave linear propagation model, the Lamb wave propagation direction is divided evenly into several sectors, and Lamb wave propagation direction within a certain sector is considered as propagation in a quasi-isotropic material to resolve the anisotropy of composite materials. The experiments on composite fiber laminate specimens show that the proposed method can accurately locate simulated single delamination damage and has less artifact interference compared to the DAS method.

Keywords: Lamb wave · Damage localization · Sparse reconstruction · Composite fiber laminate

1 Introduction

Composites have the advantages of light weight and long life compared to traditional isotropic sheets. They are widely used in various industrial production as well as in daily life. However, in the production process of carbon fiber composite panels, there will be damage defects due to process errors and other factors. And in the process of use, there will also be damage due to aging, impact, etc. Which will result in a significant reduction in structural integrity and strength. Ultrasonic Lamb wave is a type of guided wave; it has excellent application prospects to evaluate the structural health of plate-like structures since it can travel a long distance with less signal loss while being sensitive to structural damage [1, 2].

There are many algorithms that can process Lamb wave signals to obtain damage location images, most well-known imaging algorithm are time-of-flight (TOF)-based

W. Fan et al. (Eds.): AsiaSim 2022, CCIS 1713, pp. 175–186, 2022.
https://doi.org/10.1007/978-981-19-9195-0_15

methods [3, 4]. A typical one of such method is the delay-and-sum (DAS) imaging method, which uses the TOF and wave velocity to locate the damage positions [5], however, in composite materials, the propagation velocity of Lamb waves is different in each direction due to anisotropy, which will lead to different TOF in each direction and cause errors in damage localization.

In recent years, the Compressed Sensing (CS) theory-based sparse reconstruction of Lamb wave signals was introduced into damage imaging [6, 7]. In contrast to TOF-based methods, the sparse re-construction method can achieve higher damage localization accuracy and less artifact interference. Subsequently, Xu [8, 9] solves the problem of boundary noise by using an edge reflection prediction technique, and Zhang [10] proposes the adaptive BPDN algorithm to address the signal reconstruction challenges of noise and unknown sparsity, significantly improving the algorithm's efficiency. However, all these approaches are ignored the effect of anisotropy in composites; they cannot be used directly in composite materials.

In this paper, to address the problems of anisotropy in damage localization and imaging for anisotropic composites laminates by using ultrasonic Lamb wave, the Lamb wave anisotropic propagation model is proposed. In the process of building the damage dictionary, the direction-dependent phase velocity dispersion curves are used in the Lamb wave propagation model to suppress the problems of anisotropy.

2 Theory

2.1 Damage Imaging Problem as a Sparse Reconstruction Problem

Sparse reconstruction is central to the notion of CS theory [11]. For the Lamb wave-based structural health monitoring technology, the signal content of damage scattering information $m(t)$ can be represented by an N-dimensional vector of actual number s, sampled according to the Shannon–Nyquist theorem. In the CS approach, the signal m is compressed into a vector y of M-dimensional ($M \ll N$) measurement signal, i.e., vector components, using an $M \times N$ sensing matrix Φ, such that $y = \Phi m$. The fact that $M \ll N$ causes a significant data reduction also means that the sensing matrix Φ is rank-deficient; i.e., for a particular signal m_i, the exact measurements $y = \Phi m = \Phi m_i$ can be obtained from an endless number of signals m_i. It follows that, in general, m cannot be reconstructed individually from the M-dimensional measurements of y. However, supposing that the signal m is K-sparse in a particular form, the matrix Φ may be built to recover the whole value of m from the measurement vector y while significantly reducing the number of measurements [12].

Considering a given representation basis $\{\Psi_i\}_{i=1}^{N}$ for $\mathbb{R}^N$, adding the Ψ_i as columns to the $N \times N$ matrix Ψ, one can get that $m = \Psi x$, with $x \in \mathbb{R}^N$ being the representation coefficients. It means that the signal m is K-sparse in Ψ if a vector $x_k \in \mathbb{R}^N$ exists with only K ($K \ll N$) nonzero entries such that $m = \Psi x_k$.

Based on the above sparse representation framework, the sparse reconstruction can be described as follows:

$$y = \Phi m = \Phi \Psi x = Ax \tag{1}$$

In normal service, most of the structure in composite parts is undamaged, and damage occurs in a small number of discrete spots, i.e., the damage is sparse. So, we discretize the material to be measured into $M \times N$ grid points and consider each grid point as potential damage scatter source, using damage scattering signals from all potential damage sources to construct a damage dictionary A, and using the $(M \times N) \times 1$- dimensional vector x to represent the possible locations of the damage. Because of the sparseness of the damage, the vector x is K-sparse in A, and K ($K \ll N$) nonzero entries represent the actual location of the damage.

By constructing the damage matrix A, x can be solved from the measured signal y to achieve damage imaging localization.

2.2 Lamb Wave Scattering Model in Composite Fiber Laminate

Lamb waves are multi-mode guided acoustic waves with dispersion characteristics, caused by frequency-dependent velocity. In practice, Lamb wave propagation in composite materials is a complex nonlinear process that is difficult to simulate with mathematical expressions, so when building the propagation model, we simplified it to a linear propagation model. Lamb wave signals are created in various modes for any given plate thickness and excitation frequency. The dispersion characteristics of each mode vary, which means that different frequency components of a similar propagation mode will propagate at different velocities. Thus, Lamb wave dispersion can be defined as a phenomenon in which the wavenumber varies by the product of frequency and thicknesses, and the linear propagation model is defined as a function of the different frequency components of the process with a given excitation signal passing through the transform system.

Let $S(f)$ be the excitation signal in the frequency domain, after propagating a fixed distance d, the received Lamb wave waveform in the frequency domain $W(f)$ can be defined by:

$$W(f) = \text{sqrt}\left(\frac{\alpha}{d}\right) S(f) \exp(-jK_b(f)) \tag{2}$$

where b represents the different modes of Lamb wave such as S0, S1, A0, etc., $K_b(f)$ represents the frequency-wavenumber curve of the b mode, and α is the normalization constant of the distance between the excitation and reception sensor. $sqrt\left(\frac{\alpha}{d}\right)$ Accounts for the amplitude decrease caused by wave diffusion and j is an imaginary unit.

Taking the Fourier inverse transform of the signal $W(f)$, we can obtain the time domain waveform $w(t)$:

$$w(t) = \mathcal{F}^{-1}\{W(f)\} = sqrt\left(\frac{\alpha}{d}\right) \int_{-\infty}^{+\infty} S(f) \exp(-jK_b(f)d) \exp(j2\Pi ft) df \tag{3}$$

Figure 1 shows the propagation of the Lamb wave between a set of sensors, the Lamb wave is excited and received by the piezoelectric sensor attached to the upper surface of the specimen. Assuming a single-modal Lamb wave is excited at a defined location $\mathbf{p}_1$ on the plate using the frequency-domain excitation signal $S(f)$, and assuming the damage is present at position $\mathbf{q}$. The damage can be approximated as a secondary

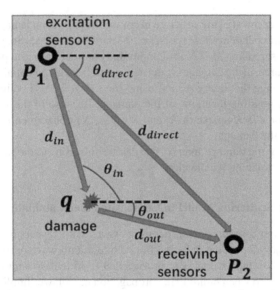

Fig. 1. Schematic diagram of Lamb wave propagation paths.

excitation source capable of completely scattering any incident wave. The waveform $W^q(f)$ measured at another location $\mathbf{p}_2$ has three components: the direct wave from $\mathbf{p}_1$ to $\mathbf{p}_2$, the damage wave from $\mathbf{p}_1$ to $\mathbf{q}$, and the scattering wave from $\mathbf{q}$ to $\mathbf{p}_2$. Ignoring geometric reflections and incoherent noise for simplicity, the measured far-field response (also in the frequency-domain) is:

$$
\begin{aligned}
W^q(f) &= W^q_{direct}(f) + W^q_{damage}(f) + W^q_{scatter}(f) = \\
&\text{sqrt}\left(\frac{\alpha}{\|\mathbf{p}_1 - \mathbf{p}_2\|_2}\right) S(f) \exp(-jK_b(f)\|\mathbf{p}_1 - \mathbf{p}_2\|_2) \\
&\text{sqrt}\left(\frac{\alpha}{\|\mathbf{p}_1 - \mathbf{q}\|_2}\right) S(f) \exp(-jK_b(f)\|\mathbf{p}_1 - \mathbf{q}\|_2) + \\
&H_{M,\theta_{in},\theta_{out}}(f)\,\text{sqrt}\left(\frac{\alpha}{\|\mathbf{q} - \mathbf{p}_2\|_2}\right) S(f) \exp(-jK_b(f)\|\mathbf{q} - \mathbf{p}_2\|_2)
\end{aligned}
\tag{4}
$$

where $\|\cdot\|_2$ represents the Euclidean distance between two points, and θ represents the angle between the propagation direction and the horizontal line; $H_{M,\theta_{in},\theta_{out}}(f)$ is the scattering coefficient of the M mode at this damage, which is a function of frequency given incoming and outgoing angles θ_{in} and θ_{out}, respectively.

The damage scattering signal $z^q(t)$ can be expressed as the measured signal in the presence of damage minus the reference signal in the absence of damage as follows, $\mathcal{F}^{-1}$ represents the Fourier inverse transform.

$$
\begin{aligned}
Z^q(f) &= W^q(f) - W^q_{direct}(f) \\
&= \text{sqrt}\left(\frac{\alpha}{\|\mathbf{p}_1 - \mathbf{q}\|_2}\right) S(f) \exp(-jK_b(f)\|\mathbf{p}_1 - \mathbf{q}\|_2) \\
&\quad H_{M,\theta_{in},\theta_{out}}(f)\,\text{sqrt}\left(\frac{\alpha}{\|\mathbf{q} - \mathbf{p}_2\|_2}\right) S(f) \exp(-jK_b(f)\|\mathbf{q} - \mathbf{p}_2\|_2)
\end{aligned}
\tag{5}
$$

The fiber laminated material comprises several single-layer fiber materials, and the frequency-wavenumber curve of Lamb waves varies by the direction of wave propagation due to the inconsistent orientation of each layer's fiber. Considering the relationships $c_p = \omega/k = 2\pi f/k$ and $c_g = d\omega/dk$, we use $C_p(f)$ to replace $C_g(f)$ under the assumption that $\omega/k = d\omega/dk$, because Lamb waves vary relatively little in frequency when selected for a specific excitation signal. Then the frequency-wavenumber curve $K_b(f)$ in Eq. (4) can be replaced by $2\pi f/C_p^b(f)$, and due to the direction-dependent $K_b(f)$, the phase velocity dispersion curve $C_p^b(f)$ also varies by the direction. So, when using Eq. (4) to simulate Lamb wave propagation in composite fiber laminate, we need to take the direction into account to ensure that the propagation model is more consistent with the real propagation of the Lamb wave.

Again, suppose that a Lamb wave is excited at location $\mathbf{p}_1$ and is measured at location $\mathbf{p}_2$ with the damage present at location $\mathbf{q}$; however, we use $C_p^{b,\theta}(f)$ to replace $K_b(f)$ and use the incoming and outgoing angles θ_{in} and θ_{out} to determine the $C_p^{b,\theta}(f)$, the signal $z^q(t)$ measured at $\mathbf{p}_2$ can be expressed as:

$$
z^q(t) = \mathcal{F}^{-1}\{\mathrm{sqrt}\left(\frac{\alpha}{\|\mathbf{p}_1 - \mathbf{q}\|_2}\right)S(f)\exp\left(-j\frac{2\pi f\|\mathbf{p}_1 - \mathbf{q}\|_2}{C_p^{b,\theta_{in}}(f)}\right)
$$

$$
+ H_{\mathrm{mod}\,es}(\theta_{in},\theta_{out},f)\mathrm{sqrt}\left(\frac{\alpha}{\|\mathbf{q} - \mathbf{p}_2\|_2}\right)S(f)\exp\left(-j\frac{2\pi f\|\mathbf{q} - \mathbf{p}_2\|_2}{C_p^{b,\theta_{out}}(f)}\right)\} \quad (6)
$$

where $C_p^{b,\theta}(f)$ represents the phase velocity dispersion curve when the propagation angle is θ and the Lamb wave mode is b, $\|\cdot\|_2$ represents the Euclidean distance between two points, and $H_{M,\theta_{in},\theta_{out}}(f)$ is the scattering coefficient of the M mode at this damage, which is a function of frequency given incoming and outgoing angles θ_{in} and θ_{out}, respectively.

3 Methodology

3.1 Damage Dictionary Construction

With the aim to reduce the calculation of the dispersion characteristic curve, we take the idea of the infinitesimal method and divide the propagation direction of Lamb waves on the composite fiber laminate into V sectors, as shown in Fig. 2. The angle size of each sector is $360°/V$. The difference between the phase dispersion curves of the Lamb waves in the sector can be neglected when V is greater than a particular value. Within a certain sector, Lamb wave can be regarded as propagation on the quasi-isotropic material and take the phase dispersion characteristic on the corner parallels of the sector as the dispersion characteristic of the whole sector. In this way, we only need to calculate the phase dispersion curve on v partitions.

Again, as shown in Fig. 1, when the potential defect is present at $\mathbf{q}$ (m^{th} row, n^{th} column in the discretized grid of the specimen), and θ_{in} is located within the ith sector, θ_{out} is located within the jth sector, the damage scattering signals $r^{m,n}(t)$ can be expressed as:

$$
r^{m,n}(t) = \mathcal{F}^{-1}\left\{\mathrm{sqrt}\left(\frac{\alpha}{\|\mathbf{p}_1-\mathbf{q}\|_2}\right)S(f)\exp\left(-j\frac{2\pi f\|\mathbf{p}_1-\mathbf{q}\|_2}{c_p^{S0,i}(f)}\right)\times\right.
$$
$$
\left. H_{modes}(\theta_{in},\theta_{out},f)\mathrm{sqrt}\left(\frac{\alpha}{\|\mathbf{q}-\mathbf{p}_2\|_2}\right)S(f)\times\exp\left(-j\frac{2\pi f\|\mathbf{q}-\mathbf{p}_2\|_2}{c_p^{S0,j}(f)}\right)\right\} \quad (7)
$$

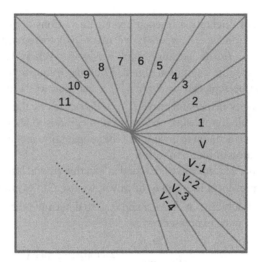

Fig. 2. Divide the specimen into V sectors.

where $C_p^{S0,i}(f)$ and $C_p^{S0,j}(f)$ represents the S0 single-mode phase velocity dispersion curve on the ith and j^{th} sector, $|| \cdot ||_2$ represents the Euclidean distance between two points, and $H_{M,\theta_{in},\theta_{out}}(f)$ is the scattering coefficient of the M mode at this damage, which is a function of frequency given incoming and outgoing angles θ_{in} and θ_{out}, respectively. While $H_{M,\theta_{in},\theta_{out}}(f)$ may be available if a very specific type of damage is expected, but in many cases, a priori knowledge of the scattering pattern of future damage is unknown and may be difficult to predict. In these cases, we setting $H_{M,\theta_{in},\theta_{out}}(f) = 1$ in the propagation model.

For damage locating and imaging, a damage dictionary A containing damage scattering signal waveforms for all potential damage sources on the interrogated area of the specimen in the simulated excitation system needs to be constructed.

- Step 1: the inspection area of the composite fiber laminate specimen is divided into M*N sufficiently fine grids, and the atoms in the damage dictionary are mapped to the grid in the specimen one-to-one correspondence.
- Step 2: Suppose that each grid in the monitored area is a potential defect, we use the damage scattering signal waveform from each grid to build the damage dictionary A.
- Step 3: For each grid, we can get $L \times (L-1)/2$ sets of signals from the sensor network with L sensors, and the atom of the damage dictionary $A_{i,j}$ corresponding to the damage scattering signal when the grid in i^{th} row and j^{th} column of the discretized monitored area is a potential damage source.

In step 3, we use the Lamb wave propagation model in Eq. (6) to simulate the damage scattering signal for each potential defect point,

Then, to be consistent with the measurement signal y, the damage scattering signal $r^{m,n}(t)$ is intercepted from the wave crest, and obtain $L*(L-1)/2$ groups of S0 single-mode damage scattering signals $r^{m,n,S0}(t)$ for cascading to form an atom $A^{m,n}$ of the damage dictionary.

$$A^{m,n} \propto \begin{bmatrix} r_1^{m,n,S0}(t) \\ r_2^{m,n,S0}(t) \\ r_3^{m,n,S0}(t) \\ \vdots \\ r_{L\times(L-1)/2}^{m,n,S0}(t) \end{bmatrix} \tag{8}$$

Finally, repeat step 3 for these $M*N$ potential damage points in the discretized grid of the specimen to obtain the damage dictionary A.

$$A = \begin{bmatrix} A^{1,1} & A^{1,2} & \dots & A^{1,N} \\ A^{2,1} & A^{2,2} & \dots & A^{2,N} \\ \vdots & \vdots & \ddots & \vdots \\ A^{M,1} & A^{M,2} & \dots & A^{M,N} \end{bmatrix} \tag{9}$$

3.2 Sparse Reconstruction Solution and Damage Localization Imaging

Finding the sparse solution x is a key step for subsequent damage localization imaging, and the sparse solution x can be obtained by solving the inverse problem of Eq. (1). Considering the vector x is sparse, the compressed perception problem can be converted into the following problem [13]:

$$min\|x\|_0 s.t. Y = Ax \tag{10}$$

Since the solution of l_0 norm requires C_N^K (N is the sparse solution dimension and K is the sparsity) possibilities for all elements in x, it is an NP-hard problem, which cannot be solved directly but can be transformed into an l_1 norm minimization problem to be solved as follows.

$$min\|x\|_1 s.t. y = Ax \tag{11}$$

Further considering the noise problem, the above representation can be rewritten as:

$$min\frac{1}{2}\|y - Ax\|_2^2 + \lambda\|x\|_1 \tag{12}$$

where λ is the regularization parameter that regulates the relationship between the allowable error $\|y - Ax\|_2^2$ and the sparsity $\|x\|_1$ of x.

Since the l_1 norm is a convex parametrization and Eq. (11) is a sparse regular least squares model containing the l_1 norm [14], it is a typical convex optimization problem, so it can be transformed into a quadratic programming problem for solving.

The BPDN [15] algorithm is a classical method for solving the above problem with high reconstruction accuracy and better robustness to measurement noise, so the spgl1 solver in MATLAB based on the BPDN algorithm is used to reconstruct the signal in this paper.

To generate a two-dimensional visualization image, the pixel value of each grid point P_k is set to x_k. The sparse vector solution x of $M*N$ dimensions solved by the sparse reconstruction corresponds. And the pixel values of all grid points in the detection area constitute a two-dimensional vector represented as an image, where the points with non-zero-pixel values correspond to the locations of the actual damage points. Therefore, the sparse vector is sparsely reconstructed from the measurement vector and the dictionary matrix, and the damage imaging results can be obtained.

4   Experimental Research and Result Analysis

Experimental testing was performed on a composite fiber laminate specimen, bonded by two composite plates of 3.49mm thickness, with a glue layer thickness of 0.02mm. In the glue layer between the two composite plates, we made simulated delamination damage. Each composite plate consists of eight layers of one-way laminates. The material parameters of the single-layer are given in Table 1, and the fiber direction of the laminate composite is [0°/45°/ 90°/-45°/45°/90°/-45°/0°], the geometric dimension is 300mm × 300mm × 7mm, as shown in Fig. 3(a).

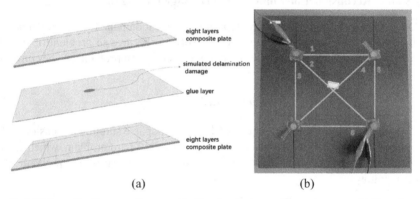

(a)   (b)

Fig. 3. (a)Schematic diagram of composite fiber laminate specimen structure, (b) Experimental specimen and sensor layout

The sensor array consisting of four PZT discs, each with a diameter of 10mm and a thickness of 0.5mm, is arranged on the upper surface of the plate. In the experiment, four PZTs are used in turn as the excitation source and the remaining three as the receiving source to collect the lamb wave signal, with a total of six sensing paths for the four sensors, so six sets of signals are collected for each plate.

The measured damage scattering signal y can be obtained by subtracting the baseline from the response signals recorded on a specimen containing damage, and using the velocity difference between S0 and A0 mode of the received Lamb wave signal, the S0 mode is extracted by intercepting the signal before the peak of the first wave packet to eliminate the influence of Lamb wave mode mixing, as shown in Fig. 4.

When constructing the damage dictionary A, the final number of partitions V is selected as 24, i.e., $15°$ per partition to better balance positioning accuracy and computation cost., considering the computational time and space cost of the propagation model. After obtaining the measured signal y and the damage dictionary A, we were able to perform damage imaging by inverse solving Eq. (1).

Table 1. Material mechanical properties of the carbon fiber layer

E1/GPa	E2/GPa	E3/GPa	G12/GPa	G13/GPa	G23/GPa	υ_{12}	υ_{13}	υ_{23}
230	15	15	24	24	5.03	0.2	0.2	0.25

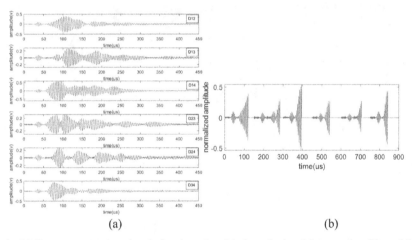

(a)   (b)

Fig. 4. (a) Damage scattering signals for six paths, (b) six path signal interception S0 modal and cascade to form the measurement signal y

To verify the method's superiority in this paper, we used three different layouts in our experiments, as shown in Fig. 5, and a comparison test with the DAS method [16] was done for the above three layouts. The damage location imaging results for the three layouts are shown in Fig. 6. From the results, one can see that the results obtained by the DAS method have significant background noise. The dynamic range of each grid point pixel (the difference between the maximum and minimum pixel value) is small, resulting in the damage localization area being too large, and the localization error of the DAS method is higher than the sparse reconstruction method in this paper due to the effect of anisotropy. In addition, it cannot eliminate the effect of composite anisotropy, which leads to significant location errors and interference from artifacts. The sparse reconstruction imaging method proposed in this paper has a smaller damage localization area and higher resolution of damage imaging. The errors between the estimated damage position and the actual damage position for DAS and the sparse reconstruction imaging method are given in Table 2.

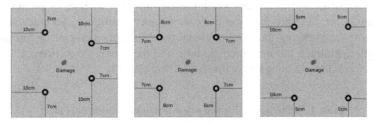

Fig. 5. Three types of sensor layouts (The gray square area represents the interrogated area of the specimen with a side length of 30 cm, black concentric circles represent four piezoelectric transducers, and the red area represents the location of the damage.)

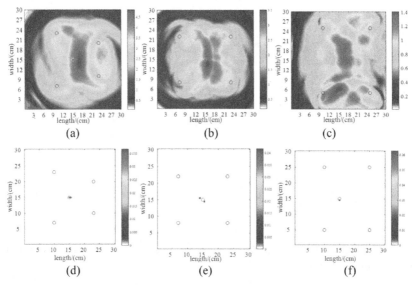

Fig. 6. Comparison of damage localization imaging results between the DAS method and this paper's method: (a) (b) (c) represents the DAS damage localization imaging results, (d)(e)(f) represents the corresponding results of the methods in this paper (In the DAS imaging results, the blue circle represents the sensor location; the white circle represents actual damage location; the area with the highest color pixel value represents the likely location of the damage; the horizontal and vertical coordinates represents the length and width of the experimental specimen.)

Table 2. Damage positioning error

Different layouts	Layout 1	Layout 2	Layout 3
Actual damage location (cm)	(15.0,15.0)	(15.0,15.0)	(15.0,15.0)
Estimated damage location(cm)/DAS	(16.9,13.6)	(15.6,13.7)	(17.3,15.4)
Estimated damage location(cm)/Sparse Reconstruction	(15.5,15.0)	(15.0,15.0)	(15.0,14.5)
Error (cm)/ DAS	2.3	1.4	2.3

5 Conclusion

This article proposed a sparse reconstruction imaging method based on Lamb wave anisotropy propagation model, which converted the composite damage localization and imaging problem to a sparse reconstruction problem by the overcomplete damage dictionary based on the sparsity of the damage location. The Lamb wave anisotropy propagation model is constructed to develop the overcomplete damage. Compared to the classic DAS method, the proposed method produces damage localization results with less background noise, less artifact interference, and higher damage positioning.

Acknowledgment. This work was supported by the National Natural Science Foundation of China (Grant No. 61671285).

References

1. Giurgiutiu, V., Santoni-Bottai, G.: Structural health monitoring of composite structures with piezoelectric-wafer active sensors. AIAA J. **49**(3), 565–581 (2011)
2. Santhanam, S., Demirli, R.: Reflection and transmission of fundamental Lamb wave modes obliquely incident on a crack in a plate. In: 2012 IEEE International Ultrasonics Symposium (2012)
3. Xu, C., Yang, Z., Tian, S., Chen, X.: Lamb wave inspection for composite laminates using a combined method of sparse reconstruction and delay-and-sum. Compos. Struct. **223**, 110973 (2019)
4. Su, C., et al.: Damage localization of composites based on difference signal and lamb wave tomography. Materials. **13**(1), 218 (2020)
5. Wang, C.H., Rose, J.T., Chang, F.-K.: A synthetic time-reversal imaging method for structural health monitoring. Smart Mater. Struct. **13**(2), 415 (2004)
6. Levine, R.M., Michaels, J.E.: Model-based imaging of damage with Lamb waves via sparse reconstruction. J. Acoust. Soc. Am. **133**(3), 1525–1534 (2013)
7. Levine, R.M., Michaels, J.E.: Block-sparse reconstruction and imaging for lamb wave structural health monitoring. IEEE Trans. Ultrason. Ferroelectr. Freq. Control **61**(6), 1006–1015 (2014)
8. Xu, C., Yang, Z., Deng, M.: Weighted structured sparse reconstruction-based Lamb wave imaging exploiting multipath edge reflections in an isotropic plate. Sensors. **20**(12), 3502 (2020)

9. Zhang, H., et al.: Adaptive sparse reconstruction of damage localization via Lamb waves for structure health monitoring. Computing **101**(6), 679–692 (2019). https://doi.org/10.1007/s00607-018-00694-0

10. Xu, C., Yang, Z., Qiao, B., Chen, X.: A parameter estimation based sparse representation approach for mode separation and dispersion compensation of Lamb waves in isotropic plate. Smart Mater. Struct. **29**(3), 035020 (2020)

11. Donoho, D.L.: Compressed sensing. IEEE Trans. Inf. Theory **52**(4), 1289–1306 (2006)

12. Candès, E.J., Wakin, M.B.: An introduction to compressive sampling. IEEE Signal Process. Mag. **25**(2), 21–30 (2008)

13. Harley, J.B., Moura, J.M.: Data-driven matched field processing for Lamb wave structural health monitoring. J. Acoust. Soc. Am. **135**(3), 1231–1244 (2014)

14. Chen, S.S., Donoho, D.L., Saunders, M.A.: Atomic decomposition by basis pursuit. SIAM Rev. **43**(1), 129–159 (2001)

15. Figueiredo, M.A., Nowak, R.D., Wright, S.J.: Gradient projection for sparse reconstruction: application to compressed sensing and other inverse problems. IEEE J. Sel. Topics Signal Process. **1**(4), 586–597 (2007)

16. Liu, L., Xia, Q., Cao, S., Ma, S., Liu, Y.: Damage detection of composite plate based on an improved DAS algorithm by time difference due to anisotropy. J. Vibroeng. **22**(8), 1747–1757 (2020)

5G Wireless Network Digital Twin System Based on High Precision Simulation

Zhongqiu Xiang[1]([✉]), Zhiqing Wang[2], Kai Fu[2], Xuemin Huang[1], Fan Chen[1], Pei Zhao[1], Shumin Jiang[1], Yantao Han[3], Wenzhi Li[3], and Feng Gao[1]

[1] China Mobile Group Design Institute Co., Ltd., Beijing, China
13811307563@139.com
[2] China Mobile Group Sichuan Co., Ltd., Chengdu, China
[3] China Mobile Group Co., Ltd., Beijing, China

Abstract. For a long time, the wireless signals of the communication network are invisible and intangible, which brings difficulties in recognizing and maintaining the network. We can achieve mathematical modeling and quantitative analysis of 5G wireless signals with the help of the ray tracing propagation model. Furthermore, we can build a 5G wireless network digital twin system through the high-precision simulation calculation and GIS presentation in the horizontal and vertical dimensions. In order to ensure the accuracy of the simulation, the coefficient correction of the ray tracing model is carried out by using the combination of the Lagrangian function and the KKT boundary condition. The 5G wireless network digital twin system can be precisely obtained through the corrected propagation model. Based on the digital twin system, the weak coverage areas of 5G network can be accurately located and the weak coverage buildings can also be accurately identified. In this way, we can carry out engineering parameter optimization and 5G indoor base station planning even in off-site work. The 5G wireless network digital twin system can achieve the goals of cost reduction and accurate resource allocation, which is of great guiding significance for engineering practice.

Keywords: 5G wireless network · Digital twin system · Ray tracing · Model coefficient correction · Engineering parameter optimization · 5G indoor base station planning

1 Introduction

Digital twin, as one of the core technologies in the age of intelligence, is a new break-through in simulation technology. It is a bridge connecting the physical world and virtual space, and assists in realizing the foresight of the physical world [1]. Digital twin makes full use of data such as physical models, sensor updates, and operation history, and integrates multi-disciplinary, multi-physical, multi-scale, and multi-probability simulation processes [2]. It can complete the mapping in the virtual space, so as to reflect the whole life cycle process of the corresponding physical equipment.

As an important part of new infrastructure, 5G covers the most fields, covers the widest range, and has the strongest economic driving effect. 5G network construction is the focus of attention of all countries in the world. For a long time, wireless electromagnetic signals are invisible and intangible. We cannot intuitively and quantitatively grasp the wireless network situation, and we have not built a digital system with a full life cycle. Building a digital twin system of 5G networks through simulation can not only objectively reflect the network operation, but also assist in solving the problems foreseen in the network.

This paper uses refined propagation model correction to carry out accurate mathematical modeling and simulation. The digital twin system of 5G wireless network is constructed through communication system simulation, and a solution for the whole life cycle of 5G wireless network is formed. It can analyze the existing problems of outdoor and indoor 5G networks. After complex iterative calculations, we seek solutions to help the planning and construction of 5G networks.

2   Communication System Simulation

2.1   Fundamentals of Communication System Simulation

Communication system simulation can predict the coverage quality and capacity performance of the network through complex computational processes. Generally, communication system simulation includes key links such as project creation, parameter configuration, path loss calculation, and simulation result output [3]. The simulation flow of the communication system is shown in the follow (see Fig. 1).

Fig. 1. Communication system simulation process

In the above process, the core link is the path loss calculation. The propagation process of the wireless electromagnetic signal is simulated by the propagation model, and the loss in the propagation path is calculated to calculate the coverage radius of the base station. Therefore, in the simulation of the entire communication system, the most important thing is to choose the propagation model.

2.2   Wireless Communication Propagation Model

The key influencing factor of the communication system simulation is the propagation model. The refined propagation model generally adopts the ray tracing model. In this paper, the LiShuttle ray tracing model developed in China is used to simulate the communication system. The propagation model is affected by factors such as geographical environment and engineering parameters. Before actual use, the propagation model should be calibrated according to the actual local test data.

Classification of Wireless Propagation Models
Traditional propagation models are divided into empirical models and deterministic models.

The empirical model is an empirical formula that is fitted by mathematical methods based on a large number of field strength test results, and generally does not require specific information about the relevant environment [4]. In the research field of wireless propagation, engineers and researchers have proposed many empirical models for various propagation environments (including rural, urban, mountainous, etc.), and most of these models are suitable for macro cells with higher antenna heights and larger coverage areas, and the parameters required by each model can be obtained according to the environmental data in the statistical sense (for example, the average height of buildings, the average street width, etc.) to obtain more accurate prediction results. Commonly used empirical models are SPM model, Okumura-Hata model and COST-231 Hata model.

The deterministic model is a method of directly applying electromagnetic theory to the specific field environment. In urban, mountainous and indoor environments, deterministic wireless propagation prediction is an extremely complex electromagnetic problem. Common methods include ray tracing, using geometric diffraction theory and other methods. The ray tracing model can better simulate the influence of buildings and streets in the city on wireless propagation. Ray tracing is a technique for predicting the propagation characteristics of radio waves in mobile communication and personal communication environments, and can be used to identify all possible ray paths between transmission and reception in multipath channels. Once all possible rays have been identified, the amplitude, phase, delay and polarization of each ray can be calculated according to the wave propagation theory, and then combined with the antenna pattern and system bandwidth, the coherent synthesis of all rays at the receiving point can be obtained result. Common ray tracing models include: LiShuttle model in China, Volcano model in France, Planet model in the United States, etc.

Correction of Ray Tracing Propagation Models
The LiShuttle ray tracing model can accurately simulate the direct, reflection, diffraction, transmission and other phenomena of electromagnetic waves in space propagation. It emits 360 rays in the horizontal and vertical dimensions, and each ray is calculated according to the cell type of the base station and the landform information. Its spatial propagation path finally generates the RSRP of the wireless signal through multipath combining [5].

After the path search in the calculation area is completed, the path loss of each grid in the area needs to be calculated, and the calculation will consider base station parameters, terminal parameters, terrain features, multipath loss, etc. The final path loss is calculated according to the following calculation formula.

$$PL = K_0 + K_{near} \log(d_{3d}) + 20 \log_{10}(f_c) + \Delta K_{farf}(d_{2d}) + \partial_{ref} PL_{ref} + \partial_{diff} PL_{diff}$$

$$(1)$$

where K_0 is the fading constant, K_{near} is the near-field coefficient, K_{far} is the far-field coefficient, ∂_{ref} is the reflection coefficient, and ∂_{diff} is the diffraction coefficient.

Direct Coefficient Correction

The calibration of the direct parameter is divided into the calibration stage of
K_{near} and K_0, and this stage uses all the data for calibration. In the ΔK_{far} correction stage, the far-field data is used for correction, and the relationship with K_{far} in the original correction formula is:

$$K_{far} = K_{near} + \Delta K_{far} \tag{2}$$

By finding the corrected ΔK_{far}, K_{far} can be obtained from this relationship.

Let $y = PL_{near} - 20\log_{10}f_c$, $x = \log(d_{3d})$, then the optimization objective is as follows

$$\min \sum_{i=1}^{N} (y_i - K_0 - K_{near}x_i)^2 \tag{3}$$

$$s.t. K_0 >= 0 \tag{4}$$

$$K_{near} >= 0 \tag{5}$$

The near-field corrected Lagrangian function is:

$$L(K_0, K_{near}, \lambda_1, \lambda_2) = \sum_{i=1}^{N} (y_i - K_0 - K_{near}x_i)^2 - \lambda_1 * K_0 - \lambda_2 * K_{near} \tag{6}$$

A. If $K_0 < 0$, then $K_0 = 0$, $\lambda_1 > = 0$, $\lambda_2 = 0$
    We can get the calculation result:

$$K_{near} = \frac{\sum_{i=1}^{N} x_i y_i}{\sum_{i=1}^{N} x_i^2} \tag{7}$$

$$\lambda_1 = \frac{2\sum_{i=1}^{N} x_i y_i \sum_{i=1}^{N} x_i - \sum_{i=1}^{N} 2y_i \sum_{i=1}^{N} x_i^2}{\sum_{i=1}^{N} x_i^2} \tag{8}$$

If $K_{near} < 0$, then discard the answer.
B. If $K_{near} < 0$, then $K_{near} = 0$, $\lambda_1 = 0$, $\lambda_2 > = 0$

We can get the calculation result:

$$K_0 = \frac{\sum_{i=1}^{N} y_i}{N} \tag{9}$$

$$\lambda_2 = \frac{2\sum_{i=1}^{N} y_i \sum_{i=1}^{N} x_i - 2N \sum_{i=1}^{N} x_i y_i}{N} \tag{10}$$

If Knear < 0, then discard the answer.

If there is a valid solution for the above edge solution, the corresponding solution of the minimum optimization function is obtained by comparison. If all of them are invalid, the solution fails, the original value is kept, no processing is performed, and the calibration of the next parameter is entered.

The far-field corrected Lagrangian function is:

$$L(\Delta K_{far}, \lambda_1, \lambda_2) = \sum_{i=1}^{N} (y - \Delta K_{far}x)^2 - \lambda_1 * \Delta K_{far} \tag{11}$$

KKT boundary conditions is:

$$\sum_{i=1}^{N} 2(y - \Delta K_{far}x_i) + \lambda_1 = 0 \tag{12}$$

$$\lambda_1 \Delta K_{far} = 0 \tag{13}$$

$$\lambda_1 \geq 0 \tag{14}$$

$$a \geq 0 \tag{15}$$

If $\lambda_1 = 0$,

$$\Delta K_{far} = \frac{\sum_{i=1}^{N} y_i}{\sum_{i=1}^{N} x_i} \tag{16}$$

If $\Delta K_{far} > 0$, the solution is successful, and exit the solution.

Solve for edge cases if no valid constraint solution is obtained $a = 0$, $\Delta K_{far} = 0$, $\lambda_1 > = 0$, then $\Delta K_{far} = 0$.

Diffraction Coefficient Correction

The calibration of the direct parameter is divided into the calibration stage of K_{near} and K_0, and this stage uses all the data for calibration.

The optimization objective is as follows:

$$\min \sum_{i=1}^{N} (y - \partial_{diff}x)^2 \tag{17}$$

$$s.t. 0 <= \partial_{diff} <= 1 \tag{18}$$

It should be noted, x is diffraction loss value, y is the true diffraction loss value.

The lagrangian function is:

$$L(\alpha_{diff}, \lambda_1, \lambda_2) = \sum_{i=1}^{N} (y - \alpha_{diff}x)^2 - \lambda_1 * \alpha_{diff} + \lambda_2 * (\alpha_{diff} - 1) \tag{19}$$

KKT boundary conditions is:

$$\sum_{i=1}^{N} 2(y - \partial_{diff}x_i) + \lambda_1 - \lambda_2 = 0 \tag{20}$$

$$\lambda_1 \partial_{diff} = 0 \tag{21}$$

$$\lambda_2 \left(\partial_{diff} - 1 \right) = 0 \tag{22}$$

$$\lambda_2 >= 0 \tag{23}$$

$$\lambda_1 >= 0 \tag{24}$$

$$0 <= \partial_{diff} <= 1 \tag{25}$$

If $\lambda_1 = 0, \lambda_2 = 0$:

$$\partial_{diff} = \frac{\sum_{i=1}^{N} y_i}{\sum_{i=1}^{N} x_i} \tag{26}$$

If $0 < \partial_{diff} < 1$, the solution is successful, exit the solution.
Solve for edge cases if no valid constraint solution is obtained a = 0 and a = 1:

A. $\partial_{diff} = 0, \lambda_2 = 0$
B. $\partial_{diff} = 1, \lambda_1 = 0$

If there are valid solutions for the above two, compare the two solutions and then compare to get the solution corresponding to the minimum value. Throws an exception if both solutions are invalid.

Reflection Coefficient Correction
The optimization objective is as follows:

$$\min \sum_{i=1}^{N} (y - \partial_{ref} x)^2 \tag{27}$$

$$s.t. 0 <= \partial_{ref} <= 1 \tag{28}$$

It should be noted, x is reflection loss value, y is the true reflection loss value.
The lagrangian function is:

$$L\left(\alpha_{ref}, \lambda_1, \lambda_2\right) = \sum_{i=1}^{N} (y - \alpha_{ref} x)^2 - \lambda_1 * \alpha_{ref} + \lambda_2 * \left(\alpha_{ref} - 1\right) \tag{29}$$

KKT boundary conditions is:

$$\sum_{i=1}^{N} 2(y - \partial_{ref} x_i) + \lambda_1 - \lambda_2 = 0 \tag{30}$$

$$\lambda_1 \partial_{ref} = 0 \tag{31}$$

$$\lambda_2 \left(\partial_{ref} - 1 \right) = 0 \tag{32}$$

$$\lambda_2 \geq 0 \tag{33}$$

$$\lambda_1 \geq 0 \tag{34}$$

$$0 \leq \partial_{ref} \leq 1 \tag{35}$$

If $\lambda_1 = 0, \lambda_2 = 0$:

$$\partial_{ref} = \frac{\sum_{i=1}^{N} y_i}{\sum_{i=1}^{N} x_i} \tag{36}$$

If $0 < \partial_{ref} < 1$, the solution is successful, exit the solution.
Solve for edge cases if no valid constraint solution is obtained a = 0 and a = 1:

A. $\partial_{ref} = 0, \lambda_2 = 0$
B. $\partial_{ref} = 1, \lambda_1 = 0$.

If there are valid solutions for the above two, compare the two solutions and then compare to get the solution corresponding to the minimum value. Throws an exception if both solutions are invalid.

Correction of Wireless Propagation Model in a Capital City

According to the above-mentioned correction principle of the refined propagation model, an important typical grid of a provincial capital city is selected to correct the propagation model. The area of Grid 86 is about 3.49 km², and there are 434 cells in the current network (expanded by three layers). After data processing and fitting calculation, the typical signal level difference is about 3 dBm, and the regional average RSRP difference is 1.45 dBm. In each level interval, the fit between drive test and simulation is good, and the level interval with the largest proportion (RSRP $\geq$ −93 dBm) differs by 0.33% (Fig. 2).

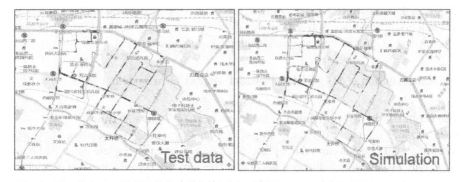

Fig. 2. Comparison of test and simulation (after coefficient correction) GIS rendering results

The proportion of each level interval is shown in the Table 1.

Table 1. Statistical comparative analysis of test and simulation RSRP

Range of RSRP (dBm)	The proportion of test data	The proportion of simulation data	Difference
RSRP – 110	0.00%	0.27%	−0.27%
−110 – RSRP <-105	0.00%	0.18%	−0.18%
−105 – RSRP <-100	0.17%	0.42%	−0.25%
−100 – RSRP < -95	1.38%	1.00%	0.38%
−95 – RSRP <-90	4.01%	2.18%	1.83%
−90 – RSRP <-85	9.19%	5.75%	3.44%
−85 – RSRP <-80	15.13%	8.28%	6.85%
−80 – RSRP <-75	19.59%	16.28%	3.31%
−75 – RSRP <-70	19.41%	27.64%	−8.23%
−70 – RSRP	31.13%	37.99%	−6.86%

3 5G Wireless Network Digital Twin System Construction and Application Research

The wireless propagation model is greatly affected by the geographical environment. The corrected model can accurately reflect the local wireless signal propagation characteristics. Therefore, we can build an accurate 5G wireless network digital twin system through high-precision ray tracing simulation. Therefore, we can build an accurate 5G wireless network digital twin system through high-precision ray tracing simulation, which can achieve accurate prediction and judgment. Using the corrected propagation model, fine-grained and accurate simulation of planar and 3D regions is performed. As a result, a digital twin system including an all-round 5G wireless network both outdoors and indoors is constructed. Based on this digital twin system, weak coverage analysis and building 5G indoor base station planning can be realized.

3.1 Construction of 5G Wireless Network Digital Twin System

The Digital Twin of Two-Dimensional Plane of 5G Wireless Network
Based on the corrected ray tracing propagation model, a refined simulation of the 5G wireless signal plane is carried out plane. Furthermore, it can complete the digital twin of 5G wireless signal two-dimensional plane. This area is a key grid area of a provincial capital city. The configuration parameters of 4 times reflection and 2 times diffraction are used to simulate the D frequency band of 5G NR. After the simulation, the signal coverage strength can be obtained. The average RSRP of this area is −84.15 dBm (Fig. 3).

Fig. 3. 2D planar digital twin results of 5G wireless signals

The Digital Twin of Three-Dimensional Space of 5G Wireless Network

On the basis of regional two-dimensional coverage prediction, a unified building database is constructed according to the building information in the three-dimensional electronic map, which can carry out three-dimensional building simulation. The method adopted in this scheme is to superimpose the antenna gain where the floor is located on the basis of the road loss of the ground floor, so as to realize the RSRP calculation of each grid of each floor height of each building. After that, based on the 3D open source technology

Fig. 4. 3D spatial digital twin results of 5G wireless signals

framework Cesium, 3D GIS rendering is performed. An example of 3D GIS rendering in a provincial capital is shown in the Fig. 4.

3.2 Application Research of 5G Wireless Network Digital Twin System

Weak Coverage Review of 5G Networks

In the planning and construction of 5G network, the digital twin system of 5G wireless network based on simulation plays an important role in application and has great guiding significance. Through the 5G wireless network digital twin system, the coverage of 5G planned sites can be reflected, and the coverage effect of the network can be predicted at the early stage of planning. And it can calculate the weak coverage area-level optimization and adjustment suggestions according to the mathematical modeling of wireless electromagnetic signals and geographical environment (Fig. 5).

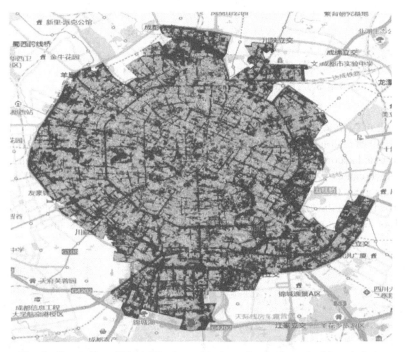

Fig. 5. Simulation of 700 MHz planned base station in a provincial capital city

This area is the core urban area of a provincial capital city. There are 3,742 700 MHz base station cells planned, and the planned area is 164.28 km^2. After the analysis and calculation of the 5G wireless network digital twin network, the coverage rate of this area reached 96.7%, and the specific calculation indicators are shown in the Table 2.

Through the mathematical modeling and calculation analysis of the RSRP value of the grid and the working parameters of the base station cell, the statistics and analysis

Table 2. Metrics for 2D plane digital twin computing

Area name	Average RSRP (dBm)	Number of base stations	Average station spacing (m)	Percentage of RSRP ≥ −88 dbm and SINR ≥ −3 db (%)	Downlink edge RSRP (dBm)	Downlink edge SINR (dB)
Central urban area	−69.51	865	468	96.7	−91.35	21.82

of the weak coverage of the area can be carried out. Further, optimization suggestions for weak coverage areas can be output. According to the calculation and analysis of the 5G wireless network digital twin network, we found 26 weak coverage areas. At the same time, based on the simulation calculation and mathematical analysis of the digital twin network, multi-dimensional information of the weak coverage area can be output, including the center latitude and longitude, whether there is a building block, geographic location, weak coverage area, weak coverage length, main service area, etc. Most importantly, it can give advice on the adjustment of engineering parameters for planning base stations, which can greatly assist in 5G network planning. Some of the results are shown in the Table 3.

Table 3. 5G weak coverage review and adjustment suggestions

Road name	Building shelter	Weak coverage area (m^2)	Weak coverage length (m)	Recommendations for optimization	Main service community
Road name-1	Yes	425	61.3	Adjust azimuth	Cell-1
Road name-2	Yes	275	60	Reduce station spacing	Cell-2
Road name-3	Yes	225	40	Reduce station spacing	Cell-3
Road name-4	Yes	200	29.1	Adjust azimuth	Cell-4
Road name-5	Yes	200	45	Reduce station spacing	Cell-5
Road name-6	Yes	200	32.4	Increase transmission power	Cell-6
Road name-7	Yes	175	25	Reduce station spacing	Cell-7
Road name-8	Yes	175	10	Adjust azimuth	Cell-8

(*continued*)

Table 3. (*continued*)

Road name	Building shelter	Weak coverage area (m^2)	Weak coverage length (m)	Recommendations for optimization	Main service community
Road name-9	No	150	13.6	Adjust azimuth	Cell-9

5G Indoor Base Station Planning

Through the 5G wireless network digital twin system, three-dimensional coverage prediction of buildings can be performed. It can calculate the 5G wireless signal coverage strength RSRP of each grid of each building. Then, the priority of 5G building construction is determined according to the strength of coverage. If the coverage of a building is poor, the construction of 5G indoor base stations will be given priority. Finally, the planning list of 5G indoor base stations for buildings can be output (Table 4).

Table 4. 5G indoor base station planning scheme in a provincial capital city

Building name	Height	Bottom area	Total area	Percentage of RSRP ≥ −114 (dBm)	Average RSRP ≥ −114 (dBm)	Percentage of RSRP ≥ −110 (dBm)	Average RSRP ≥ −110 (dBm)	Coverage level	Primary cell
Building 1	21	1672.3	11706.1	95.1	−89.6	90.1	−88.4	First	Cell-11
Building 2	18	174.8	1048.8	100	−82.3	100	−82.3	Second	Cell-12
Building 3	18	174.2	1045.2	100	−74.7	100	−74.7	Second	Cell-13
Building 4	18	352.7	2116.2	100	−82.5	100	−82.5	Second	Cell-14
Building 5	21	537.8	3764.6	87.1	−99	74.2	−96.8	First	Cell-15
Building 6	6	362.6	725.2	100	−79.7	96.2	−78.5	Second	Cell-16
Building 7	21	548.5	3839.5	57.6	−93.2	57.6	−93.2	First	Cell-17
Building 8	21	549.9	3849.3	66.3	−97	57.2	−94.5	First	Cell-18
Building 9	6	263.5	527	92.6	−90.1	77.8	−86.1	First	Cell-19

4   Conclusion

Through refined propagation model correction and 5G communication system simulation, the digital twin system of 5G wireless network can be accurately constructed. This system can not only visualize the multi-dimensional attributes of 5G wireless signals, but also accurately locate the problems existing in 5G networks, which has important guiding significance for the whole life cycle of 5G network planning, construction, maintenance, and optimization.

References

1. Ashwin, A., Martin, F., Vishal, S.: Digital twin: from concept to practice. J. Manag. Eng. **38**(3) (2022)
2. Marco, P.G., et al.: A digital twin approach for improving estimation accuracy in dynamic thermal rating of transmission lines. Energies **15**(6), 2254 (2022)
3. Xiang, Z.: Simulation design and implementation of precoding technology in 3D MIMO. Beijing University of Posts and Telecommunications (2017). (in Chinese)
4. Elumalai, P.K., Velmurugan, N.: Multiobjective optimization based on self-organizing particle swarm optimization algorithm for massive MIMO 5G wireless network. Int. J. Commun. Syst. **34**(4) (2021)
5. Xiang, Z., et al.: System simulation and coverage analysis of 5G NR communication system based on 700 MHz frequency band. In: Wang, X., Wong, K.-K., Chen, S., Liu, M. (eds.) AICON 2021. LNICSSITE, vol. 396, pp. 125–137. Springer, Cham (2021). https://doi.org/10.1007/978-3-030-90196-7_12

A Hardware-in-the-Loop Microgrid Simulation Method Based on TwinCAT3—Take Black Start as An Example

Haiqi Zhao[1,2](✉), Shufeng Dong[1], Lingchong Liu[1,2], Runzhe Lian[1], Mingyang Ge[1], and Kunjie Tang[1]

[1] College of Electrical Engineering, Zhejiang University, Hangzhou 310027, Zhejiang, China
zhaohaiqi@zju.edu.cn

[2] Polytechnic Institute, Zhejiang University, Hangzhou 310015, Zhejiang, China

Abstract. Due to the flexibility and dispatchability, the microgrid system has attracted more and more attention of scholars with large penetration of new energy. However, as the structure of the microgrid system becomes more and more complex, the existing microgrid simulation methods are outdated. Especially for complex microgrid systems, one second of simulation can take a dozen or even hundreds of times longer. In response to the above problem, a microgrid simulation method based on TwinCAT3 is proposed. The microgrid system is modeled by MATLAB/Simulink, Then, the model is converted into a TwinCAT3 model through the TE1400 component, and downloaded to the industrial computer for simulation. The experiment verifies the accuracy and efficiency of the TwinCAT3-based microgrid simulation method.

Keywords: TwinCAT3 · Microgrid · Hardware-in-the-loop simulation · Industrial computer

1 Introduction

Against the global warming backdrop, where countries are taking strong initiatives to reduce CO_2 emissions, microgrids are gaining more and more attention from researchers because of their flexibility and dispatchability, and their ability to operate in both networked and islanded mode with the larger grid.

With the large-scale access of renewable energy and power electronics, the microgrid is more prone to problems such as voltage fluctuations in actual operation, so the requirements for microgrid simulation are getting higher and higher. At present, academia has introduced hardware-in-the-loop simulation technology to conduct simulation research on microgrids, in order to solve the problems

Supported by the National Natural Science Foundation of China (52077193): Research on Autonomous and Coordinated Operation of Multi-Area Interconnected Integrated Energy Systems.

that traditional digital simulation has large errors in renewable energy simulation and the cost of physical simulation experiments is high. The most used simulation platforms are MATLAB, RT-LAB, RTDS, StarSim, etc. The researchers also use these platforms to model the microgrid, reference [1,2] established a signal-level microgrid hardware-in-the-loop simulation platform, using RTDS to build a microgrid model, and then connected to an external energy management system through I/O ports to develop and test control strategies, but due to the power output capability of the RTDS is relatively weak, so a peripheral digital power amplifier is required to amplify the RTDS signal. Reference [3] models the multi-energy complementary microgrid based on the RT-LAB platform to verify the stability of the control strategy, but its simulation parameters are relatively ideal, so it cannot fully reflect the actual working conditions. Reference [4] uses LabVIEW and an industrial computer to connect to build a microgrid system including photovoltaic cells and charging piles, but it does not consider the delay of data transmission. Reference [5] simulates the centralized energy storage microgrid system based on the Star Sim platform. In addition to the fact that the simulation environment is relatively ideal, the above-mentioned simulation methods all have the problem of long simulation time. It may take several minutes or even dozens of minutes to simulate the microgrid system for one second. Therefore, finding new hardware-in-the-loop simulation tools to improve the simulation speed is an urgent problem to be solved.

TwinCAT3 is a control software based on PC platform and Windows operating system developed by Beckhoff in Germany. It can turn industrial PC or embedded PC into a powerful PLC and motion controller, which can be installed on the production site to control various production equipment in real time. Compared with the traditional PLC, the CPU, memory and memory resources of the Beckhoff industrial computer supported by TwinCAT3 have been improved by orders of magnitude, and it supports multi-core CPU operations, making it possible to centrally control large-scale systems, and the computing speed has been greatly improved, too [6].

In this paper, the microgrid black start is taken as an example, the microgrid simulation model is downloaded to the Beckhoff industrial computer, and the TwinCAT3 software is used to build a complete simulation control system. Through the example, the accuracy and reliability of the method are verified. It is proved that this method can improve the efficiency of the simulation of the scene in the laboratory, and can modify the control more efficiently to improve the usability of the control strategy in the real scene.

2   Construction of Microgrid Equipment Model

The hardware-in-the-loop simulation case presented in this paper is a microgrid which is built for drilling rigs. This microgrid, consisting of 8 gas engines, 2 energy storage batteries and 1 variable load, is a hybrid AC/DC mini-grid with two AC bus voltages of 400V and 600V, the specific equipment parameters are shown in Table 1.

The composition of the microgrid is shown in Fig. 1.

Table 1. Parameters of microgrid scene devices

Device name	Rated power	Number of device
Energy storage	500 kW	2
Gas turbine	330 kW	8

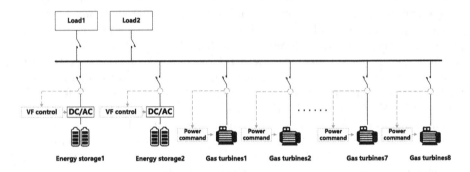

Fig. 1. The composition of the microgrid

2.1 Equipment Modelling

Energy Storage

Energy storage, as an important way of energy storage, has gradually received the attention of the academic community, and some scholars have now focused on the feasibility of using wind storage systems or optical storage systems to complete the grid black start, in addition to the research on the mechanism of energy storage with gas turbines to complete the black start, energy storage with its rapid response, for different application needs, flexible selection of energy storage power and capacity and more in the integrated energy system configuration of the advantages of With the advantages of rapid response, flexible selection of storage power and capacity for different applications and more configurations in integrated energy systems, energy storage is gradually becoming the ideal black-start power source for recovery from microgrid failures.

The energy storage battery in this case is responsible for establishing the required frequency and voltage of the system during the black start phase and smoothing out load fluctuations, for which VF control is used. It is a high-capacity lithium battery, which is converted into an industrial frequency alternating current through an inverter circuit and a filter to the busbar.

The phase-locked loop synchronises the inverter-side output current with the network-side voltage by obtaining the phase of the network-side voltage. The current controller reference value (i_{do} and i_{qo}) is obtained by sampling the three-phase voltage and current on the inverter side and inputting them into the voltage loop after abc/dq conversion. The signal is then fed into the current loop and the output value is converted by dq/abc to send a modulating signal to the modulation module, which generates the drive signal for the three-phase full-bridge inverter circuit.

The circuit structure and control structure of the energy storage is shown in Fig. 2.

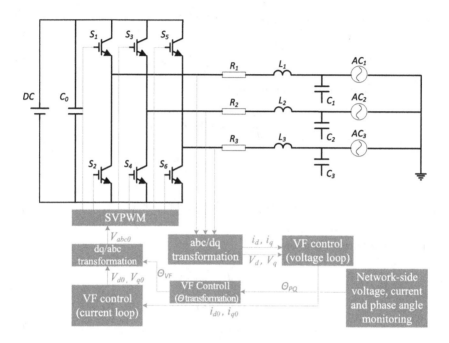

Fig. 2. The circuit structure and control structure of energy storage

Gas Turbine

The gas turbine power generation system consists of a micro gas turbine, a permanent magnet synchronous generator, a rectifier, an inverter and a filter. The micro gas turbine directly drives the built-in high speed generator and the high frequency AC power is converted to industrial frequency AC power for transmission to the load or AC grid through the SPWM modulated rectifier and inverter via VF and PQ control.

In this case, the gas turbines is controlled by giving a power command so that it outputs the power required to meet the load. In this case the gas turbines is numerically modelled to further increase the speed of the simulation. The model is based on the active power command (P_{ref}), power factor (θ), frequency (f) and system voltage (U) as inputs and is based on the function.

Calculating reactive power command:

$$Q_{ref} = P_{ref} \cdot \tan(\theta) \tag{1}$$

Calculating Three-phase active current:

$$i_{a\_d} = \frac{\sqrt{2}P_{ref}}{3U} \cdot sin\,(wt) \tag{2}$$

$$i_{b\_d} = \frac{\sqrt{2}P_{ref}}{3U} \cdot sin\left(wt - \frac{2\pi}{3}\right) \tag{3}$$

$$i_{c\_d} = \frac{\sqrt{2}P_{ref}}{3U} \cdot sin\left(wt - \frac{4\pi}{3}\right) \tag{4}$$

Calculating Three-phase reactive current:

$$i_{a\_q} = \frac{\sqrt{2}Q_{ref}}{3U} \cdot sin\left(wt - \frac{\pi}{2}\right) \tag{5}$$

$$i_{b\_q} = \frac{\sqrt{2}Q_{ref}}{3U} \cdot sin\left(wt - \frac{2\pi}{3} - \frac{\pi}{2}\right) \tag{6}$$

$$i_{c\_q} = \frac{\sqrt{2}Q_{ref}}{3U} \cdot sin\left(wt - \frac{4\pi}{3} - \frac{\pi}{2}\right) \tag{7}$$

synthesizing three-phase currents:

$$i_a = i_{a\_d} + i_{a\_q} \tag{8}$$

$$i_b = i_{b\_d} + i_{b\_q} \tag{9}$$

$$i_c = i_{c\_d} + i_{c\_q} \tag{10}$$

In order to simulate the operating conditions of the gas turbine more closely, the model is set up with five states: stopping, received a start signal but not reach rated speed, reach rated speed but not close, closed and operation, and preparing to stop. With these states, experimenters can better simulate the work of the gas turbine and achieve a more realistic control effect.

Variable Load
In the case, the variable load is equated to a reverse current source, and given the values of the active and reactive loads, the same calculation as in Eq. (1)–(10) is used to obtain the output current values, except that the currents need to be reversed.

The load is connected to the busbar via a transformer and forms a microgrid with the gas turbine and energy storage batteries.

2.2 From MATLAB/Simulink to TwinCAT3

The TE1400 is a TwinCAT3 project plug-in that uses Simulink Coder to generate C/C++ code from models in Simulink, and is further exported as a TcCOM module with input and output, this module can be instantiated in a TwinCAT3 project and the parameters of the module can also be modified. Finally, the module is executed in real time in the TwinCAT3.

With TE1400, users can easily convert a model from Simulink into a Twin-CAT3 model, which is also presented in a visual form in TwinCAT3. After this, closed-loop control can be achieved using a simple language.

The steps for converting the simulink model to TwinCAT3 are [6]:

(1) Changing the path of matlab to the TwinCAT3 installation folder and running *SetupTwinCATTarget.p* in the TE1400 folder to make matlab select the compiler for compiling the simulink model.
(2) Changing the solver to a fixed step size in the settings of the simulink model, and changing the code generation file type to *TwinCAT.tlc*. Then, setting the relevant items and finally click on apply.
(3) Clicking on *build* to start compiling the model and waitting for it to complete.
(4) Creating a new TwinCAT3 project, and importing the converted module entries and assigning them to the PlcTask module to complete the configuration in TwinCAT3.

After successful conversion you can see the visualised model in TwinCAT3 as shown in Fig. 3.

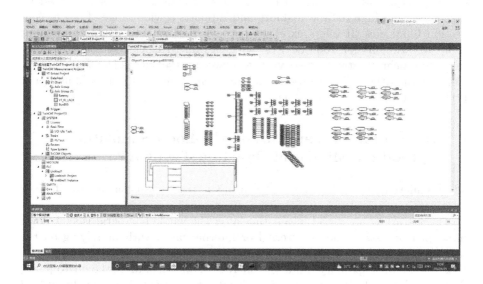

Fig. 3. The visualised model in TwinCAT3

3   Construction of the Simulation Environment

As shown in Fig. 4, the complete simulation environment consists of two parts: the Beckhoff industrial computer, which is used to model the microgrid, and the TwinCAT3 software, which implements the control of the industrial computer model.

The industrial computer used in this case is a Beckhoff control cabinet industrial PC, model C6640-0050, which is equipped with an i7-7700 (4-core, 3.6 GHz) CPU. As an industrial control machine, the C6640-0050 offers the following advantages [7]:

(1) The C6640-0050 complies with the principles of industrial design. The graphic and EtherCAT adapters can already be used directly on the board and all sockets are distributed on the upper side of the machine, allowing the connection cables to be taken directly to the wiring channels and facilitating the installation of other control cabinet equipment next to the machine.
(2) The C6640-0050 is highly scalable. The C6640 series can ensure long-term compatibility with new PC components. The motherboard, processor, memory or hard disk can be upgraded.

In the TwinCAT3, the microgrid model converted from the Simulink model is imported, and a simple code is added to complete the configuration of the complete microgrid control system.

EtherCAT is a real-time industrial Ethernet technology introduced by Beckhoff Automation in 2003. It is also a real-time Ethernet-based industrial fieldbus communication protocol and an international standard. It is characterised by high speed and high data availability and supports a wide range of device connection topologies. It offers excellent network performance. Data throughput of up to 10 kb/ms for 1500 devices, and distributed clock technology ensures that the synchronisation time deviation between these axes is less than $1\,\mu s$ [8]. Through the EtherCAT, the users can implement it on the TwinCAT3 interface, control and monitor the industrial computer.

4   Verification of Hardware-in-the-Loop Simulation Method for Microgrid Based on TwinCAT3

The simulation method based on Twincat3 is tested using a microgrid black start case.

Black start of microgrid means that after the whole microgrid is shut down into full black state due to external or internal fault, without relying on the help of large grid or other microgrid, only by starting the micro power source with black start capability inside the microgrid, and then drive the micro power source without black start capability inside the microgrid, gradually expand the recovery range of the system, and finally achieve the restart of the whole microgrid.

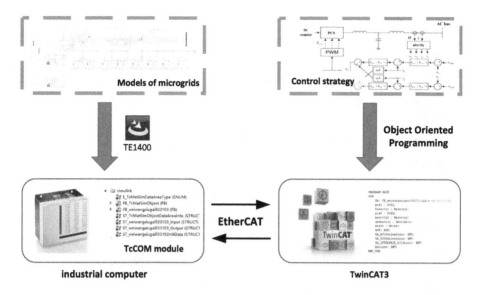

Fig. 4. The complete simulation environment

This section compares the accuracy and speed of simulations using simulink and Twincat3 based simulation methods respectively. Both MATLAB/Simulink and TwinCAT3 run on a computer with the following parameters: 12th Intel(R) Core(TM) i7-7700 CPU, 16.0 GB RAM.

The accuracy of the two simulation methods is first compared. Putting in some load immediately after the system frequency is established will lead to a load power shock to the system. As shown in Fig. 5.

In Fig. 5, curve named *Simulation method* 1 is the load curve measured by simulation method based on TwinCAT3, and the anther curve named *Simulation method* 2 is the load curve measured by simulation method based on MATLAB/Simulink. After comparison, it can be seen that the 2 simulation methods show almost identical simulation results during load input, which shows that on a millisecond time scale, the accuracy of the simulation method based on Twincat3 is similar to that of simulations using Simulink and its accuracy can be verified.

The simulation speed is then compared to the actual time taken by the gas turbine from the start of power output to an output of 200kW, and compared to the simulation time. The result is shown in Table 2.

Table 2. Comparison of simulation times between the two simulation methods

	Simulation method 1	Simulation method 2
Simulation speed	4.07 p.u	1 p.u
Efficiency gains	307%	

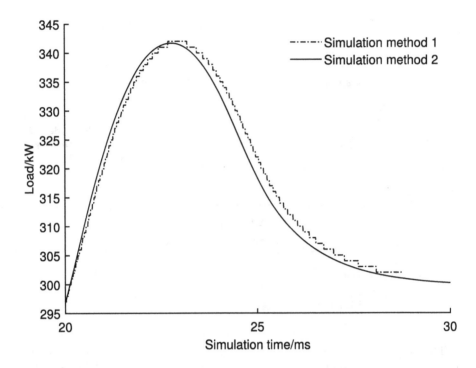

Fig. 5. Load curves measured by two simulation methods

In Table 2, the speed of the simulation is calculated by taking the time elapsed on the real time scale compared to the time elapsed on the simulation time scale and taking the reciprocal, normalised to the simulation method based on MATLAB/Simulink. The larger the value, the faster the simulation method.

The result shows that if the simulation speed of the simulink simulation method is set at 1 p.u, the simulation method based on Twincat3 achieves a speed of 4.07 p.u, an increase of more than 3 times, and the efficiency of the simulation method is verified.

Then the stability of the proposed microgrid simulation method is verified in real engineering simulations, as shown in Fig. 6, which shows the performance of the energy storage battery in the microgrid mentioned in Chap. 2 during a long time simulation. Curve 1 shows the output of the energy storage battery and curve 2 shows the load curve of the platform. It can be seen that the energy storage battery tracks the short-time fluctuations of the load well throughout the simulation time, which shows that the simulation method proposed in this paper is stable not only in the short-time simulation but also in the long-time simulation.

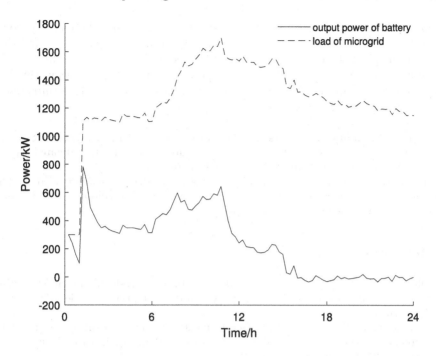

Fig. 6. Energy storage battery output versus load curve

From the above discussion, the following conclusions can be drawn:

(1) The accuracy of the simulation method based on $TwinCat3 + Industrialcomputerof Beckhoff$ is comparable to that of the simulation method based on MATLAB/Simulink.
(2) The microgrid simulation method based on $TwinCat3 + Industrialcomputerof Beckhoff$ is significantly faster than the simulation method based on MATLAB/Simulink, which has a positive effect on the rapid verification of control strategies in practical engineering.
(3) The stability of the method proposed in this paper can also be verified in long time simulations.

5   Conclusion

This paper propose the $TwinCat3 + Industrialcomputerof Beckhoff$ simulation method and improve the speed and realism of microgrid simulations through a series of methods: Using EtherCAT communication, it can support millisecond communication, up to 1,500 devices, simultaneously at a speed and scale of 10 kb/ms. In addition, this paper improves the realism of the module simulation by modelling the operating state of the gas turbine. In general, energy storage, gas turbines and variable loads in microgrid are first modelled in MATLAB/Simulink and then converted into a TwinCAT3 model using the TE1400 component of

TwinCAT3 and downloaded to the industrial computer, and the model is controlled using TwinCAT3. Through the case, the accuracy, the stability and the simulation speed of the simulation method based on Twincat3 is verified.

With comparison, the simulation speed of microgrid strategy simulation can be significantly increased by using the $TwinCat3 + Industrial\ computer of Beckhoff$ simulation method.

Acknowledgements. This work was supported by the National Natural Science Foundation of China (52077193): Research on Autonomous and Coordinated Operation of Multi-Area Interconnected Integrated Energy Systems.

In addition to the authors, we would like to acknowledge the help provided by Bin Nan and others in the Smart Grid Operation and Optimization Laboratory of Zhejiang University during the writing of this paper.

References

1. Oh, S.J., Yoo, C.H., Chung, I.Y., et al.: Hardware-in-the-loop simulation of distributed intelligent energy management system for microgrids. Energies **6**(7), 3263–3283 (2013)
2. Jeon, J., Kim, J., Kim, H., et al.: Development of hardware in the loop simulation system for testing operation and control functions of microgrid. IEEE Trans. Power Electron. **25**(12), 2919–2928 (2010)
3. Du, H., Zhao, Y.W.: Energy management system for microgrid based on RT-LAB. Electron. Des. Eng. **25**(10), 44–47+52 (2017)
4. Zhang, W.L., Yang, Y., Mao, Y.F., Yang, Y.J.: Design of PV micro-grid monitoring system based on LabVIEW. Chin. J. Power Sources **44**(7), 1017–1020+1069 (2020)
5. Yang, S.B.D., Chen, L.L., Li, B.: Research on control strategy of microgrid under centralized energy storage. Distrib. Util. **38**(6), 99–110 (2021)
6. Chen, L.J.: TwinCAT 3.1 from Entry to Mastery, 1st edn. China Machine Press, Beijing (2020)
7. C6640-0050. https://www.beckhoff.com.cn/zh-cn/products/ipc/pcs/c6xxx-control-cabinet-industrial-pcs/c6640-0050.html. Accessed 24 June 2022
8. Park, S.M., Kwon, Y., Choi, J.Y.: Time synchronization between EtherCAT network and external processor. IEEE Commun. Lett. **25**(1), 103–107 (2021)

Day-Ahead Scheduling of PV Consumption in Distribution Networks Based on Demand Response of Multiple Types of Customer-Side Loads

Lei Wang, Lu Zhang$^{(\boxtimes)}$, Jinming Zhang, Wei Tang, and Xiaohui Zhang

China Agricultural University, Beijing 100083, China
754573278@qq.com

Abstract. The crisis of energy depletion is a common problem faced by countries around the world, and renewable energy is gradually gaining popularity among researchers, and the proportion of new energy penetration in distribution networks is increasing. However, the resulting problem of new energy consumption is a major challenge. For a more in-depth analysis of the problem, This paper analyzes the method to solve the voltage crossing limit of distribution network nodes caused by distributed power sources based on the demand response dispatching model with the participation of multiple types of customer-side loads. Firstly, the user-side load models of distributed power generation, residential users, commercial users and electric vehicle users are established respectively, and then by constructing a customer satisfaction evaluation model, taking into account the economy and satisfaction of multiple types of customers participating in demand response, and finally verifies the validity of the model by simulation analysis. The validity of the model is verified by calculate examples simulation analysis.

Keywords: Distribution network · New energy consumption · Demand response

1 Introduction

Distributed energy development is an important means to achieve the goal of "double carbon" [1], with the widespread access of distributed energy in the distribution network, to a certain extent, exacerbated the voltage fluctuations and network losses in some areas of the power grid [2], however the production characteristics of residential and commercial activities make the load electricity consumption law cannot be well matched with the characteristics of photovoltaic power generation, This lead to abandonment of light, the safe and stable operation of the power grid cannot be guaranteed, limiting the local regional new energy consumption capacity, so it is urgent to carry out research on the issue of demand response on the user side of the load to consume distributed photovoltaic.

For the problem of renewable energy consumption in distribution networks, Literature [3] proposed a hierarchical scheduling optimization scheme including microgrids, load aggregators, and residential customers to improve the local consumption rate of distributed power sources such as wind power and solar power. Literature [4] used battery and air conditioner demand response resources to consume and track PV power generation, and established a multi-time scale scheduling optimization scheme under the condition that the economy and the deviation of PV power prediction values have sufficient demand-side resources to consume. In addition, the effect of demand-side response on new energy consumption is also obvious. In the literature [5], particle swarm algorithm and fuzzy control theory are used to optimize the control of residential consumers' electricity load, and in the literature [6], the characteristics of various resources in the micro-energy network are considered and demand response is used as the basis for joint optimal scheduling to further improve the economic operation of the system. In [7], two types of demand response programs, price, and incentive are incorporated into the optimal scheduling of microgrids, and a multi-objective optimal scheduling model that takes into account customer satisfaction, operating costs, and scenery consumption is established. The literature [8] establishes a forward scheduling model considering new energy generation and demand response and solves the model using the alternating direction multiplier method with the total cost minimization as the objective function. Literature [9] established a quadratic programming and dynamic programming-based electric vehicle charging power control model and reduced the network loss by correcting the node voltage. Literature [10] proposes a fast and orderly charging control method to reduce network loss by analyzing the relationship between distribution network feeder network loss, distribution network load rate and load fluctuation variance.

Therefore, in order to promote the consumption of high percentage of new energy and realize the safe and economic operation of distribution network, this paper considers the impact of PV access on the distribution network tide, the difference of electricity consumption satisfaction of different types of users during peak hours and the magnitude of demand response cost based on the distribution network day-ahead dispatching model, solves the problem of voltage fluctuation caused by high percentage of PV access to distribution network, and also calculates the equilibrium point between users' electricity consumption demand and grid demand response cost based on the subordination degree function's user satisfaction evaluation model, which has some practical engineering value.

2   User-Side Load Modeling

In order to effectively study the demand response process of customer-side scheduling, this paper classifies loads into four modeling categories according to the nature of the loads, namely distributed PV model, residential customer load modeling, commercial customer load modeling, and electric vehicle charging load modeling, where the latter three categories are adjustable loads. Since the scenario in this paper is more about the demand response modeling for urban residents' consumption, and the demand response scheduling of industrial loads has been studied in more detail, this paper does not model and analyzes industrial loads.

2.1 Distributed PV Output Load Modeling

The output power per unit time of the PV power generation model is greatly influenced by meteorological factors such as irradiation, temperature, and humidity, among which the influence of irradiation intensity is the most obvious. In the photovoltaic power generation unit in a photovoltaic power plant, the light intensity at a certain moment t obeys Beta distribution, and its probability density function can be obtained as shown in Eq. (1).

$$f(\lambda(t)) = \frac{\Gamma(\alpha_t^S + \beta_t^S)}{\Gamma(\alpha_t^S) + \Gamma(\beta_t^S)} \left(\frac{\lambda(t)}{\lambda_{\max}(t)}\right)^{\alpha_t^S - 1} \left(1 - \frac{\lambda(t)}{\lambda_{\max}(t)}\right)^{\beta_t^S - 1} \tag{1}$$

where $\lambda_{\max}(t)$ is maximum light intensity at time t; $\alpha_t^S - 1$ and $\beta_t^S - 1$ are beta distribution parameters.

At this time the overall conversion efficiency of photoelectric conversion at a fixed level near set the conversion efficiency for η. If the total area of photovoltaic panels in the photovoltaic power station for M, can get its t moment of the total active power output for

$$P_V(t) = \lambda(t)M\eta \tag{2}$$

From this, the distributed PV power output probability density function can be obtained as

$$f(P_V(t)) = \frac{\Gamma(\alpha_t^S + \beta_t^S)}{\Gamma(\alpha_t^S) + \Gamma(\beta_t^S)} \left(\frac{P_V(t)}{P_{\max}(t)}\right)^{\alpha_t^S - 1} \left(1 - \frac{P_V(t)}{P_{\max}(t)}\right)^{\beta_t^S - 1} \tag{3}$$

where $P_{\max}(t)$ is the maximum output of distributed PV at moment t.

2.2 Residential Customer Load Model

In this paper, residential loads are divided into levelizable and non-levelizable loads. A levelizable load can change its operating hours during the dispatch cycle according to dispatch instructions, such as a washing machine. Non-level shiftable loads are also called rigid loads. Rigid loads are those that have no flexible time and do not participate in scheduling, such as lamps.

In this paper, the power of residential appliances is distributed to n small scheduling cycles Δt, and the power distribution of appliances on n scheduling cycles is used to model the residential load, so the residential load at moment t can be expressed by Eq. (4).

$$\begin{cases} P_{t,r}^T = \sum_{i=1}^{m} \sum_{t=t_0}^{T} P_{t,r}^i S_i^T \\ S_i^T = \{0, 1\} \end{cases} \tag{4}$$

where m is the number of residential appliances; T is the total dispatch period (in this paper, a day is divided into 24 parts, n is 24, and the unit dispatch period is 1 h); $P_{t,r}^i$ is the active power of appliance i at moment t; and S_i^T is the operating state matrix of appliance i in period T.

2.3 Commercial User Load Model

Commercial users mainly include shopping malls office buildings and hotels etc. Commercial users can only increase or decrease the demand for electricity because it is difficult to produce a load shifting effect due to the relatively fixed demand for electricity and load time. Therefore, a load of commercial customers $P_{t,b}$ at time t can be expressed in Eq. (5).

$$P_{t,b} = P_b + \alpha \cdot \Delta p_{b,t}, \quad -1 \le \alpha \le 1 \tag{5}$$

where is the commercial base load, $\Delta p_{b,t}$ is the amount of commercial load that can be increased or decreased, and α is the adjustment factor.

2.4 Electric Vehicle Charging Load Model

With the gradual increase of electric vehicle users, the research on electric vehicle charging load participation demand response is also increasing day by day. In this paper, assuming that the car habits of electric vehicle users are consistent with those of ordinary fuel vehicle users, we can obtain that their daily mileage obeys a log-normal distribution, that is $S \sim Log - N\left(\mu_s, \sigma_s^2\right)$; Its probability density function is shown in Eq. (6).

$$f_t(x) = \begin{cases} \frac{1}{\sigma_s\sqrt{2\pi}} \exp\left(-\frac{(x-\mu_S)^2}{2\sigma_S^2}\right) & \mu_S - 12 < x < 24 \\ \frac{1}{\sigma_s\sqrt{2\pi}} \exp\left(-\frac{(x-\mu_S)^2}{2\sigma_S^2}\right) & 0 < x < \mu_S - 12 \end{cases} \tag{6}$$

The electric vehicle return moment t_0 obeys the normal distribution, that is $t_0 \sim N\left(\mu_t, \sigma_t^2\right)$, Its probability density function is shown in Eq. (7).

$$f(x) = \begin{cases} \frac{1}{\sigma_s\sqrt{2\pi}} \exp\left(-\frac{(x-\mu_S)^2}{2\sigma_S^2}\right) & \mu_S - 12 < x < 24 \\ \frac{1}{\sigma_s\sqrt{2\pi}} \exp\left(-\frac{(x-\mu_S)^2}{2\sigma_S^2}\right) & 0 < x < \mu_S - 12 \end{cases} \tag{7}$$

In this paper, we use Monte Carlo-based sampling to calculate the required charging capacity of electric vehicles in the region, as shown in Fig. 1.

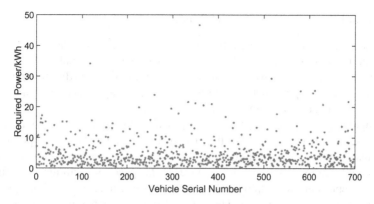

Fig. 1. Intra-regional electric vehicle charging based on Monte Carlo sampling

3   Day-Ahead Dispatch Model of a Distribution Network Based on Customer-Side Demand Response

In this paper, the objective function of the day-ahead load dispatch is constructed in terms of grid security, economic operation and customer satisfaction, which is the concerns of grid companies.

3.1   Demand Response Incentive Costs

The incentive cost required for demand response scheduling of the three types of adjustable loads in the paper can be expressed as

$$f_{\cos t} = \sum_{t=1}^{T} c_1 \left| \Delta p_t^{resdient} \right| + c_2 \left| \Delta p_t^{commerce} \right| + c_3 \left| \Delta p_t^{E-V} \right| \tag{8}$$

where c_1, c_2, c_3 are the dispatch costs per unit of load for the three types of loads; $\Delta p_t^{resident}$, $\Delta p_t^{commerce}$ and Δp_t^{E-V} are the residential load, commercial load and electric vehicle charging load at time t, respectively.

3.2   Customer Satisfaction Assessment and Demand Response Balance Point Modeling

The load of the distribution network is characterized by time-series changes, and the system operation state is bound to be more complicated if a high proportion of distributed power sources and reactive power compensation devices are connected. In order to ensure that the voltage of each node of the distribution network is within a reasonable range, the node voltage target function is set as

$$U_{\min}^* \le U_i^* \le U_{\max}^* \tag{9}$$

where U_i^* is the per-unit value of node voltage.

3.3 Demand Response Day Dispatch Model

In order to avoid the load shifting after demand response to affect the customers' electricity experience, this paper introduces electricity satisfaction q_t^s indicators to evaluate the satisfaction of customers participating in demand response load dispatching, because the electricity consumption habits of different customer types vary from time to time, and customer satisfaction changes with different time periods.

In implementing the demand response program, the grid company will not only consider the incentive cost but also the customer satisfaction, so it is hoped to find a compromise solution that takes into account the safe and economic operation of the grid and does not harm the customer's experience. In this paper, we introduce an affiliation function to represent the coupling relationship between load dispatching ratio μ and customer satisfaction q_t^s at time t. Thus, we calculate that when μ corresponds to the equilibrium point of demand response, the customer satisfaction is the highest and the incentive cost of demand response is the lowest. In practice, the customer's satisfaction is only affected when the load is moved out, so the model only evaluates the satisfaction when the load is moved out.

As shown in the Fig. 2 below, the degree of affiliation is defined as the satisfaction of demand response users for each time period. Define the load dispatch ratio as the ratio of load shifted out to the current load at that moment. At moment t, when the load dispatch ratio x is between $0 < x < a$, as the load dispatching ratio increases, the incentive received by users will also increase, so user satisfaction will rise, and when it rises to λ, it reaches a stable value; when the load dispatch ratio is between $a < x < b$, the degree of affiliation basically remains the same, and remains at λ. The user no longer increases his satisfaction because of the increase in incentive; when the load dispatch ratio $x > b$, the affiliation decreases because the scheduling volume has touched the user's important load at this time, and the user's satisfaction then decreases.

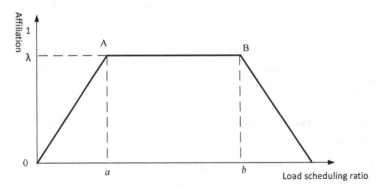

Fig. 2. Affiliation function curve

Assuming that user satisfaction rises and falls at the same rate, the functional relationship between the two is established as shown in Eq. (10).

$$q_t^s(x) = \begin{cases} \frac{\lambda}{a}x, 0 < x \le a \\ \lambda, a < x \le b \\ -\frac{\lambda}{a}x + \frac{(a+b)\lambda}{a}, b < x \end{cases} \tag{10}$$

From the analysis, it can be concluded that the demand response equilibrium point is between points A and B.

3.4 Binding Conditions

(1) Voltage constraint

$$U_{i,t}^2 = U_{i,t}^2 - 2(r_{ij}P_{ij} + x_{ij}Q_{ij,t}) + (r_{ij}^2 + x_{ij}^2)I_{ij,t}^2 \tag{11}$$

(2) Ohm's law constraint

$$p_{jin,t} = \sum_{k \in \sigma} P_{jk,t} - \sum_{l \in \eta} \left(P_{il,t} - I_{ij}^2 r_{ij}\right) \tag{12}$$

$$q_{jin,t} = \sum_{k \in \sigma} Q_{jk,t} - \sum_{l \in \eta} \left(Q_{il,t} - I_{ij}^2 x_{ij}\right) \tag{13}$$

(3) Current constraint

$$I_{ij,t}^2 = \frac{P_{ij,t}^2 + Q_{ij,t}^2}{U_{i,t}^2} \tag{14}$$

(4) Reactive power compensation constraint

Connected to the static reactive power compensation device (SVC), its reactive power compensation amount is continuously adjustable, so the SVC operation constraint is

$$Q_{i,min}^{svc} \le Q_{i,t}^{svc} \le Q_{i,max}^{svc} \tag{15}$$

where i, j are the number of the node; $U_{i,t}$ is voltage at node i at moment t; $p_{jin,t}, q_{jin,t}$ are injected active and reactive power at node i at moment t; σ, η are incoming and outgoing branch collections; $r_{ij} + x_{ij}$ is the impedance of branch i,j; $P_{ij,t}, Q_{ij,t}$ are the active and reactive power at the head end of branch i, j respectively Rate; $Q_{i,t}^{svc}$ is amount of reactive power compensation at moment t.

3.5 Demand Response Day Dispatch Model

The customer-side load demand response dissipates a high percentage of distributed PV by scheduling residential, commercial, and electric vehicle loads for different periods. The scheduling model in this paper is shown in Eq. (16), where only the load shifting in, shifting out and increasing, decreasing are considered, and the total load shifting in, shifting out increasing, decreasing in a scheduling cycle is zero, and the change in load does not exceed 30% of the load at that moment.

$$
\begin{cases}
P_{after} = \sum_{t=1}^{T} \left(P_t^{resident} - \Delta p_t^{resident} \right) + \left(P_t^{commerce} - \Delta p_t^{commerce} \right) + \left(P_t^{E-V} - \Delta p_t^{E-V} \right) \\
\Delta p_t^{resident} + \Delta p_t^{commerce} + \Delta p_t^{E-V} = 0 \\
\left| \Delta p_t^{resident} \right| \leq 0.3 P_t^{resident} \\
\left| \Delta p_t^{commerce} \right| \leq 0.3 P_t^{commerce} \\
\left| \Delta p_t^{E-V} \right| \leq 0.3 P_t^{E-V}
\end{cases}
$$

(16)

where P_{after} is the load size after scheduling; $P_t^{resident}$, $P_t^{commerce}$ and P_t^{E-V} are the residential load, commercial load and electric vehicle charging load at moment t;

3.6 Model Solving Method

The distribution network day-ahead economic dispatch model developed above is a mixed integer linear programming problem that can be solved in MATLAB using the commercial solver GUROBI. The modeling and solving process is shown in Fig. 3.

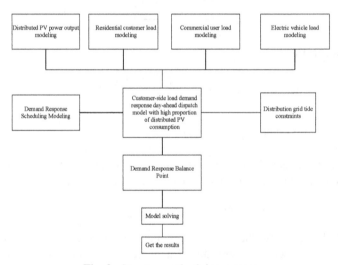

Fig. 3. Structure and solving process

4 Algorithm Simulation

4.1 Algorithm Parameters

To verify the effectiveness of the proposed model and algorithm in solving the dispatching and consumption of distributed PV by customer-side loads before the demand response day, the IEEE 33-node distribution network system is simulated with a voltage level of 12.66 kV and a reference power of 10 MVA, Ignore reactance value and the topology diagram is shown in Fig. 4. The upper and lower voltage limits are 1.05 and 0.95, respectively, and the dispatch period is T = 24 h, with a unit dispatch time of 1h. PV output is modeled using typical daily parameter loads.

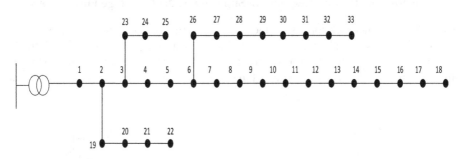

Fig. 4. IEEE 33-bus system

To simplify the calculation, this example considers connecting adjustable loads at nodes 3, 19, 23, and 26, and connecting photovoltaics at node 5, with a photovoltaic penetration rate of 75%. Moreover, the distribution of various types of loads in different areas of the distribution network may be inconsistent, so the proportion of residential users, commercial users, and electric vehicle charging loads at each node is set as shown in Table 1.

Table 1. The load ratio of each node

Node serial number	The proportion of various loads
3	Resident load: commercial load: New energy taxi charging load = 1:0:0
19	Resident load: commercial load: New energy taxi load = 0:1:0
23	Resident load: commercial load: New energy taxi load = 0:0:1
26	Resident load: commercial load: New energy taxi load = 1:1:1

The dispatch cost per kW for residential load, commercial load, and electric vehicle load is $C_1:C_2:C_3 = 0.2:0.5:0.3$. The sensitive hours of electricity consumption for residential customer load are 19:00–21:00, for commercial customer load are 11:00–12:00 and 19:00–20:00, and for New energy taxi charging are 19:00–23:00. For the customer's sensitive time period, it can be considered that the scheduling load that will make the user satisfaction decreases, so the a = b = 0 of the user satisfaction evaluation model during the sensitive time period.

In this section, the simulation of distribution network day-ahead dispatch operation is carried out according to the set objective function for three operation scenarios to analyze their operation.

Scenario1: Demand response without considering demand response.

Scenario2: Demand response day-ahead scheduling with incentive cost consideration.

Scenario3: Demand response day-ahead scheduling considering user satisfaction.

4.2 The Role of Demand Response in Suppressing Node Voltage Fluctuations

The distribution network load is characterized by time-series changes, and the system operation state is bound to be more complicated if a high proportion of distributed power sources and reactive power compensation devices are connected. The node voltage distribution of scenario one, two and three are shown in the Fig. 5 when the PV is generating at noon.

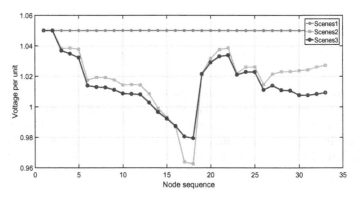

Fig. 5. Node voltage values for three scenarios at noon

It can be found that regardless of the consideration of incentive cost or user satisfaction. Scenarios II and III both solve the voltage crossing limit problem generated by distributed PV access very well. To comprehensively analyze the voltage fluctuations of all nodes of the distribution network throughout the day during demand response, the calculated voltages of the 33 nodes of Scenario III over a 24-h period are represented in Fig. 6.

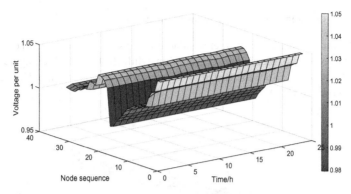

Fig. 6. Voltage distribution of all nodes in Scenario 3 throughout the day

As can be seen from the figure, the demand response day-ahead scheduling ensures that the voltages of all nodes are within the allowed range for the whole time.

4.3 Analysis of Incentive Cost Results for Scenario 2

The first analysis to reach the minimum demand response incentive cost can consider the extreme case of scheduling only the residential load with the minimum incentive cost, and the scheduling cost is ¥3941.5, and the scheduling results are shown in Fig. 7.

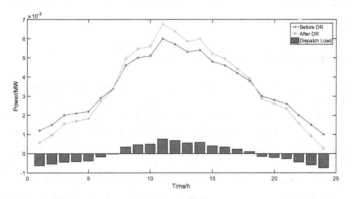

Fig. 7. Demand Response Scheduling Results for Scenario 2 with Comparison

In the above figure, it can be found that such dispatching scheme does not consider the customer satisfaction of residential loads during the electricity sensitive hours, especially during the electricity sensitive hours of 19:00–21:00 for residential customer loads, and still moves the loads outward, which is obviously difficult to implement in the actual situation.

4.4 Analysis of User Satisfaction Results for Scenario 3

Scenario 3 mainly considers customer satisfaction and satisfies the demand of customers during sensitive hours of electricity consumption, while the load dispatching during other hours is sufficient to satisfy the distribution network tide constraints. The scheduling results are shown in Fig. 8.

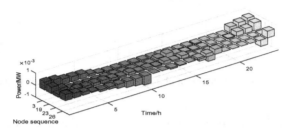

Fig. 8. Scenario 3 Demand Response Load Scheduling Results

From the figure, it can be found that the load is not dispatched during the sensitive hours of 19:00–21:00 for residential customer loads, 11:00–12:00 and 19:00–20:00 for commercial customer loads, and 19:00–23:00 for New energy taxi charging. This type of dispatching scheme does not perform load dispatching, so that customers are not affected by demand response during the sensitive hours of electricity consumption. Thus achieving a high level of satisfaction in electricity consumption, but with incentive costs of ¥8312.4.

4.5 Demand Response Balance Point Analysis

Customers participating in demand response generally have a certain cooperative relationship with the grid and are equipped with high-density collection devices such as smart meters, and the parameters for determining their customer characteristics can be obtained based on user research and analysis of collection device measurement data. In this paper, it is considered that the parameters a and b in the customer satisfaction assessment model can be obtained directly through customer research, and then the balance point of demand response for each type of customer can be found. Taking a user load as an example, according to the demand response equilibrium point model in this paper, it is known that a = 0.3 and b = 0.7 in a certain demand response. The scheduling load, incentive cost and user satisfaction curve are shown in Fig. 9.

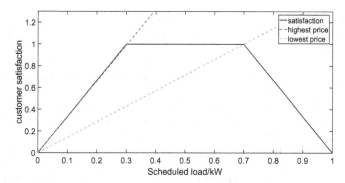

Fig. 9. Satisfaction and schedulable load

As can be seen from the figure, the equilibrium point price of demand response is between 1.42 and 3.3.

5 Conclusion

The contributions of this paper are as follows:

(1) This paper solves the nodal voltage crossing limit problem caused at high percentage of PV penetration into the distribution network by establishing a day-ahead dispatching model based on demand response.
(2) Considering the incentive cost required for different types of customers to respond to demand response, this paper constructs a customer satisfaction evaluation model based on the subordination function, and analyzes the equilibrium point between customer satisfaction and demand response cost by optimizing the scheduling scheme, which provides a feasible strategy for the implementation of demand response in the power grid.

References

1. Zhigang, Z., Chongqing, K.: Challenges and prospects of constructing new power system under carbon neutrality target. China Elect. Eng. Report **42**(08), 2806–2819 (2022). https://doi.org/10.13334/j.0258-8013.pcsee.220467
2. Hamza-Ebtsam, A., Sedhom-Bishoy, E., Badran-Ebrahim, A.: Impact and assessment of the overvoltage mitigation methods in low-voltage distribution networks with excessive penetration of PV systems: a review. Int. Trans. Elect. Energy Syst. **31**(12), 1–16 (2021)
3. Ying, C., Jianfeng, L., Jinxia, G.: Wind-solar consumption scheduling optimization model for load aggregators. Renew. Energy **36**(4), 563–567 (2018)
4. Yang, S., Jingye, Z., Lei, W., et al.: Research on multi-time scale scheduling of photovoltaic consumption considering forecast deviation. Power Eng. Technol. **37**(1), 58–64 (2018)
5. Ruelens, F., Claessens, B., Quaiyum, S., et al.: Reinforcement learning applied to an electric water heater : form theory to practice. IEEE Trans. Smart Grid. 1 (2015)

6. Liang, S., Jian, S., Cheng, Z., et al.: Day-ahead scheduling and quantitative evaluation of micro-energy network considering demand-side response. Chinese J. Solar Energy **42**(9), 461–469 (2021)
7. Tian Biyuan, X., Haiqi, Z.X., et al.: Day-ahead optimal dispatch method for microgrid considering electricity price incentive demand response. Elect. Demand Side Manage. **22**(6), 45–50 (2020)
8. Tang, C., Zhang, F., Zhang, N., et al.: Day-ahead economic dispatch of power systems considering randomness and demand response of renewable energy. Autom. Electric Power Syst. **43**(15), 18–25+63 (2019)
9. Clement-Nyns, K., Haesen, E., Driesen, J.: The impact of charging plug-in hybrid electric vehicles on a residential distribution grid. IEEE Trans. Power Syst. **25**(1), 371–380 (2010)
10. Sortomme, E., Hindi, M.M., MacPherson, S.D.J., et al.: Coordinated charging of plug-in hybrid electric vehicle to minimize distribution system losses. IEEE Trans. Smart Grid. **2**(1), 186–193 (2011)

A Simulink-Based Control Method for Energy Storage Assisted Black-Start

Shuang Huang[1]([✉]), Runzhe Lian[2], and Haiqi Zhao[2]

[1] Wuhan Second Ship Design and Research Institute, Wuhan 430205, China
huangshuang0709@126.com
[2] Zhejiang University, Hangzhou 310000, China

Abstract. To improve the black start capability of microgrids, this paper proposes a control strategy of energy storage assistance. First, it explores the advantages and feasibility of energy storage devices in a black start. Then, it figures out a method to realize the establishment and maintenance of both voltage and frequency of a microgrid system through VF (voltage and frequency) control. Second, this paper puts forward a control strategy of energy storage assisted black start. Specifically, with the energy storage battery as the black start power source, after the systecy3m self-check, the battery automatically outputs power to the system and establishes the voltage and frequency through VF control. Then, according to the actual loads, the gas turbines are connected to the system after the system stabilizes. Finally, based on the Simulink platform, a microgrid with energy storages, combustion turbines and loads is built to verify the feasibility and effectiveness of the energy storage assisted black start strategy. Consequently, the simulation results prove that the strategy can significantly improve the black start speed and enhance the stability of the microgrid system.

Keywords: Black-start · Energy Storage · VF control

1 Introduction

With the construction of new power systems, the reliability and stability of power grids are challenged by the continuous access of new energy devices such as photovoltaic and wind power. Especially for small-scale microgrids, it is easy to cause power outages after some natural disasters and human misoperation. After a system is down and stopped, black start control is needed to revive the whole system. Black start means that a whole system is not supported by other systems, and generation equipment with self-start capability in the local system drives the units without self-starting capability through transmission lines, thus the whole system is restored gradually. In a new microgrid, the self-starting capability of both traditional gas turbines and new energy generation equipment such as photovoltaic can not meet the black start requirements. Various types of energy storage devices are ideal for black start power supply because of their good dynamic performance and stable power output capability [1, 2].

This paper firstly analyzes the black start capability of energy storage, and the problem of the control method in the process of microgrid black start. Then this paper designs

W. Fan et al. (Eds.): AsiaSim 2022, CCIS 1713, pp. 225–234, 2022.
https://doi.org/10.1007/978-981-19-9195-0_19

a control strategy of storage assisted microgrid black start. Finally, this paper builds a simulation model of microgrid composed of gas turbines, energy storage and loads based on MATLAB/Simulink platform to verify the effectiveness and feasibility of the proposed black start strategy.

2   Architecture of Energy Storage Assisted Black-Start

2.1   Microgrid System Structure

According to a small microgrid system of an actual project, this paper designs a 400-600 V two voltage levels low voltage microgrid system, as shown in Fig. 1. The microgrid system consists of eight 330 kW gas turbines, two 500 kW energy storage sources and one variable load. The gas turbines and energy storage batteries are connected to 400 V bus, and the load is connected to 600 V bus.

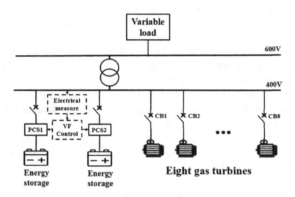

Fig. 1. Microgrid system architecture diagram

330 kW gas turbines: conventional power generation equipment, used as a power output during the normal operation phase of the microgrid, with the three-phase AC output connected directly to the 400 V bus.

500 kW energy storage device: Li-ion battery is selected as the energy storage battery, including battery pack, energy inverter and PQ-VF control module, etc. The energy storage battery can switch between PQ control and VF control modes according to the actual demand, and the control command is issued by the control system. The three-phase AC output of the energy storage power supply is connected to the 400 V bus via a transformer.

Variable load: consists of a 150 kW fixed load and a variable load. The load is connected to a 600 V bus, and the two voltage levels of the bus are connected by a transformer.

2.2   Analysis of Black Start Capability of Energy Storage Devices

In a microgrid system, the selection of black start power is the key to success. Compared to traditional black start sources, energy storages have the following advantages [3–5]:

(1) The VF control of storage inverters can quickly establish and maintain the voltage and frequency in a microgrid;

(2) They have sufficient capacity and good dynamic performance, which can quickly track the change of loads and withstand the power shock when the non-black start power supply starts;

(3) In addition to black-start power, they also support other functions such as peak-shaving and valley-filling.

Therefore, the energy storage system is chosen as the black start power source in this paper.

2.3 Energy Storage Assisted Black Start Strategy

In traditional large grids, black start control is often manually operated according to established procedures; however, in microgrids, manual operation is difficult to adapt to the actual situation considering their small system size and a large number of future applications. Therefore, an automatic microgrid black start strategy needs to be designed, for which energy storage assisted black start strategy is proposed in this paper [6].

The flow of the energy storage assisted black start strategy is as follows.

1) System self-inspection. To avoid the phenomenon of failure shutdown due to insufficient capacity of energy storage batteries and large loads in the early stage of a black start of microgrid, it is necessary to check each piece of equipment in the system. Self-check includes battery remaining capacity check, removal of all loads, gas turbine closing condition check, etc. To ensure that the system does not have problems such as insufficient energy storage power.

2) Start energy storage devices. The VF control of energy storage devices establishes the voltage and frequency of microgrid systems. The rated power of energy storage devices is required to meet the no-load loss of the AC bus of the microgrid, while its output can withstand the power shock generated by the non-black start power in the system during start-up.

3) Putting in loads. To quickly restore power supply to important loads, put in part of the important loads under the premise of considering the rated capacity of energy storage devices; during the input process, ensure that the system voltage and frequency fluctuate within the rated range to avoid destabilization of the whole system.

4) Access to non-black start sources such as gas turbines. At this time, the microgrid system already has a certain capacity, and it is necessary to gradually start the non-black start sources after a comprehensive assessment of the capacity matching in the microgrid system, to expand the system's power generation capacity.

5) Increase loads. With the connection of each power generation unit, the system power supply capacity increases, and loads can be increased gradually according to load characteristics and system power generation capacity.

3 Microgrid System Modeling and Control Implementation

3.1 Microgrid Simulink Simulation Model Building

To verify the energy storage assisted black start strategy, this paper builds a microgrid simulation model using Simulink as shown in Fig. 2 below, based on the microgrid system architecture in Sect. 2.1.

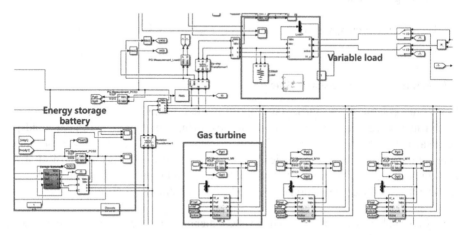

Fig. 2. Microgrid simulation model

Energy Storage Battery. The simulation model of the energy storage battery is shown in Fig. 3, which is mainly composed of dc power supply, SOC (state of charge) calculation module, inverter, LC filter and PQ-VF control module. Energy storage batteries input active power P, reactive power Q and PQ-VF control signal, and output three-phase AC power, battery SOC and remaining capacity.

VF Control: VF control establishes the voltage and frequency of the microgrid system during black start, and adjusts the amplitude and frequency of the voltage according to the set values [7].

The principle of VF control is shown in Fig. 4, in which the DC power is inverted to AC by the SPWM-controlled three-phase inverter, and then connected to an AC side bus through the LC three-phase filter circuit. The SPWM generation module mainly includes phase-locked loop and park conversion, current loop control, and voltage loop control. First, an input voltage is calculated by the phase-locked loop to get the phase angle, and the voltage and current are converted to components in dq0 coordinates. Then by voltage loop control, the set dq component reference voltage is input, after which the reference current values i_{dref} and i_{qref} of the dq component are obtained by PI control. After that, since the output reference current has a dq axis coupling, and at the same time is mainly controlled by the dq axis voltage, PI control decoupling calculation is needed, and its decoupling calculation equation is as follows.

$$\begin{cases} u_{dref} = u_d + \left(k_p + k_i/s\right)\left(i_{dref} - i_d\right) - \omega L i_q \\ u_{qref} = u_q + \left(k_p + k_i/s\right)\left(i_{qref} - i_q\right) - \omega L i_d \end{cases} \tag{1}$$

transformation of
coordinates   voltage loop

current loop

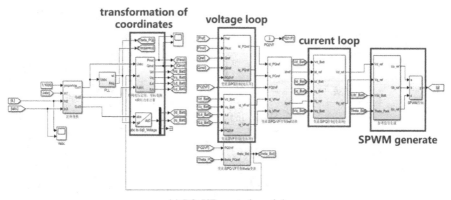

SPWM generate

(a) PQ-VF control module

Energy storage
battery   Energy storage
converter

LC filter

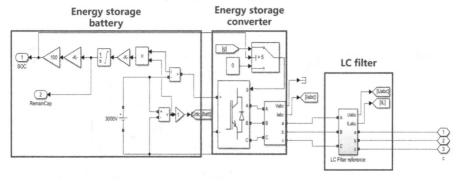

(b) Energy storage battery and converter module

Fig. 3. Simulation model of energy storage battery

After getting the voltage under dq coordinate system, finally, the three-phase reference voltage under abc coordinate system is obtained by park inverter conversion, and the switching signal controlling the three-phase bridge inverter is obtained after SPWM modulation.

PQ Control: The role of PQ control is to regulate the active and reactive power output of the energy storage battery to track the corresponding set values in real time. When the frequency and voltage of the microgrid are supported by other power sources (e.g. gas turbine), the energy storage battery does not require frequency and voltage control, and PQ control is used to achieve specific power output. [8].

The PQ control schematic is shown in Fig. 5, where the left side of the inverter is the DC input, which is inverted to AC after the three-phase inverter, and then connected to the AC side through the LC three-phase filter circuit. i_{Labc} and u_{abc} are the filtered three-phase current and the grid three-phase voltage, respectively.

The principle of PQ control is similar to that of VF control, which mainly includes phase locked loop and park conversion, current loop control and power control. Compared with VF control the difference is in the voltage loop control replaced by power

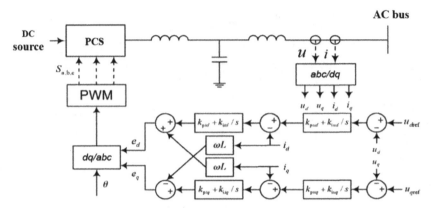

Fig. 4. VF control schematic

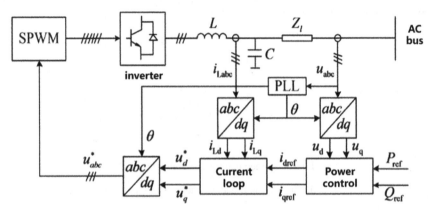

Fig. 5. PQ control schematic

control, the input is set active power and reactive power, after which the reference current values idref and iqref of the dq component are obtained by PI control.

Gas Turbine. The gas turbine simulation model inputs the active power P, bus voltage and enables control signal. Then, it calculates the current output value based on the active power and bus voltage, and outputs AC power by the controlled AC current source, and is connected to the 400 V bus after the LC filter circuit and circuit breaker.

At the same time, to better simulate the state change during the black start of the gas turbine, a gas turbine state control module is added to realize the setting of the gas turbine start time, rated speed time and shutdown time. The control module inputs superior control signals and gas turbine status, and outputs enable and break signals to the gas turbine module. In this module, the gas turbine status is divided into shutdown, start-up acceleration, reaching rated speed pending shutdown, shutdown operation, shutdown hot standby, and shutdown at reduced speed.

3.2 Black Start Control Strategy Implementation

Design the black start control module according to the energy storage assisted black start control strategy in Sect. 2.3, which is as follows.

1) System self-check. The remaining capacity of storage batteries is checked, and the minimum required capacity of storage batteries for a black start is calculated according to the start-up and closing time of gas turbines. In this paper, the rated capacity of storage batteries is set to 552,000 kW·h, and the remaining capacity of the storage batteries before a black start is greater than 30%. Check the working status of loads and gas turbines
2) Start energy storage devices. Energy storage devices are switched to VF control mode, and the energy storage device start signal is issued.
3) Input loads. The microgrid simulation model in this paper sets a fixed load of 300 kW.
4) Access to gas turbines. After the voltage and frequency of the microgrid system are established, the first gas turbine start signal is issued, and the time to start the gas turbine until it reaches the rated speed and closes is set to 5 min, with an output power of 300 kW.
5) Normal operation of the system. After the first gas turbine is connected, the power has been balanced with the fixed load, and the remaining gas turbines are subsequently connected to the microgrid according to the actual demand, while the energy storage devices switch the charging and discharging mode according to the source load.

4 Simulation Results Analysis

4.1 Simulink Simulation Results Analysis

To verify the feasibility and effectiveness of the energy storage assisted black start strategy proposed in this paper, it is tested on the Simulink simulation model of the microgrid shown in Fig. 3. It is assumed that eight gas turbines in the system go into a full black state due to failure or other reasons, the fixed load is 300 kW, and energy storage batteries are used as the black start sources. The parameters are set as follows: the rated power of every energy storage battery is 500 kW, the rated capacity is 552000 kWh, the initial SOC value is 0.5; the rated power of the gas turbine is 330 kW, the rated voltage is 400 V, the frequency is 50 Hz, and the starting and closing time is 5 min; the voltage loop PI parameters for VF control are Kpud = 10, Kiud = 5000, Kpuq = 10 The current loop PI parameters are Kpud = 5, Kiud = 2.5, Kpuq = 5, Kiuq = 2.5.

The simulation duration is set to 5s, and the main observations during the simulation are the bus voltage, frequency, two energy storage devices and the power of the load. The simulation results are shown in Fig. 6.

Simulation result analysis: The energy storage assisted black start can quickly establish the voltage and frequency of the microgrid system, and it can complete the black start task about 0.02 s after the black start starts. It is about 10 times faster than the response of conventional diesel generators as black start sources [9]. From the bus voltage and frequency, it can be seen that there is a certain fluctuation at the beginning of the black start, but it will stabilize immediately with less fluctuation; and the single storage output

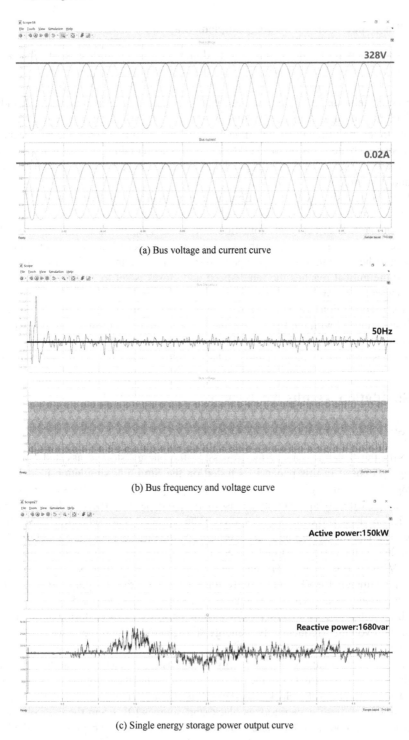

(a) Bus voltage and current curve

(b) Bus frequency and voltage curve

(c) Single energy storage power output curve

Fig. 6. Simulation results of energy storage assisted black start

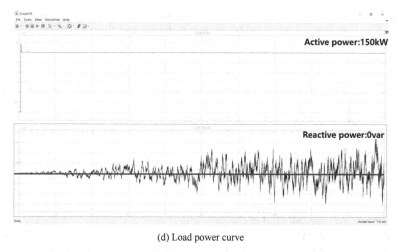

(d) Load power curve

Fig. 6. (*continued*)

power is stabilized at 150 kW after a small spike at the beginning, and the load's output power is similar and can be stabilized at 300 kW quickly. It can meet the black start requirements of microgrid system.

4.2 Practical Engineering Applications

In practical engineering applications, the above microgrid simulation model and black-start control strategy have been fully tested. The strategy can complete the system black start task quickly and stably. No unexpected situation occurs in the one-year engineering application, which shows that the strategy has high reliability.

5 Conclusion

This paper proposes an energy storage assisted black start strategy. By analyzing the characteristics of microgrid and the advantages of energy storage devices as a black start power source, a black start scheme with faster response speed and better stability is proposed. Based on the Simulink platform, the microgrid model and black start control module are built, and the simulation implementation of the energy storage assisted black start strategy is completed. The simulation analysis results show that the strategy can realize the black start of microgrid quickly and stably, and the stability of the system has been greatly improved.

References

1. Li, L., Wang, Y., Zhao, Y.: The energy storage system control research based on black-start. In: 2014 China International Conference on Electricity Distribution (CICED), pp. 1472–1476. IEEE, Shenzhen (2014)

2. Meng, Q., Mou, L., Xu, X., Zhu, G.: Power distribution control strategy of energy storage system in black start process. Elect. Power Autom. Equip. **34**(03), 59–64 (2014)
3. Li, X.: Research on black-start strategy of lsolated microgrid. Electrotechn. Elect. (Jiangsu Elect. Appar.) **2019**(06), 1–4, 20 (2019)
4. Cul, H., Wang, D., Yang, B., Wu, F., Zhu, T., Zhao, J.: Coordinated control strategy of wind-solar-storage power station supporting black start of power grid. Elect. Power Constr. **41**(09), 50–57 (2020)
5. Asheibi, A., Shuaib, S.: A case study on black start capability assessment. In: 2019 International Conference on Electrical Engineering Research & Practice (ICEERP), pp. 1–5. IEEE, Sydney (2019)
6. Wang, A., Gang, H., Qiu, P., Luo, X., Pang, X., Zhang, C.: Self-starting analysis of new energy system with wind power and energy storage. In: 2018 37th Chinese Control Conference (CCC), pp. 7422–7427. IEEE CPP, Wuhan (2018)
7. Lu, H., Yuan, X., Zhang, L.: Improved VF control strategy for flexible access to microgrid. Elect. Power Sci. Eng. **38**(02), 1–8 (2022)
8. Yu, Y., Wang, M., Zhang, R., Feng, L., Zhou, J.: Direct PQ control strategy for grid-connected inverter of energy storage system based on backstepping control. Energy Storage Sci. Technol. **37**(10), 11–17 (2021)
9. Liu, M., Liu, G., Sun, Z., Liang, S., Qiu, X.: Selection and simulation of black-start diesel generating set in regional power grid. In: 2018 China International Conference on Electricity Distribution (CICED), pp. 1774–1777. IEEE, Tianjin (2018)

Double-Layer Control Strategy for Power Distribution of Energy Storage System Based on AOE and Simulation Analysis

Lingchong Liu[1,2(✉)], Shufeng Dong[2], Kaicheng Lu[2], Mingyang Ge[2], Bin Nan[2], and Haiqi Zhao[1,2]

[1] Polytechnic Institute, Zhejiang University, Zhejiang 310015, Hangzhou, China
`22160028@zju.edu.cn`
[2] College of Electrical Engineering, Zhejiang University, Zhejiang 310027, Hangzhou, China
`dongshufeng@zju.edu.cn`

Abstract. In the context of dual carbon, the power distribution strategy for energy storage systems considering SOC (state of charge) balance and the difficulty of implementing control strategies is of great significance for slowing down battery aging and allowing more users to participate in the dual carbon goal. Aiming at the problem of power distribution in energy storage system, a double-layer control strategy for power distribution in energy storage system based on AOE (activity on edge) is proposed. First of all, the selection of charge and discharge battery clusters is realized based on the relationship between the SOC and the charge and discharge capacity in the upper layer. In addition, the optimal distribution of power is realized in the lower layer with the goal of SOC balance of battery packs. Then, the strategy can be implemented by writing Excel configuration files based on AOE which has the advantages of less difficulty in use, simple writing, intuitive control process and high calculation efficiency. Finally, the effectiveness of the strategy is verified through case simulation analysis.

Keywords: Energy storage system · AOE · Power distribution · SOC balance

1 Introduction

With the implementation of the dual carbon goals, new energy sources represented by wind power and photovoltaics are playing an increasingly important role. Their volatility and intermittency bring a huge impact on the grid, which is not conducive to the safe and stable operation of the grid. The rapid development of large-scale energy storage technology can effectively improve the grid's ability to absorb new energy and improve the security, economy and flexibility of grid operation [1]. Therefore, it is of great academic significance and engineering application value to study the power distribution

This work was supported by the Science and Technology Program of State Grid Corporation of China (5100-202256008A-1-1-ZN).

strategy of energy storage system and coordinate operation with new energy to meet the requirements of different dispatch commands.

Some scholars have conducted research on the power distribution of energy storage systems. Reference [2] used triple bidirectional DC/DC circuit to split multiple parallel batteries into each battery pack, and distributed the given power to each battery pack equally. Reference [3] proposed a real-time power distribution method based on the first-order Butterworth filter, but did not analyze the optimal distribution among lithium battery packs. Reference [4] proposed a fuzzy power distribution and coordinated control method for DC microgrid with electricity/hydrogen composite energy storage, and implemented the control strategy based on TMS320F28335 DSP and real-time simulation RT-LAB hardware-in-the-loop experimental platform. But the implementation process was complicated and difficult to learn.

This paper takes the power distribution strategy of energy storage system as the research object. Firstly, considering the selection of charging and discharging battery clusters and the SOC balance, a double-layer control strategy for power distribution of energy storage system is proposed. Secondly, considering the difficulty of implementing the control strategy, the proposed control strategy is realized by writing Excel configuration files based on AOE. Finally, taking a typical energy storage power station as an example, the simulation results verify the effectiveness of the strategy.

2   Double-Layer Control Strategy for Power Optimization Distribution of Energy Storage System

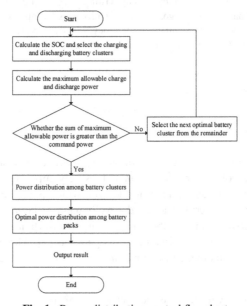

Fig. 1. Power distribution control flowchart

Due to different charging and discharging work state of each energy storage battery cluster, SOC is different in the energy storage system. In order to reduce the number of charge-discharge cycles, prevent over-charge and over-discharge, and maintain the safe and stable operation of the battery cluster, this paper proposes a double-layer control strategy for power optimization distribution of the energy storage system based on AOE. The control flow chart is shown in Fig. 1. The upper layer takes the selection of the optimal working battery clusters as the objective function, and selects the optimal charge and discharge battery cluster according to the relationship between the command power and the real charge and discharge power of the energy storage battery clusters. The lower layer takes the SOC balance as the objective function to realize the power distribution among battery packs.

2.1 Upper Layer Control Strategy - Battery Clusters Selection Layer

Taking the command power $P_{ref} > 0$, that is, the energy storage battery cluster in the discharging working state as an example, the upper-layer control strategy is described in detail below.

Step 1: Select the discharge battery clusters. Considering the large number of battery clusters in the energy storage system, in order to reduce the number of discharges, it is particularly important to select the optimal discharge battery clusters according to the command power and the discharge capacity of the battery cluster. Since the discharge capacity is inversely proportional to the SOC, the principle of selecting battery clusters can be described as [5]:

$$max(SOC_1, SOC_2, \cdots, SOC_n) \tag{1}$$

Step 2: Discharge power judgment. The maximum allowable discharge power of the energy storage battery cluster is determined by its own discharge characteristics and remaining power, which can be described as [6]:

$$P_{D\_limit}(t) = min \{ P_{D\_max}, \frac{[E(t-1) - E_{min}]\eta_D}{\Delta t} \} \tag{2}$$

where $P_{D\_limit}(t)$ is the maximum allowable discharge power of the energy storage battery cluster at time t. P_{D_max} is the maximum discharge characteristic determined by the material of the energy storage battery cluster. $E(t-1)$ is the remaining power of the energy storage battery cluster at the last moment. E_{min} is the minimum allowable remaining power of the energy storage battery cluster. η_D is the discharge efficiency. Δt is the calculation window time. Determine the relationship between the maximum allowable discharge power $P_{D\_limit}(t)$ and the command power P_{ref} of the selected optimal discharge battery cluster, when $P_{D\_limit}(t) > P_{ref}$, the selected optimal battery cluster is discharged. When $P_{D\_limit}(t) > P_{ref}$, repeat step 1, select the next target battery cluster from the remaining battery clusters, and compare the sum of the maximum allowable discharge power with the command power until the command power is satisfied.

Step 3: Determine the number of selected discharge battery clusters according to the above steps, and divide the power equally among the energy storage battery clusters.

The calculation formula is:

$$P_D = \frac{P_{ref}}{m} \tag{3}$$

When the command power is $P_{ref} \leq 0$, the energy storage battery cluster is in the charging working state, which is basically similar to the above process and is not be repeated here.

2.2 Lower Control Strategy - Power Optimization Distribution Layer

According to the obtained optimal discharge battery clusters and the power assumed by each energy storage battery cluster, the lower-level control strategy takes the minimum SOC balance as the objective function within the energy storage battery cluster, that is, between the energy storage battery packs, to achieve optimal power distribution. Which can be expressed as:

$$\min \sum_{i=1}^{n} (SOC_i^0 - \frac{P_{Di}\Delta t}{E_i} - SOC_{ave})^2 \tag{4}$$

where SOC_i^0 is the current SOC of the ith energy storage battery pack. P_{Di} is the discharge power of the ith energy storage battery pack. E_i is the remaining power of the ith energy storage battery pack. SOC_{ave} is the average SOC of all energy storage battery packs. Constraints:

1) Discharge power constraint:

$$0 \leq P_{Di} \leq P_{D\_limit}(t) \tag{5}$$

2) SOC constraint:

$$SOC_{i,min} \leq SOC_i \leq SOC_{i,max} \tag{6}$$

3) Total output constraint:

$$\sum_{i=1}^{n} P_{Di} = P_{ref} \tag{7}$$

When the command power changes, the double-layer control strategy of power optimization and distribution will be cyclically executed in the set window time, so as to meet the requirements of the dispatch command and coordinate the operation of the new energy [7].

3 Implementation of Control Strategy Based on AOE

In order to verify the proposed double-layer control strategy for power optimization distribution of the energy storage system, this paper takes the energy storage power station as an example, uses Simulink to establish the simulation model of the energy storage power station, and implements the control strategy based on AOE to realize the simulation analysis.

AOE is a weighted directed acyclic graph composed of multiple event-driven processes, whose nodes represent events and edges represent actions. When a node satisfies the condition and is triggered, the event represented by its corresponding edge is executed [8]. The energy storage power optimal distribution control strategy based on AOE can be realized only by writing the Excel configuration file, which has many advantages such as low difficulty in use, simple programming, intuitive control process, high calculation efficiency. The Excel configuration file includes measurement point configuration, AOE configuration and communication channel configuration. They are described in detail as follows [9].

3.1 Measurement Point Configuration

The measurement point configuration is mainly used to configure the parameters of the input and output signals of the controller, and define the name of the measurement point, whether it is discrete, whether to calculate the point and default values.

The measurement point of the case includes the system active command power issued by the main control system, the start signal, the SOC and the real power. The measurement point configuration file is shown in Fig. 2 below.

	A	B	C	D	E	F	M	N	O	P	Q	R	S	T
	serial number	point number	point name	alias	whether discrete	whether calculated point	upper limit	lower limit	maximum allowable change	minimal allowable change	whether real time points	whether SOC	defaults	note
2	1	1001	measurement point 1	GEN_TOTAL_P_POINT	FALSE	FALSE	99999	-99999	0	0	FALSE	FALSE	1	system active command
3	2	1002	measurement point 2	P_POINT_G1	FALSE	FALSE	99999	-99999	0	0	FALSE	FALSE	1	G1 active command
4	3	1003	measurement point 3	BAMS1_SOC_POINT	FALSE	FALSE	99999	-99999	0	0	FALSE	FALSE	0	soc
5	4	1004	measurement point 4	BAMS2_SOC_POINT	FALSE	FALSE	99999	-99999	0	0	FALSE	FALSE	0	soc
6	5	1005	measurement point 5	BAMS3_SOC_POINT	FALSE	FALSE	99999	-99999	0	0	FALSE	FALSE	0	soc
7	6	1006	measurement point 6	BAMS4_SOC_POINT	FALSE	FALSE	99999	-99999	0	0	FALSE	FALSE	0	soc
8	7	1007	measurement point 7	BAMS5_SOC_POINT	FALSE	FALSE	99999	-99999	0	0	FALSE	FALSE	0	soc
9	8	1008	measurement point 8	BAMS6_SOC_POINT	FALSE	FALSE	99999	-99999	0	0	FALSE	FALSE	0	soc
10	9	1009	measurement point 9	BAMS7_SOC_POINT	FALSE	FALSE	99999	-99999	0	0	FALSE	FALSE	0	soc
11	10	1010	measurement point 10	BAMS8_SOC_POINT	FALSE	FALSE	99999	-99999	0	0	FALSE	FALSE	0	soc
12	11	1011	measurement point 11	BAMS1_P_POINT	FALSE	FALSE	99999	-99999	0	0	FALSE	FALSE	0	power
13	12	1012	measurement point 12	BAMS2_P_POINT	FALSE	FALSE	99999	-99999	0	0	FALSE	FALSE	0	power

Fig. 2. Measurement point configuration file

3.2 AOE Configuration

AOE configuration is to describe the control strategy by defining the basic information, variables, events and actions of AOE. This case includes a total of 6 nodes and 5 actions.

The above control strategy is represented by AOE, as shown in Fig. 3. Node 1 is a Simulink communication event node, which is an event trigger type. When the energy storage system starts to work, the node is triggered to start executing the control strategy. Node 2 represents the completion of the calculation of intermediate variables. When considering the reduction of SOH due to changes in the total charge and discharge mileage, the current SOH of each energy storage and the actual capacity reduction caused by the loss of energy storage operating life are calculated and updated. Node 3 represents the optimal battery pack selection event. Node 4 represents the event that the optimization calculation is completed. In the optimization solution method built into the controller, the value of all optimization variables is 0 when the optimal solution does not exist. When there is an optimal solution, the value of flag is 1, so the binary variable flag can be set as the judgment condition for whether there is a solution to the optimization problem. The node type is set to switch. When a node event is triggered, action 4 is executed; otherwise, action 5 is executed after a timeout occurs. Node 5 represents the event that the optimal solution has been obtained for the optimization problem. The occurrence of this event indicates that the optimal solution has been found and issued, and the AOE strategy ends. Node 6 indicates that the optimization problem has no optimal solution.

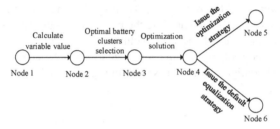

Fig. 3. Control strategy AOE topology

	A	B	C	D	E	F
1	AOE ID	whether to enable	name	triggering conditions	trigger condition parameters	variable initial value
2	5003	TRUE	Secondary Frequency Regulation	Event_Drive		deadband:100;SOCref:0.55;dt:0.25;last_cmd:100;w:0.0000005;max_power:4500
3						
4	AOE ID	Variable definitions				
5						
6						
7	AOE ID	node ID	name	node type	timeout time (ms)	expression
8	5003	1	start node	Condition	1000	active1==1&&P_POINT_G1 != last_cmd
9	5003	2	Intermediate volume calculation completed	Condition	15000	1
10	5003	3	The optimization solution ends	Switch	1000	flag == 1
11	5003	4	There is an optimal solution	Condition	1000	1
12	5003	5	There is no optimal solution	Condition	1000	1

Fig. 4. AOE configuration file

Action 1 is to calculate variable value which is executed after the communication event node is triggered, and calculates the SOC of the energy storage battery cluster. Action 2 is the selection of optimal battery clusters. According to the SOC of the energy storage battery clusters and the objective function, the charging and discharging battery cluster is optimally selected. Action 3 is to solve the power optimization distribution problem of the energy storage battery pack. When there is an optimal solution, perform action 4 to issue an optimization strategy, and when there is no optimal solution, perform action 5 to issue a default equalization strategy. When there is no optimal solution, it means that the command power is unreasonable, or the capacity of the energy storage is insufficient. The AOE configuration file is shown in Fig. 4 below.

3.3 Communication Channel Configuration

The communication channel configuration is used to describe the communication mode between the controller and the simulation model of the Simulink energy storage power station, so as to realize the information exchange between the energy storage battery and the controller. According to the attribute of the measuring point in the measuring point configuration file, the register information and address for saving the measuring point value are given. The communication channel configuration file includes two parts: communication mode configuration and measurement point register configuration. It supports Modbus, IEC104, MQTT and serial port as communication modes [10, 11]. Communication mode configuration includes port number, baud rate, parity bit, data bit and stop bit. The measuring point register configuration is used to give the address information for storing the measuring point data, including serial number, register type, starting address, data type, new request flag, polling cycle and point number. In this case, the controller is used as a slave station and Modbus is used to realize the communication between the controller and the simulation model [12]. The measurement point data are all of the EightByteFloat data type, occupying 4-bit registers, and then complete the address allocation according to the data type. The communication channel configuration file is shown in Fig. 5 below.

	A	B	C	D	E	F	G	H	I	J	K
1	channel name	server test channel	connection name	test channel 1	no.	register type	start address	data type	new request flag	polling period	point no.
2	number of connections 1	1	number of measuring points	79	1	HOLDING	1	EightByteFloat	FALSE	2000	1001
3	service port	502	client IP	192.168.0.139	2	HOLDING	5	EightByteFloat	FALSE	2000	1002
4			client port	9999	3	HOLDING	9	EightByteFloat	FALSE	2000	1003
5			slave id	1	4	HOLDING	13	EightByteFloat	FALSE	2000	1004
6			letter of agreement	XA	5	HOLDING	17	EightByteFloat	FALSE	2000	1005
7			the upper limit of the number of registers read at one time	125	6	HOLDING	21	EightByteFloat	FALSE	2000	1006
8			the upper limit of the number of switches for one reading	2000	7	HOLDING	25	EightByteFloat	FALSE	2000	1007
9			the upper limit of the number of registers that can be written at one time	120	8	HOLDING	29	EightByteFloat	FALSE	2000	1008
10			the upper limit of the number of switches in one write	1968	9	HOLDING	33	EightByteFloat	FALSE	2000	1009
11			polling Period (ms)	5000	10	HOLDING	37	EightByteFloat	FALSE	2000	1010
12			timeout (ms)	1000	11	HOLDING	41	EightByteFloat	FALSE	2000	1011
13					12	HOLDING	45	EightByteFloat	FALSE	2000	1012

Fig. 5. Communication channel configuration file

4   Simulation Results Analysis

Set the specific parameters of the simulation model: the rated capacity of the battery pack is 10000 Wh, the maximum allowable charge and discharge power is 12000 W, and the maximum and minimum SOC values are 0.9 and 0.1, respectively.

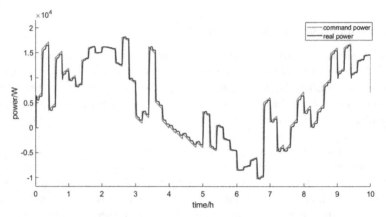

Fig. 6. Command power and real output power of energy storage power station

By running the controller and the simulation model, the command power and the total real output power of the energy storage power station are obtained as shown in Fig. 6. It can be seen from the figure that the real output power of the energy storage power station can accurately follow the change of the command power. The output power fluctuates slightly due to the characteristics of the energy storage battery, but the fluctuation value is within 200 W.

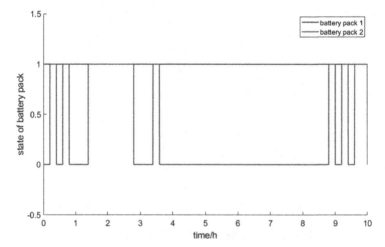

Fig. 7. Working state change of the energy storage battery pack

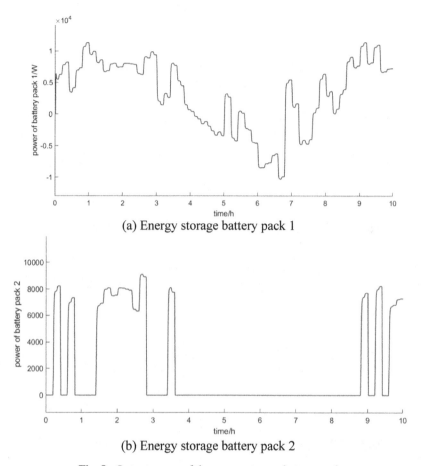

(a) Energy storage battery pack 1

(b) Energy storage battery pack 2

Fig. 8. Output power of the energy storage battery pack

Figures 7 and 8 show the changes in the working state and output power of the energy storage battery cluster. The working state is represented as "1", and the shutdown state is represented as "0". Among the energy storage battery clusters 1, 2, and 3, the energy storage battery cluster 1 has the largest SOC. It can be seen from the figure that according to the upper-layer control strategy, when the command power is satisfied, the energy storage battery cluster 1 is in the working state, and the battery clusters 2 and 3 are in the shutdown state.

Figure 9 shows the SOC changes of the double-layer control strategy of energy storage battery packs. Compared with the control strategy of power equalization distribution in Fig. 10, it can be seen from the figure that the lower control strategy realizes the reasonable distribution of power among the energy storage battery packs. After 2h, the SOC reaches the basic equilibrium. Therefore, the strategy in this paper realizes the SOC balance among the energy storage battery packs, reduces the difference of SOC, avoids overcharge and overdischarge, and slows down the aging of the energy storage battery.

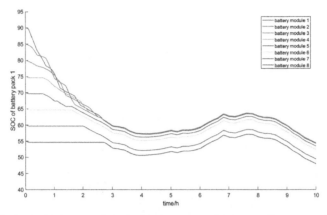

Fig. 9. SOC changes of the double-layer control strategy of battery pack 1

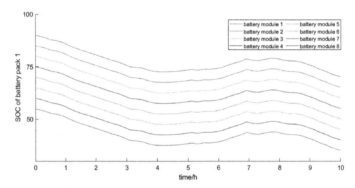

Fig. 10. SOC changes of the equalization control strategy of battery pack 1

5   Conclusions

In this paper, a double-layer control strategy for power distribution of energy storage system based on AOE is proposed, and the effectiveness of the strategy is verified through simulation analysis. The main conclusions are as follows.

1) The optimal charging and discharging battery cluster selection is used as the upper-layer control strategy to realize the intermittent charging and discharging tasks of the energy storage battery clusters. It does not require all battery clusters to undertake the charging and discharging tasks in real time. The number of charging and discharging is effectively reduced.
2) With power optimization distribution as the lower control strategy, the SOC balance among energy storage battery packs is realized, which avoids overcharging and over discharging of energy storage batteries and slows down aging.
3) The control strategy is realized based on AOE, and the control strategy can be written through Excel. The realization process is simple and flexible, and has many advantages such as high versatility, high calculation efficiency, and low learning

difficulty, which is conducive to allowing more industrial users to participate in the dual carbon goal.

Subsequently, the control strategy is optimized from the perspectives of energy storage cost and the balance of energy storage battery utilization times on the optimal distribution of energy storage power.

References

1. Feng, X., Zhang, S., Zhu, T., et al.: Large-scale energy storage selection analysis based on interval type-2 fuzzy multiple attribute decision making method. High Volt. Eng. **47**, 4123–4136 (2021)
2. Ding, M., Lin, G., Chen, Z., et al.: A Control Strategy for Hybrid Energy Storage Systems. Proc. CSEE. **32**, 1–6+184 (2012)
3. Li, X., Liu, J.: Real-time power distribution method adopting second-order filtering for hybrid energy storage system. Power Syst. Technol. **43**, 1650–1657 (2019)
4. Zhang, X., Pei, W., Mei, C., et al.: Fuzzy power allocation strategy and coordinated control method of islanding DC microgrid with electricity/hydrogen hybrid energy storage systems. High Voltage Eng. **48**, 958–968 (2022)
5. Li, H., Fu, B., Yang, C., et al.: Power Optimization Distribution and Control Strategies of Multistage Vanadium Redox Flow Battery Energy Storage Systems. Proc. CSEE. **33**, 70–77+16 (2013)
6. Qiu, Y., Li, X., Chen, W., et al.: Power distribution of vanadium redox battery energy storage system based on P-AWPSO algorithm. High Voltage Eng. **46**, 500–510 (2020)
7. Faisal, M., Hannan, M.A., Ker, P.J., et al.: Fuzzy-based charging-discharging controller for lithium-ion battery in microgrid applications. IEEE Trans. Ind. Appl. **57**, 4187–4195 (2021)
8. Dong, S., Tang, K., Liu, L., et al.: The teaching method of energy storage control experiment based on Simulink and low-code controller. Energy Stor. Sci. Technol. **1**, 1–13 (2022)
9. Kang, X., Zhang, X., Wu, S.: Research on modeling technology of virtual maintenance process based on Petri Net and AOE network. In: 2013 Fourth International Conference on Digital Manufacturing and Automation, pp. 1090–1093 (2013)
10. S. Tamboli, M. Rawale, R. Thoraiet, et al, "Implementation of Modbus RTU and Modbus TCP communication using Siemens S7–1200 PLC for batch process," 2015 International Conference on Smart Technologies and Management for Computing, Communication, Controls, Energy and Materials (ICSTM), pp. 258–263, 2015
11. Silva, C.R.M., Silva, F.A.C.M.: An IoT gateway for Modbus and MQTT integration. In: 2019 SBMO/IEEE MTT-S International Microwave and Optoelectronics Conference (IMOC), pp. 1–3 (2019)
12. Gonzalez, L., Calderon, A.J., Portalo, J.M.: Innovative multi-layered architecture for heterogeneous automation and monitoring systems: application case of a photovoltaic smart microgrid. Sustainability **13**, 2234–2258 (2021)

Design of Oceanic Eddy Simulation Platform for Autonomous Ocean Vehicles

Wenhao Gan[1]($\boxtimes$), Dalei Song[1,2], and Xiuqing Qu[1]

[1] College of Engineering, Ocean University of China, Shandong, China
gwh@stu.ouc.edu.cn
[2] Institute for Advanced Ocean Study, Ocean University of China, Shandong, China

Abstract. Real-time and dynamic monitoring of oceanic eddies by ocean vehicles is an essential part of the layout of the national ocean observation network. In this paper, the mesoscale eddy environment is constructed, and a near-real-time simulation platform for ocean observation is established with the autonomous surface underwater vehicle (ASUV) as the observer. Then, a cluster observation strategy based on the virtual leader structure and ocean characteristic estimation method is proposed, so that vehicles can search and track the eddy centre and front through autonomous perception. In the end, the simulation experiments of eddy observation with multi-vehicle are carried out, which further verifies the practicability of the simulation platform and the validity of the observation strategy.

Keywords: Autonomous ocean vehicle · Mesoscale eddy · Simulation platform · Cooperative observation

1 Introduction

Oceanic eddies are ubiquitous, and their formation, evolution, and termination have essential effects on various parameters in the ocean environment, biology, and chemistry. Among them, the mesoscale eddy is a kind of oceanic eddy with a spatial scale of 50–500 km and a life scale of 10–100 days, which has the characteristics of random movement and closed circulation. In the last decade, the three-dimensional characteristics of mesoscale eddies have attracted extensive attention of oceanographers [1].

In recent years, the autonomous ocean vehicle (AOV), a mobile observation platform, has been introduced into the ocean observation. With its characteristics of autonomy, controllability, and intelligence, the dynamic and continuous observation of small and medium-sized ocean phenomena has become possible [2–4]. And the ocean fusion data provided by single or multiple satellite altimeters can also be used when vehicles are deployed in the field to improve the observation efficiency. For the general observation strategies for these phenomena, such

This work was partially supported by the Fundamental Research Funds for the Central Universities (Grant No. 201962010).

W. Fan et al. (Eds.): AsiaSim 2022, CCIS 1713, pp. 246–260, 2022.
https://doi.org/10.1007/978-981-19-9195-0_21

as mesoscale eddy centres, fronts, isotherms, etc., pieces of literature [4,5] made some summaries. On the premise of obtaining estimation or gradient estimation of the characteristic field, the vehicle can be driven to the desired extremum (or isoline) by moving along the gradient direction (or the vertical direction of the gradient). At the same time, literatures [6–8] made some summaries for the spatial and temporal distribution, thermohaline distribution, current field distribution, and the motion pattern of mesoscale eddies in the certain regions. And in literature [9], the rationality of eddy observation using autonomous ocean vehicles was proved. Moreover, in recent years, multi-agent systems (MAS) and swarm robotics are developing rapidly [10], which provide a reliable methodology and inspiration for the ocean observation.

Based on the above analysis, in order to meet the requirements of high purpose, real-time performance and high spatio-temporal resolution of mesoscale eddies observation, autonomous surface underwater vehicles (ASUVs) will be used as the platform to observe their central and frontal (an area with maximum flow speed) characteristics. The main innovations of this paper are as follows:

* Based on previous studies on the characteristics of oceanic eddies, this paper constructs a high-fidelity eddies simulation platform, which can be used for the design and verification of observation algorithms.
* A cluster observation strategy for multi-ASUV is proposed. By analyzing the real-time observation data collected by vehicles, the information of local ocean characteristic field can be estimated, and the movement direction of the vehicle group can be planned.
* The virtual leader strategy is combined with the artificial potential field method to coordinate the navigation of each vehicle. The whole observation process carries on this cycle, which lays the foundation for realizing the fine three-dimensional observation of mesoscale eddies.

The organizational structure of this paper is as follows: Sect. 2 describes the construction of the simulation platform. In Sect. 3, the observation framework of multi-ASUV is constructed. In Sect. 4, the simulation experiments and analysis of the eddy centre and front observation are carried out, and Sect. 5 is the conclusion.

2   Platform Architecture

The structure of the simulation platform is shown in Fig. 1. As can be seen, the platform is mainly composed of three parts:

(1) Model unit. ① Vehicle modelling: realizing the motion simulation of the vehicle at sea surface/undersea. ② Eddy modelling: realizing the simulation of the eddy motion and eddy characteristics. ③ Ocean modelling: realizing the visual simulation of the sea surface/undersea and providing the interface for environmental disturbance for the vehicle.

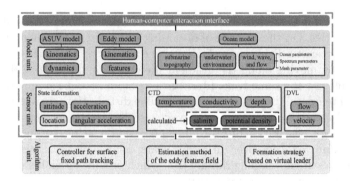

Fig. 1. The design diagram of the simulation platform.

(2) Sensor unit. This unit is mainly responsible for modelling vehicle-borne sensors and interacting with the model unit to obtain real-time information about vehicles and surrounding environment.
(3) Algorithm unit. This unit is mainly responsible for the planning and controlling the vehicle group in the eddy observation task.

2.1 Vehicle Modelling

This paper uses the autonomous surface underwater vehicle (ASUV) as the observation platform to carry out the simulation research. The vehicle model and its workflow are shown in Fig. 2.

Fig. 2. ASUV solid model and its workflow.

The vehicle combines the profile detection function of the Argo buoy and the propulsion function of the autonomous underwater vehicle (AUV). The horizontal navigation in the sea surface (AUV function) and the vertical profile observation (Argo function) can be achieved with the ASUV. And by changing the position of the battery pack, the center of gravity of the vehicle can be greatly adjusted to realize the adjustment of AUV and Argo modes. For motion simulations (kinematic and dynamic simulations) of ASUV, the corresponding modeling derivation is described in the literature [11]. And Fig. 3 shows the motion visual effect on the platform.

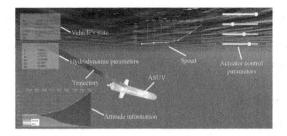

Fig. 3. ASUV dynamics simulation and the visualization of its state parameters.

The vehicle is equipped with multiple measuring equipments, such as the Acoustic Doppler Current Profiler (ADCP), which measures the current velocity at various depths, and the Conductivity, Temperature, and Depth measurement (CTD), which further calculates the corresponding salinity and potential density of sea water.

2.2 Conventional Ocean Environment

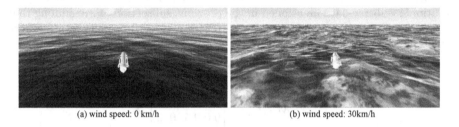

(a) wind speed: 0 km/h   (b) wind speed: 30km/h

Fig. 4. Motion state of the ASUV under different wind speeds.

When the vehicle is navigating on the sea surface, it will be disturbed by the waves, winds, and ocean current, which will lead to changes of the position and posture of the vehicle at any time [12]. Given that the current wave simulation is usually completed by using the wave spectrum and wave functions [13], the Phillips spectrum and 3D Gerstner function are used to complete the construction of the sea surface in this platform. By controlling the wind speed ω_s, the wave situation under different sea conditions can be simulated.

When the ASUV is in the AUV mode, its buoyancy can be expressed as $B = mg = \rho g \nabla$, where m represents the vehicle's mass, g is the gravitational constant, ρ is the density of seawater, and ∇ is the volume of the body submerged by seawater. Because ∇ changes as the waves rise and fall, B needs to be updated in real-time. In addition, waves will also provide additional disturbing forces and moments for the ASUV, which can be calculated by using the wave velocity v_{wave}

and wave normal n_{wave} in the position of the vehicle's body. In this way, the motion state of ASUV under different wind speeds can be simulated, as shown in Fig. 4. And for the ocean current, it is simulated by applying the eddy geostrophic flow to the kinematic model of ASUV.

2.3   Eddy Modelling

Eddy Characteristics. Compared with the ideal equation simulation, the eddy environment constructed from the observed data can give more convincing results. In literature [1], Sun et al. used the SSHA data and Argo profiles to reconstruct the three-dimensional structure of mesoscale eddies in the Kuroshio Extension (KE) region and reveal the distribution of related physical parameters. In this simulation platform, the distribution information of temperature, salinity, potential density, and geostrophic flow anomalies mentioned about the composite cyclonic eddy is used to build the eddy environment.

To match these data and ensure the authenticity of experiments, the platform also stores several groups of historical motion data of the mesoscale composite cyclonic eddy in the KE region provided by the AVISO website [14].

Eddy Kinematic Modeling. The motion of the eddy can be described using a constant velocity (CV) model and with the help of its position, velocity, and rotational velocity data mentioned in the previous subsection [4]. Assuming that X_k is the state vector composed of the position r_k and velocity $\dot{r}_k$, then the eddy kinematic model can be expressed as follows:

$$X_{k+1} = M \cdot X_k = \begin{bmatrix} 1 & 0 & 0 & T & 0 & 0 \\ 0 & 1 & 0 & 0 & T & 0 \\ 0 & 0 & 1 & 0 & 0 & T \\ 0 & 0 & 0 & 1 & 0 & 0 \\ 0 & 0 & 0 & 0 & 1 & 0 \\ 0 & 0 & 0 & 0 & 0 & 1 \end{bmatrix} X_k \tag{1}$$

$$X_k = \begin{bmatrix} r_k \\ \dot{r}_k \end{bmatrix} = (x_k, y_k, \psi_k, \dot{x}_k, \dot{y}_k, \dot{\psi}_k)^{\text{T}}$$

where k represents the k-th time-step, T represents the step-length, set as 0.02h; x_k and y_k represent the position of eddy centre in the inertial coordinate system, and ψ_k represent the rotation angle of the eddy.

3   Observation Algorithm

The sampling data obtained by a single vehicle is relatively limited, and the sampling process may be affected by a variety of factors, such as sampling errors caused by equipments, ocean space-time variation errors, etc., which will lead to inaccurate estimations of the characteristic field. Therefore, the research in this paper will adopt the method of synchronous array observation, and the navigation process of multi-ASUV can be divided into two stages:

Stage I: the searching stage. The multi-ASUV navigates on the sea surface in AUV mode, continuously collects the eddy information, and maintains communication with the shore station.

Stage II: the tracking stage. The multi-ASUV constantly switches between AUV and Argo mode, uses the AUV mode to approach the eddy centre (or front) and the Argo mode to sample the vertical information of the eddy.

3.1 Multi-ASUV Control Method

For ocean navigation tasks, the following points should be considered:

(1) Due to the complexity of ocean environment, the cooperative control algorithm used should have strong engineering practicability.
(2) The stability of maritime communications is uncertain, so the algorithm's reliability is critical.
(3) The algorithm should not generate control commands frequently to reduce the energy consumption as much as possible.

Based on the above three points, a multi-vehicle cooperative control method based on the virtual leader and artificial potential field is proposed in this paper for the oceanic eddy observation task. The so-called virtual leader is used to guide followers to the target points, which can avoid the problem of group communication failure effectively. When the velocity and posture of the leader are determined, each follower can adjust its velocity and posture according to the offset to the leader, as shown in Fig. 5.

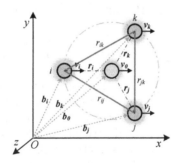

Fig. 5. Representation of multi-ASUV system in the inertial coordinate system. (Color figure online)

where the yellow circle is the virtual leader, which is the reference point of the system, and the green circle is the follower, which is the ASUV used in this paper. And b_i is the position of the i-th vehicle in the inertial coordinate system, v_i is its velocity, and satisfies $\dot{b}_i = v_i$. The control force acting on the vehicle is defined as μ and satisfies $\dot{v}_i = \mu$. Then, a coordinate system with the virtual leader as the origin and no rotation is established. Assuming that the leader is moving forward at the velocity v_0, the position of the follower i in this coordinate system can be expressed as r_i and the velocity is $\dot{r}_i = v_i - v_0$.

Followers interact with each other by forces, which are derived from the potential field [15]. The system uses $\boldsymbol{f}_I$ to represent the force between the follower and any adjacent follower, and the corresponding potential field is V_I:

$$V_I = \begin{cases} a_I \left(\ln (r_{ij}) + \frac{d_0}{r_{ij}} \right), 0 < r_{ij} < d_1 \\ a_I \left(\ln (r_{ij}) + \frac{d_0}{d_1} \right), r_{ij} \geq d_1 \end{cases} \tag{2}$$

$$\text{and } \boldsymbol{f}_I = -\nabla_{r_{ij}} V_I = \begin{cases} (f_{Ix}, f_{Iy}), 0 < r_{ij} < d_1 \\ (0,0), r_{ij} \geq d_1 \end{cases}$$

At the same time, the follower will also be affected by the force $\boldsymbol{f}_h$ of the virtual leader, and the corresponding potential field is V_h:

$$V_h = \begin{cases} a_h \left(\ln (h_{ik}) + \frac{h_0}{h_{ik}} \right), 0 < h_{ik} < h_1 \\ a_h \left(\ln (h_1) + \frac{h_0}{h_1} \right), h_{ik} \geq h_1 \end{cases} \tag{3}$$

$$\text{and } \boldsymbol{f}_h = -\nabla_{h_{ik}} V_h = \begin{cases} (f_{hx}, f_{hy}), 0 < h_{ik} < h_1 \\ (0,0), h_{ik} \geq h_1 \end{cases}$$

where a_I and a_h represent control gains; d_1 and h_1 represent influence ranges of potential fields; d_0 and h_0 represent the minimum radii of potential energy field; r_{ij} represents the distance between the follower i and j; h_{ik} represents the distance between the follower i and the virtual leader k. And the direction of $\boldsymbol{f}_h$ is the same as the direction of $\boldsymbol{r}_{ij}$.

In addition, each follower will be affected by an additional dissipative force $\boldsymbol{f}_{v_i} = -a_v \dot{\boldsymbol{r}}_i$, which is used to control the follower to maintain the desired speed. Where a_v represents the control gain and it is greater than 0. In general, the total control force received by the follower i can be expressed as:

$$\boldsymbol{\mu}_i = -\sum_{j \neq i}^{n} \nabla_{r_i} V_I (r_{ij}) - \sum_{k=0}^{m-1} \nabla_{r_i} V_h (h_{ik}) + \boldsymbol{f}_{v_i} \tag{4}$$

where n and m represent the number of followers and virtual leaders, respectively.

For the observation task in this paper, to form the equilateral triangle formation as shown in Fig. 5, it is necessary to set a expected distance between the virtual leader and followers, and between each follower, and select relatively large d_1 and h_1 to ensure that each vehicle is within the range of potential field. Finally, The expression of the whole formation system can be obtained:

Virtual leader : $\dot{\boldsymbol{b}}_0 = \boldsymbol{v}_0$.
Follower 1 : $\dot{\boldsymbol{r}}_1 = \boldsymbol{v}_1$,
$\dot{\boldsymbol{v}}_1 = -\nabla_{r_{12}} V_I (r_{12}) - \nabla_{r_{13}} V_I (r_{13}) - \nabla_{h_{10}} V_h (h_{10})$
$\quad -a_{12} (\boldsymbol{v}_1 - \boldsymbol{v}_2) - a_{13} (\boldsymbol{v}_1 - \boldsymbol{v}_3) - a_{10} (\boldsymbol{v}_1 - \boldsymbol{v}_0)$.
Follower 2 : $\dot{\boldsymbol{r}}_2 = \boldsymbol{v}_2$,
$\dot{\boldsymbol{v}}_2 = -\nabla_{r_{21}} V_I (r_{12}) - \nabla_{r_{23}} V_I (r_{13}) - \nabla_{h_{20}} V_h (h_{20})$
$\quad -a_{21} (\boldsymbol{v}_2 - \boldsymbol{v}_1) - a_{23} (\boldsymbol{v}_2 - \boldsymbol{v}_3) - a_{20} (\boldsymbol{v}_2 - \boldsymbol{v}_0)$.
Follower 3 : $\dot{\boldsymbol{r}}_3 = \boldsymbol{v}_3$,
$\dot{\boldsymbol{v}}_3 = -\nabla_{r_{31}} V_I (r_{31}) - \nabla_{r_{32}} V_I (r_{32}) - \nabla_{h_{30}} V_h (h_{30})$
$\quad -a_{31} (\boldsymbol{v}_3 - \boldsymbol{v}_1) - a_{32} (\boldsymbol{v}_3 - \boldsymbol{v}_2) - a_{30} (\boldsymbol{v}_3 - \boldsymbol{v}_0)$.

$$\tag{5}$$

3.2 Characteristic Field Estimation Method

With the clustering strategy in place, it is also needed to specify the velocity v_0 of the virtual leader at each step. Vehicle sampling is a discrete process, the sampling area is large, and there may be some data deviations during the sampling process. Therefore, the sampling data of multi-ASUV is processed by the moving average method, and the window size is n. In this way, three processed data $F_1(k)$, $F_2(k)$ and $F_3(k)$ can be obtained at the same step k. Meanwhile, since the position of each vehicle can be obtained by GPS, the estimated characteristic value of the triangular characteristic field surrounded by the vehicles can be obtained by the area interpolation method, as shown in Eq. (6).

$$F(k) = \frac{F_1(k)S_{23} + F_2(k)S_{13} + F_3(k)S_{12}}{2\left(S_{12} + S_{23} + S_{13}\right)} \tag{6}$$

where $F(k)$ is the estimated characteristic of a point in the triangular characteristic field, S_{ij} represents the area size of the characteristic field surrounded by the follower i, j, and the point to be estimated.

3.3 Overall Observation Process

Before the mission, the location of the latest eddy centre/front S can be obtained from satellite data. Then we can place three ASUVs and set the initial position of the virtual leader as P. Then, the navigation strategy of the ASUV can be divided into two methods: "surface fixed path tracking" and "autonomous ocean characteristic searching". For the former, it is mainly used for navigation when the eddy characteristics are not obvious, and its main workflow is as follows:

a Calculate the direction of the virtual leader $\theta = \tan^{-1}((S_y - P_y)/(S_x - P_x))$.
b Set leader's speed as the economic speed v_f, then $v_0 = [v_f\cos(\theta), v_f\sin(\theta)]^T$.

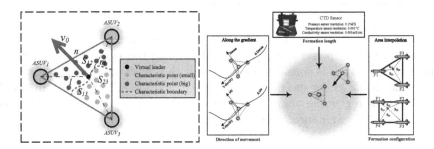

Fig. 6. Schematic diagram of strategy design for multi-ASUV eddy observation.

And for the latter, it is used to guide the vehicle to reach target sampling points accurately, and its main workflow is as follows:

a The eddy information is collected by three vehicles and processed by the moving average method.

b The area interpolation method is used to get the characteristic variation of the area surrounded by vehicles.

c The increasing/decreasing direction of the potential density is θ_{dens}, and the accelerating/slowing direction of the flow speed is θ_{uv}. The overall characteristic change is defined as the weighted sum of partial characteristics: $\theta_{\text{all}} = \omega_1 \theta_{\text{dens}} + \omega_2 \theta_{\text{uv}}$, where $\omega_1 + \omega_2 = 1$.

d The direction of characteristic gradient determines the moving direction of the virtual leader, and its speed is set as the economic speed v_f, then $\boldsymbol{v}_0 = [v_f \cos(\theta_{\text{all}}), v_f \sin(\theta_{\text{all}})]^{\text{T}}$.

Where the interpolation characteristics can be divided into two categories, as shown in the left subgraph in Fig. 6.

Then the overall observation process can be obtained, as shown in the right subgraph in Fig. 6, and the pseudocode is shown in Algorithm 1. Among them, d_t represents the distance threshold, which is defined according to the Euclidian distance between the vehicle and the preset observation point.

Algorithm 1. Observation strategy for oceanic eddy

1: **for** each day **do**
2:     Update the location point S based on satellite data
3:     **for** each step **do**
4:         **if** $\|S - P\|_2 > d_t$ **then**
5:             Adopted "**surface fixed path tracking**" method
6:         **else**
7:             Adopted "**autonomous ocean characteristic searching**" method.
8:         **end if**
9:         Multiple ASUVs form a formation according to the cooperative control method and follow the virtual leader.
10:     **end for**
11: **end for**

Meanwhile, when the potential density/flow speed values in adjacent K samples are within the preset ranges ($F_{\text{sam}} \in F_{\text{pre}}$), and the signs of their gradients change frequently, it is consider the feature at that position is worth to sample. Then, the virtual leader speed v_0 will be set to 0 and the ASUV group will switch to Argo mode for profile sampling. Otherwise, the ASUV group will continue to navigate on the water surface by using AUV mode.

4   Experiment and Analysis

By analyzing the eddy statistics, the following strategies are adopted:

(1) Observation of the eddy centre: searching for the area with the maximum potential density anomalies, $\omega_1 = 1, \omega_2 = 0$.

(2) Observation of the front: searching for the area with the maximum flow anomalies, $\omega_1 = 0, \omega_2 = 1$.

In order to meet the observation requirements of large range of characteristic fields, d_t is set as twice the eddy radius ($d_t = 2R = 200$ km). The formation length L of the multi-ASUV is set as 10 km, the economic speed v_f is 2 knots, the sampling interval Δt is 0.1 h, and the window size n is 10. The predicted characteristic range of the eddy center is represented by the potential density anomaly, i.e., $F_{\text{pre}} = 0.15 \sim 0.3$ kg/m^3; The predicted characteristic range of the front is represented by the flow anomaly, i.e., $F_{\text{pre}} = 0.15 \sim 0.3$ m/s; The sampling is set every 10 times as a cycle, and the condition for switching Argo mode is set as the sign change frequency of characteristic gradient is greater than 3 times/cycle, which is designed to ensure the characteristic at the current position is close to the local abnormal maximum as far as possible.

4.1   Validation Experiment

The feasibility of the method will be tested using the observation data in a static eddy environment (fixed position, no rotation). Three experiments are carried out for eddy centre search and front search, respectively, and the placement positions of the ASUV group in each experiment are randomly selected within the eddy boundary. The experimental results are shown in Fig. 7, the corresponding performance parameters are shown in Table 1 and 2.

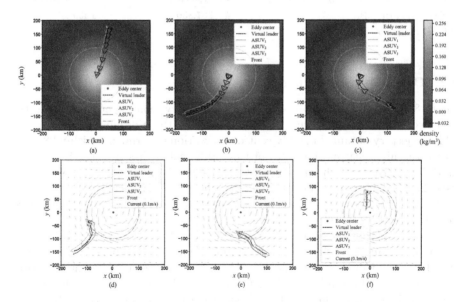

Fig. 7. Formation searching experiments. (a), (b) and (c): searching for the eddy center. (d), (e) and (f): searching for the eddy front.

It can be seen that, by combining the proposed multi-vehicle cooperative control and characteristic field estimation method, the ASUV group can search the eddy center/front accurately and ensure the final search error within a small range on the premise of selecting multiple parameters reasonably. Next, a complete set of dynamic eddy searching and tracking experiments will be simulated to further verify the reliability and practicability of the proposed method.

Table 1. Performance parameters of the static eddy center searching experiments.

Parameters	Value (a)	Value (b)	Value (c)
ASUV1 place point (km)	(62.723,171.691)	(−142.177, −146.309)	(132.123, −121.309)
ASUV2 place point (km)	(45.2,175.281)	(−159.7, −142.719)	(114.6, −117.719)
ASUV3 place point (km)	(49.639,166.081)	(−155.261, −151.919)	(119.039, −126.919)
Navigation time (h)	99.961	119.316	128.401
Final search error (km)	2.436	2.3436	4.560

Table 2. Performance parameters of static front searching experiments.

Parameters	Value (a)	Value (b)	Value (c)
ASUV1 place point (km)	(−137.677, −152.209)	(108.123, −164.809)	(−1.377,21.691)
ASUV2 place point (km)	(−155.2, −148.619)	(90.6, −161.219)	(−18.9,25.281)
ASUV3 place point (km)	(−150.761, −157.819)	(95.039, −170.419)	(−14.461,16.081)
Navigation time (h)	85.490	87.850	46.439
Final range error (km)	2.516	18.7329	23.453

4.2 Comprehensive Experiment

Figure 8 shows the visualization of the eddy observation experiment using the designed simulation platform. Figure 9 shows the three groups' formation and eddy trajectory obtained in eddy centre observation experiments within 28 days, where the historical motion data of three different eddies in the KE region is used. Figure 10 shows the tracking error and state changes of the ASUV. Table 3 lists the specific performance parameters, where ME represents the mean value of navigation error in the observation process, and SE represents the standard deviation of the error.

These experiments show that the ASUV group can complete the searching and tracking task based on the proposed method. In the searching stage, the distance error between the vehicle and the observation point decreases monotonously, which indicates that the mulit-ASUV can estimate the local characteristic field well and find a relatively short path to get close to the eddy centre. In the tracking stage, the formation track basically overlaps with the eddy track, and the tracking error is always kept within a small range. And when the tracking error is small, the vehicle could automatically switch to the Argo mode to

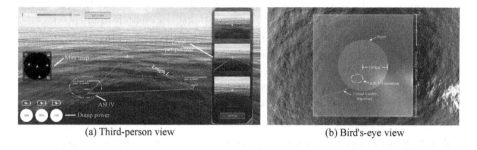

(a) Third-person view   (b) Bird's-eye view

Fig. 8. Visualization of formation observation using the simulation platform.

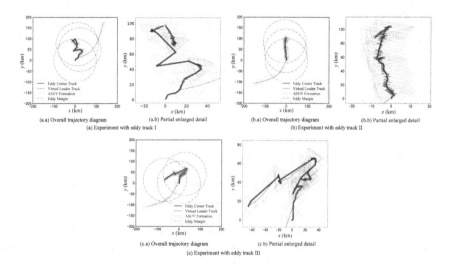

Fig. 9. Formation observation experiments for dynamic eddy center.

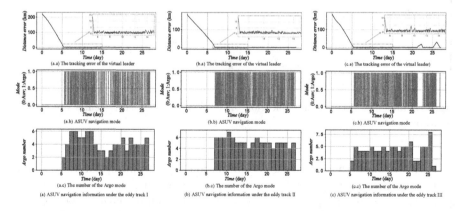

Fig. 10. Tracking error and the state changes of the mulit-ASUV.

profile sampling. In this way, the validity of eddy data collected by the ASUV group can be fully guaranteed. And compared with the widely used methods of manually setting the observation path and manually controlling the vehicle in the past, the tracking accuracy of the proposed method is greatly improved, and it is higher than the method using only satellite data (the time accuracy is 1 day, the average daily error of trajectory prediction of the eddy center is about 8 km, and the 7-day prediction error is about 18.2 km [16]).

Table 3. Performance parameters of dynamic eddy center observation experiments.

Parameters	Value (track I)	Value (track II)	Value (track III)
ASUV1 place point (km)	(147.403,180.528)	(−140.320, −109.517)	(−143.238, −111.963)
ASUV2 place point (km)	(121.635,183.526)	(−164.522, −101.974)	(−165.473, −105.243)
ASUV3 place point (km)	(130.907,166.169)	(−157.843, −120.563)	(−158.96, −124.267)
Search Time (h)	101.219	102.622	125.153
ME (km)	1.160	1.049	1.354
SE (km)	0.802	0.734	1.467
Argo times	93	104	94

4.3 Parameter Analysis

To analyze the influence of different parameter selection on experimental performance, eight groups of eddy centre observation experiments (10 tests for each group) for 28 consecutive days are carried out under different formation length L and sampling interval Δt, and the data of track III is uniformly used. The changes of tracking error in the first two experiments are shown in Fig. 11, and the total statistical results are shown in Table 4.

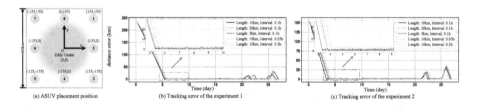

(a) ASUV placement position   (b) Tracking error of the experiment 1   (c) Tracking error of the experiment 2

Fig. 11. Influence of parameter selection on tracking error in eddy center observation experiments.

It can be seen that reducing the formation length appropriately is helpful to improve the tracking results, and the reasonable length can be set by analyzing the dynamic properties of the vehicle, the communication period, and the accuracy of vehicle-borne sensors. At the same time, the length should not be too small to avoid excessive similarity of the sampling data, which will lead to inaccurate estimation of local characteristic fields. A reasonable selection of sampling interval is also essential. If the interval is too small, it may lead to frequent manoeuvring of the vehicle, thus increasing the frequency of the Argo times. If the interval is too large, the vehicle may miss sampling key features.

Table 4. Comparison of experimental performance of eddy center observation under different parameters.

Formation length (km)	Sampling interval (h)	Searching time (h)	ME (km)	SE (km)	Argo times
10	0.1	88.153	1.266	1.455	100
5	0.1	88.412	1.0345	1.468	126
20	0.1	100.682	1.920	1.527	75
10	0.05	88.467	1.011	1.891	129
10	0.2	88.711	1.752	1.251	81

5 Conclusion

In this paper, a high-fidelity oceanic eddy simulation platform is constructed to meet the requirement of synchronous array observation of mesoscale eddy with multi-ASUV, and a multi-vehicle cooperative control method based on the virtual leader and artificial potential field is proposed, as well as an ocean characteristic estimation strategy based on real-time sampling. Sufficient experiments show that under the action of the proposed method, the ASUV group can complete the observation task of ocean eddy while maintaining low tracking error. Our method can be regarded as an effective supplement to previous observation methods, which improves the tracking accuracy of the observer and the validity of the sampling data. At the same time, those experiments also show that the simulation platform has strong practicability in various stages of the vehicle test, application, strategy presentation, and algorithm evaluation.

References

1. Sun, W., Dong, C., Wang, R., et al.: Vertical structure anomalies of oceanic eddies in the K uroshio E xtension region. J. Geophys. Res. Oceans **122**(2), 1476–1496 (2017)
2. Ren, Q., Fei, Y.U., Shuo, L.I., et al.: Analysis on the experimental data acquired from the sea tests for domestic underwater glider in the South China sea in 2014. J. Ocean Technol. **036**, 52–57 (2017)
3. Zhang, Y., Kieft, B., Hobson, B.W., et al.: Autonomous tracking and sampling of the deep chlorophyll maximum layer in an open-ocean eddy by a long-range autonomous underwater vehicle. IEEE J. Oceanic Eng. **45**(4), 1308–1321 (2019)

4. Zhao, W., Yu, J., Zhang, F., et al.: Tracking moving mesoscale eddies with underwater gliders under autonomous prediction and control. Control. Eng. Pract. **113**, 104839 (2021)

5. Zhang, S., Zhang, A., Yu, J.: Ocean observing with underwater glider in South China Seas. In: 2015 IEEE International Conference on Cyber Technology in Automation, Control, and Intelligent Systems (CYBER), pp. 1109–1114. IEEE (2015)

6. Hwang, C., Wu, C.R., Kao, R.: TOPEX/Poseidon observations of mesoscale eddies over the Subtropical Countercurrent: kinematic characteristics of an anticyclonic eddy and a cyclonic eddy. J. Geophys. Res. Oceans **109**(C8) (2004)

7. Yang, G., Wang, F., Li, Y., et al.: Mesoscale eddies in the northwestern subtropical Pacific Ocean: Statistical characteristics and three-dimensional structures. J. Geophys. Res. Oceans **118**(4), 1906–1925 (2013)

8. Matsuoka, D., Araki, F., Sasaki, H.: Event detection and visualization of ocean eddies simulated by ocean general circulation model. Int. J. Model. Simul. Sci. Comput. **10**(03), 1950018 (2019)

9. Zong, Z., Xiong, X.J., Liu, Y.H., et al.: The method of mesoscale eddy observation using underwater lider. Adv. Mar. Sci. **2** (2018). (in Chinese)

10. Xin, B., Zhang, J., Chen, J., et al.: Overview of research on transformation of multi-AUV formations. Complex Syst. Model. Simul. **1**(1), 1–14 (2021)

11. Song, D., Gan, W., Yao, P., et al.: Guidance and control of autonomous surface underwater vehicles for target tracking in ocean environment by deep reinforcement learning. Ocean Eng. **250**, 110947 (2022)

12. Nie, Y., Luan, X., Gan, W., et al.: Design of marine virtual simulation experiment platform based on Unity3D. In: Global Oceans 2020: Singapore-US Gulf Coast, pp. 1–5. IEEE (2020)

13. Chen, L., Jin, Y., Yin, Y.: Ocean wave rendering with whitecap in the visual system of a maritime simulator. J. Comput. Inf. Technol. **25**(1), 63–76 (2017)

14. Zhao, W.T., Yu, J.C., Zhang, A.Q., et al.: Dynamic characteristic detection of mesoscale eddies based on SLA data. J. Mar. Sci. **3**, 62–68 (2016)

15. Leonard, N.E., Fiorelli, E.: Virtual leaders, artificial potentials and coordinated control of groups. In: Proceedings of the 40th IEEE Conference on Decision and Control (Cat. No. 01CH37228), vol. 3, pp. 2968–2973. IEEE (2001)

16. Wang, X., Wang, X., Yu, M., et al.: MesoGRU: deep learning framework for mesoscale eddy trajectory prediction. IEEE Geosci. Remote Sens. Lett. **19**, 1–5 (2021)

Analysis of Influencing Factors of Ground Imaging and Positioning of High-Orbit Remote Sensing Satellite

Wenyu Zhu[✉], Xinyan Liu, and Zuowei Wang

Beijing Institute of Control Engineering, Beijing 100190, China
Pisuke1203@163.com

Abstract. With the continuous development of remote sensing satellites, it has become the core task of optimizing the application efficiency of high-orbit remote sensing satellites to improve the accuracy of imaging positioning. Aiming at the situation that the payload camera and the star sensor are installed on the same board, a ground imaging positioning model of remote sensing satellite is proposed, and the distance between the theoretical and actual imaging points is used as the evaluation index of positioning accuracy. The influence of key error factors such as the attitude and orbit parameters and the installation angle of the payload camera on the positioning accuracy are simulated and analyzed. The analysis results show that the satellite orbit amplitude angle error and the ascending node equinox error have the greatest impact on the positioning accuracy, and the load camera X and Y axis installation angle error has the second highest impact on the positioning accuracy. It is necessary to comprehensively consider various error factors in order to effectively improve the ground imaging positioning accuracy of high-orbit remote sensing satellites.

Keywords: Remote sensing satellite · Imaging localization; simulation model · Attitude orbit control · Positioning accuracy · Analysis of influencing factors

1 Introduction

Remote sensing satellites, artificial satellites which are used as remote sensing platforms in outer space, can cover the entire Earth or any designated area within a specified time. With the rapid development of aerospace technology, geostationary orbit satellites can realize continuous remote sensing of a designated area on the surface of the Earth, and have advantages of rapid response to key events and near real-time observation of small and medium-scale targets. It has become an important development direction in the field of remote sensing satellite [1, 2]. With the launch of a number of high-resolution satellites, such as France's SPOT-5 and America's QuickBird, WorldView and GeoEye, satellite operators have publicly announced the positioning capabilities of their satellite systems, typically within 50 m. Typical advanced satellites represented by the United States and France, such as WorldView-3/4, GeoEYE-1, PleIADES-1, etc., have positioning accuracy of 5 m and 3M without control points. However, typical high

W. Fan et al. (Eds.): AsiaSim 2022, CCIS 1713, pp. 261–272, 2022.
https://doi.org/10.1007/978-981-19-9195-0_22

resolution optical remote sensing satellite positioning accuracy is around 50M in China, and there is a certain gap with the international.

Regarding the imaging positioning of remote sensing satellites, Zhang Renwei proposed a remote sensing image geometric positioning model for obtaining the data of remote sensing points by scanning the intersection of the optical axis line of sight and the surface of the Earth [3]. Ling Qiong calculated the coordinates of the imaging point of the space-borne camera in the ground-fixed coordinate system based on the satellite's orbit determination data and high-precision attitude data, and proposed a method to eliminate the deviation of the vertical line of the Earth [4]. Man Yiyun proposed to use the inner azimuth elements of the optical remote sensor and the outer azimuth elements of the satellite platform's attitude and orbit to jointly establish a physical geometric model based on the collinear equation to achieve accurate geometric positioning of remote sensing images [5]. Regarding the pointing accuracy of the on-board payload, Pan Bo proposed a pointing model of the on-board antenna using a homogeneous transformation matrix instead of a directional cosine array which considers the on-board payload installation displacement and installation angles and their errors [6].

Regarding the geometric positioning error of remote sensing satellites, Li Deren divided it into inner and outer azimuth element errors [7]. The inner azimuth element describes the mutual positional relationship between the photographic center of the payload camera and the image, including the focal length and the image principal point in the image. The external orientation element refers to the space position and attitude of the photography beam at the moment of photographing, including the orbital term and attitude term.

Studies have established simplified models of remote sensing satellite positioning, from the perspective of imaging precision of remote sensing satellite is analyzed, in this paper, from the Angle of satellite on-orbit control precision, optical remote sensing satellite as the research object, based on optical remote sensing camera load, puts forward a model of remote sensing satellite ground imaging localization, aiming at each key error affects positioning surface imaging simulation analysis, Finally, some suggestions are given to improve the ground imaging positioning accuracy of high orbit remote sensing satellite.

2   Ground Imaging Positioning Modeling

The composition of the ground imaging system of the remote sensing satellite is shown in Fig. 1. The system consists of the satellite body, the payload camera and the star sensor. The payload camera and the star sensor are connected by a common plate installation method, and are connected with the satellite body, and move according to their instructions.

In order to clearly describe the ground imaging system of remote sensing satellites and their errors, the following coordinate systems are established: the geocentric equatorial inertial coordinate system $O_I X_I Y_I Z_I$, the Earth-fixed coordinate system $O_E X_E Y_E Z_E$, the orbital coordinate system $O_O X_O Y_O Z_O$, the star sensor measurement coordinate system $O_S X_S Y_S Z_S$, and the satellite payload camera coordinate system $O_C X_C Y_C Z_C$.

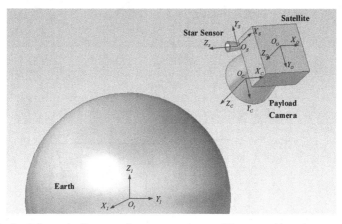

Fig. 1. Schematic diagram of remote sensing satellite ground imaging system

Define the point of the center of the Earth as point O, the point of the satellite as point S, the satellite position vector $\boldsymbol{r} = \overrightarrow{OS}$, define the direction vector of the optical axis of the satellite payload camera as $\boldsymbol{u}$, and define the line in the direction of $\boldsymbol{u}$ to intersect the surface of the Earth at point E, which is the ground of the remote sensing satellite. The imaging anchor point, $\boldsymbol{H} = \overrightarrow{SE} = H \cdot \boldsymbol{u}$, H is the distance between point S and point E. Define vector $\boldsymbol{\rho} = \overrightarrow{OE}$. The geometric relationship is given by

$$\boldsymbol{\rho} = \boldsymbol{r} + \boldsymbol{H} \tag{1}$$

The coordinate vector $\boldsymbol{r}_I$ of the direction vector $\boldsymbol{r}$ in the geocentric inertial coordinate system can be obtained from the six roots of the satellite's orbit [8],

$$r_i = C_{IO}r_O = C_{OI}^T r_O \tag{2}$$

where $r_O = \begin{bmatrix} 0 & 0 & -r \end{bmatrix}^T$, is the coordinate vector of r in the orbital coordinate system. $r = a(1 - e^2)/(1 - e\cos u)$ is the distance between the center of the Earth and the satellite. a, e, i, Ω, ω and f are orbital elements. $u = \omega + f$ is the orbital argument. Δi, $\Delta\Omega$ and Δu are the errors of orbital inclination, right ascension of ascending node and argument of orbit respectively.

The coordinate vector $\boldsymbol{u}_I$ of the direction vector u in the geocentric inertial coordinate system can be obtained from the satellite attitude parameters [9, 10],

$$u_I = C_{IS}u_S \tag{3}$$

where $\widehat{\boldsymbol{q}}_{IS} = \begin{bmatrix} \widehat{q}_1 & \widehat{q}_2 & \widehat{q}_3 & \widehat{q}_4 \end{bmatrix}^T$, $\widehat{q}_i = q_i + \Delta q_i (i = 1, 2, 3, 4)$. $\begin{bmatrix} q_1 & q_2 & q_3 & q_4 \end{bmatrix}^T$ is the satellite attitude quaternion obtained by the star sensor, which $\begin{bmatrix} q_1 & q_2 & q_3 \end{bmatrix}^T$ is the vector part, and q_4 is the scalar part. Δq_1, Δq_2, Δq_3 and Δq_4 is the satellite attitude quaternion error.

Reference construction of satellite payload camera and star sensor is an important means to improve the ground imaging and positioning accuracy of remote sensing satellites. In engineering, the main method to realize reference construction is to install the two plates together, as shown in Fig. 2.

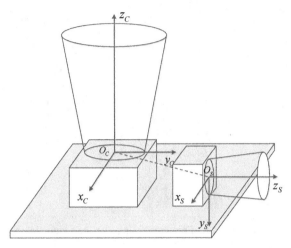

Fig. 2. Schematic diagram of mounting load camera and star sensor on board

The coordinate vector $\mathbf{u}_S$ of the direction vector $\mathbf{u}$ in the measurement coordinate system of the star sensor can be obtained from the installation displacement and installation angles of the star sensor relative to the satellite payload camera,

$$\begin{bmatrix} \mathbf{u}_S \\ 1 \end{bmatrix} = \begin{bmatrix} \mathbf{C}_{SC} & \mathbf{L}_{SC} \\ \mathbf{0} & 1 \end{bmatrix} \begin{bmatrix} \mathbf{u}_C \\ 1 \end{bmatrix} \tag{4}$$

where $\mathbf{C}_{SC}$ is the transformation matrix of the satellite sensor measurement coordinate system relative to the satellite payload camera coordinate system, $\mathbf{L}_{SC} = \begin{bmatrix} L_{SCx} + \Delta L_{SCx} & L_{SCy} + \Delta L_{SCy} & L_{SCz} + \Delta L_{SCz} \end{bmatrix}^T$. L_{SCx}, L_{SCy} and L_{SCz} are the displacements of the star sensor measurement coordinate system relative to the satellite payload camera coordinate system in x_C, y_C, z_C directions. ΔL_{SCx}, ΔL_{SCy} and ΔL_{SCz} are the displacement errors. φ_{SC}, θ_{SC} and ψ_{SC} are the installation angles of the star sensor relative to the satellite payload camera, and the rotation sequence is ZYX. $\Delta\varphi_{SC}$, $\Delta\theta_{SC}$ and $\Delta\psi_{SC}$ is the installation angles errors. $\mathbf{u}_C = \begin{bmatrix} 0 & 0 & 1 \end{bmatrix}^T$ is the coordinate vector of the direction vector $\mathbf{u}$ in the coordinate system of the satellite payload camera.

The geometric relationship is solved in the geocentric inertial coordinate system, and the earth is regarded as a rotating ellipsoid with OZ_I as the axis, that is, the elliptic surface obtained through the intersection of the plane of OZ_I axis and the globe have the same characteristics. Referring to the geometric relationship, the surface equation can be expanded as

$$\frac{(Hu_{Ix} + r_{Ix})^2 + (Hu_{Iy} + r_{Iy})^2}{R_e^2} + \frac{(Hu_{Iz} + r_{Iz})^2}{R_p^2} = 1 \tag{5}$$

where R_e is the equatorial radius of the Earth, and R_p is the polar radius of the Earth.

Coordinate vector $\boldsymbol{\rho}$ in the geocentric inertial coordinate system [6]

$$\boldsymbol{\rho}_I = \mathbf{r}_I + \mathbf{H}_I = \mathbf{r}_I + H \cdot \mathbf{u}_I \tag{6}$$

where H is obtained by substituting the components of r_I and u_I into Eq. (8)

Coordinate vector ρ in Earth-fixed coordinate system

$$\rho_E = C_{EI}\rho_I = C_{IE}^T\rho_I \tag{7}$$

where $C_{IE} = C_3(G(t))$, $G(t)$ is the vernal equinox angle of Greenwich.

To calculate the geocentric longitude and latitude of remote sensing satellite ground imaging positioning points on the surface of the Earth

$$\lambda = \arctan\left(\frac{\rho_{Ey}}{\rho_{Ex}}\right) \tag{8}$$

$$\delta = \arcsin\left(\frac{\rho_{Ez}}{|\rho_E|}\right) \tag{9}$$

The geocentric latitude of the imaging positioning point obtained based on the above steps does not eliminate the vertical line deviation. It is necessary to convert the geocentric latitude δ into geographic latitude δ'

$$\delta' = \arctan\left(\frac{\tan\delta}{(1-h)^2}\right) \tag{10}$$

where $h = (R_e - R_p)/R_e$ is the flattening of the Earth. Then, the geographic latitude and longitude of the imaging anchor point $\left[\lambda\ \delta'\right]^T$ can be calculated.

3 Analysis of Ground Imaging Positioning Accuracy of High-Orbit Remote Sensing Satellites

For high-orbit remote sensing satellites, the key error terms of ground imaging positioning can be divided into satellite orbit error and payload camera attitude error [11, 12], as shown in Fig. 3. The satellite attitude determination error is expressed as attitude quaternion error. The assembly error is mainly represented by the installation displacement error of the load and the thermal deformation error is represented by the load installation angles errors.

Combining the above (1)–(11), the geographic latitude and longitude of the imaging location can be summarized as

$$\begin{bmatrix} \lambda \\ \delta' \end{bmatrix} = \begin{bmatrix} f_1(O, q, \psi, E) \\ f_2(O, q, \psi, E) \end{bmatrix} \tag{11}$$

where O is the orbit parameter vector, q is the attitude parameter vector of star sensor, ψ is the installation parameter vector of star sensor and payload camera, E is the error vector containing all the errors of orbit, attitude and payload camera installation parameters.

Regardless of all errors, $E = 0$, the geographic latitude and longitude of the remote sensing satellite ground imaging positioning point is $\left[\lambda_1\ \delta_1'\right]^T$.

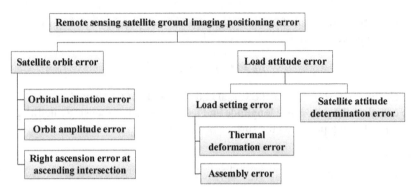

Fig. 3. Analysis of positioning error of remote sensing satellite ground imaging

Considering all errors E, the geographic latitude and longitude of the remote sensing satellite ground imaging location point is $\left[\lambda_2 \; \delta'_2\right]^T$.

Geographic latitude and longitude error of remote sensing satellite ground imaging positioning can be expressed as

$$\Delta\lambda = \lambda_2 - \lambda_1 \tag{12}$$

$$\Delta\delta' = \delta'_2 - \delta'_1 \tag{13}$$

The final remote sensing satellite ground imaging positioning accuracy can be expressed as the distance between the theoretical imaging point and the actual imaging point on the surface of the Earth [13, 14]. The larger the distance, the lower the positioning accuracy.

$$\Delta L_E = \sqrt{\left(\frac{\Delta\lambda}{360°} \cdot 2\pi R_e\right)^2 + \left(\frac{\Delta\delta'}{180°} \cdot \left[\pi R_p + 2\left(R_e - R_p\right)\right]\right)^2} \tag{14}$$

4 Simulation Analysis of Key Factors Affecting Imaging Positioning Accuracy

Based on the ground imaging positioning model of remote sensing satellite, the influence of orbit parameters, attitude parameters, installation parameters of star sensor and payload camera on positioning accuracy is studied. For the remote sensing satellite in the geostationary orbit, the orbital inclination is about 0°, the orbital eccentricity is about 1, which is a nearly circular orbit. The distance between the satellite and the center of the earth r can be simplified to a fixed value of 42164 km.

Taking a geostationary orbit remote sensing satellite as an example, the parameter values of the simulation model are shown in Table 1, and the error values of each link are shown in Table 2. Since the distance between the satellite and the ground is much larger than the installation size between the payload camera and the star sensor, the influence of the installation displacement error is small and can be ignored.

Table 1. Simulation model parameters

Model parameters	Numerical value
Right ascension of ascending node Ω	$105.0400°$
Perigee Angle ω	$60.1402°$
Satellite attitude quaternion **q**	$[-0.0000, -0.0000, 0.7924, 0.6100]^{T}$
Payload camera installation angles $\begin{bmatrix} \varphi_{SC} & \theta_{SC} & \psi_{SC} \end{bmatrix}^{T}$	$[90°, 0°, 0°]^{T}$
Payload camera installation displacement $\begin{bmatrix} L_{SCx} & L_{SCy} & L_{SCz} \end{bmatrix}^{T}$	$[0\ m, 1\ m, 0\ m]^{T}$

Table 2. Error of each link

Error source	Difference	Error source	Difference
Orbital inclination error Δi	$0.001°$	Load camera installation angle error $\Delta \varphi_{SC}$	$0.001°$
Argument of orbit error Δu	$0.001°$	Load camera installation angle error $\Delta \theta_{SC}$	$0.001°$
Right ascension of ascending node error $\Delta \Omega$	$0.001°$	Load camera installation angle error $\Delta \psi_{SC}$	$0.001°$

Set the orbital inclinations of high orbit remote sensing satellites as $0°$, $2.5°$ and $5°$ respectively, and the kinematics equations are solved. The influence of orbit parameter error, attitude quaternion error and loading camera installation angles errors on the latitude, longitude and positioning accuracy of the ground imaging point are calculated.

4.1 Influence of Orbital Parameter Errors

4.1.1 Influence of Right Ascension of Ascending Node Error

Because the right ascension of ascending node error of satellite directly affects the geographic longitude of the ground imaging anchor point, the influence is equal proportion transmission, and it has no influence on the geographic longitude and latitude of the ground imaging anchor point. When the right ascension of ascending node error is $\Delta \Omega = 0.001°$, the influence of the imaging positioning accuracy of one orbit is a fixed value, which is about 111.32 m.

4.1.2 Influence of Argument of Orbit Error

Figure 4 shows the variation trend of the positioning accuracy of the positioning point affected by the argument of orbit error $\Delta u = 0.001°$ during the satellite orbital orbit under different orbital inclinations. The positioning error distance fluctuates periodically, the minimum positioning error distance is about 1113.2 m, and the fluctuation amplitude increases with the increase of the orbital inclination. The larger the orbital inclination, the larger the longitude and latitude errors, and the lower the positioning accuracy. This is because with the increase of orbital inclination, the satellite is no longer strictly in the geostationary orbit directly above the earth's equator, and the trajectory of the subsatellite point is in the shape of "8". The ground imaging site has a southern and northern displacement in the process of orbiting the orbit of the satellite, causing the error of the fixed point of $\Delta u = 0.001°$ to have longitude and latitude, and the latitude component increases with the Angle of the orbital.

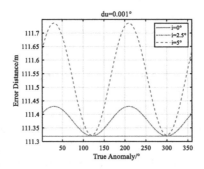

Fig. 4. Positioning accuracy with different orbital inclinations affected by Δu during 1 orbit

4.1.3 Influence of Orbital Inclination Error

Under different orbital inclination angles, the change trends of longitude and latitude errors of positioning points and positioning accuracy affected by argument of orbit error $\Delta i = 0.001°$ in one week of satellite orbit overlap. The positioning error distance fluctuates periodically from 0.0785 m to 111.8467 m with a period of 180°, and the fluctuation amplitude is not affected by the change of orbital inclination.

It is shown in Table 3 that the maximum longitude and latitude error of the anchor point, the positioning accuracy and the changes of each error term affected by the orbital inclination error $\Delta i = 0.001°$ during one orbit of the satellite under different orbital inclination angles. As the orbital inclination increases from 0° to 5°, the longitude error caused by the argument of orbit error increases, while the latitude error changes little, and the maximum positioning accuracy error increases.

Table 3. Table captions should be placed above the tables.

Orbital inclination/°	Maximum longitude error/°	Maximum positioning degree error change/10^{-8}°	Maximum latitude error/10^{-4}°	Maximum positioning latitude degree error variation/10^{-7}°	Top positioning accuracy/m	Minimum positioning accuracy/m	Minimum positioning accuracy change/10^{-3} m
0	0.0010	0.0001	0.0000	0.1755	111.3195	111.3195	0.0001
0.001	0.0010		0.0002		111.3195	111.3195	
2.5	0.0010	0.0763	0.4387	0.1754	111.3205	111.4255	0.0849
2.501	0.0010		0.4389		111.3205	111.4256	
5	0.0010	0.1533	0.8766	0.1749	111.3235	111.7447	0.1707
5.001	0.0010		0.8768		111.3235	111.7449	

4.2 Influence of Attitude Quaternion Error

Through the simulation experiment, when the value of orbital inclination i changes, the positioning accuracy of the satellite in one orbit is invariable due to the error of each element of the satellite attitude quaternion. The attitude quaternion errors are set to 0.0001, 0.0005, 0.001 and 0.002 respectively, and the calculation results of the positioning accuracy are shown in Fig. 5.

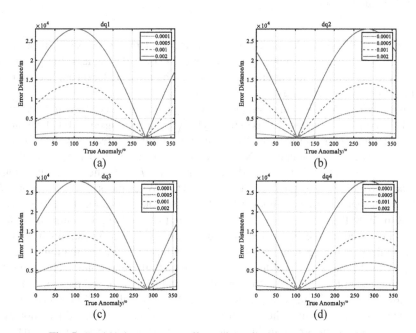

Fig. 5. Positioning accuracy affected by attitude error during 1 orbit

When the orbital inclination $i = 0°$, the positioning accuracy fluctuates periodically with the orbit of the satellite affected by the attitude quaternion error. Taking the error term Δq_1 as an example, as shown in Fig. 6(a), while $\Delta q_1 = 1 \times 10^{-4}$, the positioning error distance fluctuates between 0.2425 m and 1403.6 m, and the fluctuation range is 1403.4 m. While $\Delta q_1 = 1 \times 10^{-3}$, the positioning error range fluctuates between 2.425 m and 14036 m, and the fluctuation range is 14034 m. The distance of positioning error increases proportionally with the increase of quaternion error of satellite attitude.

4.3   Influence of Installation Angles Errors of Payload Camera

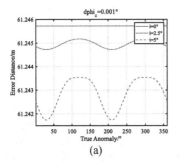 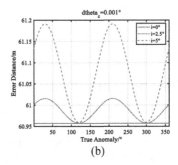

(a)   (b)

Fig. 6. Positioning accuracy of the loaded camera installation error with different orbital inclinations during 1 orbit

Since the installation angles of the payload camera relative to the star sensor around Z axis does not affect the pointing of the optical axis of the camera, the positioning accuracy of the satellite during orbit is not affected by the installation angle error of the payload camera around Z axis.

Figure 6(a) shows the variation curve of the positioning accuracy affected by the installation angle error of the X-axis of the payload camera during one orbit of the satellite under different orbital inclinations. The positioning accuracy affected by the installation angle error of the loading camera X axis $\Delta \varphi_{SC} = 0.001°$ fluctuates periodically with the orbit of the satellite. When the orbital inclination is $i = 0°$, the highest positioning accuracy is 61.2457 m and the fluctuation is about 0m. When the orbital inclination is $i = 5°$, the highest positioning accuracy is 61.2417 m, and the fluctuation amplitude is about 0.0018 m. With the increase of orbital inclination, the satellite positioning accuracy is slightly improved and the fluctuation amplitude increases. Figure 6(b) shows the variation curves of the positioning accuracy of the satellite affected by the installation angle error of the payload camera's Y-axis during one orbit under different orbital inclinations. The positioning accuracy affected by the installation angle error of the payload camera Y axis $\Delta \theta_{SC} = 0.001°$ fluctuates periodically with the orbit of the satellite, and the highest accuracy is about 60.96 m. When the orbital inclination is $i = 0°$, the fluctuation range is about 0 m. When the orbital inclination is $i = 5°$, the fluctuation range is about 0.2321 m. The amplitude of fluctuation increases with the increase of orbital inclination.

5   Conclusions

In this paper, a remote sensing satellite ground imaging positioning model based on the co-location of the payload camera and the star sensor is given, and the distance between the theoretical imaging point and the actual imaging point on the Earth's surface is taken as the evaluation index of remote sensing satellite ground imaging positioning accuracy. The influence of key error factors such as orbit parameters, attitude parameters and load camera installation angle on the ground imaging positioning accuracy is analyzed through simulation calculations. The analysis results show that the influence of satellite orbit amplitude angle error and ascending node equinox error on imaging positioning accuracy is about one hundred meters, which is the most significant, and the influence of load camera X and Y axis installation angle error on imaging positioning accuracy is the next most significant, which is about ten meters. Therefore, it is necessary to choose an orbit with the smallest possible inclination angle and improve the satellite orbiting accuracy, reduce the thermal deformation error of the load camera installation, and obtain the satellite attitude with a high precision star sensor in order to effectively improve the ground imaging and positioning accuracy of high-orbiting remote sensing satellites. The model and analysis method proposed in this paper have certain reference value for improving the effectiveness of high-orbiting satellites in other fields of application and further improving the imaging and positioning accuracy of low-orbiting remote sensing satellites.

References

1. Wang, M., et al.: High-precision on-orbit geometric calibration of Gaofen-4 geostationary satellite. J. Survey. Map. **46**(1), 9 (2017). (in Chinese)
2. Wang, M., et al.: On-orbit geometric calibration and geometric quality assessment for the high-resolution geostationary optical satellite GaoFen4. ISPRS J. Photogram. Remote Sens. **125**, 63–77 (2017)
3. Renwei, Z.: Satellite Orbit Attitude Dynamics and Control. Beihang University Press, Beijing (1998). (in Chinese)
4. Ling, Q., et al.: A new attitude control accuracy evaluation method for agile remote sensing satellites. Spacecr. Eng. **28**(6) (2019). (in Chinese)
5. Yiyun, M., Zhijun, J.: Error analysis of optical remote sensing satellite plane positioning accuracy. Aerosp. Return Remote Sens. **42**(1) (2021) (in Chinese)
6. Bo, P., Zhang, D.: Modeling and analysis of satellite-borne antenna pointing accuracy. Spacecr. Eng. **20**(5) (2011). (in Chinese)
7. Li, D., Wang, M.: A review of high-resolution optical satellite mapping technology. Space Return Remote Sens. **41**(2) (2020). (in Chinese)
8. Zhenduo, L., Yongjun, L.: Satellite Attitude Measurement and Determination. National Defense Industry Press, Beijing (2013). (in Chinese)
9. Markley, F.L., Mortari, D.: Quaternion attitude estimation using vector observations. J. Astronaut. Sci. **48**(2) (2000)
10. Zuowei, W., Xiaoxiang, L., Shuhua, Z.: Attitude and Orbit Control Technology of Geostationary Satellite. China Aerospace Press, Beijing (2021). (in Chinese)
11. Roques, S., et al.: Satellite attitude instability effects on stereo images. In: IEEE International Conference on Acoustics. IEEE (2004)

12. Huang, Y., et al.: Research and discussion on high precision image location technology in large scale surveying and mapping mission. In: 2018 Eighth International Conference on Instrumentation & Measurement, Computer, Communication and Control (IMCCC) (2018)
13. Xia, Z., et al.: Study on geometric performance assessment method of high resolution optical remote sensing satellite imagery. In: International Symposium on Advanced Optical Manufacturing & Testing Technologies: Large Mirrors & Telescopes. International Society for Optics and Photonics (2016)
14. Guoliang, T., Qiaolin, H., Hongyan, H., Zhongqiu, X.: Analysis of geometric positioning accuracy evaluation method of remote sensing satellite image. Space Return Remote Sens. **38**(05), 106–112 (2017). (in Chinese)

Mass Characteristics Identification and Intelligent Automatic Balancing Technology of Three-Axis Air-Bearing Testbed

Xu Xu[1], Zhang Jierui[1], Qin Jie[2], Li Zening[2], Ma Guangcheng[1], and Xia Hongwei[1($\boxtimes$)]

[1] School of Astronautics, Harbin Institute of Technology, Harbin 150001, China
hxia@hit.edu.cn

[2] Shanghai Aerospace Control Technology Institute, Shanghai 201109, People's Republic of China

Abstract. Mass characteristic identification and balancing of the centre of mass are the core key technologies of the full physical trial based on the three-axis air-bearing testbed. According to the above point of view, this paper proposes a mass characteristics identification method and an intelligent automatic balancing technology of a large three-axis air-bearing testbed. First, we establish the identification model of the three-axis air-bearing testbed based on the Euler angle. Next, an online identification algorithm of mass characteristics of the three-axis air-bearing testbed is investigated based on the asymptotic least squares method. Utilizing the mass characteristics obtained by the online identification algorithm, an intelligent automatic balancing algorithm for the three-axis air-bearing testbed is proposed. Finally, a physical trial is carried out to illustrate the benefits and effectiveness of the online identification algorithm and the intelligent automatic balancing algorithm. The physical trial results show that the intelligent automatic balancing algorithm proposed in this paper can be applied to most three-axis air-bearing testbed systems and the accuracy of the identification algorithm is better than 3.8%, which has high stability and application value.

Keywords: Three-axis air-bearing testbed · Mass characteristics · Asymptotic least squares · Online identification algorithm · Intelligent automatic balancing algorithm · Physical trial

1 Introduction

With the development of science and technology, the structure of spacecraft is becoming more and more complex. It is difficult to achieve the desired results during spacecraft development with digital and semi-physical simulation [1]. Only the full physical simulation method can simulate microgravity on the ground, and carry out the test and analysis of the whole process of the space mission. Among them, the air-bearing method is the most widely used and the most mature technology for the full physical simulation test method of the spacecraft on the ground [2, 3].

W. Fan et al. (Eds.): AsiaSim 2022, CCIS 1713, pp. 273–287, 2022.
https://doi.org/10.1007/978-981-19-9195-0_23

In order to simulate the microgravity and undamped space environment on the ground, the rotation centre and the centre of mass of the three-axis air-bearing testbed must be coincident; otherwise, there will generate the eccentric moment of gravity, and the ground simulation test of the space mission cannot be carried out. Therefore, the balance adjustment technology is the prerequisite for the ground simulation test using the three-axis air-bearing testbed. Besides, the mass characteristic parameter identification technology is essential to the high-precision automatic balancing technology [4].

There are some studies in the field of mass characteristics identification. Wang and Hou proposed a satellite mass characteristics identification method using a thruster combination as excitation, which decouples the mass centre and inertia matrix to reduce the identification error [5, 6]. Aiming at the large-angle maneuvering problem of spacecraft, He et al. proposed an on-orbit identification method for flexible spacecraft rotational inertia based on Kalman filtering for modal state estimation of flexible attachment vibration [7]. However, most of the above methods are only based on theory and digital simulation, lacking the full physical test and the guidance for practical applications.

In practical engineering, the most widely used method for balancing the three-axis air-bearing testbed is the manual method. The advantage of the manual method is that the operation is relatively simple, and there is no need to design a complex algorithm. However, the balance accuracy is challenging to meet the requirements of ground tests. Therefore, an intelligent balance adjustment algorithm is urgently needed in engineering to further reduce the eccentric moment of gravity of the three-axis air-bearing testbed to improve the reliability of the ground full physical simulation test. Liu compensated the constant interference moment by introducing the tilting moment, so as to realize the automatic balance of the air-floating table [8], but they only carried out digital experiments and lacked guidance for practical application. Chen proposed an automatic balancing algorithm based on the compound pendulum model [9], but the test object was a small 30 kg air-floating testbed, and it was relatively easy to adjust the balance. Xiang introduced a tilting moment to compensate for the constant interference moment of the table body [10], but this method of adjusting the base is a manual balancing method with low precision.

Based on the above analysis and the existing problems, we propose a high-efficiency online identification method based on the asymptotic least squares method for a large three-axis air-bearing testbed with a load of 500 kg, and an intelligent adjustment method of the centre of mass is given according to the identification results.

The organization of this paper proceeds as follows: Sect. 2 presents the design scheme of the three-axis air-bearing testbed system; the mass characteristic identification algorithm of the triaxial air flotation table is given in Sect. 3; Sect. 4 presents the intelligent automatic balancing algorithm for the centre of mass of the three-axis air-bearing testbed; practical experiments results and the analysis are given in Sect. 5, and Sect. 6 concludes the paper.

2   Three-Axis Air-Bearing Testbed Simulation System

The three-axis air-bearing testbed simulation system is divided into an on-testbed integrated test system and a monitoring system outside the testbed.

The on-testbed integrated test system consists of a three-axis air-bearing testbed with an instrumentation platform, an on-testbed management and control unit, a power supply module, a balancing module, and a data transmission module. The instrumentation platform mainly provides the hardware installation interface for each component of the on-testbed integrated test system; the on-testbed management and control unit is primarily responsible for the control and management of the on-testbed simulation test equipment; the power supply module is used to supply power to all electrical equipment on the on-testbed integrated test system; the balancing module is used to adjust the mass characteristics of the three-axis air-bearing testbed before the test; the data transmission function module is used to collect all experimental data on the stage and the operational status of each equipment and send it to the off-stage monitoring system; the data transmission module contains all the experimental data and the functional level of each piece of equipment on the testbed and sends them to the monitoring system outside the testbed.

The monitoring system outside the testbed consists of a data transmission module, a safety auxiliary module and monitoring software. The data transmission module is responsible for receiving the experimental data sent from the data transmission module on the testbed; the safety auxiliary module is used to limit the motion of the three-axis air-bearing table and support it; the monitoring software is used to display, analyze, store and playback the experimental data, and command the on-testbed integrated test system (Figs. 1 and 2).

Fig. 1. Model diagram of the three-axis air-bearing testbed.

In practical engineering, the rotation centre and the centre of mass of the three-axis air-bearing testbed must be coincident; otherwise, the eccentric moment of gravity will lead to the failure of the ground simulation test. Therefore, the balance adjustment technology is the prerequisite for the ground simulation test using the three-axis air-bearing testbed. Besides, the mass characteristic parameter identification technology is essential to the high-precision automatic balancing technology.

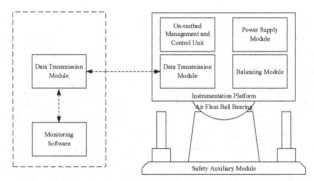

Fig. 2. System composition diagram.

3 Algorithm for the Identification of Mass Characteristics of the Three-Axis Air-Bearing Testbed

3.1 Mathematical Model

Coordinate System Definition. Different coordinate systems need to be defined to facilitate the attitude description of the three-axis air-bearing testbed in the study of the dynamics model of a three-axis air-bearing testbed. The coordinate systems required in this paper are defined as follows:

(1) **Geographic Coordinate System**
   The origin of the geographic coordinate system $O\ x_0\ y_0\ z_0$ is the rotation centre of the three-axis air-bearing testbed; the z_0 axis is vertically upward, the x_0 axis points to the geographic east direction, and the y_0 axis points to the geographic north direction, forming a right-handed orthogonal coordinate system.
(2) **Ontology Coordinate System**
   The origin of the ontology coordinate system $O\ x\ y\ z$ is the rotation centre of the three-axis air-bearing testbed. In the initial state, the ontology coordinate system coincides with the geographic coordinate system. When the attitude movement occurs, the ontology coordinate system starts to rotate and no longer overlaps with the geographic coordinate system (Fig. 3).

Kinetics Analysis. Define T_{Gb} as the gravitational interference moment of the three-axis air-bearing testbed in the ontology coordinate system, M is the mass of the instrumentation platform, r_p is the distance vector from the rotation centre to the centre of mass, and g_b is the projection of the gravitational acceleration vector in the ontology coordinate system, then we have:

$$T_{Gb} = M \cdot (r_p \times g_b) = M r_p \times (A_{Gb} \cdot g) \tag{1}$$

$$A_{Gb} = \begin{bmatrix} c\theta c\gamma - s\varphi s\theta s\gamma & -s\gamma c\varphi & s\theta c\gamma + s\varphi s\gamma c\theta \\ s\gamma c\theta + s\varphi s\theta c\gamma & c\varphi c\gamma & s\theta s\gamma - s\varphi c\theta c\gamma \\ -s\theta c\gamma & s\varphi & c\varphi c\theta \end{bmatrix} \tag{2}$$

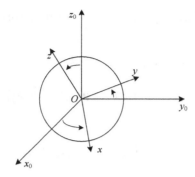

Fig. 3. Coordinate system definition diagram.

where A_{gb} is the Euler angle attitude transformation matrix in the 3–1–2 rotation sequence; φ, θ, γ are the pitch angle around the x-axis, the roll angle rotating around the y-axis, and the yaw angle rotating around the z-axis of the ontology coordinate system; $g = \begin{bmatrix} 0 & 0 & g \end{bmatrix}^T$ is the projection of the gravitational acceleration in the geographic coordinate system.

Substitute Eq. (2) into Eq. (1) and we can get:

$$T_{Gb} = A(\varphi, \theta) \cdot Mgr_p \tag{3}$$

where

$$A(\varphi, \theta) = \begin{bmatrix} 0 & a_3(\varphi, \theta, \gamma) & -a_2(\varphi, \theta, \gamma) \\ -a_3(\varphi, \theta, \gamma) & 0 & a_1(\varphi, \theta, \gamma) \\ a_2(\varphi, \theta, \gamma) & -a_1(\varphi, \theta, \gamma) & 0 \end{bmatrix} \tag{4}$$

$$a_1(\varphi, \theta, \gamma) = \sin\theta \cos\gamma + \sin\varphi \sin\gamma \cos\theta \tag{5}$$

$$a_2(\varphi, \theta, \gamma) = \sin\theta \sin\gamma - \sin\varphi \cos\theta \cos\gamma \tag{6}$$

$$a_3(\varphi, \theta, \gamma) = \cos\varphi \cos\theta \tag{7}$$

Then the dynamic equation of the three-axis air-bearing testbed can be obtained as follows:

$$T_{Gb} + T = J \cdot \dot{\omega} \tag{8}$$

where T is the moment excitation, J is the rotational inertia matrix, and ω is the projected angular velocity of the rotation of the three-axis air-bearing testbed in the inertial coordinate system.

Online Identification Algorithm Based on Asymptotic Least Squares.
The dynamic equation of the three-axis air-bearing testbed is given in the previous subsection, and the online identification algorithm of mass characteristics based on the asymptotic least squares is presented in this subsection according to the equation.

According to the Eq. (8), the dynamic difference equation of the three-axis air-bearing testbed can be obtained:

$$W(k+1) = W(k) + dt \cdot J^{-1}[T(k+1) + T_{Gb}(k+1)] \tag{9}$$

$$T(k) = J_f \cdot \left[A(k) - W_f(k) \times W(k)\right] \tag{10}$$

where

$$W(k) = \left[\omega_x(k)\ \omega_y(k)\ \omega_z(k)\right]^{\mathrm{T}} \tag{11}$$

$$T(k) = \left[T_x(k)\ T_y(k)\ T_z(k)\right]^{\mathrm{T}} \tag{12}$$

$$A(k) = \left[\alpha_x(k)\ \alpha_y(k)\ \alpha_z(k)\right]^{\mathrm{T}} \tag{13}$$

$$W_f(k) = \left[\omega_1(k)\ \omega_2(k)\ \omega_3(k)\right]^{\mathrm{T}} \tag{14}$$

and ω_x, ω_y, ω_z are the angular velocity of each axis; T_x, T_y, T_z are the input torque, which is provided by the momentum wheel; dt is the sampling time of the system; α_x, α_y and α_z are the accelerations of the momentum wheels on the three axes, respectively; ω_1, ω_2 and ω_3 are the rotational speeds of the momentum wheels on the three axes of the three-axis air-bearing testbed, respectively.

Definition:

$$\begin{cases} a_1(k) = a_1(\varphi(k), \theta(k), \gamma(k)) \\ a_2(k) = a_2(\varphi(k), \theta(k), \gamma(k)) \\ a_3(k) = a_3(\varphi(k), \theta(k), \gamma(k)) \end{cases} \tag{15}$$

According to Eq. (10), the asymptotic least squares recurrence equation can be established with unidentified parameters ε_1, ε_2, ε_3, r_x, r_y and r_z:

$$\varphi^T(k) = [\omega(k-1), T_x(k), T_y(k), T_z(k), a_3(k), \\ -a_2(k), a_1(k), -a_3(k), a_2(k), -a_1(k)] \tag{16}$$

$$\theta(k) = [1, \varepsilon_1, \varepsilon_2, \varepsilon_3, \mu_1, \mu_2, \mu_3, \mu_4, \mu_5, \mu_6] \tag{17}$$

$$K(k+1) = \frac{P(k)\varphi(k+1)}{\lambda + \varphi^T(k+1)P(k)\varphi(k+1)} \tag{18}$$

$$P(k+1) = \frac{1}{\lambda}[I - K(k+1)\varphi^T(k+1)]P(k)] \tag{19}$$

$$\theta(k+1) = \theta(k) + K(k+1)[\omega(k+1) - \varphi^T(k+1)\theta(k)] \tag{20}$$

Recursive operations are performed on the x, y and z axes respectively, and the moment of inertia matrix can be obtained:

$$J = (\Theta \cdot T)^{-1} \tag{21}$$

where

$$\boldsymbol{\Theta} = \left[\theta_x^{\mathrm{T}}(k)\ \theta_y^{\mathrm{T}}(k)\ \theta_z^{\mathrm{T}}(k) \right]^{\mathrm{T}} \tag{22}$$

$$\boldsymbol{T} = \left[\mathbf{0}_{3\times1}\ \boldsymbol{E}_{3\times3}\ \mathbf{0}_{3\times6} \right]^{\mathrm{T}} \tag{23}$$

At the same time, the distance vector from the centre of rotation of the three-axis air-bearing testbed to the centre of mass can be obtained as follows:

$$
\boldsymbol{r}_p = \left[r_x\ r_y\ r_z \right]^{\mathrm{T}}
$$
$$
= \begin{bmatrix} \frac{1}{6}\left(\frac{\mu_{4x}}{\varepsilon_{2x}} + \frac{\mu_{4y}}{\varepsilon_{2y}} + \frac{\mu_{4z}}{\varepsilon_{2z}} + \frac{\mu_{5x}}{\varepsilon_{3x}} + \frac{\mu_{5y}}{\varepsilon_{3y}} + \frac{\mu_{5z}}{\varepsilon_{3z}} \right) \\ \frac{1}{6}\left(\frac{\mu_{1x}}{\varepsilon_{1x}} + \frac{\mu_{1y}}{\varepsilon_{1y}} + \frac{\mu_{1z}}{\varepsilon_{1z}} + \frac{\mu_{6x}}{\varepsilon_{3x}} + \frac{\mu_{6y}}{\varepsilon_{3y}} + \frac{\mu_{6z}}{\varepsilon_{3z}} \right) \\ \frac{1}{6}\left(\frac{\mu_{2x}}{\varepsilon_{1x}} + \frac{\mu_{2y}}{\varepsilon_{1y}} + \frac{\mu_{2z}}{\varepsilon_{1z}} + \frac{\mu_{3x}}{\varepsilon_{2x}} + \frac{\mu_{3y}}{\varepsilon_{2y}} + \frac{\mu_{3z}}{\varepsilon_{2z}} \right) \end{bmatrix} \tag{24}
$$

The algorithm flow is shown in Table. 1.

Table 1. Online identification algorithm based on least squares method.

Online Identification Algorithm Based On Least Squares Method
Initialize θ
While $k < N$, **do**
Record the value of $\omega(k\text{-}1)$, $T_x(k)$, $T_y(k)$, $T_z(k)$, $a_1(k)$, $a_2(k)$ and $a_3(k)$
Initialize λ
Generate observation vector $\varphi^T(k)$
Calculate $K(k+1)$, $P(k+1)$
Update the vector of parameters to be identified θ
Set $k=k+1$
End while

4 The Intelligent Automatic Balancing Algorithm of the Centre of Mass

In the previous section, the moment of inertia matrix and the identification method of the position of the centre of mass are given. In this section, the intelligent automatic balancing algorithm of the centre of mass is given based on the identification results of the centre of mass (Fig. 4).

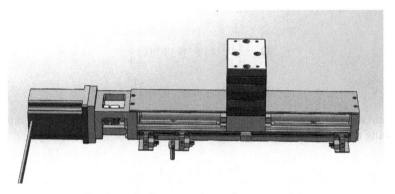

Fig. 4. Model diagram of the balancing module.

4.1 Manual Balancing of the Centre of Mass

From Sect. 2, the distance vector r_p from the centre of rotation of the three-axis air-bearing testbed to the centre of mass is obtained. Then determine the mass M of the three-axis air-bearing testbed and the mass m of each slider on the three axes of the balancing module, and then the target position vector r_{ba} of the three sliders in the manual balancing stage of the centre of mass can be calculated.

$$r_{ba} = -\frac{M}{m} \cdot r_p \tag{25}$$

4.2 Intelligent Automatic Delicate Balancing of the Centre of Mass

After manual balancing adjustment of the centre of mass, the three-axis air-bearing testbed has a state of centre of mass balance. The intelligent automatic delicate balancing of the centre of mass is divided into two stages: the x, y-axis plane centre of mass intelligent balancing and the z-axis centre of mass intelligent balancing.

The X, y-axis Plane Centre of Mass Intelligent Balancing.

Choose the incremental Proportion Integration Differentiation (PID) method to control the three-axis air-bearing testbed to the target attitude $f_1 = \{\theta_x = 0; \theta_y = 0\}$; Define $e_1(k)$ as the steady-state error of the x, y-axis angle of the control process; Define $r_{ba\_precise}$ as the target position vector of the slider of the x, y-axis balancing module, and we have:

$$r_{ba\_precise} = \rho \cdot \frac{e_1(k)}{m} \tag{26}$$

When $\lim\limits_{k \to \infty} e_1(k) = 0$, $r_{ba\_precise} = 0$. The x, y-axis plane centre of mass intelligent balancing ends, where ρ is the intelligent balancing coefficient. If the coefficient is too large, $e_1(k)$ will fail to converge; if it is too small, the convergence speed of $e_1(k)$ will decrease.

The z-axis Centre of Mass Intelligent Balancing.

Choose the incremental PID method to control the three-axis air-bearing testbed to the target attitude $f_2 = \{\theta_x = 5°; \theta_y = 0\}$; Let $e_2(k)$ be the steady-state error of the x-axis angle of the control process; Let $r_{z\_precise}$ be the target position of the slider of the z-axis balancing module, there are:

$$\begin{cases} r_{z\_precise} = q_1 \cdot \frac{e_2(k)}{m} & e_2(k) > 0 \\ r_{z\_precise} = q_2 \cdot \frac{e_2(k)}{m} & e_2(k) < 0 \end{cases} \tag{27}$$

After the control process steps into a steady-state, when $e_2(k) > 0$, it indicates that the centre of mass is still below the centre of rotation; when $e_2(k) < 0$, it indicates that the centre of mass has moved above the centre of rotation. In the moment, the centre of mass should be moved to below the rotation centre as soon as possible to ensure the stability of the three-axis air-bearing testbed.

When $\lim_{k \to \infty} e_2(k) = 0^+$, $r_{z\_precise} = 0$, the z-axis centre of mass intelligent balancing ends, where q_1 and q_2 are intelligent balancing coefficients, and $q_2 > q_1$. Similarly, if the coefficients are too significant, the oscillation of $e_2(k)$ will fail to converge; if it is too small, the convergence speed of $e_2(k)$ will decrease (Table 2).

Table 2. Intelligent automatic balancing algorithm of the centre of mass.

Intelligent Automatic Balancing Algorithm of the Centre of Mass
Set $f_1 = \{\theta_x = 0; \theta_y = 0\}$
While $\lvert e_1(k)\rvert \geq \varepsilon$, **do**
$r_{ba\_precise} = \rho \cdot \dfrac{e_1(k)}{m}$
End While
Set $f_2 = \{\theta_x = 5^\square; \theta_y = 0\}$
While $e_2(k) < 0$, **do**
$r_{z\_precise} = q_2 \cdot \dfrac{e_2(k)}{m}$
End While
While $0 < e_2(k) < \varepsilon$, **do**
$r_{z\_precise} = q_1 \cdot \dfrac{e_2(k)}{m}$
End while

5 Trials and Analysis

The third and fourth sections give the identification algorithm and intelligent automatic balancing algorithm of the mass characteristics of the three-axis air-bearing testbed. This section verifies the feasibility of the algorithm through practical trials.

5.1 Mass Characteristic Identification Trial

During the trial, the momentum wheels on each axis provide the torque input of the system, and then the characteristic mass parameters are identified through the torque information and the attitude data collected by the inertial components.

The actual moment of inertia matrix of the three-axis air-bearing testbed used for the ground trial is:

$$J = \begin{bmatrix} 391.370 & 1.004 & -7.130 \\ 1.004 & 398.567 & -2.205 \\ -7.130 & -2.205 & 391.936 \end{bmatrix} \tag{28}$$

Let momentum wheels output square waves of torque on the x, y and z-axis; the amplitude of the square waves is $\pm$ 100 mNm, and the period is 14s, 16s and 18s, respectively. The sampling time is 0.02s, and the results of the moment of inertia of each axis are shown in Fig. 5, Fig. 6, and Fig. 7, and the results of the position of the centre of mass are shown in Fig. 8, Fig. 9, and Fig. 10.

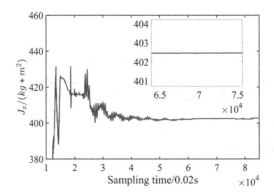

Fig. 5. J_x identification result chart.

The moment of inertia of each axis are identified as $J_x = 402.4$ kg·m^2, $J_y = 408.1$ kg·m^2 and $J_z = 406.7$ kg·m^2, respectively. The errors from the actual moment of inertia are 2.8%, 2.4% and 3.8%, respectively.

The identified centre of mass positions of the three-axis air-bearing testbed are $r_x = -0.0049$ m, $r_y = 0.0043$ m and $r_z = -0.0151$ m, respectively.

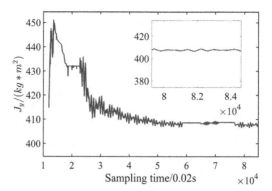

Fig. 6. J_y identification result chart.

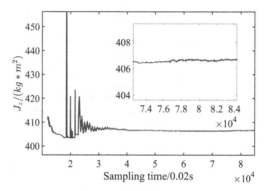

Fig. 7. J_z identification result chart.

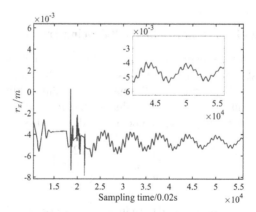

Fig. 8. r_x identification result chart.

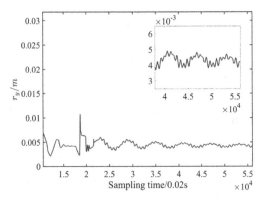

Fig. 9. r_y identification result chart.

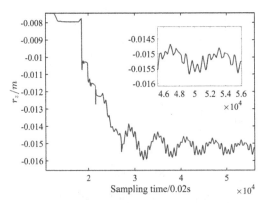

Fig. 10. r_z identification result chart.

5.2 Centre of Mass Intelligent Automatic Balancing Trial

In the previous section, the identification results of the centre of mass were given through the mass characteristic identification trial. It is known that the mass of the three-axis air-bearing testbed is $M = 1371.25$ kg, and the mass of each slider of the balancing module is $m = 30$ kg. According to the above results, the target position vectors of the three sliders in the manual balancing stage can be obtained as follows:

$$r_{ba} = -\frac{M}{m} \cdot r_p = \begin{bmatrix} 0.224 \\ -0.197 \\ 0.691 \end{bmatrix} \tag{29}$$

Obviously, the z-axis has exceeded the adjustment range of the slider of the balancing module. At this time, the counterweight below the three-axis air-bearing testbed should be appropriately reduced to make the centre of mass move up correctly, and then repeat the process of the centre of mass identification and manual balancing of the centre of mass until the centre of mass of the x-axis and y-axis is within ± 0.001 m, and the centre of mass of the z-axis is near -0.005 m.

The angle, angular velocity and rotational speed of the momentum wheel of each axis in the balancing algorithm are given as follows.

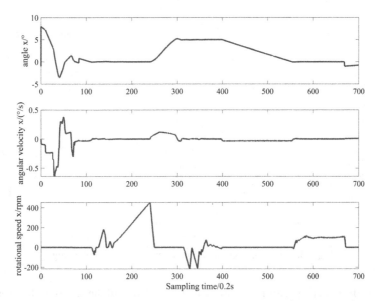

Fig. 11. x-axis data graph of intelligent automatic balancing.

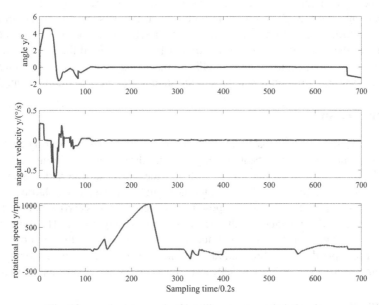

Fig. 12. y-axis data graph of intelligent automatic balancing.

We can conclude from Fig. 11 that the centre of mass intelligent balancing of the x, y-axis plane is performed in the first 100 sampling points. According to the intelligent

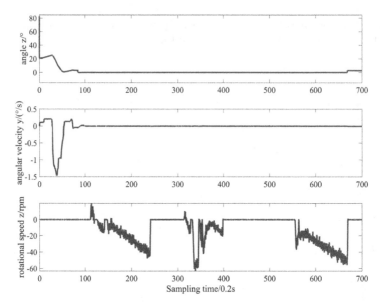

Fig. 13. z-axis data graph of intelligent automatic balancing.

algorithm, the slider's position is continuously adjusted until the steady-state error tends to zero. At this time, the centre of mass intelligent balancing of the x and y-axis plane is completed, and the intelligent adjustment of the z-axis is entered (Fig. 12, 13).

When entering the 250th sampling point, the x-axis is started to be controlled to the target attitude angle of $5°$. According to the intelligent algorithm, the position of the slider of the balancing module is continuously adjusted until the steady-state error tends to 0^+, the z-axis centre of mass intelligent balancing stage ends, and the three-axis air-bearing testbed is re-controlled back to the state where the x-axis attitude angle is 0.

6 Conclusion

In this paper, the automatic balancing method of the three-axis air-bearing testbed has been investigated. The identification model of the three-axis air-bearing testbed was established based on the Euler angle. Utilizing the mass characteristics obtained by the online identification algorithm, an intelligent automatic balancing algorithm for the three-axis air-bearing testbed was proposed. An experiment was carried out to illustrate the benefits and effectiveness of the developed control algorithm. The experimental results have shown that the intelligent automatic balancing algorithm proposed in this paper can be applied to most three-axis air-bearing testbed systems, and the identification algorithm's accuracy is better than 3.8%, which has high stability and application value.

References

1. Spencer, M., Chernesky, V., Baker, J.: Bifocal relay mirror experiments on the NPS three axis spacecraft simulator. AIAA Guidance, Navigation, and Control Conference And Exhibit, pp. 71–5 (2002)

2. Schwartz, J.L., Peck, M.A., Hall, C.D.: Historical review of air-bearing spacecraft simulators. J. Guid. Control. Dyn. **26**(4), 513–522 (2003)
3. Prado, J., Bisiacchi, G.: Dynamic balancing for a satellite attitude control simulator. Instrumentation & Development (2000)
4. Scharf, D.P., Lawson, P.R.: Flight-like ground demonstration of precision formation flying spacecraft. In: Proceedings Of Spie - The International Society For Optical Engineering, vol. 6693, p. 669307–12 (2007)
5. Wang, S.T., Cao, X.B.: Online mass-property identification algorithm research for satellite. Chinese Control Conference, vol. 01, pp. 587–591, Harbin (2006)
6. Hou, Z., Wang, Z., Zhang, Y.: Research on identification of mass characteristics for spacecraft combination based on thrusters. Aero-space Control **33**(01), 54–60 (2015)
7. He, X., Tan, S., Wu, Z.: On-orbit identification of the moment of inertia for a spacecraft with flexible appendages during a large-angle maneuver. Journal of Astronautics **38**(09), 927–935 (2017)
8. Liu, Z., Guo, J., Zhang, K.: Dynamic modeling and simulation for automatic balancing system of triaxial air bearing table, based on the Software Simulink. Microcomputer Information **25**(22), 138–139+170 (2009)
9. Chen, Z., Luo, Z., Wu, Y.: Research on the automatic balance system of three-axis air bearing test-bed. Computer Simulation **39**(04), 218–222 (2022)
10. Xiang, D., Yang, Q., Bao, G.: Research on analysing and compensation of the steady disturbing torque of the three axis air bearing table. Journal of Astronautics **30**(02), 448–452 (2009)

Simulation of Microstructure Evolution
of Ti-3Al-2Fe Alloy as Fabricated by VAR

Ling Ding[1]([✉]), Jiuyang Bai[1], Weiye Hu[1], Hui Chang[2], and Fuwen Chen[2]

[1] Nanjing Chenguang Group Co., Ltd, Nanjing 210006, China
dingling2013@njtech.edu.cn
[2] Tech Institute for Advanced Materials and College of Materials Science and Engineering,
Nanjing Tech University, Nanjing 210009, China

Abstract. In this paper, the microstructure evolution of Ti-3Al-2Fe alloy as fabricated by Vacuum arc remelting(VAR) is studied by simulation combined with experiments. Different parts of the molten pool are affected by the temperature gradient and cause a difference in morphology. The center and non-edge regions of the molten pool (zone1 and znoe2) are mainly equiaxed crystals, and the dendrite width is small. The edge region of the molten pool (zone3) is dominated by columnar crystals, and the dendrite width is large. At the solid-liquid interface, the mass fraction of Al in the solid phase gradually decreases with time, while the mass fraction distribution of Fe increases. At the interface, the partition coefficient of Al decreases slightly and then increases with time, while that of Fe increases first and then decreases.

Keywords: Titanium alloy · Simulation · Dendrite · Microsegregation

1 Introduction

Vacuum arc remelting(VAR) is a refining process that removes impurities and pores from ingots [1]. It is suitable for the production of titanium alloy, nickel base superalloy and other aerospace materials. It is widely used in titanium alloy due to the advantages of low cost and fast smelting. Since the melting and solidification processes are performed in a closed environment at high temperatures, the physical parameters in the processes are difficult to measure experimentally. Therefore, simulation methods are used to obtain relevant data, and the finite element method and phase field method are commonly used methods. The finite element method is used to predict macroscopic physical fields such as flow, strain/stress distribution, temperature distribution [2–4], while the phase field method can simulate the evolution of microstructures, such as nuclei formation and dendrite growth. If a region of appropriate size is selected from the macroscopic finite element model and the physical field data within this range are imported into the phase field model for further microscopic simulation, the cross-scale coupling of temperature field and microstructure evolution can be realized to a certain extent, which is closer to the actual process than the single-scale microstructure simulation.

Segregation may form during VAR solidification. Kawakami et al. studied the segregation coefficients of certain elements in commercial titanium alloys by spot analysis of as-cast alloys [5]. It was found that the values of the segregation coefficients were determined different due to the strong interactions between the elements in these systems. Kondrashov et al. devoted a simple thermal model for VAR, including solutions to nonlinear heat conduction equations with typical VAR nonlinear boundary conditions [6]. The finite difference simulation of the initial equation of the model is obtained by the finite volume method and the model can fit the experimental data well under various remelting conditions. In recent years, many studies have been done on the simulation of solidification process based on the finite element method and phase field method. Kelkar et al. proposed a comprehensive calculation model for predicting VAR process performance [7]. The calculation model included the electromagnetic field, turbulence in the metal pool, magnetic stirring, heat transfer, macrosegregation and inclusion behavior during the entire melting process, and the model had been used to analyze the actual process of Ti-6Al-4V alloys. Zhang et al. studied the tensile properties and deformation resistance of 3 mm thick commercial pure titanium strip with abrasive pits on the surface by the finite element method [8], Yi et al. verified the mechanical properties of TiAl alloy casting by finite element simulation, but no microstructure simulation was involved [9]. Kundin et al. studied the microstructure formation of Inconel 718 supperalloy by phase-field modeling, and the effect of growth velocity on simulation was discussed [10]. Choo et al. tried to couple the phase field method with the plastic model, which further increased the application scope of phase field method, but there was no research on the VAR process [11].

In this work, Ti-3Al-2Fe alloy (an alpha + beta titanium alloy, smelted twice by VAR) was used to simulate the microstructure evolution during VAR process, due to alloying effect of Fe and Al in titanium alloys and the significance in industrial production. Al is the most widely used alloying element in titanium alloys because it is the only common metal that increases the phase transition point and dissolves in both the alpha and beta phases. Fe is a stronger β-eutectoid alloying element than Cr in titanium alloy, which has a strong influence on the phase transition point [12]. According to the procedure of microstructure simulation, firstly, the macroscopic finite element model of VAR should be established to simulate the temperature field distribution of the molten pool in the VAR process. Then the parameters such as the temperature gradient and cooling rate at the edge of the mushy zone obtained by finite element simulation were introduced into the phase field to simulate microstructure.

2 Model Description

2.1 Model of Finite Element

In this work, the finite element method is used to simulate the temperature field in the ingot VAR melting process. The main assumptions of the model are based on the following conditions:

(1) The ingot is cylindrical, and all parameters have axis symmetry;
(2) For the crystal growth analysis in the non-steady state, the time step is automatically adjusted;
(3) The model takes into account the main physical phenomena in the whole process, namely melt flow, heat transfer and macrosegregation of major alloying elements.

This work refers to the classic VAR model [13–16], modifies the governing equations and boundary conditions, and establishes a simplified VAR model, as shown in Fig. 1.

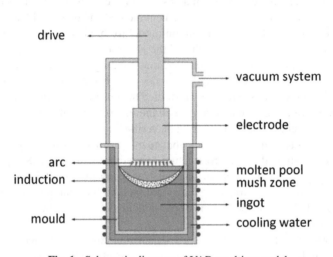

Fig. 1. Schematic diagram of VAR smelting model

In this page, the energy control equation is as follows:

$$\frac{\partial}{\partial t}(\rho C_p T) + \frac{\partial}{\partial x}(\rho V_x C_p T) + \frac{\partial}{\partial y}(\rho V_y C_p T) = \frac{\partial}{\partial x}\left(K\frac{\partial T}{\partial x}\right) + \frac{\partial}{\partial y}\left(K\frac{\partial T}{\partial y}\right) + Q_V \quad (1)$$

K- thermal conductivity, T- temperature, C_p- specific heat, Q_V- volumetric heat source.

There are two types about thermal boundary conditions of temperature field. One is the boundary condition of the heat convection that occurs at the wall and bottom of the crucible. At this time, Newton's law of cooling is adopted:

$$q = \alpha(T_f - T_w) \quad (2)$$

α- convective heat transfer coefficient, T_f- cooling water temperature, T_w- temperature of the water-cooled copper crucible.

The second is the boundary condition of heat input at the top of the molten pool. During solidification, the latent heat generated by phase transformation affects the top temperature T of molten pool. T can be calculated according to formula (3):

$$T = T_L + \Delta T(J, D_i) \quad (3)$$

T- surface temperature of the molten pool, T_L- liquidus temperature, $\Delta T(J, D_i)$-melting overheating, J- current density, D- diameter of the ingot.

2.2 Phase Field Model

As the temperature change calculated according to simulation is less than 0.1 K, the following assumptions are made:

(1) It is considered that the diffusion coefficients of Al and Fe in the solid phase and the liquid phase do not change in this temperature range.
(2) It is considered that the diffusion of Al and Fe have no influence on each other.
(3) it is considered that the phase diagram data in the whole process remain invariant.

Dendritic growth and grain growth models were established using micress 6.2 software. The dendrite growth model has a mesh size of 1000 × 1000, a cell resolution of 0.2 μm, and a minimum time step of 1×10^{-6}s. The initial condition is considered to be 1 for the initial grain. A rectangular area set at the bottom of the model, having a length of 200 μm and a thickness of 2 μm. The grain growth model has a grid size of 1000 × 1000, a cell resolution of 0.5 um, and a minimum time step of 1×10^{-6}s. Set 8 initial grain levels to randomly generate grains according to grain radius and distribution density. Based on the principle of minimum free energy, the phase field equation of micress is used:

$$\dot{\varphi}(\vec{x}, t) = \mu \left[\sigma \left\{ \nabla^2 \varphi - \frac{(1 - \varphi)(1 - 2\varphi)\varphi}{\eta^2} \right\} + \frac{1}{\eta} \Delta G \varphi (1 - \varphi) \right] \tag{4}$$

x- space, t- time, ΔG- Gibbs free energy, φ- order parameter of phase field, σ- interfacial energy, η- interfacial thickness, and μ- interfacial mobility.

The simulated interface energy can use common interface energy [17]. The phase diagram data required for the simulation is directly extracted from the Thermo-Calc TTTi3 database. The solid phase diffusion coefficient of the Al is calculated from the MOBTI1 database, and the liquid phase diffusion coefficient of Al is estimated. Since there is no diffusion data of Fe in the MOBTI1 database, a kinetic database containing Fe is prepared by Chen Y's study of β phase diffusion kinetics of Ti-Al-Fe alloy [18], and the data obtained is imported into micress to calculate the solid phase diffusion coefficient of Fe. The liquid phase diffusion coefficient of Fe is derived from the solid phase diffusion coefficient of Fe with reference to Kundin's study [19] (Table 1).

Table 1. Partial physical parameters [17–19]

Physical parameters	Ti-3Al-2Fe
Interface energy σ (J/cm**2)	25
Al Liquid diffusion coefficient D_l (cm^2/s)	5×10^{-5}
Al Solid diffusion coefficient D_s (cm^2/s)	2×10^{-7}
Fe Liquid diffusion coefficient D_l (cm^2/s)	1×10^{-4}
Fe Solid diffusion coefficient D_s (cm^2/s)	2×10^{-5}
Molar volume V(cm^3/mol)	10.3
Calculated temperature T (K)	1885
Anisotropic strength η	0.05

3   Material and Methods

The Ti-3Al-2Fe sample (smelted twice by VAR to obtain a cylindrical ingot of about 30 kg with a diameter of 160 mm, furnace cooling) with the following chemical composition (wt.%) was studied: 2.9 Al, 2.1 Fe, 0.1B and balance titanium. A 10 mm thick flat plate was cut by wire electrode cutting in the middle of the ingot. As shown in Fig. 2, three 10 * 10 mm squares were cut from the flat plate along the isotherm. The samples were electrolytic polished (using HClO$_4$: C$_2$H$_5$OH = 3: 57 electrolyte) and quickly washed in alcohol and distilled water.

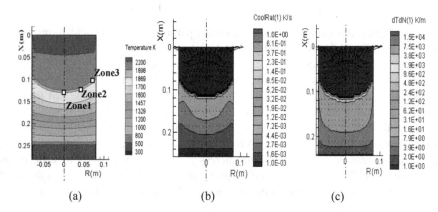

Fig. 2. Calculation of temperature field of Ti-3Al-2Fe alloy ingot, (a) distribution of temperature field and selected position, (b) distribution of cooling rate, (c) distribution of temperature gradient

The metallographic photographs of different positions were observed with Optical Microscope (OM, Carl Zeiss, Jena, Germany). Line scan and spot scan images of grain boundary of Ti-3Al-2Fe alloy were obtained by Scanning electron microscope(SEM, JEOL, Tokyo, Japan) equipped with an energy dispersive spectrometer (EDS) analysis system.

4   Results and Discussions

4.1   Temperature Field Simulation of Ti-3Al-2Fe Alloy

The macroscopic temperature field calculated by Melt Flow-VAR software is shown in Fig. 2(a). The melting point of the alloy was calculated by Thermo-Calc to be 1885 K. It is considered that the temperature in the region above 1885 K is the molten pool, and the temperature is from 1700 K to 1885 K as the position of the mushy zone. As shown in Fig. 2, the obtained temperature field data was analyzed to get a temperature gradient and cooling rate distribution map of the ingot in the middle of smelting. At this time, the liquid metal near the bottom of the crucible solidifies first. The molten metal entering the crucible can't solidify immediately, and gradually form a flat shallow molten pool, due to the small contact area between the ingot and the crucible wall. The distribution trend of cooling rate and temperature gradient is similar near the boundary of mushy zone and molten. In the range of the molten pool, the cooling rate and temperature gradient are zero because the solidification has not yet occurred. Along the direction of solidification, the cooling rate and temperature gradient increase as a whole, reaching the extreme value near the bottom edge of ingot, the maximum cooling rate is about 1 K/s, and the maximum temperature gradient is close to 1.5×10^4 K/m.

The initial position of solid-liquid phase transformation should be at the boundary of the mushy zone and the molten pool. An isotherm should be chosen at the boundary, and three areas are selected at the bottom/middle/edge of the isotherm to simulate the dendrite growth in the solidification process with the phase field model. The areas in the mushy zone boundary are all 0.04 mm2 (corresponding to zone 1, zone 2, zone 3 in Fig. 2(a), respectively). The temperature gradient G and the cooling rate R in the normal direction corresponding to the three positions are selected to calculate the dendrite growth rate V. The solidification parameters of the different phase field calculation domains in Table 2 are obtained, and the data transmission from the macroscopic temperature field to the phase field is completed to realize the cross-scale coupling between macroscopic simulation and phase field simulation.

Table 2. Solidification parameters at different locations

Zone	Temperature gradient G(K/m)	Cooling rate R (K/s)	Growth rate V(mm/s)	
1	2341	0.43		0.184
2	3954	0.41		0.104
3	7642	0.52		0.068

4.2   Phase Field Simulation of Microstructures

As shown in Fig. 3(a), solidification is in the initial stage, and undulations are formed at the solid-liquid interface, resulting in more fine grains. As time goes on, the fine grains gradually grow up. Due to the competitive growth between the grains, some of the small grains are swallowed by the large ones. When the crystal grains continue to grow, some of the crystal grains merge with each other, and the grain width becomes larger, and fine secondary dendrites appear between the coarse primary dendrites. Comparing Fig. 3(c) with Fig. 3(g), the dendrite morphology at zone1 and zone2 is similar, and the growth rate is very close. Since the temperature gradients at these two places are relatively close, it can be considered that the influence of the temperature gradient is negligible in a small range, and columnar crystals growing along the temperature gradient direction are formed at both places.

It can be seen from Fig. 3(i) that the interface undulation at the initial stage of zone3 is small, and the number of grains at the initial stage of solidification is significantly less than that of zone1 and zone2. In the process of continuous grain growth, there is still competition in grain growth, but due to the small initial grain size, the competitive growth trend at zone3 is not as strong as that of zone1 and zone2. The grains can grow freely to a greater width, as shown in Fig. 3 (j-l). Since the temperature gradient at zone3 is much larger than zone1 and zone2, the width of the primary dendrite increases as the temperature gradient is larger.

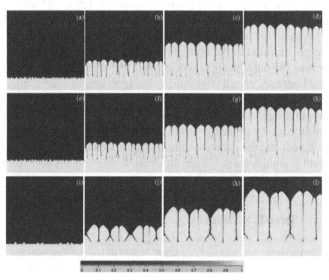

Fig. 3. Dendrite Morphology at Zone 1 (a 0.01s, b 0.02s, c 0.03s, d 0.04s), Zone 2(e 0.01s, f 0.02s, g 0.03s, h 0.04s), Zone 3(i 0.01s, j 0.02s, k 0.03s, l 0.04s)

Figure 4(a-d) show the grain growth at the zone1 location. Random nucleation under the influence of undercooling, due to the small temperature gradient, the nucleation preferentially grows into equiaxed grains. The equiaxed grains compete in the growth process, the grains are larger in the regions with lower nucleation density, and the triangular grain boundaries are formed by extrusion in the regions with higher density. As time goes by, the solid phase content in the area gradually increases. Under the influence of the temperature gradient, some grains grow into columnar grains along the temperature gradient direction, as shown in Fig. 4(d). In the zone 2 position, since the temperature gradient is similar to that of zone1, grain growth is also changed from equiaxed crystals to columnar crystals, and dendrites grow faster along the temperature gradient direction.

Compared to zone1 and zone2, zone3 is closer to edge of crucible. The bottom of the simulation area is set as the edge of the crucible in the simulation. It is considered that the undercooling is larger at the position near the bottom, and it is easier to nucleate. As shown in Fig. 4(i), dendrites grow rapidly along the direction of the temperature gradient under the influence of a large temperature gradient. Dendritic growth in other directions was significantly inhibited, and the grains appeared to be more pronounced columnar crystals. In the later stage of grain growth, the residual liquid phase between the grain boundaries gradually disappears. Due to the larger temperature gradient of the zone 3 location, the original grain is less. The surface of the grains is large, hence the area of grain boundaries is reduced, which is consistent with the experimental results.

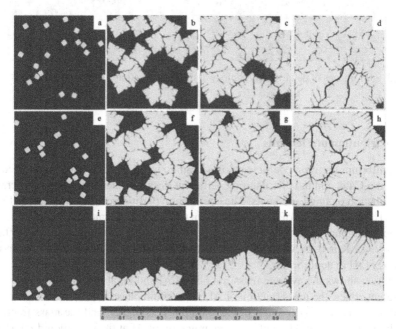

Fig. 4. Grain growth at Zone1 (a 0.005s, b 0.02s, c 0.03s, d 0.04s), Zone2 (e 0.005s, f 0.02s, g 0.03s, h 0.04s), Zone3 (i 0.005s, j 0.02s, k 0.04s, l 0.08s)

As shown in Fig. 5, since the ingot cooling process uses furnace cooling, the matrix is the beta phase grain with acicular alpha phase particles of the Widmanstatten structure and the grain size at zone3 is large. At zone 1, the molten metal is directly nucleated and grown, preferentially forming equiaxed grains. The transition from equiaxed to columnar crystals gradually occurs along the direction of the temperature gradient, and the dendrites with the same or similar orientation as the temperature gradient are more developed. Zone 2 is similar to Zone 1 and also forms an equiaxed crystal structure. Most of the dendrites are more developed in the direction of the temperature gradient and exhibit a stronger orientation. In the zone 3 position, part of the grains nucleate on the crucible wall, forming a fine-grained zone and gradually growing into columnar crystals with the same or similar orientation as the temperature gradient. The orientation of dendrite growth in Zone3 is the most obviously, dendrites that deviate significantly from the direction of the temperature gradient are slower and inhibited. The solidified structure near the center of the ingot is composed of equiaxed grains, and the dendrite growth has a certain orientation. The edge of the ingot is formed into a solidified structure in which columnar crystals and equiaxed crystals are combined, and the orientation of the dendrite growth is more pronounced [20, 21].

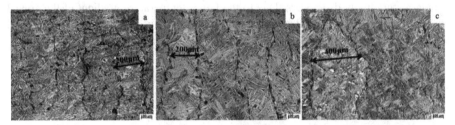

Fig. 5. OM images of dendrite morphology at different locations (a zone1,b zone2,c zone3)

4.3 Distribution of Elements

The equilibrium partition coefficients of Al and Fe in Ti-3Al-2Fe alloy were calculated by Thermo-Calc, and the distribution coefficient discussed here specifically referred to the distribution of single elements. The equilibrium partition coefficient of Al at 1885 K is 1.27. The mass fraction of the solid phase in the solid-liquid interface should be higher than the liquid phase mass fraction in equilibrium. The equilibrium partition coefficient of Fe at 1885 K is 0.35, which is opposite to the trend of solid and liquid phase mass fractions of Al in equilibrium.

Figure 6 shows the average mass fraction of Al and Fe at the solid-liquid interface in different zones, the mass fraction of Al in the liquid phase is low, the lowest value is 2.27 wt%, and the mass fraction in the solid phase is higher, and the mass fraction in the solid phase is up to 3.08 wt%. The segregation of Al at the solid-liquid interface is not strong and conforms to the phase diagram. However, the mass fraction of Fe in the liquid phase is higher, the highest value is 6.04 wt%, while the lowest in the solid phase is only 1.61 wt%.

At the solid-liquid interface, the mass fraction of Al in the solid phase in different regions gradually decreases with time. At 0.005 s, the three zones are all 3.08, at 0.055 s, zone 1 and zone 2 are 3.01 and zone 3 are 3.03, which are close to the nominal composition as a whole. However, the mass fraction distribution of Fe increases with time. At 0.005 s, the three zones are all 1.45. At 0.055 s, the zone 1 and zone 2 are close to 1.94, while the zone 3 is only 1.85, which is much different from the nominal composition. The change trend of mass fraction of the two elements in the liquid phase is similar to that in the solid phase as a whole, except that the slope of change in the solid phase is first high and then low, while in the liquid phase it is the opposite. Because the diffusion rate of the two elements in the liquid phase is higher than that in the solid phase, the change range of the mass fraction in the liquid phase is significantly larger than that in the solid phase. In particular, the mass fraction of the Fe element at the beginning of the simulation was almost three times that at the end of the simulation due to the lower partition coefficient.

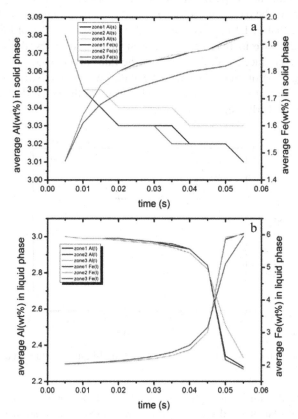

Fig. 6. The average mass fraction of Al and Fe at the solid-liquid interface in different zones(a solid phase, b liquid phase)

Figure 7 shows the distribution coefficients of Al and Fe at the solid-liquid interface in different zones. The distribution coefficient of Al as a whole increases with time and is opposite to Fe. According to the study of J. C. Brice [22], the solute partition coefficient k at the solid-liquid interface is related to the growth rate.

$$k = \frac{\alpha}{\beta} + \frac{f(1 - k_0)}{\beta v} \tag{5}$$

v- diffusion rate of the solute, α- adhesion coefficient of the solute at the solid-liquid interface, β- dissociation constant of the adsorbed solute, k_0- equilibrium partition coefficient, and $k_0 = \alpha/\beta$, f- interface movement rate (that is, the growth rate of the crystal).

According to Eq. (5), when $k_0 < 1$, the value of k increases with the growth rate of the crystal. Therefore, at the initial stage of simulation, the partition coefficient of Fe at zone3 is lower than that at zone1 and zone2. As the solidification process progresses, the growth rate of the crystal decreases gradually. Similarly, when $k_0 > 1$, k value decreases with the increase of crystal growth rate. While the partition coefficient of Al at zone3 is slightly higher than that at zone1 and zone2. Since the k_0 of Al is closer to 1, the effect of crystal growth rate on the partition coefficient is relatively small.

At the interface, the partition coefficient of Al decreases slightly and then increases with time, while that of Fe increases first and then decreases. According to Hughel's study [23], the value of α increases with increasing mass fraction. As the mass fraction of Fe in solid-liquid interface increases with time, the value of α increases, which contribute to the increasing of partition coefficient of Fe at the beginning of simulation. Similarly, the partition coefficient of Al decreased slightly, due to the small decreasing of mass fraction of Al in the early stage.

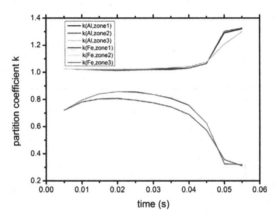

Fig. 7. The partition coefficient of Al and Fe at the solid-liquid interface in different zones

According to the simulation results, after 0.045 s, the content of liquid phase was much lower than that of solid phase, and the movement rate of solid-liquid interface rapidly decreased, and its influence on the partition coefficient at the interface decreased. As shown in Fig. 8, at 0.055 s, the solid-liquid interface movement in the simulation region nearly stopped, and the remaining liquid phase between the solid phases was shown in red. At this point, the interface partition coefficient was close to k_0. Therefore, after 0.045 s, the interface partition coefficients of the three zones get crossing due to their rapid proximity (Fig. 6).

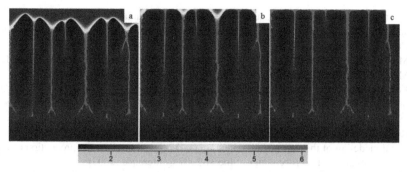

Fig. 8. Mass fraction of Fe in zone 3 at the end of simulation (a 0.45 s, b 0.50 s, c 0.55 s)

5 Conclusion

In this paper, the finite element method is used to simulate the temperature field distribution in the VAR process, and the temperature gradient and cooling rate in different regions near the molten pool during solidification are obtained. The temperature field data is brought into the microscopic phase field model to obtain the desired microstructure. The specific conclusions are as follows:

(1) Different parts of the molten pool are affected by the temperature gradient and cause a difference in morphology. The center and non-edge regions of the molten pool (zone1 and zone2) are mainly equiaxed crystals, and the dendrite width is small. The edge region of the molten pool (zone3) is dominated by columnar crystals, and the dendrite width is large.
(2) At the solid-liquid interface, the mass fraction of Al gradually decreases with time, while the mass fraction distribution of Fe increases. The mass fraction of Al decreases from 3.08 to 3.01 in solid phase and from 3 to 2.3 in liquid phase. The mass fraction of Fe increases from 1.6 to 1.9 in solid phase and from 2.1 to 6 in liquid phase.
(3) At the interface, the partition coefficient of Al decreases slightly and then increases with time, while that of Fe increases first and then decreases. The partition coefficient of Al increases from 1 to 1.3, and that of Fe decreases from 0.7 to 0.3.

References

1. Zhang, J.: Numerical investigation of segregation evolution during the vacuum arc remelting process of Ni-based superalloy ingots. Metals **11**, 2046 (2021)
2. Karimi-Sibaki, E., Kharicha, A., Abdi, M., et al.: A Numerical study on the influence of an axial magnetic field (AMF) on vacuum arc remelting (VAR) process. Metall. and Mater. Trans. B. **2**, 3354–3362 (2021)
3. Yang, Z., Cao, J., Yu, W., et al.: Effects of microstructure characteristics on the mechanical properties and elastic modulus of a new Ti–6Al–2Nb–2Zr–0.4B alloy. Materials Science and Eng. A **820**, 141564 (2021)

4. Günnemann, S., Kremer, H., Laufkötter, C., Seidl, T.: Simulation on solidification structure and shrinkage porosity (hole) in tc4 ingot during vacuum arc remelting process. Special Casting & Nonferrous Alloys **32**, 418–421 (2012)
5. Mitchell, A., Kawakami, A., Cockcroft, S.: Segregation in titanium alloy ingots. High Temp. Mater. Processes (London) **26**, 59–78 (2007)
6. Kondrashov, E., Musatov, M., Maksimov, A.Y., Goncharov, A., Konovalov, L.: Calculation of the molten pool depth in vacuum arc remelting of alloy vt3-1. J. Eng. Thermophys. **16**, 19–25 (2007)
7. Kelkar, K.M., Patankar, S.V., Mitchell, A., Kanou, O., Fukada, N., Suzuki, K.: In Computational modeling of the vacuum arc remelting (var) process used for the production of ingots of titanium alloys. In: Proceedings of the Ti-2007 Conference (2007)
8. Zhang, J., Yu, W., Dong, E., Zhang, Z., Shi, J., Gong, G.: Study on grinding and deformation fracture control of cold rolled titanium strip. Metals **10**, 323 (2020)
9. Jia, Y., Xiao, S., Tian, J., Xu, L., Chen, Y.: Modeling of tial alloy grating by investment casting. Metals **5**, 2328–2339 (2015)
10. Kundin, J., Mushongera, L., Emmerich, H.: Phase-field modeling of microstructure formation during rapid solidification in inconel 718 superalloy. Acta Mater. **95**, 343–356 (2015)
11. Choo, J., Sun, W.: Coupled phase-field and plasticity modeling of geological materials: From brittle fracture to ductile flow. Comput. Methods Appl. Mech. Eng. **330**, 1–32 (2018)
12. Lütjering, G., Williams, J.C.: Titanium. Springer, Heidelberg (2007)
13. Beaman, J.J., Felipe, L.L., Williamson, R.L.: Modeling of the vacuum arc remelting process for estimation and control of the liquid pool profile. J. Dyn. Syst. Meas. Contr. **136**, 031007 (2014)
14. Senkevich, K.S., Pozhoga, O.Z.: Experimental Investigation of hydrogen absorption by commercial high alloyed Ti2AlNb-based alloy in cast and rapidly solidified state. Vacuum **3**, 110379 (2021)
15. Patel, A., Fiore, D.: On the modeling of vacuum arc remelting process in titanium alloys. IOP Conference Series: Materials Science and Eng. **143**, 012017 (2016)
16. Yang, Z., Wang, L., Gortschakow, S.: Numerical simulation and experimental investigation of transient anode surface temperature in vacuum arc. J. Physics, D. Applied Physics: A Europhysics J. **54**, 505201 (2021)
17. Guo, J., Li, X., Su, Y., Wu, S., Li, B., Fu, H.: Phase-field simulation of structure evolution at high growth velocities during directional solidification of ti55al45 alloy. Intermetallics **13**, 275–279 (2005)
18. Chen, Y., Li, J., Tang, B., Kou, H., Segurado, J., Cui, Y.: Computational study of atomic mobility for bcc phase in ti–al–fe system. Calphad **46**, 205–212 (2014)
19. Kundin, J., et al.: Phase-field modeling of eutectic ti–fe alloy. Comput. Mater. Sci. **63**, 319–328 (2012)
20. Cruz, K.S., Meza, E.S., Fernandes, F.A.P., Quaresma, J.M.V., Casteletti, L.C., Garcia, A.: Dendritic arm spacing affecting mechanical properties and wear behavior of al-sn and al-si alloys directionally solidified under unsteady-state conditions. Metall. and Mater. Trans. A. **41**, 972–984 (2010)
21. Gentry, S.P., Thornton, K.: Simulating recrystallization in titanium using the phase field method. IOP Conference Series: Materials Science and Eng. **89**, 012024 (2015)
22. Brice, J.: The variation of interface segregation coefficients with growth rate of crystals. J. Cryst. Growth **10**, 205–206 (1971)
23. Hughel, T., Bolling, G.: Solidification. Amer. Soc. Metal, Metals Park, OH (1971)

An Improved RRT* Path Planning Algorithm in Dynamic Environment

Jianyu Li, Kezhi Wang, Zonghai Chen, and Jikai Wang(✉)

University of Science and Technology of China, Hefei, Anhui, China
wangjk@ustc.edu.cn

Abstract. With the increasingly diverse usage scenarios of mobile robots, the path planning of mobile robots in dynamic environments has become a hot issue. Aiming at the slow planning speed of RRT* algorithm in dynamic environment, this paper proposes an improved RRT* algorithm: efficient dynamic rapidly-exploring random tree star (ED-RRT*). In a static environment, the algorithm uses goal-biased sampling to reduce the randomness of the RRT* sampling method, and sets the sampling step size to be adaptive to improve the planning speed of the initial path. When the dynamic environment encounters obstacles and needs to be re-planned, the local target points are given based on the initial path and environmental information to quickly complete the local re-planning. The path information planned in the static environment is used as the prior information to improve the efficiency of the replanning algorithm. Finally, in the simulation environment based on OpenCV2, the ED-RRT* algorithm is compared with the improved RRT* algorithm. The experimental results show that ED-RRT* has faster dynamic performance and fewer nodes.

Keywords: Mobile robot · Dynamic environment · Path planning · RRT*

1 Introduction

In recent years, Mobile robots have been applied in many fields, such as Search-and-Rescue [1], healthcare services [2], autonomous delivery [3], and so on. Path planning has always been one of the hot issues in the field of robotics. In static environment, there are already many typical algorithms like A* [4], Dijkstra's algorithm [5], improved A* [6]. However, due to the fact that obstacle movement is irregular in dynamic environment, it is still challenging to plan a collision-free, safe, efficient and high-quality path.

Because of the speed and robustness in finding the path to the target, sampling-based planning algorithms are popular global path planning approaches. RRT [7], which quickly explores the configuration space through random sampling, has high computational efficiency and probabilistic completeness. Although the generality of the RRT algorithm can solve the path planning problem of the robot in different dimensions, it cannot guarantee asymptotic optimality. In environments with narrow channels, the convergence rate of the RRT algorithm is slow.

W. Fan et al. (Eds.): AsiaSim 2022, CCIS 1713, pp. 301–313, 2022.
https://doi.org/10.1007/978-981-19-9195-0_25

In order to solve the shortcomings of the traditional RRT algorithm, scholars have made improvements to the RRT algorithm. The RRT-connect algorithm [8] is an algorithm based on the RRT algorithm, which simultaneously grows two rapid-exploration random trees from the starting point and the ending point to search the state space. Bidirectional rapid-exploration random trees speed up algorithm convergence. Urmson and Simmons [9] proposed a heuristic function that makes the expansion of random trees biased to speed up the convergence of the algorithm. Karaman and Frazzoli proposed the RRT* algorithm [10] to find the optimal path of the RRT algorithm. By reselection of parent nodes and rewire function, RRT* algorithm can find an optimal or near-optimal path. Rapidly-exploring random trees star fixed nodes (RRT*FN) [11] proposed by reduces the memory usage of the computing center by limiting the maximum number of nodes in the tree. In order to solve the problem of large RRT* sampling space and long sampling time, Gammell and Srinivasa proposed the Informed-RRT* algorithm [12]. The algorithm generates an ellipse sampling space determined by the starting point, the target point, and the current path length, and by heuristic sampling in this ellipse region, it accelerates the convergence to the optimal solution.

Although the algorithms mentioned above have been optimized to a certain extent, none of them utilizes environmental information. In order to introduce environmental information to guide RRT, some researchers have explored the method of combining RRT with artificial potential field [13]. The algorithm switches to RRT planning when the artificial potential field method falls into a local minimum, and switches back to the artificial potential field method when it jumps out of the local minimum. The smart rapidly-exploring random tree star (RRT*-Smart) [14] has made improvements on the basis of RRT*, mainly by optimizing the path. RRT*-Smart is exactly the same as RRT* in the first stage, but after finding a feasible path from the start point to the end point it starts to optimize the path. The optimization process starts from the leaf node, constantly looking for whether it can directly connect to the parent node without collision. If it can be directly connected, there will be one more straight line and one less curve. These algorithms implicitly utilize environmental information, but they are improved based on RRT or RRT*. They are dedicated to searching for the optimal path and have poor real-time performance, so they are not suitable for dynamic path planning with high real-time requirements.

Previous work on re-planning is incremental replanning algorithm, such as D* [15] and D* Lite [16]. The D* Lite algorithm is a reverse search. At first, according to the known environmental information, the unknown part is regarded as a free space, and the global optimal path from the target point to the starting point is planned. At this time, a path field is established to provide the basis for the optimal approach to the target point incrementally. The D* Lite algorithm can be well applied to unknown environment for path planning. Due to the idea of incremental planning, it has fewer re-planning times. However, when the space is relatively large, the number of grid nodes to be maintained in the reverse search process increases sharply, which increases the time complexity of the search. Qi et al. [17] proposed a real-time dynamic path planning method combining artificial potential field method and goal-biased RRT algorithm. The algorithm offsets the random sampling points towards the target point, uses the constructed gravitational potential field and repulsive potential field to make the random tree approach the target while staying away from obstacles, and has an adaptive growth step in different local environments. It greatly reduces the repeated calculation of the algorithm in the local minimum region, and uses a local re-planning strategy for dynamic environments. However, this algorithm still has the problem of long search time and long replanning path.

In terms of above issues, we propose an RRT*-based algorithm termed as ED-RRT* algorithm. ED-RRT* consists of two stages, the first stage is the global path planning, and the second stage is the re-planning for the dynamic environment. In the first stage, the idea of goal-biased sampling is used to make the sampling points converge quickly. This ensures that a safe path can be planned as an initial path in a relatively short time. When the robot moves according to the global path and encounters dynamic obstacles, the algorithm will enter the re-planning stage. Taking points on the global path as new target points, replanning is performed within a certain distance to avoid collisions.

To summarize, the main contributions of this work are as follows:

(1) **Global Path Planning:** The initial environment is regarded as a static environment for initial planning, and goal-biased sampling is introduced to reduce the randomness of the RRT* algorithm. Apply an adaptive step size method to increase exploration efficiency. Such a sampling strategy can not only guide the random tree to grow toward the target, but also has the characteristic of preferentially growing a branch, which is beneficial to improve efficiency and quickly plan a path.
(2) **Path Re-planning:** The local replanning strategy is used to further improve the real-time performance of the algorithm in dynamic environments. The robot senses the movement of obstacles through sensors. When re-planning is required, the initially planned path is used as a path cache, and a certain node of the initial path is selected as a new target point to re-plan locally.

The remainder of the paper is organized as follows. The related work is reviewed in Sect. 2. An overview of ED-RRT* algorithm is provided in Sect. 3. Section 4 compares the ED-RRT* algorithm with other method under simulated scenarios. Section 5 presents our conclusions.

2 Related Work

2.1 Task Definition

Denote S as the configuration space, which represents the space where the mobile robot needs to do path planning. The space that collides with obstacles is defined $S_{obs} \subset S$, and the space that does not collide with obstacles is defined as free space $S_{free} = S \backslash S_{obs}$, Let the starting position of the mobile robot be $S_{start} \in S_{free}$ and the target area be $S_{goal} \subset S_{free}$. Therefore, the feasible path between the starting position and the target area can be defined as a set $\sigma : [0, 1] \rightarrow S_{free}$ with $\sigma(0) = S_{start}$ and $\sigma(1) \in S_{goal}$.

The path cost is defined as $c : S_{free} \rightarrow R$, where R represents the set of non-negative real numbers. Finding a path with a lower path cost is a goal of path planning, and the minimum path cost function $\sigma *$ is defined as follows:

$$\sigma^* = \arg\min_{\sigma \in S}\{c(\sigma)|\sigma(0) = S_{start}, \sigma(1) = S_{goal}, \forall x \in [0, 1], \sigma(x) \in S_{free}\} \quad (1)$$

2.2 RRT* Algorithm

Below is the pseudo code for RRT*.

Algorithm 1: RRT* Algorithm

```
1:  Input: S, S_start, S_goal
2:  Output: A path σ from S_start to S_goal
3:  initializeTree(S)
4:  insertRootNode()
5:  for i = 1 to n do
6:      q_rand ←SampleRandomNode(S)
7:      q_nearest ←FindNearestNeighbor()
8:      q_new ←Steer(q_nearest, q_rand, stepsize)
9:      if ObtacleFree(q_nearest, q_new)
10:         q_min ←SelectParent()
11:         InsertNode(q_min, q_new)
12:         Rewire(Q_near, q_min, q_new)
```

RRT* algorithm is much similar in comparison to RRT. RRT* starts by building a tree using random samples from the robot's operating space and adds new samples to the tree. However, there are two notable differences with respect to RRT: new edge addition and additional step to optimize path cost through rewiring.

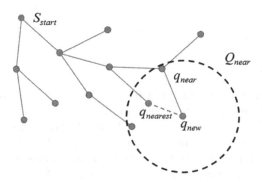

Fig. 1. Node growth progress of RRT*

Figure 1 shows the process of adding a new node in the RRT* algorithm. After RRT* finds the node $q_{nearest}$ closest to q_{new}, it does not immediately add edge $(q_{nearest}, q_{new})$ to the expansion tree, but takes q_{new} as the center and r as the radius to find all potential parent node sets. As shown by the dotted circle in the figure, Q_{near} represents the search area to find potential parent nodes. For every potential parent node in the boundary, as shown by the red dotted line in the figure, a path is drawn starting from the newly added node to the nearby node. If the path is obstacle free and the total cost of this path is lower than the cost of current path, the previous edge in the tree will be deleted and the new edge will be added in the tree.

3   ED-RRT* Path Planning Algorithm

Global path planning is an improved RRT* algorithm that combines goal-biased sampling and adaptive step size. Global path planning is an improved RRT* algorithm that combines biased sampling and adaptive step size. Compared with the RRT* algorithm, bias sampling is introduced to guide the growth direction of the tree, thereby saving the time spent on path planning. Change the fixed step size to an adaptive step size, which can increase the efficiency of exploring the environment. In order to adapt to the dynamic environment, it is important to improve the planning efficiency.

3.1   ED-RRT* Algorithm Framework

ED-RRT* runs three threads for global path planning, local path replanning and visualization. The description and relationship of these three threads are shown in the following figure (Fig. 2):

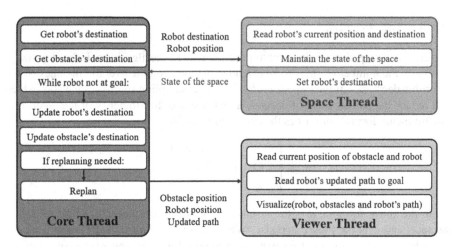

Fig. 2. Framework of the ED-RRT* algorithm

After the initial path is planned, space thread is responsible for maintaining the state information in the environment and reading the current position and target of the robot.

At the same time, it also needs to let the robot drive according to the planned path. Core thread needs to obtain the position and movement of the robot and obstacles in real time.

If the robot does not reach the goal, it is necessary to determine whether the robot will collide with a dynamic obstacle before the next node along the current path. When there is a possibility of collision, it will enter the re-planning stage. Viewer thread is a simulation and visualization tool for the ED-RRT* algorithm, which obtains the position of the robot, the position of obstacles and the path of the robot from the core thread. After rendering, they are presented in the simulation environment built by this article.

3.2 Goal-Biased Sampling

The generation of nodes in the RRT* algorithm is completely random, which causes the exploration process to be non-directional. In turn, a lot of time is spent on path exploration in invalid areas. The sampling method of goal-bias can guide the growth direction of the tree, so that the exploration process is carried out towards the goal point. The generation probability formula for each direction of random nodes is as follows:

$$q_{rand} = \begin{cases} \mu \times q_{goal} + (1 - \mu) \times q_{rand} & obstacle = 0, n > \alpha \\ q_{goal} & obstacle = 0, n \leq \alpha \\ q_{rand} & obstacle = 1 \end{cases} \quad (2)$$

When there is no obstacle nearby, that is, when obstacle $= 0$, the random nodes will tend to be chosen as points near the goal point. This tendency depends on the size of μ, and the larger the μ is, the higher the tendency will be. n represents a random number, and α represents the probability of directly selecting the goal point as a random node. When obstacles are detected, new nodes are randomly generated.

3.3 Adaptive Step Size

The choice of step size is important to the efficiency of the algorithm. If the step size is too small, the number of steps to be extended is large, and the overall efficiency of the algorithm decreases. If the step size is too large, the sampling success rate is low, which will reduce the performance of obstacle avoidance and the efficiency of the algorithm. The following formula describes the choice of step size:

$$\tau = \begin{cases} \lambda & d \leq 0.75d_s \\ 1.5\lambda & 0.75d_s < d \leq d_s \\ 2\lambda & d > d_s \end{cases} \quad (3)$$

In the above formula, d is the distance from nearby obstacles, d_s is the set safety distance, and λ is the preset step size. When the expansion tree is far away from the obstacle, a fixed step is used for expansion to speed up the convergence speed of the algorithm; when the expansion tree is expanded to the vicinity of the obstacle, a small step is used for expansion.

3.4 Path Re-planning

Re-planning begins with the determination of whether or not the moving obstacle is blocking the path. In the detection range of the sensor, the method of judging whether the dynamic obstacle will block the path is shown in Fig. 3.

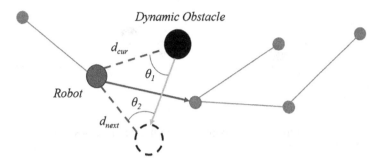

Fig. 3. Schematic diagram of judging whether dynamic obstacles block the path (Colour figure online)

In Fig. 3, the red dots represent the current position of the robot. The black point represents the current position of the dynamic obstacle. At the next moment, the robot will follow the blue trajectory to the next green node, and the dynamic obstacle will follow the yellow trajectory to the position of the dotted circle. The distance between the robot's current moment and the dynamic obstacle is d_{cur}. The distance between the position of the robot at the current moment and the dynamic obstacle at the next moment is d_{next}. θ_1 and θ_2 respectively represent the angle between the two lines.

First, it is judged whether the distance between d_{cur} and d_{next} is less than the safe distance d_{safe}. If so, it is considered that a collision will occur. If not, judge the magnitude of angles θ_1 and θ_2. When the angles of A and B are both less than 90°, dynamic obstacles cross the robot's path. In this case we think there will also be a collision. When there is a possibility of a collision, the algorithm re-plans the path.

The replanning process is divided into the following three steps:

(1) **Set Up Neighborhood for Replanning:** Any sensor has a corresponding sensing range. In the ED-RRT* algorithm, a circle with robot as the center and radius R simulates the range covered by the sensor. If there are many dynamic obstacles in the sensing range, the neighborhood for replanning will be smaller, otherwise it will be larger.

(2) **Set Up a Local Goal:** Using the initial planned path can save the time of re-planning, so when re-planning, only the nodes on the initial path are selected as the local goal. Figure 4 shows the setting of replanning regions and local goals under different numbers of dynamic obstacles. When there are many dynamic obstacles around the robot, the local goal will be set closer to the robot. If the re-planned

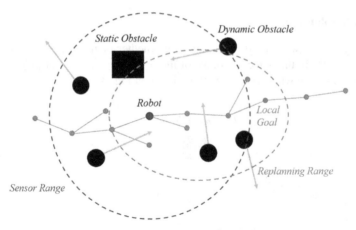

(a) In scenarios with many dynamic obstacles

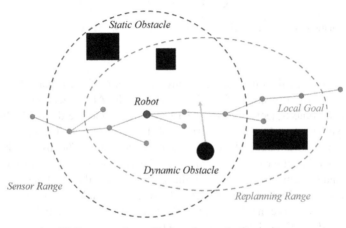

(b) In scenarios with few dynamic obstacles

Fig. 4. In different scenarios, the selection of neighborhood and local goal

path is too long, it will still be blocked by other dynamic obstacles and needs to be re-planned.

(3) **Rewire:** After the local goal is established, the tree will be pruned locally. ED-RRT* algorithm will plan a new local path according to the location of the local goal. After the robot reaches the local goal, it still follows the initial path. As shown in Fig. 5, the green dotted line is part of the initial path. When replanning, this part of the path is discarded, and the blue path is replanned to reach the local goal. Compared with direct re-planning, this local re-planning utilizes the initial path information to save computation time.

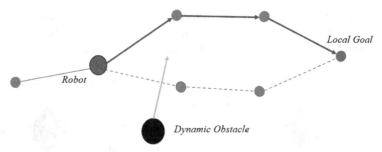

Fig. 5. The rewire progress of ED-RRT* (Colour figure online)

4 Experimental Results

This paper builds a simulation environment based on OpenCV2 to realize the simulation and visualization of the path planning algorithm. We compare our algorithm with improved RRT* algorithm in three different scenarios, and the results are shown in

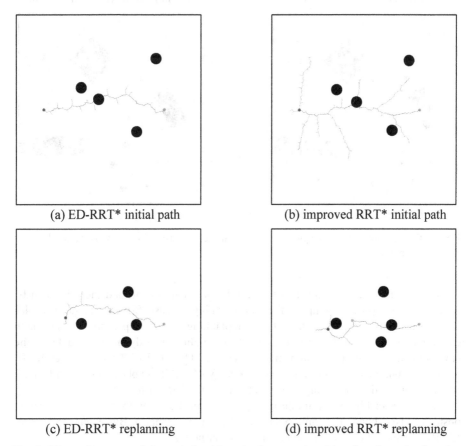

(a) ED-RRT* initial path

(b) improved RRT* initial path

(c) ED-RRT* replanning

(d) improved RRT* replanning

Fig. 6. The performance of the two algorithms in the scenario with a few dynamic obstacles (Colour figure online)

Fig. 6, Fig. 7 and Fig. 8. The red dots in these figures represent the mobile robot, the black circles represent dynamic obstacles, the yellow dots represent the local goal, the gray track represents the RRT, and the blue track represents the planned path.

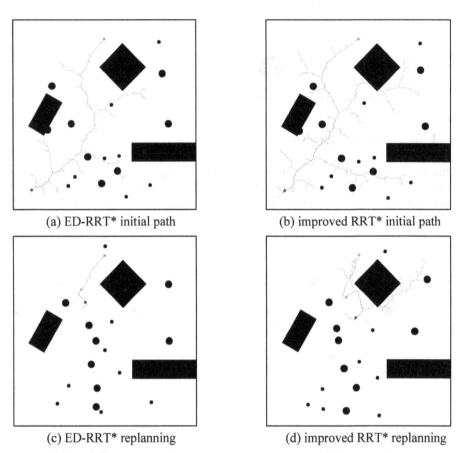

(a) ED-RRT* initial path	(b) improved RRT* initial path
(c) ED-RRT* replanning	(d) improved RRT* replanning

Fig. 7. The performance of the two algorithms in the scenario with many dynamic obstacles (Colour figure online)

The obstacles are expanded in the simulation environment, and the robot can be regarded as a particle to simplify the path planning process. Since dynamic obstacles are not treated as obstacles in the initial path planning, the initial path may pass through dynamic obstacles. As shown in Fig. 6, when the number of dynamic obstacles in the environment is small, the initial path lengths planned by ED-RRT* and improved RRT* are similar, but the number of nodes generated by ED-RRT* exploration is significantly smaller. During the replanning phase, both algorithms perform well.

As shown in Fig. 7, the initial path planning in the scenario with many dynamic obstacles also obtains the same results as the simple scenario. But when replanning, ED-RRT* plans shorter paths and explores fewer nodes.

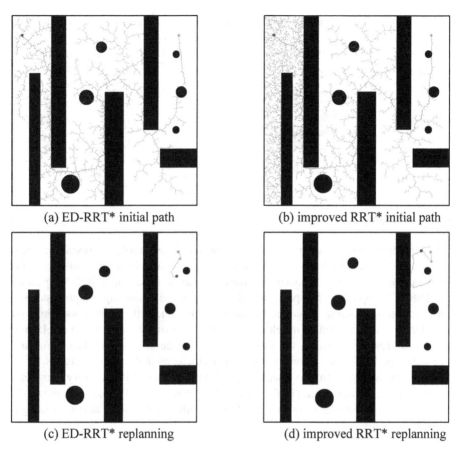

(a) ED-RRT* initial path   (b) improved RRT* initial path

(c) ED-RRT* replanning   (d) improved RRT* replanning

Fig. 8. The performance of the two algorithms in a maze scenario with dynamic obstacles (Colour figure online)

As shown in Fig. 8, in a maze scenario with dynamic obstacles, improved RRT* needs to expand more nodes to explore feasible paths. Since ED-RRT* utilizes the goal-biased sampling method, it will explore towards the target point. In the re-planning stage, ED-RRT* can also plan a shorter path and improve the operating efficiency of the algorithm.

Table 1 presents the average path lengths and average running times of ED-RRT* and improved RRT* under the three scenarios, respectively.

As shown in Table 1, in Scenario 1, the average path length planned by ED-RRT* is 7.9% shorter and the running time is 9.4% shorter than improved RRT*. In Scenario 2, the average path length planned by ED-RRT* is 21.2% shorter and the running time is 24.7% shorter than improved RRT*. in Scenario 3, the average path length planned by ED-RRT* is 16.0% shorter and the running time is 20.7% shorter than improved RRT*.

Table 1. Comparison of the average path length and average running time of the two algorithms

Scenario	Algorithm	Path Length	Running time/s
Scenario 1	ED-RRT*	58	8.51
	Improved RRT*	63	9.39
Scenario 2	ED-RRT*	89	14.78
	Improved RRT*	113	19.62
Scenario 3	ED-RRT*	237	28.05
	Improved RRT*	282	35.37

5 Conclusion

Path planning in dynamic scenarios is one of the key issues in robotics research. Due to its randomness, the sampling-based algorithm will lead to an increase in the amount of computation, and it performs poorly in dynamic scenarios. Based on the RRT* algorithm, the ED-RRT* proposed in this paper introduces goal-biased sampling, adaptive step size and local reprogramming. The goal-biased sampling guides the exploration direction. The adaptive step size method accelerates the efficiency of exploration, and re-planning in local sub-regions increases the dynamic performance of the algorithm. Compared with improved RRT*, this algorithm has higher efficiency in dynamic scenarios.

Although the algorithm has relatively good performance, it is necessary to increase the trajectory optimization part of the back-end for practical problem, so that the trajectory can meet the kinematic constraints of the mobile robot. This is also something to consider in future work.

Acknowledgments. This work was supported by the Natural Science Found of China (Grant No. 62103393).

References

1. Punith, K.M.B., Sumanth, S., Savadatti, M.A.: Internet rescue robots for disaster management. Int. J. Wirel. Microwave Technol. **11**(2), 13–23 (2021)
2. Vasquez, A., et al.: Deep detection of people and their mobility aids for a hospital robot. In: 2017 European Conference on Mobile Robots (ECMR). IEEE (2017)
3. Alatise, M.B., Hancke, G.P.: A review on challenges of autonomous mobile robot and sensor fusion methods. IEEE Access **8**, 39830–39846 (2020)
4. Hart, P.E., et al.: A formal basis for the heuristic determination of minimum cost paths. IEEE Trans. Syst. Sci. Cybern. **4**(2), 100–107 (2007)
5. Dijkstra, E.W.: A note on two problems in connexion with graphs. Numer. Math. **1**(1), 269–271 (1959)
6. Li, Z., Shi, R., Zhang, Z.: A new path planning method based on sparse A* algorithm with map segmentation. Trans. Inst. Meas. Control. **44**(4), 916–925 (2022)
7. LaValle, S.M., Kuffner, J.J.: Randomized kinodynamic planning. Int. J. Robot. Res. **20**(5), 378–400 (2001)

8. Kuffner, J.J., LaValle, S.M.: RRT-connect: an efficient approach to single-query path planning. In: Proceedings of the 2000 ICRA. Millennium Conference. IEEE International Conference on Robotics and Automation. Symposia Proceedings (Cat. No. 00CH37065), vol. 2, pp. 995–1001 (2000)

9. Urmson, C., Simmons, R.: Approaches for heuristically biasing RRT growth. In: Proceedings of the IEEE/RSJ International Conference on Intelligent Robots and Systems. IEEE, Las Vegas, USA, vol. 2, pp. 1178–1183 (2003)

10. Karaman, S., Frazzoli, E.: Sampling-based algorithms for optimal motion planning. Int. J. Robot. Res. **30**(7), 846–894 (2011)

11. Adiyatov, O., Varol, H.A.: Rapidly-exploring random tree based memory efficient motion planning. In: 2013 IEEE International Conference on Mechatronics and Automation, pp. 354–359. IEEE (2013)

12. Gammell, J.D., Srinivasa, S.S., Barfoot, T.D.: Informed RRT*: optimal sampling-based path planning focused via direct sampling of an admissible ellipsoidal heuristic. In: Proceedings of the 2014 IEEE/RSJ International Conference on Intelligent Robots and Systems. IEEE (2014)

13. Fang, Z., Qidan, Z., Guoliang, Z.: Path optimization of manipulator based on the improved rapidly-exploring random tree algorithm. J. Mech. Eng. **47**(11), 30–35 (2011)

14. Nasir, J., Islam, F., Malik, U., et al.: RRT*-smart: a rapid convergence implementation of RRT*. Int. J. Adv. Rob. Syst. **10**(7), 299–311 (2013)

15. Stentz, A., et al.: The focussed d* algorithm for real-time replanning. IJCAI **95**, 1652–1659 (1995)

16. Koenig, S., Likhachev, M.: Improved fast replanning for robot navigation in unknown terrain. In: Proceedings 2002 IEEE International Conference on Robotics and Automation (Cat. No.02CH37292), vol. 1, pp. 968–975 (2002)

17. Xi, Y., et al.: A real-time dynamic path planning method combining artificial potential field method and biased target RRT algorithm. J. Phys: Conf. Ser. **1**, 2021 (1905)

Lidar Localization Method for Mobile Robots Based on Priori Pose Compensation

Guangpu Zhao, Jikai Wang, Liang Chen, and Zonghai Chen[✉]

University of Science and Technology of China, Hefei 230026, China
{zhaoguangpu9807,wangjk,cltogether}@mail.ustc.edu.cn,
chenzh@ustc.edu.cn

Abstract. Accurate and robust map-based localization is crucial for autonomous mobile robots. In this paper, we build a set of lidar localization framework, from initialization to pose tracking, to achieve real-time localization of mobile robots on priori map. To tackle the problem that large localization errors happen when mobile robot is turning fastly, we proposes a priori pose compensation method to improve it. By extracting the global features of the point cloud and performing pre-alignment, we provide an accurate a prior pose for further point cloud registration to improve the accuracy of robot localization in turns. We tested the proposed localization approach on a mobile robot platform equipped with a lidar sensor in campus scenario. The experimental results show that our method can reliably and accurately localize the mobile robot in the campus scenario and operate online at the lidar sensor frame rate to track the robot pose.

Keywords: Mobile robots · Lidar localization · Pre-alignment · Pose compensation

1 Introduction

Precise localizaion is a fundamental capability required by most autonomous mobile systems. With a localizaion system [1], a mobile robot or autonomous vehicle is able to estimate its pose in map based on observations obtained with onboard sensors. Autonomous navigation of mobile robots [2] requires accurate and reliable LiDAR-based global localization [3], especially when GPS is limited or when GPS cannot provide accurate localization results. Most autonomous mobile systems have a 3D LiDAR sensor to sense the environment and directly provide 3D range measurements [4], and solving the localization problem of mobile robots based on such sensors is popular in the area of autonomous driving. However, there are less work focusing on the global localization problem [5], more work mainly focused on designing representations [6] for matching between sensor readings and the map [7]. Learning the representation of 3D point cloud is explored in objects detection [8], but inapplicable in localization problem. In summary, the critical challenge for map-based localization we consider is the lack of real-time effective global pose estimation method [9].

© The Author(s), under exclusive license to Springer Nature Singapore Pte Ltd. 2022
W. Fan et al. (Eds.): AsiaSim 2022, CCIS 1713, pp. 314–324, 2022.
https://doi.org/10.1007/978-981-19-9195-0_26

In this paper, our contribution is mainly reflected in two aspects. First, in the first aspect, we build a localization framework to solve the 3D Lidar-based global localization problem, by converting the global localization problem into two parts: initial localization on priori map and pose tracking [10]. Firstly, the point clouds collected by laser in the campus scene are stitched into a global point cloud map according to Groundtruth, which performing the localization of the mobile robot on. In order to perform initial localization on the map [11], we segment the global map according to the selected grid size, extract FPFH local features [12] for the segmented point cloud and the first frame point cloud respectively, and then perform feature matching + RANSAC to iteratively obtain the poses and matching errors. The pose with the lowest matching error is selected as the coarse initial pose, which is used for ndt matching [13] between the current frame point cloud and map, and the fine matching pose is obtained as the system initial localization result. The subsequent pose tracking is NDT matching based on scan to map [14] for each frame of the point cloud, and UKF(Unscented Kalman Filter) [15] is introduced to provide a predicted initial value for the next matching based on the previous localization result. The current matching result is applied as the observation of the filter to update the predicted initial value, and the updated result is used as the global pose of the current frame point cloud. Another aspect of our contribution is to propose a priori pose compensation method. For the problem of large localization errors that occur in turns [16], we propose a priori positional compensation method. By extracting the global features [17] of the current frame and the previous point cloud to perform pre-alignment on the yaw angle, we provide a compensation pose for the initial value predicted by the UKF, which reduces error in the initial prediction value and thus achieves a great improvement on the final localization accuracy. The rest of this paper is organized as follows: Sect. 2 describes our localization framework. In Sect. 3, we introduce our priori pose compensation method. we evaluate the accuracy of our method on campus scene dataset in Sect. 4. Section 5 concludes with a brief discussion on our methods.

2 System Overview

The global localization framework proposed in this paper consists of two parts: initial localization and pose tracking. The framework of the localization system is shown in Fig. 1, and the system hardware composition is shown in Fig. 2.

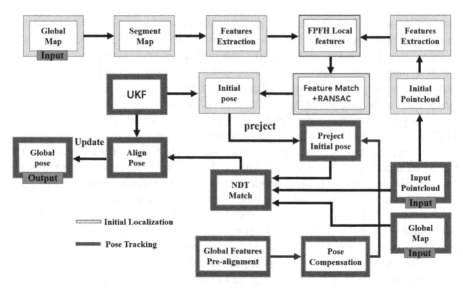

Fig. 1. Global localization system framework. It includes initial localization and pose tracking.

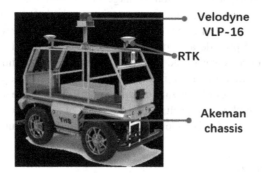

Fig. 2. Mobile robot platform hardware architecture

2.1 System Hardware

The framework proposed in this paper is validated on a campus scene dataset collected by a mobile robotics platform equipped with Velodyne VLP-16 LiDAR, using RTK measurements which can Provide accurate centimeter level localization accuracy as Groundtruth. The VLP-16 measurement range is up to 100 m with an accuracy of $\pm$ 3 cm. It has a vertical field of view (FOV) of $30°$ ($\pm15°$) and a horizontal FOV of $360°$. The 16-channel sensor provides a vertical angular resolution of $2°$. The horizontal angular resolution varies from $0.1°$ to $0.4°$ based on the rotation rate. Throughout the paper, we choose a scan rate of 10 Hz, which provides a horizontal angular resolution of $0.2°$. The localization algorithm mentioned in this paper was validated on a laptop equipped with an AMD Ryzen 5800 h CPU and was able to run in real time on a mobile robotics platform.

2.2 Initial Localization

We segment the priori global point cloud map according to the selected grid size (40 m × 40 m) for the purpose of feature matching with the current query point cloud to estimate the optimal initial poses. In the feature extraction phase, some query points are selected uniformly in the point cloud, and the simplified point feature histogram (SPFH) of each point in the k-neighborhood of the query points is calculated separately by parameterizing the spatial differences between the query points and the neighboring points. The individual SPFH in the neighborhood are weighted and counted into the final fast point feature histogram (FPFH). The FPFH feature is invariant to the 6-dimensional pose of the point cloud and is robust to different sampling densities or noise levels in the neighborhood. In the feature matching phase, FPFH is used to synthetically describe the feature information. The local correlation of feature information is established between the matched point clouds, and then the initial coarse matching is achieved by sampling consistent initial transformation (RANSAC) according to the correspondence of feature point pairs. Finally, the matching is performed using NDT based on the obtained initial values to achieve the precise transformation relationship and matching error. The transformation relation with the lowest matching error is selected as the global poses of the query point cloud, i.e., the initial pose of the proposed localization method. Figure 3 shows the result of successful initialization.

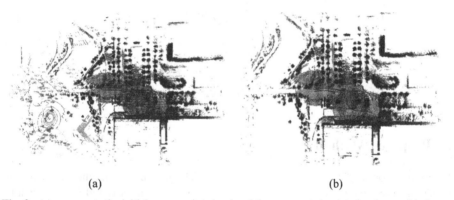

(a)	(b)

Fig. 3. (a) represents the initial query point cloud and the segmented point cloud map. (b) shows the result of successful initialization.

2.3 Pose Tracking

After the successful initialization of the localization system, the subsequent pose tracking is a continuous NDT matching based on scan to map, and the state of the entire system is maintained by the Unscented Kalman filter (UKF). Compared to EKF, which approximates the nonlinear distribution with Taylor series near the Gaussian mean point to obtain an approximate Gaussian distribution, UKF extracts a series of representative points including the mean point from the original Gaussian distribution, and adds different weights to the representative points to bring them into the nonlinear equation and approximates around them to obtain better approximation results. Combining the localization methods of this paper, UKF considers the position, velocity and quaternion information as the mean of the maintained Gaussian distribution. For the localization of the current frame, as shown in Eqs. (1) and (2), UKF adds some process noise to the localization result of the previous frame to perform an update of the mean and variance of the maintained distribution, which generates a predicted initial value as the initial pose of NDT matching between the current frame and the map.

$$\mu' = \sum_{i=0}^{2n} \omega^{[i]} g(\chi^{[i]}) \tag{1}$$

$$\Sigma' = \sum_{i=0}^{2n} \omega^{[i]} \left(g\left(\chi^{[i]}\right) - \mu' \right) \left(g\left(\chi^{[i]}\right) - \mu' \right)^T + R_t \tag{2}$$

As shown in Eq. (3), (4), (5), (6), where K, T is the gain of the filter. UKF updates the mean and variance once again with the point cloud matching result as the observation of the filter, and the updated result is considered as the global pose of the final current frame pose. In this way, the iterative calculation achieves the real-time estimation of the global pose during the motion of the robot.

$$z = h(\chi) \tag{3}$$

$$\hat{z} = \sum_{i=0}^{2n} \omega^{[i]} z^{[i]} \tag{4}$$

$$\mu = \mu' + K(z - \hat{z}) \tag{5}$$

$$\Sigma = (I - KT)\Sigma' \tag{6}$$

3   Proposed Method

Like the majority of laser localization methods, The localization method built in this paper has a relatively large localization error when the robot turns and is not robust to rotation. For this problem, after analysis, we found that the predicted initial value given by UKF has a large deviation especially in the rotation angle, when the robot is turning, and the final localization result has the same error tendency as the predicted initial value. Therefore, this paper proposes a priori pose compensation method that extracts the global features of the current frame and the previous frame point cloud when the robot is in a turn. A priori pose is pre-calculated by aligning the global features to compensate the predicted initial value of UKF to make it closer to the real value, so that we can get more accurate localization results during the turn.

3.1   Pre-alignment of Yaw Angle

In contrast to local features, global feature representations tend to have better rotational robustness. Therefore, it can reflect the variations of robot's viewpoint well and is often used in the position recognition of lidar point clouds. Since the ground of campus scenes tends to be flat, the pitch and roll angles do not vary much. Therefore, this paper uses the rotation undeformation of global features to pre-calculate the variation of yaw angle between last and current frames. As shown in Fig. 4, similar to scancontext [17], the point cloud is projected to the x_y plane, divided into different bins according to orientation and radius, and the whole point cloud is encoded into a matrix, where each row represents a ring and each column represents a sector, preserving the absolute geometric structure of the point cloud. The height information of all point clouds in each bin is encoded using an eight-bit binary code, which is linearly discretized into 8 intervals in the order of height, for which the value of the interval is 1, otherwise it is 0. Finally, the 8-bit binary code is converted to decimal numbers as the value of each bin. In this paper, since the mobile robot is moving slowly and the rotation angle is low, the number of rings is taken as 20 and the number of sectors is 360. Since the last and current frames are definitely at the same position, the yaw angle alignment is performed by continuously moving the matrix by column while calculating the similarity of the matrix at each column movement, and the number of column movements corresponding to the maximum similarity is the yaw angle change between the last and current frame point clouds.

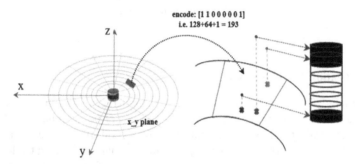

Fig. 4. Schematic diagram of generating global descriptor

3.2 Priori Pose Compensation

The localization method (without pose compensation) in this paper shows fine localization performance when the robot is moving in a straight line, so we simply need to apply the compensation pose pre-calculated when turning. How to determine when the robot has been rotated? We recover the Euler angles from the localization results of the previous two frames of the current frame. Referring to the algorithm logic of evo to ensure that the recovered Euler angles are continuous. Then we calculate the absolute value of the difference of the yaw angle between the two frames and make a threshold judgment on the absolute value. If the absolute value is above a certain threshold (The threshold value in this paper is $1.0°$), it is considered to be in a rotating state and the compensation pose is applied. It is not necessary if below the threshold.

4   Experiment

We validate the localization method of this paper on a campus scene dataset collected by a mobile robot platform equipped with Velodyne VLP-16 LIDAR, and Online verification was also performed. Qualitatively and quantitatively, we analyze the necessity and feasibility of positional compensation, and the improvement of localization accuracy.

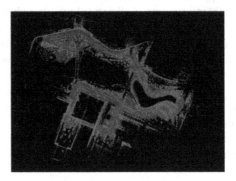

Fig. 5. Global point cloud map

4.1   Localization Error Analysis

As shown in Fig. 5, we use the ground truth of the campus scene dataset to stitch the point clouds as a global priori map. We re-collected two sets of point cloud data, denoted as trajectory1 and trajectory2, within the map area. We re-collected two sets of point cloud data, denoted as trajectory1 and trajectory2, within the map area. Using the localization method in this paper (without compensation pose) to verify the localization accuracy. we can see that trajectory1 (Fig. 6a) fails to localize at the later turn, and trajectory2 (Fig. 6b), although successfully localized, also shows poor localization performance during the turn.

We further analyze the localization success of trajectory2 to explore the reasons for its poor localization performance. We compare the predicted initial values given by UKF during localization with the localization results, and we can observe that the localization results are consistent with the predicted initial values (Fig. 6c and Fig. 6d), i.e., the localization results depend strongly on the accuracy of the initial values. Therefore, our idea is to compensate the predicted initial value to make it as close to the ground truth as possible to improve the final localization result. To investigate whether the error in the initial value is affected by the rotation, we take the delta yaw of the localization result between the before and after frames as the horizontal coordinate, and the yaw difference between the predicted initial value and the ground truth as the vertical coordinate. The results show that the yaw compensation required to make the predicted initial value close to the ground truth is very large when the delta yaw is large, i.e. the robot is in turn. Therefore it validates the correctness of our idea of providing compensation to the predicted initial value at the turn.

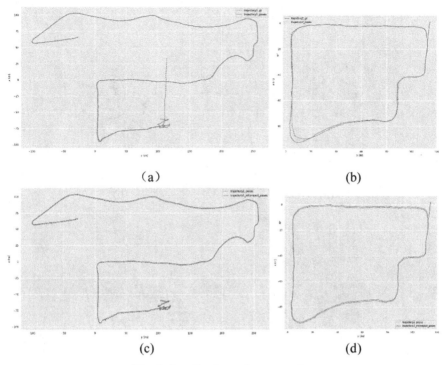

(a)   (b)

(c)   (d)

Fig. 6. Localization trajectory results.

4.2 Compensation Pose Verification

To verify the accuracy of the initial poses recovered from the global features we constructed, we selected a set of point clouds with reverse closure from the dataset of the campus scene. The above global features are extracted from the point clouds, and the variation of the yaw angle between two frames of the point clouds is calculated by the matrix similarity metric as the initial poses for NDT matching of the point clouds. No initial positional alignment of the point cloud is provided for NDT registration for comparison. The matching results are shown in Fig. 7 below. We further calculated root mean square error (RMSE) for the other closed-loop frames in the sequence for comparison, and the results are shown in the following Figs. 7. It can be obviously observed that the initial pose recovered by global features has a great improvement on the point cloud matching performance.

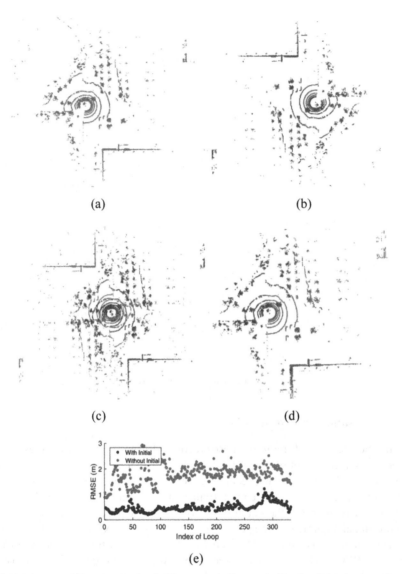

Fig. 7. (a) represents Query pointcloud. (b) represents Detected pointcloud. (c) shows registration without initial. (d) shows registration with initial value. (e) shows the RMSE of the matching result

4.3 Localization Results with Compensation Pose

We apply the pose compensation to the localization method proposed in this paper and re-test trajectory1 and trajectory2 on the global a priori map. The experimental results are shown in Fig. 8 and Table 1 below. Ensure the accuracy of the RTK Localization results as the groundtruth, the selected operation tracks are located in the open space of the buildings. We apply the pose compensation to the localization method proposed in this paper and re-test trajectory1 and trajectory2 on the global a priori map. The experimental results are shown in Fig. 8 and Table 1 below. The localization error is the average result of three times of localization. Trajectory1 can successfully locate where the original localization failed, and has fine localization accuracy, and trajectory2 has a great improvement in localization accuracy, especially at the turns. It can verify the improvement of localization performance with the priori compensation pose method proposed in this paper.

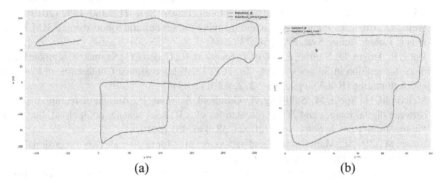

(a)   (b)

Fig. 8. Localization results with compensation pose

Table 1. Comparison of localization result

Trajectory	Method	Mean error (m)	RMSE (m)	Time (ms)
Trajectory1	Original_localization	5.9952	20.1840	8 ms
	Improved_localization	0.2920	0.5219	10 ms
Trajectory2	Original_localization	1.6171	2.4739	8 ms
	Improved_localization	0.4902	0.6987	10 ms

5   Conclusion

This paper proposes a lidar localization method based on a priori compensation pose. We conducted offline and online tests in a campus scenario using a mobile robot equipped with Velodyne VLP-16 LIDAR. The global localization of the mobile robot on the priori map can be achieved in real time through initial localization and pose tracking. By pre-calculating compensation pose through global features during rotation, the localization accuracy has been greatly improved. In the future, we will focus on how to improve the localization accuracy of mobile robots during rotation.

Acknowledgments. This work was supported by the National Natural Science Found of china (Grant No. 62103393).

References

1. Gao, Y., Liu, S., Atia, M.M., Noureldin, A.: INS/GPS/LiDAR integrated navigation system for urban and indoor environments using hybrid scan matching algorithm. Sensors **15**, 23286–23302 (2015)
2. Nakhaeinia, D., Tang, S.H., Noor, S.M., Motlagh, O.: A review of control architectures for autonomous navigation of mobile robots. Int. J. Phys. Sci. **6**, 169–174 (2011)
3. Yin, H., Wang, Y., Ding, X., Tang, L., Huang, S., Xiong, R.: 3d lidar-based global localization using siamese neural network. IEEE Trans. Intell. Transp. Syst. **21**, 1380–1392 (2019)
4. Weiss, U., Biber, P.: Plant detection and mapping for agricultural robots using a 3D LIDAR sensor. Robot. Auton. Syst. **59**, 265–273 (2011)
5. Dubé, R., Dugas, D., Stumm, E., Nieto, J., Siegwart, R., Cadena, C.: Segmatch: segment based place recognition in 3d point clouds. In: 2017 IEEE International Conference on Robotics and Automation (ICRA), pp. 5266–5272. IEEE (2017)
6. Meierhold, N., Spehr, M., Schilling, A., Gumhold, S., Maas, H.: Automatic feature matching between digital images and 2D representations of a 3D laser scanner point cloud. Int. Arch. Photogramm. Remote Sens. Spat. Inf. Sci. **38**, 446–451 (2010)
7. Bosse, M., Zlot, R.: Map matching and data association for large-scale two-dimensional laser scan-based slam. Int. J. Robot. Res. **27**, 667–691 (2008)
8. Li, B., Zhang, T., Xia, T.: Vehicle detection from 3d lidar using fully convolutional network (2016). arXiv preprint arXiv:1608.07916
9. Choi, S., Choi, S., Kim, C.: MobileHumanPose: toward real-time 3D human pose estimation in mobile devices. In: Proceedings of the IEEE/CVF Conference on Computer Vision and Pattern Recognition, pp. 2328–2338 (2021)
10. Xiao, B., Wu, H., Wei, Y.: Simple baselines for human pose estimation and tracking. In: Proceedings of the European Conference on Computer Vision (ECCV), pp. 466–481 (2018)
11. Liu, G.-X., Shi, L.-F., Chen, S., Wu, Z.-G.: Focusing matching localization method based on indoor magnetic map. IEEE Sens. J. **20**, 10012–10020 (2020)
12. Rusu, R.B., Blodow, N., Beetz, M.: Fast point feature histograms (FPFH) for 3D registration. In: 2009 IEEE International Conference on Robotics and Automation, pp. 3212–3217. IEEE (2009)
13. Akai, N., Morales, L.Y., Takeuchi, E., Yoshihara, Y., Ninomiya, Y.: Robust localization using 3D NDT scan matching with experimentally determined uncertainty and road marker matching. In: 2017 IEEE Intelligent Vehicles Symposium (IV), pp. 1356–1363. IEEE (2017)
14. Fu, H., Ye, L., Yu, R., Wu, T.: An efficient scan-to-map matching approach for autonomous driving. In: 2016 IEEE International Conference on Mechatronics and Automation, pp. 1649–1654. IEEE (2016)
15. Huang, G.P., Mourikis, A.I., Roumeliotis, S.I.: A quadratic-complexity observability-constrained unscented Kalman filter for SLAM. IEEE Trans. Rob. **29**, 1226–1243 (2013)
16. Choi, S., Hong, D.: Position estimation in urban U-turn section for autonomous vehicles using multiple vehicle model and interacting multiple model filter. Int. J. Automot. Technol. **22**, 1599–1607 (2021)
17. Kim, G., Kim, A.: Scan context: egocentric spatial descriptor for place recognition within 3d point cloud map. In: 2018 IEEE/RSJ International Conference on Intelligent Robots and Systems (IROS), pp. 4802–4809. IEEE (2018)

A Method of Square Root Central Difference Kalman Filter for Target Motion Analysis

Yi Zheng[1,3](✉), Mingzhou Wang[1], Youfeng Hu[2], Yunchuan Yang[1,3], and Xiangfeng Yang[1,3]

[1] Xi'an Precision Machinery Research Institute, Xi'an 710000, China
z_yi25@163.com
[2] Kunming Branch of Xi'an Precision Machinery Research Institute, Kunming 650118, China
[3] Science and Technology on Underwater Information and Control Laboratory, Xi'an 710000, China

Abstract. The central differential Kalman filter (CDKF) and square root central differential Kalman filter (SR-CDKF) are widely used methods for addressing nonlinear problems in target motion analysis. However, for the bearings-only target motion analysis by single observer, sometimes the CDKF and the SR-CDKF experience the divergence problem or even filter interruption due to the observer's mobility and system noise. To solve these problems, SR-CDKF is improved, and an adaptive singular value decomposition square root center difference Kalman filter (ASVDSR-CDKF) is proposed. The covariance square root update method based on the singular value decomposition method is deduced. The error discriminant statistics and adaptive factors are constructed. When the disturbance is too large, the adaptive factor is automatically adjusted and the square root updating form is selected. The simulation of bearings only target motion analysis is carried out under three different conditions, and the performance of CDKF, SR-CDKF, square root unscented Kalman filter (SR-UKF) and the proposed ASVDSR-CDKF method are compared. The simulation results demonstrate that the proposed method not only has high accuracy, but is also more stable than the other three methods.

Keywords: Center difference Kalman filter · Adaptive filtering · Target motion analysis · Bearing only · Nonlinear filtering

1 Introduction

Recognition and tracking of underwater targets are very important for the safe operation of submarines, ships, torpedoes and, unmanned underwater vehicles. To facilitate this, it is necessary to conduct a passive analysis on the target motion state of the underwater vehicle without actively radiating energy. Underwater bearings-only target motion analysis (BOTMA) is a process of estimating the target motion parameters (such as position and velocity) based on the measurements of bearings [1]. Underwater bearings-only target motion analysis can be applied to sensor networks [2, 3] or to motion observers [4]. Compared with a multi-sensor setup, single observer bearings-only target motion analysis needs fewer observers and has strong concealment, and has wider application across a diverse set of scenarios [5–7] e.g. close-range tasks or when a torpedo attacks a moving target. In this paper, we consider BOTMA at a close range using a single observer with a passive bearing sensor. Due to increased bearing jitter when the observer is closer to the target, the filtering is more prone to divergence or interruption. At the same time, a complication occurs: the potential for target location and velocity may be not fully observable, which makes it more difficult to estimate the state of target [8–10].

The BOTMA is a nonlinear problem which can be effectively solved by and nonlinear filtering. Extended Kalman filter (EKF) [11] is a nonlinear filtering method that was applied to bearings-only target motion analysis. The nonlinear function is linearized by a Taylor series expansion in order to estimate the target state in the nonlinear system. Since EKF is only a first-order approximation of nonlinear functions, the linearization error is large, which results in poor accuracy and stability of filtering [12, 13]. Another method to achieve nonlinear filtering is the sampling method, which includes deterministic sampling and random sampling. A common example of random sampling is the particle filter (PF) [14] and its class method [15, 16]. This kind of method is not constrained by the linear and Gaussian assumptions of the model and can, in theory, have higher accuracy. Nonetheless, the disadvantages of PF include the large amount of calculation and the possible problems of particle degradation or particle depletion, which limits this method's use in engineering applications. Representatives of deterministic sampling methods include the: unscented Kalman filter (UKF) [17], cubature Kalman filter (CKF) [18], and central differential Kalman filter (CDKF) [19, 20]. Among them, UKF is the most widely used method [21–23]. UKF is based on unscented transformation, where at least second-order approximation can be reached, and its computational complexity has the same order as that of EKF. However, UKF requires three parameters, α, β and λ, which need to be selected according to the nonlinearity of the motion or observational model. The selection of parameter values will affect the estimation accuracy of the filter. Considering that the nonlinear degree of the model cannot be accurately estimated, it is difficult to find the optimal values of these parameters. Square root unscented Kalman filter (SR-UKF) based on UKF is also used in many fields [24, 25]. SR-UKF method has been used in target tracking of underwater bearing-only and bearing-Doppler measurements with good results [26]. Therefore, SR-UKF is also included in the simulation as a comparison method.

CDKF is also a kind of deterministic sampling method. The sampling points are selected in a different way than that of UKF. Based on the interpolation formula of sterling polynomials, CDKF estimates nonlinear functions in the form of central difference, where at least second-order approximation can be reached. The accuracy of CDKF is equivalent to that of UKF, and only one parameter h needs to be adjusted. CDKF has good application in target tracking [27, 28] and navigation systems [29]. Using CDKF as a basis, a square root central difference Kalman filter (SR-CDKF) algorithm is proposed [30]. The square root (SR) of covariance is being used instead of covariance to participate in the recursive operation, which improves the filtering accuracy and provides good results in many scenarios [31, 32]. However, the observability of the system is poor and the measurement noise has a great influence on BOTMA by the single observer. Therefore, it is possible that the filter is divergent and the covariance matrix is not positive definite after decomposition, which makes the filter unstable or even incapable of working [8–10]. This phenomenon becomes more common when the target distance is close.

The objective of this paper is to solve the problem of filter instability in the BOTMA by the single observer in the case of close distances and small maneuvers. In this research, we propose an improved method of SR-CDKF: adaptive singular value decomposition square root central difference Kalman filter (ASVDSR-CDKF). The paper is organized as follows. Section 2 contains a description of the BOTMA problem by the single observer. Then, we present the application steps of SR-CDKF in BOTMA. In Sect. 3, an ASVDSR-CDKF method is proposed. It adds an adaptive factor based on SR-CDKF, uses the error discriminant statistics constructed by the residuals in order to judge the disturbance to the system, and uses the singular value decomposition (SVD) method to update the square root of covariance when the disturbance is large. Section 3-A derives a covariance square root update method based on SVD. An adaptive factor for SR-CDKF is constructed in Sect. 3-B. The steps of the ASVDSR-CDKF method are described in Sect. 3-C. In Sect. 4, the simulations are carried out in three different cases. Through the simulation results, we compare the performance of CDKF, SR-CDKF, SR-UKF, and ASVDSR-CDKF in terms of error and filtering stability. The concluding remarks are presented in Sect. 5.

2   Problem Description and Filtering Method

2.1   Bearings-Only Target Motion Analysis by Single Observer

The movement of the target is non-maneuverable in many cases underwater. Therefore, the nearly constant velocity (NCV) model [11, 33] is suitable for the underwater BOTMA. In this research, we consider a single non-maneuverable target and a single poor-maneuvering observer in a two-dimensional space. The system model in the Cartesian coordinate system is demonstrated in Fig. 1.

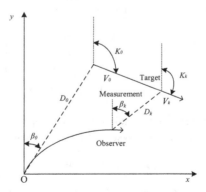

Fig. 1. An overview of the target and observer (bearings only) geometry.

In Fig. 1, K_k is the direction of the target motion, V_k is the target velocity, $\beta_0, \beta_1, \cdots\cdots, \beta_k$ are the target bearings corresponding to the time $t_0, t_1, \cdots\cdots, t_k$, and D_k is the distance between the target and the observer at time t_k. $[D_k, K_k, V_k, \beta_k]$ can determine the motion trajectory of the target for the target moving in a straight line at a uniform velocity.

The target state can be indicated as: $X_k = [x_k, \dot{x}_k, y_k, \dot{y}_k]^T$. $[x_k, y_k]$ is the coordinate of the target's position; $\dot{x}_k$ and $\dot{y}_k$ are the velocity components of the target. Thus, the following formula can be obtained:

$$
\begin{cases}
x_k = D_k \sin \beta_k + x_{o,k} \\
y_k = D_k \cos \beta_k + y_{o,k} \\
\dot{x}_k = V_k \sin K_k \\
\dot{y}_k = V_k \cos K_k
\end{cases}
\tag{1}
$$

where $[x_{o,k}, y_{o,k}]$ is the coordinate of the observer, which is known. $X_k = [x_k, \dot{x}_k, y_k, \dot{y}_k]^T$ can also determine the trajectory of the target. The motion parameters of the target can be obtained by acquiring the target motion state at the complete time. Target motion analysis is the process of solving the parameter $[x_k, \dot{x}_k, y_k, \dot{y}_k]$ or $[D_k, K_k, V_k, \beta_k]$ or its subset. In BOTMA, the sensor of the observer can only obtain the measurement of bearings. When β_k is known, the target trajectory can be uniquely determined according to $[D_k, K_k, V_k]$.

For the NCV model, the target moves in a straight line with the uniform velocity, and the state equation of the system can be expressed as follows:

$$
X_{k+1} = \Phi X_k + w_k
\tag{2}
$$

where Φ is the 4×4 dimensional state transition matrix, w_k is the zero-mean Gaussian process noise with variance matrix Q. For the NCV model:

$$
\Phi = \begin{bmatrix}
1 & \Delta t & 0 & 0 \\
0 & 1 & 0 & 0 \\
0 & 0 & 1 & \Delta t \\
0 & 0 & 0 & 1
\end{bmatrix}
\tag{3}
$$

$$
Q = \delta_w^2 \begin{bmatrix}
\Delta t^3/3 & \Delta t^2/2 & 0 & 0 \\
\Delta t^2/2 & \Delta t & 0 & 0 \\
0 & 0 & \Delta t^3/3 & \Delta t^2/2 \\
0 & 0 & \Delta t^2/2 & \Delta t
\end{bmatrix}
\tag{4}
$$

where δ_w^2 is the process noise intensity and Δt is the sampling interval.

The measurement equation of the system can be represented as follows:

$$
Z_k = f(X_k) + v_k
\tag{5}
$$

where v_k is the zero-mean Gaussian noise with variance R. For BOTMA by single observer, R is 1×1 dimensional measurement noise variance matrix. In Eq. (5), $f(X_k) = \arctan[(y_k - y_{o,k})/(x_k - x_{o,k})]$.

Equations (2) and (5) constitute the system state space model of BOTMA, which is a typical nonlinear filtering problem. The observer maneuver is only a necessary condition for ensuring the observability rather than a sufficient condition, and the observer maneuver plays a key role in the filtering effect. However, in the case of close pursuit of the target, it is easy for the observer to lose the target in a strong maneuver, which can result in either poor maneuvering or no maneuvering. Therefore, it is more challenging for bearings-only target motion analysis by a single observer compared with other tracking systems.

2.2 Square Root Central Differential Kalman Filter

It is assumed that the system state variables obey Gaussian distribution, and the mean and the covariance are known. The core idea of CDKF is to use the sterling's polynomial interpolation formula to expand the nonlinear equation in the form of central difference without calculating the Jacobian matrix of the function. The second-order Stirling polynomial interpolation formula expanded at $x = \overline{x}$ of the nonlinear function $f(x)$ is as follows:

$$f(x) \approx f(\overline{x}) + f_D'(\overline{x})(x - \overline{x}) + f_D''(\overline{x})(x - \overline{x})^2/2! \tag{6}$$

where, $f_D'(\overline{x})$ is the first-order difference operator, $f_D''(\overline{x})$ is the second-order difference operator:

$$f_D'(\overline{x}) = [f(\overline{x} + h\delta) - f(\overline{x} - h\delta)]/2h \tag{7}$$

$$f_D''(\overline{x}) = [f(\overline{x} + h\delta) + f(\overline{x} - h\delta) - 2f(\overline{x})]/h^2 \tag{8}$$

Among them, h is the half step of the central difference, and its value can determine the distribution interval of the sampling points. The optimal value of h suitable for Gaussian distribution is $\sqrt{3}$. δ is a zero-mean random variable with the same covariance as x. Equation (6) can be regarded as replacing the derivative operation in Taylor series expansion with the central difference operation, as well as replacing the first and second derivatives with the first and second order central difference operators.

The central difference filter in reference [19] and the derivation of the divided difference filter in reference [20] are both based on Stirling's polynomial interpolation formula. The two filters are essentially the same, that is, CDKF. CDKF approximates the distribution function of state variables by deterministic weighted sampling points (sigma points), and estimates the mean and covariance of state variables after nonlinear transformation by the nonlinear transformation of sampling points. The number of sigma sampling points required to construct the L-dimensional state vector is 2L + 1. Sigma points have the same mean, variance, and high-order center distance with the real state vector. UKF is similar to CDKF and belongs to sigma point Kalman filtering as well. Their sigma point sampling methods and corresponding weights are similar, but their covariance matrices are different. On the basis of sigma point Kalman filtering, the square root of the covariance matrix is used to replace covariance for calculation, which has lower computational complexity, stronger numerical stability, and smoother filtering results. SR-UKF and SR-CDKF are based on UKF and CDKF respectively. Cholesky decomposition and QR factorization are used to calculate the square root. The covariance is replaced by the square root of covariance to increase the numerical stability of the filtering result and make it smoother. The difference in estimation accuracy between CDKF and UKF is negligible. The CDKF uses only a single scalar parameter h, as opposed to the three that the SR-UKF uses. Therefore, the CDKF was chosen as a typical nonlinear Gaussian filter to be studied in this paper.

The SR-CDKF method based on CDKF can be used to solve BOTMA problem. The specific steps can be summarized as follows:

Step 1: When k = 0, initialize

$$\hat{X}_0 = E(X_0) \tag{9}$$

$$S_0 = \text{Chol}\{ E[(\hat{X}_0 - X_0)(\hat{X}_0 - X_0)^T]\} \tag{10}$$

where Chol{} represents the Cholesky factorization, also known as the square root method. $S = chol\{P\}$ factorizes symmetric positive definite matrix P into an upper triangular R that satisfies $P = S^T S$.

When $k = 1, 2, \cdots$, steps 2 through 8 are used.

Step 2: Calculate sigma points and corresponding weights

$$\begin{cases} \boldsymbol{\chi}_{0,\ k-1} = \hat{X}_{k-1}^-, i = 0 \\ \boldsymbol{\chi}_{i,k-1} = \hat{X}_{k-1}^- + hS_{k-1(i)}, i = 1 \cdots L \\ \boldsymbol{\chi}_{i,k-1} = \hat{X}_{k-1}^- - hS_{k-1(i-L)}, i = L \cdots 2L \end{cases} \tag{11}$$

$$\begin{cases} w_i^{(m)} = (h^2 - L)/h^2 \\ w_i^{(m)} = 1/2h^2, i = 1 \cdots 2L \\ w^{(c1)} = 1/4h^2 \\ w^{(c2)} = (h^2 - 1)/4h^2 \end{cases} \tag{12}$$

where the subscript (i) represents the i-th column of the matrix.

Step 3: Time update

$$\boldsymbol{\chi}_{i,k|k-1} = \boldsymbol{\Phi} \boldsymbol{\chi}_{i,k|k-1}, i = 0, 1 \cdots 2L \tag{13}$$

$$\hat{X}_k^- = \sum_{i=0}^{2L} w_i^{(m)} \boldsymbol{\chi}_{i,k|k-1} \tag{14}$$

where $\boldsymbol{\Phi}$ is shown in (3).

Step 4: QR factorization of covariance matrix

$$S_k^- = \text{QR}\left\{ \begin{bmatrix} \sqrt{w_{c1}}\left(\boldsymbol{\chi}_{1:L,k|k-1} - \boldsymbol{\chi}_{L+1:2L,k|k-1}\right)^T \\ \sqrt{w_{c2}}\left(\boldsymbol{\chi}_{1:L,k|k-1} + \boldsymbol{\chi}_{L+1:2L,k|k-1} - 2\boldsymbol{\chi}_{0,k|k-1}\right)^T \\ \sqrt{Q}^T \end{bmatrix} \right\} \tag{15}$$

where Q is the covariance matrix of process noise and QR{} represents the QR factorization of a matrix. The QR factorization expresses an m-by-n matrix A as $\mathbf{A} = \boldsymbol{QR}$ Here, Q is an m-by-m unitary matrix, and R is an m-by-n upper triangular matrix. In this article, QR{} returns the upper triangular matrix of the QR factorization.

Step 5: Sigma point set update

$$
\begin{cases}
\chi_{0,\ k-1} = \hat{X}_{k-1}^-, i = 0 \\
\chi_{i,k-1} = \hat{X}_{k-1}^- + hS_{k(i)}^-, i = 1 \cdots L \\
\chi_{i,k-1} = \hat{X}_{k-1}^- - hS_{k(i-L)}^-, i = L \cdots 2L
\end{cases}
\tag{16}
$$

Step 6: Measurement and covariance update

$$
\varXi_{i,k|k-1} = f(\chi_{i,k|k-1}), i = 1, 2 \cdots 2L \tag{17}
$$

$$
\hat{Z}_k^- = \sum_{i=0}^{2L} w_i^{(m)} \varXi_{i,k|k-1} \tag{18}
$$

$$
S_k^{(z)} = \mathrm{QR}\left\{ \begin{bmatrix} \sqrt{w_{c1}}\left(\varXi_{1:L,k|k-1} - \varXi_{L+1:2L,k|k-1}\right)^{\mathrm{T}} \\ \sqrt{w_{c2}}\left(\varXi_{1:L,k|k-1} + \varXi_{L+1:2L,k|k-1} - 2\varXi_{0,k|k-1}\right)^{\mathrm{T}} \\ \sqrt{R}^{\mathrm{T}} \end{bmatrix} \right\} \tag{19}
$$

$$
P_k^{(xz)} = \sqrt{w^{(c1)}}S_k^-\left(\varXi_{1:L,k|k-1} - \varXi_{L+1:2L,k|k-1}\right)^{\mathrm{T}} \tag{20}
$$

Step 7: Target state update

$$
\kappa_k = (P_k^{(xz)} / S_k^{(z)\mathrm{T}})S_k^{(z)} \tag{21}
$$

$$
\hat{X}_k = \hat{X}_k^- + \kappa_k(Z_k - \hat{Z}_k^-) \tag{22}
$$

Step 8: Square root of posterior covariance matrix

$$
S_k = \text{cholupdate}\{S_k^-, \kappa_k S_k^{(z)}, -\} \tag{23}
$$

where cholupdate{} represents the rank 1 update to Cholesky factorization. The (23) returns the Cholesky factor of $S_k^{-\mathrm{T}}S_k^- - \kappa_k S_k^{(z)}(\kappa_k S_k^{(z)})^{\mathrm{T}}$, which is an upper triangular matrix.

3   Adaptive Singular Value Decomposition Square Root Central Differential Kalman Filter

SR-CDKF, like SR-UKF, can be used for TMA problems. However, due to the influence of observer mobility, observation error, and system noise, in the case of BOTMA by a single observer, the $S_k^{-\mathrm{T}}S_k^- - \kappa_k S_k^{(z)}(\kappa_k S_k^{(z)})^{\mathrm{T}}$ in (23) is not necessarily positive definite. So Cholesky factor of $S_k^{-\mathrm{T}}S_k^- - \kappa_k S_k^{(z)}(\kappa_k S_k^{(z)})^{\mathrm{T}}$ may not exist, so the filtering may be interrupted. Therefore, an improved SR-CDKF method is presented in this paper. This method provides two improvements based on the SR-CDKF. First, the singular value decomposition is used instead of the Cholesky factorization with the purpose of improving the robustness of the system. Second, the prediction residuals are used to construct error discriminators in order to judge the perturbations to the system. When the perturbations are too large, SVD is used instead of the square root method and the adaptive factor is used for the adjustment. We refer to the proposed method as an adaptive singular value decomposition square root central differential Kalman filter (ASVDSR-CDKF).

3.1 Covariance Square Root Update Based on Singular Value Decomposition

Firstly, the singular value decomposition is introduced. The decomposed matrix in SVD does not need to be positive definite. It is a matrix decomposition method with strong stability and high precision, and it is easy to implement on computer. If $A \in R^{m \times n}(m \geq n)$, then the singular value decomposition of the matrix A is as follows:

$$A = U \Lambda T^{\mathrm{T}} = U \begin{bmatrix} S & 0 \\ 0 & 0 \end{bmatrix} T^{\mathrm{T}} \tag{24}$$

where $U \in R^{m \times m}, T \in R^{n \times n}, \Lambda \in R^{m \times n}, S = \mathrm{diag}(s_1, s_2, \ldots \ldots s_r). s_1 \geq s_2 \geq \ldots \ldots \geq s_r \geq 0$ are the singular values of A. The covariance matrix $P = E[(\hat{X} - X)(\hat{X} - X)^{\mathrm{T}}]$ of the state equation of the single observer bearing-only target motion analysis system is a symmetric matrix. Therefore, after the singular value decomposition of P, $U = V$ is obtained, the $\sqrt{P} = U\sqrt{S}$ can be acquired.

Thus, a new update method of covariance square root matrix is proposed based on the SR-CDKF. For step 8 in the SR-CDKF, we replace it with the following SVD.

The covariance updates of the system are calculated first:

$$P_k = S_k^{-\mathrm{T}} S_k^- - \kappa_k S_k^{(z)} (\kappa_k S_k^{(z)})^{\mathrm{T}} \tag{25}$$

Then, the singular value decomposition of covariance:

$$(U, S^{(d)}, V) = svd(P_k) \tag{26}$$

$U, S^{(d)}, V$ in (26) correspond to U, Λ, T in (24) formula, respectively. Finally, the square root of the covariance matrix is:

$$S_k = U\sqrt{S^{(d)}} \tag{27}$$

SR-CDKF based on SVD can be obtained by replacing step 8 in SR-CDKF with the formulas (25)–(27). Since the singular value is not negative, $\sqrt{S^{(d)}}$ is always solvable. Even if the covariance of formula (25) is not positive definite, the singular value decomposition can be completed and S_k is calculated.

3.2 Adaptive Factor

In the process of BOTMA, the system may be disturbed, and the predicted residual error can reflect the disturbance. The predicted residual error can be expressed as follows:

$$V_k = f(\hat{X}_k^-) - Z_k \tag{28}$$

The influence of disturbance can be reduced to a certain extent by selecting prediction residuals to construct error discriminant statistics and constructing adaptive factors. The error discriminant statistics are constructed by residuals:

$$\Delta V_k = (V_k^T V_k / tr(S_{V_k}))^{1/2} \tag{29}$$

$$S_{V_k} = QR \left\{ \begin{bmatrix} \left[\sqrt{w_{c1}} \left(\boldsymbol{\Xi}_{1:L,k|k-1} - \boldsymbol{\Xi}_{L+1:2L,k|k-1} \right)^T \\ \sqrt{w_{c2}} \left(\boldsymbol{\Xi}_{1:L,k|k-1} + \boldsymbol{\Xi}_{L+1:2L,k|k-1} - 2Z_k \right)^T \\ \sqrt{R}^T \end{bmatrix} \right] \right\} \tag{30}$$

where tr $(\cdot)$ represents the trace of the matrix. The adaptive factors α_k are as follows:

$$\alpha_k = \begin{cases} 1 & |\Delta V_k| \le c_1 \\ \frac{c_1(c_2 - |\Delta V_k|)}{|\Delta V_k|(c_2 - c_1)} & c_1 < |\Delta V_k| < c_2 \\ 10^{-2} & |\Delta V_k| \ge c_2 \end{cases} \tag{31}$$

where c_1 and c_2 are empirical constants, and their values usually satisfy $1 < c_1 < 3$ and $3 < c_2 < 8$. When the values of c_1 and c_2 are too small, the effect of the adaptive factor increases, which may result in over-adjustment of the covariance matrix. When their values are too large, the adaptive factor may lose its effect. In this paper, $c_1 = 2.3$, $c_2 = 7.5$. The partial covariance and covariance decomposition matrix can be modified by using adaptive factors α_k:

$$S_k^{(z)} = S_k^{(z)} / \alpha_k \tag{32}$$

$$P_k^{(xz)} = P_k^{(xz)} / \alpha_k \tag{33}$$

After the update of each measurement, the adaptive factor can be constructed by formula (31). In the following steps, add (32) and (33) to make the adaptive correction for $S_k^{(z)}$ and $P_k^{(xz)}$, and use the modified values to calculate. Through the adaptive factor α_k, the weight ratio of model information and observation information can be adjusted to reduce the influence of abnormal disturbance.

At the same time, the error discriminant statistic constructed by residual V_k can be used to judge whether the prediction residual is too large or not to evaluate whether the system is too disturbed. When the result is $|\Delta V_k| \ge c_2$, the square root of the covariance is being updated by the formulas (25)–(27) instead of (23).

3.3 Process of ASVDSR-CDKF Method

ASVDSR-CDKF is proposed in order to overcome the issue of SR-CDKF having no output due to non-positive definite matrix. This method adds an adaptive factor to make an adaptive adjustment to $S_k^{(z)}$ and $P_k^{(xz)}$. When the system error discriminant statistic is too large, that is, $|\Delta V_k| \geq c_2$, the covariance square root is updated using the singular value decomposition method.

The ASVDSR-CDKF method consists of the following steps:

Step 1: When k = 0, initialize

$$\hat{X}_0 = E(X_0) \tag{34}$$

$$S_0 = \text{Chol}\{ E[(\hat{X}_0 - X_0)(\hat{X}_0 - X_0)^{\mathrm{T}}]\} \tag{35}$$

When $k = 1, 2, \cdots$, proceed to steps 2 through 8.

Step 2: Calculate sigma points and corresponding weights

$$\begin{cases} \chi_{0,\ k-1} = \hat{X}_{k-1}^-, i = 0 \\ \chi_{i,k-1} = \hat{X}_{k-1}^- + hS_{k-1(i)}, i = 1 \cdots L \\ \chi_{i,k-1} = \hat{X}_{k-1}^- - hS_{k-1(i-L)}, i = L \cdots 2L \end{cases} \tag{36}$$

$$\begin{cases} w_i^{(m)} = (h^2 - L)/h^2 \\ w_i^{(m)} = 1/2h^2, i = 1 \cdots 2L \\ w^{(c1)} = 1/4h^2 \\ w^{(c2)} = (h^2 - 1)/4h^2 \end{cases} \tag{37}$$

Step 3: Time update

$$\chi_{i,k|k-1} = \Phi \chi_{i,k|k-1}, i = 0, 1 \cdots 2L \tag{38}$$

$$\hat{X}_k^- = \sum_{i=0}^{2L} w_i^{(m)} \chi_{i,k|k-1} \tag{39}$$

Step 4: QR factorization of covariance matrix

$$S_k^- = \text{QR}\left\{ \begin{bmatrix} \sqrt{w_{c1}}(\chi_{1:L,k|k-1} - \chi_{L+1:2L,k|k-1})^{\mathrm{T}} \\ \sqrt{w_{c2}}(\chi_{1:L,k|k-1} + \chi_{L+1:2L,k|k-1} - 2\chi_{0,k|k-1})^{\mathrm{T}} \\ \sqrt{Q}^{\mathrm{T}} \end{bmatrix} \right\} \tag{40}$$

Step 5: Sigma point set update

$$\begin{cases} \chi_{0,\ k-1} = \hat{X}_{k-1}^-, i = 0 \\ \chi_{i,k-1} = \hat{X}_{k-1}^- + hS_{k(i)}^-, i = 1 \cdots L \\ \chi_{i,k-1} = \hat{X}_{k-1}^- - hS_{k(i-L)}^-, i = L \cdots 2L \end{cases} \tag{41}$$

Step 6: Measurement prediction

$$\Xi_{i,k|k-1} = f(\chi_{i,k|k-1}), i = 1, 2 \cdots 2L \tag{42}$$

$$\hat{Z}_k^- = \sum_{i=0}^{2L} w_i^{(m)} \Xi_{i,k|k-1} \tag{43}$$

Step 7: Residual discriminant statistics

$$V_k = f(\hat{X}_k^-) - Z_k \tag{44}$$

$$\Delta V_k = (V_k^{\mathrm{T}} V_k / \mathrm{tr}(S_{V_k}))^{1/2} \tag{45}$$

$$S_{V_k} = QR \left\{ \begin{bmatrix} \sqrt{w_{c1}} \left(\Xi_{1:L,k|k-1} - \Xi_{L+1:2L,k|k-1} \right)^{\mathrm{T}} \\ \sqrt{w_{c2}} \left(\Xi_{1:L,k|k-1} + \Xi_{L+1:2L,k|k-1} - 2Z_k \right)^{\mathrm{T}} \\ \sqrt{R}^{\mathrm{T}} \end{bmatrix} \right\} \tag{46}$$

Step 8: Adaptive factor

$$\alpha_k = \begin{cases} 1 & |\Delta V_k| \leq c_1 \\ \frac{c_1(c_2 - |\Delta V_k|)}{|\Delta V_k|(c_2 - c_1)} & c_1 < |\Delta V_k| < c_2 \\ 10^{-2} & |\Delta V_k| \geq c_2 \end{cases} \tag{47}$$

Step 9: Adaptive factor modified covariance

$$S_k^{(z)} = \frac{1}{\alpha_k} QR \left\{ \begin{bmatrix} \sqrt{w_{c1}} \left(\Xi_{1:L,k|k-1} - \Xi_{L+1:2L,k|k-1} \right)^{\mathrm{T}} \\ \sqrt{w_{c2}} \left(\Xi_{1:L,k|k-1} + \Xi_{L+1:2L,k|k-1} - 2\Xi_{0,k|k-1} \right)^{\mathrm{T}} \\ \sqrt{R}^{\mathrm{T}} \end{bmatrix} \right\} \tag{48}$$

$$P_k^{(xz)} = \frac{\sqrt{w^{(c1)}}}{\alpha_k} S_k^- (\Xi_{1:L,k|k-1} - \Xi_{L+1:2L,k|k-1})^{\mathrm{T}} \tag{49}$$

Step 10: Kalman gain and state update:

$$\kappa_k = (P_k^{(xz)} / S_k^{(z)\mathrm{T}}) S_k^{(z)} \tag{50}$$

$$\hat{X}_k = \hat{X}_k^- + \kappa_k (Z_k - \hat{Z}_k^-) \tag{51}$$

$$P_k = S_k^{-\mathrm{T}} S_k^- - \kappa_k S_k^{(z)} (\kappa_k S_k^{(z)})^{\mathrm{T}} \tag{52}$$

Step 11: If $|\Delta V_k| \geq c_2$ or P_k is not positive definite, then:

$$(U, S^{(d)}, V) = svd(P_k) \tag{53}$$

$$S_k = U\sqrt{S^{(d)}} \tag{54}$$

Step 12: If $|\Delta V_k| < c_2$ or P_k is positive definite, then:

$$S_k = \text{cholupdate}\{S_k^-, \kappa_k S_k^{(z)}, -\} \tag{55}$$

These are all the steps of ASVDSR-CDKF. The update of the covariance square root of this method is adaptive and is determined by the error identification statistics. When c_2 approaches zero, the square root update method is based on the singular value decomposition, then the ASVDSR-CDKF becomes a SR-CDKF based on a singular value decomposition proposed in Sect. 3-A of this paper. When c_2 approaches ∞, the covariance square root is always updated by (55), so the ASVDSR-CDKF actually becomes an adaptive SR-CDKF.

The flow of the ASVDSR-CDKF method is demonstrated in Fig. 2:

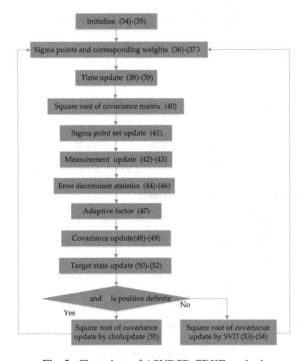

Fig. 2. Flow chart of ASVDSR-CDKF method.

4 Simulation

In order to verify the effectiveness and advantages of the proposed method in BOTMA by single observer, three close-range tracking trajectories are simulated in this section. In simulations, the measurement noise covariance is $R = 5$, the process noise intensity is $\delta_w^2 = 0.01$. We suppose that the variance of the process noise and the measurement noise are unavailable. The estimation of measurement noise covariance is $\hat{R} = 3$, and the estimation of process noise intensity is $\hat{\delta}_w^2 = 0.1$. The sampling interval is $\Delta t = 0.1$ s. There are some errors in the initial state estimation $\hat{X}_0$. The initial distance error is 5%, and the initial velocity error is 5%. The mean square error of the initial heading is $3°$. One hundred Monte Carlo experiments are performed by four methods: CDKF, SR-CDKF, SR-UKF (another common square root class method), and the proposed ASVDSR-CDKF.

The root mean square error (RMSE) is used to measure the estimation error, which is defined as follows:

$$\text{RMSE} = \sqrt{\frac{1}{n}\sum_{i=1}^{n}\left[\left(x_{true,k}(i) - \hat{x}_k(i)\right)^2 + \left(y_{true,k}(i) - \hat{y}_k(i)\right)^2\right]} \qquad (56)$$

When the filter is interrupted due to the non-positive definite matrix, the estimation error cannot be calculated. In order to ensure the fairness of error comparison, the simulation error only counts the case that all methods have solutions, that is, the filtering is not interrupted. In this way, the simulation conditions of all filters are completely consistent in error statistics.

The relative distance error can be used to judge whether the filter is divergent, which is defined as the ratio of the error distance to the real distance. The expression is as follows:

$$\text{RRE} = \sqrt{\frac{(\hat{x}_k - x_{true,k})^2 + (\hat{y}_k - y_{true,k})^2}{(x_{o,k} - x_{true,k})^2 + (y_{o,k} - y_{true,k})^2}} \times 100\% \qquad (57)$$

In this research, when the relative distance error is more than 100%, it is determined as filtering divergence. When the decomposed matrix is not positive definite and the decomposition cannot be carried out so that the filtering cannot be continue, we determine that the filtering is interrupted. When the filter is not interrupted and not judged to diverge, it is determined that the filtering is completed.

4.1 Case 1

The observer tracks the target with a polyline maneuver. The target moves east with a velocity of 50 knots in a uniform straight line. The initial position of the observer is 1000 m $60°$ north by east of the target. For the first ten seconds, the observer moves in a uniform straight line at a velocity of 50 knots, with a heading of $50°$ south by west. After 10 s, it turns to $20°$ south by west and continues to move in a straight line at a constant velocity of 50 knots. The total observation time is 20 s. The movement of both the target and the observer in case 1 is demonstrated in Fig. 3. The RMSE of the four methods is indicated in Fig. 4 and the comparison of measurement, true azimuth and estimated azimuth, is shown in Fig. 5. The comparison of distance error, velocity error, and heading error is established in Table 1.

Where ΔD is the average distance error; ΔV is the average velocity error; ΔK is the average heading error.

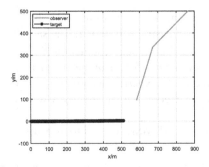

Fig. 3. Movement of target and observer in case 1: target's true trajectory (black) and observer's trajectory (green) (Color figure online)

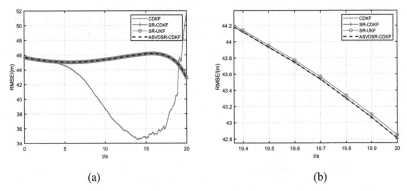

(a)   (b)

Fig. 4. RMSE comparison of CDKF (red), SR-CDKF (blue), SR-UKF (mauve) and ASVDSR-CDKF (black) in case 1: (a) The total scan; (b) the local enlarged figure of (a). (Color figure online)

Table 1. Performance comparison of four filtering methods in case 1

Filtering method	ΔD(m)	ΔV(m/s)	ΔK(°)	Completion times
CDKF	31.2957	1.7265	2.6663	100
SR-CDKF	30.2131	1.1736	1.4493	100
SR-UKF	30.2134	1.1736	1.4493	100
ASVDSR-CDKF	**30.2131**	**1.1735**	**1.4493**	**100**

It can be concluded from the completion times of filtering in Table 1 that, in this case, all four filtering methods achieve 100% complete filtering. This is because, at the end of the observation, the observer is not very close to the target (about 100 m). This means the filter disturbance is not severe and each filter can complete the filtering.

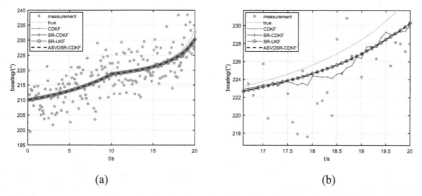

Fig. 5. The measurement (green), the true bearings (cyan) and the estimated bearings by CDKF (red), SR-CDKF (blue), SR-UKF (mauve) and ASVDSR-CDKF (black) in case 1: (a) The total scan; (b) The local enlarged figure of (a). (Color figure online)

By comparing the bearing estimation of each method with the real bearing of the target in Fig. 5, the bearing estimation results of SR-UKF, SR-CDKF and ASVDSR-CDKF methods are similar and smooth. However, the direction estimation results of the CDKF method fluctuate greatly. This is because the covariance in the form of square root reduces the influence of measurement jitter on the filtering result and makes the estimation result smoother.

From the RMSE in Fig. 4 and the average errors in Table 1, the following conclusions can be drawn. Compared with CDKF and SR-UKF method, the RMSE of SR-CDKF and ASVDSR-CDKF is smaller. The estimation errors of the two methods are almost the same, and the velocity estimation error of ASVDSR-CDKF is slightly lower than that of SR-CDKF. This is because, in case 1, the interference is small, the error discrimination statistics are small, and the adjustment effect of the adaptive factor is small. According to the simulation results, in case 1, the proposed ASVDSR-CDKF method has the lowest estimation error.

4.2 Case 2

At close range, the observer tracks the target in the angle of advance. The target moves eastward in a straight line at a speed of 50 knots. The observer is initially 250 m away from the target in the direction of 60° north by east of the target. It proceeds to track the target with a velocity of 50 knots and an advance angle of 7°. The observation time is 7 s. The movement of target and observer in case 2 is demonstrated in Fig. 6. The RMSE comparison of the four methods is shown in Fig. 7. Figure 8 indicates the comparison of measurements, true azimuth, and various methods of calculating azimuth. The number of times all four methods complete the filter (uninterrupted and non-divergent), as well as the final distance error, speed error, and heading error of the four methods are counted, is demonstrated in Table 2.

Using Fig. 7, 8 and Table 2, the errors and completion times of each method can be compared. The errors of CDKF are the largest, followed by SR-UKF. The errors of SR-CDKF and ASVDSR-CDKF are very close, and the errors of SR-CDKF are lower.

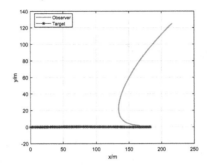

Fig. 6. Movement of target and observer in case 2: target's true trajectory (black) and observer's trajectory (green) (Color figure online)

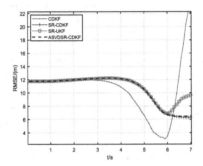

Fig. 7. RMSE comparison of CDKF (red), SR-CDKF (blue), SR-UKF (mauve), and ASVDSR-CDKF (black) in case 2 (Color figure online)

But only SR-UKF and ASVDSR-CDKF can complete the filtering 100 times, and the error of ASVDSR-CDKF is smaller.

Both SR-UKF and SR-CDKF belong to the square root form of sigma point Kalman filter. In case 1, the estimation results of the two methods are very similar. But in the last second of case 2, the RMSE of SR-UKF increases, which is different from that of SR-CDKF. This is because the covariance matrices of the two methods are different. As the distance between target and observer decreases, the influence of target position estimation error on measurement estimation increases. Because two subtraction methods are used in the calculation of the covariance matrix of SR-CDKF, it is more sensitive to the change, and its estimated value easily approaches the real value of the movement of the observer. At the same time, the value can also easily appear non-positive definite in the square root update of covariance matrix.

Where ΔD is the average distance error; ΔV is the average velocity error; ΔK is the average heading error.

In the above 100 simulations, only SR-UKF and ASVDSR-CDKF achieve a 100% completion rate. In order to count the times of divergence and interruption of all four methods and understand the influence of measurement noise covariance on filtering results, the measurement noise covariance R is adjusted to 1–5, and 1000 Monte Carlo experiments are carried out in case 2. The percentages of divergence and interruption of

Table 2. Performance comparison of four filtering methods in case 2

Filtering method	ΔD(m)	ΔV(m/s)	ΔK(°)	Completion times
CDKF	8.2428	2.3858	3.0980	89
SR-CDKF	7.2113	1.2413	1.5496	98
SR-UKF	7.3037	1.2472	1.5691	100
ASVDSR-CDKF	**7.2122**	**1.2673**	**1.5701**	**100**

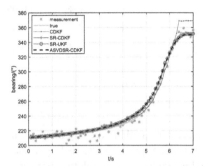

Fig. 8. The measurement (green), the true bearings (cyan) and the estimated bearings by CDKF (red), SR-CDKF (blue), SR-UKF (mauve), and ASVDSR-CDKF (black) in case 2 (Color figure online)

each method under different noises are calculated respectively, and the results are shown in Table 3. Figure 9 shows the total completion percentage of the four methods under different measurement noises.

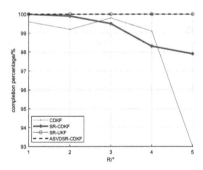

Fig. 9. Completion percentage of CDKF (red), SR-CDKF (blue), SR-UKF (mauve) and ASVDSR-CDKF (black) in case 2. (Filtering is considered complete without interruption and divergence). (Color figure online)

Combining the simulation results of Table 3 and Fig. 9, it is clear that CDKF diverges easily and has poor stability. SR-CDKF has a fatal weakness that the matrix is not positive definite in updating the square root of covariance, which leads to filter interruption. This

Table 3. Interruption or divergence percentage statistics for each method in case 2.

Variance of noise	Filtering method	CDKF	SR-CDKF	SR-UKF	ASVDSR-CDKF
1	Interruption	0.0%	0.0%	0.0%	**0.0%**
	Divergence	0.4%	0.0%	0.0%	**0.0%**
2	Interruption	0.0%	0.1%	0.0%	**0.0%**
	Divergence	0.8%	0.0%	0.0%	**0.0%**
3	Interruption	0.0%	0.5%	0.0%	**0.0%**
	Divergence	0.2%	0.0%	0.0%	**0.0%**
4	Interruption	0.0%	1.7%	0.0%	**0.0%**
	Divergence	0.9%	0.0%	0.0%	**0.0%**
5	Interruption	0.0%	2.1%	0.0%	**0.0%**
	Divergence	7.0%	0.0%	0.0%	**0.0%**

phenomenon is especially common with increasing measurement errors. Although there is no interruption in SR-UKF, the estimation error is large. The ASVDSR-CDKF method proposed in this research, has the highest completion percentage of filtering completion and lower estimation error Thus, of the four methods, ASVDSR-CDKF achieves the best performance.

4.3 Case 3

The observer meets the target in a straight line. The target moves east at velocity of 50 knots in a uniform straight line. The observer is initially 500 m away from the target in the direction of 85 ° north by east of the target. It tracks the target with a velocity of 50 knots, in a uniform straight line 82° south by west. The observation time is 10 s. The movement of the target and the observer in case 3 is demonstrated in Fig. 10. The simulation results for case 3 are established in Fig. 11, Fig. 12 and Table 4.

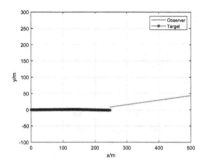

Fig. 10. Movement of target and observer in case 3: target's true trajectory (black) and observer's trajectory (green) (Color figure online)

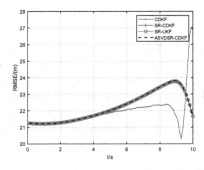

Fig. 11. RMSE comparison of CDKF (red), SR-CDKF (blue), SR-UKF (mauve), and ASVDSR-CDKF (black) in case 3 (Color figure online)

Table 4. Performance comparison of four filtering methods in case 3

Filtering method	ΔD(m)	ΔV(m/s)	ΔK(°)	Completion times
CDKF	21.5951	1.2941	1.4297	98
SR-CDKF	21.4012	1.3573	1.5794	74
SR-UKF	21.4086	1.3573	1.5710	97
ASVDSR-CDKF	**21.4003**	**1.3550**	**1.5821**	**99**

Where ΔD is the average distance error; ΔV is the average velocity error; ΔK is the average heading error.

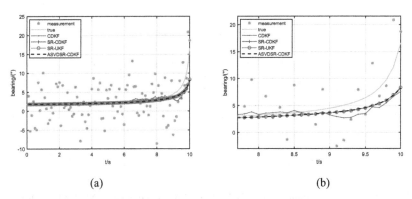

(a)   (b)

Fig. 12. The measurement (green), the true bearings (cyan) and the estimated bearings by CDKF (red), SR-CDKF (blue), SR-UKF (mauve) and ASVDSR-CDKF (black) in case 3: (a) The total scan; (b) The local enlarged figure of (a). (Color figure online)

Using the simulation results in Fig. 11, Fig. 12, and Table 4, the following conclusions can be made about case 3. SR-CDKF has the least number of completions because its covariance matrix in case 3 tends to make the rank 1 update to Cholesky factorization non-positive definite. The error of ASVDSR-CDKF is the smallest of the other three methods.

In the BOTMA by single observer, due to the ambiguity of distance, it is more difficult to estimate the target state in the face-to-face case of case 3. Among the 100 simulations, none of the methods completed filtering 100% due to the influence of movement. On the basis of case 3, the observed noise covariance R is adjusted to 1–5, and 1000 Monte Carlo tests are performed. Figure 13 shows the completion percentage of four filters. The percentage of divergence and interruption of each filter under different noises is in Table 5.

Table 5. Interruption or divergence percentage statistics for each method in case 3.

Variance of noise	Filtering method	CDKF	SR-CDKF	SR-UKF	ASVDSR-CDKF
1	Interruption	0.0%	26.8%	4.0%	**0.0%**
	Divergence	0.0%	0.0%	0.0%	**0.0%**
2	Interruption	0.0%	24.3%	3.2%	**0.0%**
	Divergence	0.0%	0.0%	0.0%	**0.0%**
3	Interruption	0.0%	26.4%	3.8%	**0.0%**
	Divergence	1.1%	1.0%	1.0%	**1.0%**
4	Interruption	0.0%	24.2%	2.8%	**0.0%**
	Divergence	0.4%	0.0%	0.0%	**0.1%**
5	Interruption	0.0%	27.4%	3.7%	**0.0%**
	Divergence	0.4%	0.4%	0.4%	**0.4%**

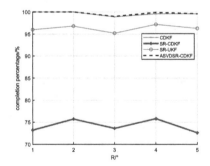

Fig. 13. Completion percentage of CDKF (red), SR-CDKF (blue), SR-UKF (mauve), and ASVDSR-CDKF (black) in case 3. (Filtering is considered complete without interruption and divergence). (Color figure online)

Combining the simulation results in Fig. 13 and Table 5, a conclusion can be drawn. In case 3, SR-CDKF has the most interruptions and is therefore the most unstable method. SR-UKF also has several filter interruptions due to a non-positive definite matrix in the rank 1 update to Cholesky factorization. Among the four methods, the proposed ASVDSR-CDKF has no filter interruption, the least divergence and the highest filter completion percentage.

Combining the simulation results of each filter in the three cases, it can be concluded that:

CDKF: The conventional CDKF method has no interruption, but is prone to divergence and has a large error. This is because its covariance does not take the form of square root and is more sensitive to disturbances to the system.

SR-UKF: The SR-UKF method has fewer filtering interruptions than SR-CDKF, but the estimation error is greater than SR-CDKF when the observer maneuvers to approach the target. This is because, compared with SR-CDKF, its covariance matrix is less sensitive to disturbances and is therefore less sensitive to changes in information caused by observer's maneuvering.

SR-CDKF: SR-CDKF has a low estimation error when the filter interrupts are excluded. But it has the most filter interruptions. For the bearings-only target motion analysis by single observer, the observation and tracking of the target cannot be redone, and the filter interruption is fatal. Therefore, SR-CDKF is not suitable for short-range BOTMA by single observer due to poor reliability.

ASVDSR-CDKF: The ASVDSR-CDKF method proposed in this paper is based on the SR-CDKF framework. As it can automatically adjust the covariance square root update based on disturbance, it inherits the advantages of low error of SR-CDKF and solves the problem of the SR-CDKF being prone to interruption. The ASVDSR-CDKF method has the highest percentage of filter completion under the three simulation conditions. Additionally, among the methods without filter interruption, the method has the lowest error. Therefore, ASVDSR-CDKF has the best stability and comprehensive performance.

5   Conclusion

In this paper, we proposed an ASVDSR-CDKF method in order to deal with the filter instability that occasionally occurs in BOTMA. Based on the SR-CDKF method, ASVDSR-CDKF adds an adaptive factor to adjust the covariance adaptively. This allows it to update the square root of covariance with the SVD instead of the Cholesky factorization when there is a large disturbance in the filtering process of SR-CDKF.

The main contributions of this paper are as follows: first, a new method for updating the square root of covariance based on the singular value decomposition for SR-CDKF is introduced. Then, by adding an adaptive factor, the new method to make the system adjust adaptively and select the square root of covariance automatically is presented, and the flow of ASVDSR-CDKF method is given. Finally, three groups of simulation experiments are carried out in order to verify the validity of the ASVDSR-CDKF method. The results show that the proposed ASVDSR-CDKF method inherits the advantages of SR-CDKF and has lower filtering error than conventional CDKF and SR-UKF. At the same time, ASVDSR-CDKF avoids the problem of filter interruptions that SR-CDKF and SR-UKF are prone to. Thus ASVDSR-CDKF is the method with the highest percentage of filter completion among the four methods.

Our future research directions are as follows. Firstly, explore the effectiveness of this method for tracking a maneuvering target. Secondly, consider optimizing this method in order to further reduce the probability of divergence in the case of close-range target tracking.

References

1. Nardone, S., Lindgren, A., Gong, K.: Fundamental properties and performance of conventional bearings-only target motion analysis. IEEE Trans. Autom. Control **29**(9), 775–787 (1984)
2. Isbitiren, G., Akan, O.B.: Three-dimensional underwater target tracking with acoustic sensor networks. IEEE Trans. Veh. Technol. **60**(8), 3897–3906 (2011)
3. Fang, X., Jiang, Z.H., Nan, L., Chen, L.J.: Noise-aware localization algorithms for wireless sensor networks based on multidimensional scaling and adaptive Kalman filtering. Comput. Commun. **101**, 57–68 (2017)
4. Dogancay, K.: 3D Pseudolinear target motion analysis from angle measurements. IEEE Trans. Signal Process. **63**(6), 1570–1580 (2015)
5. Alexandri, T., Diamant, R.: A reverse bearings only target motion analysis for autonomous underwater vehicle navigation. IEEE Trans. Mob. Comput. **18**(3), 494–506 (2019)
6. Mao, D., Fang, Y., Gao, X.: Target tracking method with bearings-only measurements based on reinforcement learning. IEICE Commun. Express **5**(1), 19–26 (2016)
7. Kim, J., Suh, T., Ryu, J.: Bearings-only target motion analysis of a highly manoeuvring target. IET Radar Sonar Navig. **11**(6), 1011–1019 (2017)
8. Jauffret, C., Perez, A., Pillon, D.: Observability: range-only versus bearings-only target motion analysis when the observer maneuvers smoothly. IEEE Trans. Aerosp. Electron. Syst. **53**(6), 2814–2832 (2017)
9. Zheng, Y., Wang, M.Z.: A sliding backward recursive EKF bearings-only target tracking method. J. Unmanned Undersea Syst. **6**(28), 663–669 (2020)
10. Badriasl, L., Arulampalam, S., Nguyen, N.H., Finn, A.: An algebraic closed-form solution for bearings-only maneuvering target motion analysis from a nonmaneuvering platform. IEEE Trans. Signal Process. **68**, 4672–4687 (2020)
11. Shalom, Y., Li, X., Thiagalingam, R.K.: Estimation with Applications to Tracking and Navigation. Wiley, New York (2001)
12. Lin, X., Kirubarajan, T., Bar-Shalom, Y., Maskell, S.: Comparison of EKF, pseudomeasurement, and particle filters for a bearing-only target tracking problem. In: Signal and Data Processing of Small Targets 2002, Orlando, FL, USA, pp. 240–250 (2002)
13. Konatowski, S., Kaniewski, P., Matuszewski, J.: Comparison of estimation accuracy of EKF, UKF and PF filters. Ann. Navig. **23**(1), 69–87 (2016)
14. Gordon, N.J., Salmond, D.J., Smith, A.F.M.: Novel approach to nonlinear/non-Gaussian Bayesian state estimate. IEEE Proc. Radar Sonar Navig. **140**(2), 107–113 (1993)
15. Zhang, H.W., Xie, W.X.: Constrained auxiliary particle filtering for bearings-only maneuvering target tracking. J. Syst. Eng. Electron. **13**(4), 684–695 (2019)
16. Tiwari, R.K., Radhakrishnan, R., Bhaumik, S.: Particle filter for underwater passive bearings-only target tracking with random missing measurements. In: European Control Association (EUCA), Limassol, Cyprus, pp. 2732–2737 (2018)

17. Julier, S.J., Uhlmann, J.K.: Unscented filtering and nonlinear estimation. Proc. IEEE **92**(3), 401–422 (2001)
18. Arasaratnam, I., Haykin, S.: Cubature Kalman Filters. IEEE Trans. Autom. Control **54**(6), 1254–1269 (2009)
19. Ito, K., Xiong, K.: Gaussian filter for nonlinear filtering problems. IEEE Trans. Autom. Control **5**, 910–927 (2000)
20. Nørgaard, M., Poulsen, N.K., Ravn, O.: New developments in state estimation for nonlinear systems. Automatica **36**(11), 1627–1638 (2000)
21. Li, L., Qin, H.: An UKF-based nonlinear system identification method using interpolation models and backward integration. Struct. Control Health Monit. **25**(4), 2129 (2018)
22. Zhao, J., Mili, L.: Robust unscented Kalman Filter for power system dynamic state estimation with unknown noise statistics. IEEE Trans. Smart Grid **10**(2), 1215–1224 (2019)
23. Costanzi, R., Fanelli, F., Meli, E., Ridolfi, A., Caiti, A., Allotta, B.: UKF-based navigation system for AUVs: online experimental validation. IEEE J. Ocean. Eng. **44**(3), 633–641 (2019)
24. Yao, Q., Su, Y., Li, L.: Application of square-root unscented Kalman filter smoothing algorithm in tracking underwater target. Adv. Eng. Res. **150**, 526–531 (2017)
25. Menegaz, H.M.T., Ishihara, J.Y.: Unscented and square-root unscented Kalman filters for quaternionic systems. Int. J. Robust Nonlinear Control **28**, 4500–4527 (2018)
26. Li, X., Zhao, C., Yu, J., Wei, W.: Underwater bearing-only and bearing-doppler target tracking based on square root unscented Kalman filter. Entropy **21**(8), 740 (2019)
27. Lou, T., Yang, N., Wang, Y., Chen, N.H.: Target tracking based on incremental center differential Kalman filter with uncompensated biases. IEEE Access **6**, 66285–66292 (2018)
28. Dai, J., Li, X., Wang, K., Liang, Y.: A novel STSOSLAM algorithm based on strong tracking second order central difference Kalman filter. Robot. Auton. Syst. **116**, 114–125 (2019)
29. Ye, W., Li, J., Fang, J., Yuan, X.: EGP-CDKF for performance improvement of the SINS/GNSS integrated system. IEEE Trans. Industr. Electron. **65**(4), 3601–3609 (2018)
30. Merwe, R.V.D.: Sigma-point Kalman filters for probabilistic inference in dynamic state-space models. Ph.D. dissertation, Dept. Elec. Comp., Oregon Health & Science Univ., Oregon, Portland (2004)
31. Liu, D., Duan, J., Shi, H.: A strong tracking square root central difference fast SLAM for unmanned intelligent vehicle with adaptive partial systematic resampling. IEEE Trans. Intell. Transp. Syst. **17**(11), 3110–3120 (2016)
32. Xie, J., Ma, J., Chen, J.: Available power prediction limited by multiple constraints for LiFePO 4 batteries based on central difference Kalman filter. Int. J. Energy Res. **42**(15), 4730–4745 (2018)
33. Li, X.R., Jilkov, V.P.: Survey of maneuvering target tracking-Part I. dynamic models. IEEE Trans. Aerosp. Electron. Syst. **39**(4), 1333–1364 (2004)

A Novel Electromagnetic Radiation Source Localization Method Based on Dynamic Data Driven Simulations

Xu Xie[✉] and Yuqing Ma

College of Systems Engineering, National University of Defense Technology,
Changsha, China
x.xie@hotmail.com

Abstract. Locating enemy targets via their electromagnetic radiation signal is vital to block and attack the enemy targets at an earlier stage. Traditional electromagnetic radiation source localization methods in literature are essentially geometric methods. Although they are simple and intuitive, they might fail to locate the source due to measurement noise. This paper proposes a novel electromagnetic radiation source localization method based on dynamic data driven simulations. In the proposed approach, we first model the spatial propagation process of the electromagnetic radiation signal emitted from the target, and then we assume a proper model for the noisy measurements. Based on the signal propagation model and the measurement model, the particle filter is employed to estimate the target position, and in the process addresses measurement and modeling errors. Identical-twin experiment is conducted to test and validate the proposed approach. The simulation results show that the proposed method can accurately locate the electromagnetic radiation source, and is robust to errors both in the model and in the data.

Keywords: Electromagnetic radiation source localization · Signal propagation modeling · Dynamic data driven simulation · Particle filters

1 Introduction

At the end of the 19th century, the discovery of wireless electromagnetic waves not only provided a new information carrier for communication, but also could be used to locate and trace the source of electromagnetic radiation. In the military field, especially the ocean security, the main threats are to prevent the invasion of enemy fleets, prevent drone reconnaissance, and prevent submarines from sneaking in, etc. If we could locate the targets by detecting their electromagnetic radiation signals, we would be able to block and attack the enemy targets at an earlier stage.

In literature, the commonly used electromagnetic radiation source localization methods can be classified into four categories, namely the Received Signal

© The Author(s), under exclusive license to Springer Nature Singapore Pte Ltd. 2022
W. Fan et al. (Eds.): AsiaSim 2022, CCIS 1713, pp. 348–359, 2022.
https://doi.org/10.1007/978-981-19-9195-0_28

Strength (RSS) method, the Time of Arrival (TOA) method, the Time Difference of Arrival (TDOA) method, and the Angle of Arrival (AOA) method [1]. The RSS method is designed based on the fact that the power of the wireless signal will decay according to a certain law as the signal propagates spatially [2]. The RSS method first records the signal power received at each measurement point, and then calculates the distances between the measurement points to the target based on a semi-empirical model, and finally exploits the triangulation theory to estimate the location of the target. The TOA method is proposed based on the principle that the product of the speed and time is equal to the travel distance [3]. The target sends electromagnetic wave signal, and each measurement point receives and records the arrival time of the signal; based on the transmission media, the travel speed can be determined, and then the travel distance can be calculated accordingly. Finally, the triangulation theory is again used to locate the target. When using the TOA method, it is necessary to ensure that the clocks among all measurement points are strictly synchronized, otherwise the localization error would be large. Due to this limitation, [4] proposes the TDOA method, which locates the target according to the difference among time instants when signal arrives at each measurement points, thereby reducing the time synchronization requirements. The AOA method uses a receiver to measure the angle of the signal emitted from the target to each measurement point, and uses geometric relations to estimate the coordinates of the target. The advantage of the AOA method is that it does not require time synchronization, but it is susceptible to be affected by wave reflection.

The electromagnetic radiation source localization methods in literature are essentially geometric methods, and their advantages are that they are simple and intuitive, but in practical applications, measurement noise inevitably exist, which will result in failures when using these methods, since the intersection point (i.e. the estimated target position) may not exist or multiple intersection points exist. Therefore, this paper proposes a novel electromagnetic radiation source localization method based on dynamic data driven simulations (DDDS). Dynamic data driven simulations are a new simulation paradigm, where the simulation is continually influenced by the real time data for better analysis and prediction of a system under study [5,6]. In a dynamic data driven simulation, the noisy observations from the system under study are continually injected into the simulation which mimics the dynamic state evolution of the system. At the meantime, the data assimilation technique [7] is exploited to combine noisy observations and (simulation) model predictions to estimate the system state. In the proposed approach, we first model the spatial propagation process of the electromagnetic radiation signal emitted from the target, and then we assume a proper model for the noisy measurements. Based on the signal propagation model and the measurement model, the particle filter [8,9] is employed to estimate the target position. Identical-twin experiment is conducted to test and validate the proposed approach. The simulation results show that the proposed method can accurately locate the electromagnetic radiation source.

The rest of the paper is organized as follows. We first overview the dynamic data driven simulations in Sect. 2. Section 3 then models the spatial propagation process of the electromagnetic radiation signal emitted from the target, after which Sect. 4 presents the electromagnetic radiation source localization method based on dynamic data driven simulations in detail. The simulation study to test the proposed approach is presented in Sect. 5, and finally, the paper is concluded in Sect. 6.

2   Dynamic Data Driven Simulation

Modeling & Simulation are a method of choice for studying and predicting dynamic behavior of complex systems. However, models inevitably contain errors, which arise from many sources in the modeling process, such as inadequate sampling of the real system when constructing the behavior database for the source system [10], or conceptual abstraction in the modeling process [11]. Due to these inevitable errors, even elaborate complex models of systems cannot model the reality perfectly, and consequently, results produced by these imperfect simulation models will diverge from or fail to predict the real behavior of those systems [12,13]. With the advancement of measurement infrastructures, such as sensors, data storage technologies, and remote data access, the availability of data, whether real-time on-line or archival, has greatly increased [12,13]. This allows for a new paradigm – *dynamic data driven simulations*, in which the simulation is continuously influenced by fresh data sampled from the real system [5].

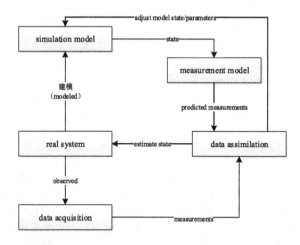

Fig. 1. A general dynamic data driven simulation [6]

Figure 1 shows a general dynamic data driven simulation, which consists of 1) a simulation model, describing the dynamic behavior of the real system; 2) a data acquisition component, which essentially consists of sensors that collect data from the real system; and 3) a data assimilation component, which carries out state estimations based on information from both measurements and the simulation [7,14]. Since the state evolution of a simulation system usually contains nonlinear and/or non-Gaussian behavior, the particle filter is always adopted to conduct data assimilation in dynamic data driven simulations. The particle filters approximate a probability density function by a set of particles and their associated importance weights, and therefore they put no assumption on the properties of the system model. As a result, they can effectively deal with nonlinear and/or non-Gaussian applications [15].

By assimilating actual data, the simulation can dynamically update its current state to be closer to the real system state, which facilitates real-time applications of simulation models, such as real-time control and analysis, real-time decision making, and understanding the current state of the real system. Besides, if the model state is extended to include model parameters, on-line model parameter calibration can be achieved together with the state estimation [16]. With more accurate model state and model parameters adjusted by assimilating real-time data, we can experiment (off-line) on the simulation model with the adjusted state and parameters, which will lead to more accurate results for follow-on simulations [6].

3   Spatial Propagation Model of Electromagnetic Radiation Signal

After the electromagnetic radiation signal is emitted from the target, it will be received by receivers after attenuating, delay and adding noise. The received signal can be described as:

$$r_i(t) = \mu_i s(t - \tau_i)e^{j2\pi f_i t} + n_i(t) \tag{1}$$

where $s(t)$ represents the signal emitted by the target at time t, and τ_i is the path propagation delay from the source to the i-th receiver; μ_i is the propagation path gain coefficient from the target to the i-th receiver; f_i is the drift value of the frequency caused by the Doppler effect; n_i is a complex noise; $r_i(t)$ is a complex number that represents the signal received by the receiver at time t.

3.1   Electromagnetic Radiation Signal $s(t)$

The electromagnetic radiation signal emitted by the target can be expressed as:

$$s(t) = |s|e^{j2\pi ft} \tag{2}$$

where f is the frequency of the signal in hertz (Hz), $|s|$ is the amplitude of the signal, and the unit is watts (W).

3.2   Propagation Path Gain Coefficient μ_i

The propagation path gain coefficient $|\mu_i|$ represents the power gain rate of the signal from the sender (i.e. the target) to the receiver. After the signal arrives at the receiver, it is filtered, and amplified by a low-noise amplifier (LNA), and finally sampled. So $|\mu_i|$ is related to the length of the propagation path and the magnification of the LNA. Due to the phase inconsistency between the sender and the receiver, there will exist an amplitude angle for μ_i. Therefore, we model the propagation path gain coefficient $|\mu_i|$ as:

$$\mu_i = |\mu_i|e^{j2\pi(f-f_r^i)} = \frac{k_i e^{j2\pi(f-f_r^i)}}{|P_i(t) - G|} \tag{3}$$

where f_r^i is the center frequency of the i-th receiver, and $2\pi(f - f_r^i)$ represents the phase difference between the sender and the receiver. $P_i(t)$ is the position of the i-th receiver, while $G = [g_x, g_y, g_z]^T$ is the position of the target. k_i is the LNA magnification in decibels (dB).

3.3   The Frequency Drift f_i

The moving receiver will cause the Doppler effect, which will incur the frequency drift. The frequency drift value of the i-th receiver can be expressed by:

$$f_i = \frac{v_i \cos(\theta_i)}{\lambda} = \frac{v_i f \cos(\theta_i)}{c} \tag{4}$$

where

$$\cos(\theta_i) = \frac{(G - P_i(t)) \cdot (P_i(t - \Delta t) - P_i(t))}{|G - P_i(t)| \times |P_i(t - \Delta t) - P_i(t)|}$$
$$G - P_i(t) = [g_x - (x_0^i + v_x^i \times t), g_y - (y_0^i + v_y^i \times t), g_z - (z_0^i + v_z^i \times t)]^T$$
$$P_i(t - \Delta t) - P_i(t) = [-v_x^i \times \Delta t, -v_y^i \times \Delta t, -v_z^i \times \Delta t]^T$$

where $[x_0^i, y_0^i, z_0^i]^T$ is the initial position of the i-th receiver, and $v_i = [v_x^i, v_y^i, v_z^i]^T$ is its velocity, and Δt is the sampling period of the signal.

3.4   The Received Signal at the i-th Receiver

Bring the electromagnetic radiation signal $s(t)$ (Eq. (2)), the propagation path gain coefficient μ_i (Eq. (3)), and the frequency drift f_i (Eq. (4)) into Eq. (1), we can get the signal received by the i-th receiver at time t:

$$r_i(t) = \mu_i s(t - \tau_i)e^{j2\pi f_i t} + n_i(t)$$
$$= \frac{k_i|s|}{|P_i(t) - G|}e^{j(2\pi(f-f_r^i)+2\pi f(t-\frac{|P_i(t)-G|}{c})+2\pi \frac{v_i f \cos(\theta_i)}{c})} + n_i(t) \tag{5}$$

where $\tau_i = \dfrac{|P_i(t) - G|}{c}$ is the signal transmission delay, and c is the speed of light.

4 The Electromagnetic Radiation Source Localization Approach

In this section, the proposed electromagnetic radiation source localization approach is presented. Since the proposed approach is based on the DDDS framework, we first formalize the system model which describes the moving of the target in Sect. 4.1, after which we assume the measurement model in Sect. 4.2 based on the spatial propagation model of electromagnetic radiation signal emitted by the target which is derived in Sect. 3. Finally, particle filters are employed to locate the target, which is described in Sect. 4.3.

4.1 System Model

The position of the target is $G = [g_x, g_y, g_z]^T$, so we define the system state as $s_k = [g_x, g_y, g_z]^T$. In this paper, we assume that the target keeps still, therefore we can define the state evolution of the target as a slow-change process:

$$s_k = s_{k-1} + \nu_{k-1}, k = 1, 2, \ldots \tag{6}$$

where the noise vector ν_{k-1} is a Gaussian white noise with an average of 0 and variances of $\sigma_{g_x}^2$, $\sigma_{g_y}^2$, $\sigma_{g_z}^2$, respectively.

4.2 Measurement Model

Suppose we have n receivers, and the measurement at the k-th step can be defined as $m_k = [r_1(k \times \Delta t), \cdots, r_n(k \times \Delta t)]^T$, where Δt is the signal sampling period of the receiver. According to the signal propagation model derived in Sect. 3, the measurement model can be defined as:

$$
\begin{aligned}
m_k &= g(s_k) + \varepsilon_k \\
&= \begin{bmatrix} \frac{k_1|s|}{|\boldsymbol{P}_1(k \times \Delta t) - \boldsymbol{G}|} e^{j(2\pi(f - f_r^1) + 2\pi f(k \times \Delta t - \frac{|\boldsymbol{P}_1(k \times \Delta t) - \boldsymbol{G}|}{c}) + 2\pi \frac{v_1 f \cos(\theta_1)}{c})} \\ \vdots \\ \frac{k_n|s|}{|\boldsymbol{P}_n(k \times \Delta t) - \boldsymbol{G}|} e^{j(2\pi(f - f_r^n) + 2\pi f(k \times \Delta t - \frac{|\boldsymbol{P}_n(k \times \Delta t) - \boldsymbol{G}|}{c}) + 2\pi \frac{v_n f \cos(\theta_n)}{c})} \end{bmatrix} + \varepsilon_k
\end{aligned}
$$

$$\tag{7}$$

where $\cos(\theta_i), |\boldsymbol{P}_i(k \times \Delta t) - \boldsymbol{G}|, i = 1, \ldots, n$ are defined in Sect. 3.3. ε_k is the measurement noise, and we assume that both the real and imaginary parts of ε_k are Gaussian white noise with mean 0 and variance σ_m^2.

4.3 Locating the Target Using Particle Filters

Based on the system model and the measurement model, we locate the target based on particle filters (see Algorithm 1). The input of the algorithm is the data sequence collected by n receivers, i.e. $data = \{[d_k^1, \cdots, d_k^n]^T\}_{k=1}^N$, where N is the number of time steps; particle filters estimate the target position as

$\{\{[g_{x,k}^i, g_{y,k}^i, g_{z,k}^i]^T\}_{i=1}^{N_p}\}_{k=1}^N$, where N_p is the number of particles. Note that at step k, the output of the particle filter is a group of particles $\{[g_{x,k}^i, g_{y,k}^i, g_{z,k}^i]^T\}_{i=1}^{N_p}$, approximating the probability distribution of the target position. Therefore, given the particles at each step k, the position of the target and the corresponding variance can be estimated:

$$[\hat{g}_{x,k}, \hat{g}_{y,k}, \hat{g}_{z,k}]^T = [\frac{1}{N_p}\sum_{i=1}^{N_p} g_{x,k}^i, \frac{1}{N_p}\sum_{i=1}^{N_p} g_{y,k}^i, \frac{1}{N_p}\sum_{i=1}^{N_p} g_{z,k}^i]^T$$

$$\sigma_{x,k}^2 = \frac{1}{N_p - 1}\sum_{i=1}^{N_p}(g_{x,k}^i - \hat{g}_{x,k})^2$$

$$\sigma_{y,k}^2 = \frac{1}{N_p - 1}\sum_{i=1}^{N_p}(g_{y,k}^i - \hat{g}_{y,k})^2$$

$$\sigma_{z,k}^2 = \frac{1}{N_p - 1}\sum_{i=1}^{N_p}(g_{z,k}^i - \hat{g}_{z,k})^2$$

4.4   Weight Computation

Given the particle state s_k^i, we can calculate $m_k^i = [r_1^i(k\Delta t), \cdots, r_n^i(k\Delta t)]^T$ according to the measurement model. Then the particle weight can be computed as:

$$w_k^i = w_{k-1}^i \times p([d_k^1, \cdots, d_k^n]^T|m_k^i) = w_{k-1}^i \times \prod_{j=1}^{n} \frac{1}{\sqrt{2\pi}\sigma_m} e^{-\frac{|r_j^i(k\Delta t) - d_k^j|^2}{2\sigma_m^2}} \quad (8)$$

Note that $r_j^i(k\Delta t)$ and d_k^j are complex numbers, and $|r_j^i(k\Delta t) - d_k^j|$ represents the modulo of the difference between two complex numbers.

5   Experimental Results and Analysis

5.1   Experiment Scenario and Parameter Setting

We use the identical-twin experiment [17] to test and validate the proposed approach. In the identical-twin experiment, a simulation is first run, and the corresponding data are recorded. This simulation is regarded as the 'real' system. Then we add noise to these data, and use these noisy data to estimate the position of the target based on the proposed approach. Finally we compare the estimated results with the ground truth data to quantify the accuracy of the estimation.

In the first simulation which is used to generate the ground truth data, the parameters of the target are set based on Table 1. Four drones are used as receivers to collect data (i.e. the amplitude and phase of the signal), and the

---

Algorithm 1: The particle filter for electromagnetic radiation source localization

---

Input: $data = \{[d_k^1, \cdots, d_k^m]^T\}_{k=1}^N$

Output: estimated location of the target: $\{\{[g_{x,k}^i, g_{y,k}^i, g_{z,k}^i]^T\}_{i=1}^{N_p}\}_{k=1}^N$

1 % the initialization step
2 **for** $i = 1 : N_p$ **do**
3   generate the i-th particle $s_0^i = [g_{x,0}^i, g_{y,0}^i, g_{z,0}^i]^T$
4   set weight $w_0^i = 1/N_p$
5 **end**
6 **for** $k = 1 : N$ **do**
7   % the sampling step for any time $k \geq 1$
8   **for** $i = 1 : N_p$ **do**
9   according to the state at time step $k - 1$, i.e. s_{k-1}^i, generate a new particle at time step k, i.e. s_k^i, using equation (6)
10   calculate m_k^i based on s_k^i using equation (7), and update the weight of the particle (see more details in section 4.4):

$$w_k^i = w_{k-1}^i \times p([d_k^1, \cdots, d_k^m]^T | m_k^i)$$

11   **end**
12   normalizes the particle weights and denote them as $\{s_k^i, w_k^i\}_{i=1}^{N_p}$
13   % The resampling step
14   resample $\{s_k^i, w_k^i\}_{i=1}^{N_p}$ using the standard resampling method which samples particles in proportion to their weights; the resampled results are again denoted as $\{s_k^i, w_k^i\}_{i=1}^{N_p}$
15   Set the weight of each particle as $1/N_p$
16   % record data for estimation
17   record each particle's estimate of the target position: $\{[g_{x,k}^i, g_{y,k}^i, g_{z,k}^i]^T\}_{i=1}^{N_p}$
18 **end**

---

relevant parameters are set based on Table 2. The drones collect the data every second (i.e. $\Delta t = 1\ s$) for a total of $100\,s$ (i.e. $N = 100$). For data noise, we set σ_m to 5% of the signal amplitude. For model errors, we set σ_{g_x}, σ_{g_y} and σ_{g_z} to 5% of the corresponding coordinates of the target position.

5.2 The Performance Indicators

In order to quantify the accuracy of the proposed localization method, we define the following performance indicators (since the z-coordinate of the target is 0, the localization error of the z-coordinate is not considered):

$$
\begin{aligned}
RMSE_x &= \sqrt{\frac{\sum_{k=1}^N (\hat{g}_{x,k} - g_x)^2}{N}} \\
RMSE_y &= \sqrt{\frac{\sum_{k=1}^N (\hat{g}_{y,k} - g_y)^2}{N}}
\end{aligned}
\tag{9}
$$

Table 1. The parameter settings of the target

Parameter	Value
Position of the target	$(-9960.3, 123686, 0)$
Signal amplitude $|s|$	200 W
The frequency of the signal f	100 MHz

Table 2. The parameter settings of the drones (i.e. the receivers used in the experiment)

Parameter	Value
Initial position of drone#1	$(-120341, 110974, 12000)$
Velocity of drone#1	$(69, 72.4, 0)$
Initial position of drone#2	$(68625, 131345, 12000)$
Velocity of drone#2	$(58.5, 81.1, 0)$
Initial position of drone#3	$(-50000, 130000, 12000)$
Velocity of drone#3	$(60.0, 40.0, 0)$
Initial position of drone#4	$(50000, 111000, 12000)$
Velocity of drone#4	$(50.0, 35.0, 0)$
LNA magnification	160 dB
Center frequency of the receiver f_r	315.425 MHz

where g_x and g_y are the true horizontal and vertical coordinates of the target, while $\hat{g}_{x,k}$ and $\hat{g}_{y,k}$ are the corresponding estimates given by the particle filtering algorithm.

5.3   The Experiment Results

In this experiment, we use $N_p = 1000$ particles. In the initialization step, we randomly generate particles, and each particle represents a guess of the position of the target. The particle initialization result is shown in Fig. 2. It can be seen that the particles are evenly distributed within the state space in which the target may exist.

As more data are assimilated, it is expected that the particles will converge to the true position of the target. We show the data assimilation results at $t = 6$ in Figs. 3a and 3b. We can see that in the sampling step (Fig. 3a), the particles that are close to the target are assigned larger weights (see the different colors); After resampling (Fig. 3b), the particles with larger weights are kept for the data assimilation at the next time step, and consequently the probability distribution

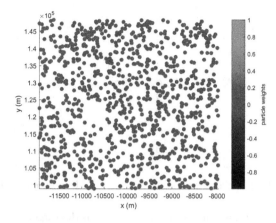

Fig. 2. The particle initialization result (the true target position is represented by a black cross)

of the estimated target position will gradually converge to the true distribution of the target (see that the area of the particle dispersion gradually shrinks).

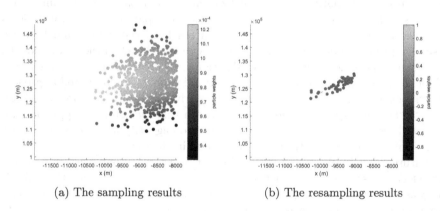

(a) The sampling results   (b) The resampling results

Fig. 3. The data assimilation results at $t = 6$ (the true target position is represented by a red cross) (Color figure online)

The estimated position of the target at different time steps is shown in Figs. 4a and 4b. Based on the estimated positions at different steps, we also calculate the performance indicators defined in Eq. (9), and the values are:

$$RMSE_x = 987.7626 \ m$$
$$RMSE_y = 6476 \ m$$

and the relative errors are $RMSE_x/g_x \times 100\% = 9.92\%, RMSE_y/g_y \times 100\% = 5.24\%$. These results show that the proposed approach can accurately locate the target.

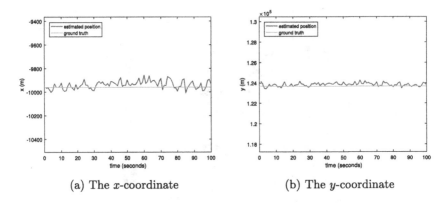

(a) The x-coordinate   (b) The y-coordinate

Fig. 4. The estimated position of the target

6   Conclusions and Future Work

Locating enemy targets via their electromagnetic radiation signal is vital to block and attack the enemy targets at an earlier stage. In this paper, we propose a novel electromagnetic radiation source localization method based on dynamic data driven simulations (DDDS). Firstly, we model the spatial propagation process of the electromagnetic radiation signal emitted from the target, and then we assume a proper model for the noisy measurements. Based on these two models, we design a particle filter based localization method to locate the target. We adopt the identical-twin experiment to test and validate the proposed approach, and the experiment results show that the relative localization errors are within 10% of the true target positions, which prove that the propose approach can accurately locate the target using noisy electromagnetic radiation signals, and is also robust to errors both in the model and in the data.

Future work is planned in the following directions. First, we need to test the effectiveness of the approach when the target is moving, since this experiment assumes that the target keeps still. Second, more realistic electromagnetic radiation process models will be researched.

Acknowledgments. This research is supported by the National Natural Science Fund of China (Grant No. 62103428) and the Natural Science Fund of Hunan Province (Grant No. 2021JJ40702).

References

1. Liu, W.: Radio direction-finding and location research in the radio monitoring. Master's thesis, Xihua University (2013)
2. Li, Z., Braun, T., Zhao, X., Zhao, Z., Fengye, H., Liang, H.: A narrow-band indoor positioning system by fusing time and received signal strength via ensemble learning. IEEE Access **6**, 9936–9950 (2018)

3. Wang, Y., Ho, K.C.: Unified near-field and far-field localization for AOA and hybrid AOA-TDOA positionings. IEEE Trans. Wireless Commun. **17**(2), 1242–1254 (2018)
4. Liu, Y., Guo, F., Yang, L., Jiang, W.: An improved algebraic solution for TDOA localization with sensor position errors. IEEE Commun. Lett. **19**(12), 2218–2221 (2015)
5. Xiaolin, H.: Dynamic data driven simulation. SCS M&S Mag. **II**(1), 16–22 (2011)
6. Xie, X.: Data assimilation in discrete event simulations. Ph.D. thesis, Delft University of Technology (2018)
7. Nichols, N.: Data assimilation: aims and basic concepts. In: Swinbank, R., Shutyaev, V., Lahoz, W.A. (eds.) Data Assimilation for the Earth System, pp. 9–20. Springer, Dordrecht (2003). https://doi.org/10.1007/978-94-010-0029-1_2
8. Arulampalam, S., Maskell, S., Gordon, N., Clapp, T.: A tutorial on particle filters for online nonlinear/non-Gaussian Bayesian tracking. IEEE Trans. Signal Process. **50**(2), 174–188 (2002)
9. Djurić, P., Kotecha, J., Zhang, J., Huang, Y., Ghirmai, T., Bugallo, M., Miguez, J.: Particle filtering. IEEE Signal Process. Mag. **20**(5), 19–38 (2003)
10. Zeigler, B., Praehofer, H., Kim, T.G.: Theory of Modeling and Simulation: Integrating Discrete Event and Continuous Complex Dynamic Systems, 2nd edn. Academic Press, Cambridge (2000)
11. Lahoz, W.A., Khattatov, B., Menard, R.: Data Assimilation: Making Sense of Observations, 1st edn. Springer, Heidelberg (2010). https://doi.org/10.1007/978-3-540-74703-1
12. Darema, F.: Dynamic data driven applications systems: a new paradigm for application simulations and measurements. In: Bubak, M., van Albada, G.D., Sloot, P.M.A., Dongarra, J. (eds.) ICCS 2004. LNCS, vol. 3038, pp. 662–669. Springer, Heidelberg (2004). https://doi.org/10.1007/978-3-540-24688-6_86
13. Darema, F.: Dynamic data driven applications systems: new capabilities for application simulations and measurements. In: Sunderam, V.S., van Albada, G.D., Sloot, P.M.A., Dongarra, J.J. (eds.) ICCS 2005. LNCS, vol. 3515, pp. 610–615. Springer, Heidelberg (2005). https://doi.org/10.1007/11428848_79
14. Bouttier, F., Courtier, P.: Data assimilation concepts and methods. Meteorological Training Course Lecture Series, ECMWF (European Centre for Medium-Range Weather Forecasts) (1999)
15. Gu, F.: Dynamic data driven application system for wildfire spread simulation. Ph.D. thesis, Georgia State University (2010)
16. Bai, F., Guo, S., Hu, X.: Towards parameter estimation in wildfire spread simulation based on sequential Monte Carlo methods. In: Proceedings of the 44th Annual Simulation Symposium, Boston, MA, USA, pp. 159–166 (2011)
17. Xue, H., Gu, F., Hu, X.: Data assimilation using sequential Monte Carlo methods in wildfire spread simulation. ACM Trans. Model. Comput. Simul. **22**(4), 23:1–23:25 (2012)

A Parallel Simulation Method for Active Distribution Network Transient Process Based on Real-Time Simulator

Haotian Ma[(✉)], Keyan Liu, and Wanxing Sheng

China Electric Power Research Institute, Haidian District, Beijing 100192, China
mahaotian1996@163.com

Abstract. When the distributed generation is integrated to power distribution system, the network structure and operating characteristics become complex and challenging. Due to the complex control system of distributed generation, the simulation time for transient process of active distribution network increases exponentially. In order to better describe the transient characteristics and satisfy the increasing simulation requirements, the parallel simulation method is proposed to simulate the active distribution network in partitions by using the active distribution network simulator (ADN-Sim). Firstly, a decoupling equivalent model for short-distance branch based on characteristic line is established, which solves the inapplicability of the distributed parameter long-line decoupling method in the distribution network, and further the transient simulation accuracy is improved. Secondly, the coupling relationship between the node voltage equation and the current loop equation in the parallel simulation is reduced by establishing the logical relationship of the network structure and the electrical relationship of the component model after partitioning, so that the decoupling and dimension reduction of the parallel simulation in the active distribution network is further realized. In addition, the main factors affecting the efficiency of parallel simulation are given, and the simulation error in the transient process of distribution network is briefly analyzed. Finally, an actual 62-node distribution network model is simulated as a typical scenario to verify the parallel simulation effect under the transient process of active distribution network loop closing operation. Through verification, it can be shown that the proposed method is suitable for the transient process of active distribution network, and the simulation speed can be improved effectively.

Keywords: Active distribution network · Parallel transient simulation · Real-time simulation · Partition decoupling

1 Introduction

As the proportion of distributed power generation connected to the distribution network gradually increases, its transient process characteristics will gradually become more complex, especially when the external operating conditions change or faults occur. From the control point of view, the transient characteristics of the distribution network will become more and more complex, and the existing coordinated control algorithms will not be able to meet the demand. Distribution network transient simulation is an important means to study the relationship between various dynamic processes and the basis for revealing various transient process problems. It has engineering application value and theoretical practical guiding significance. Therefore, distribution network transient simulation needs to be further studied.

In order to improve the simulation speed and fully reflect the transient process of the distribution network, real-time simulation is generally used to model and analyze large-scale distribution network systems. If multi-core parallel simulation is used, it is necessary to establish inter-core communication and electrical connection to decouple a system. Therefore, establishing a suitable equivalent decoupling method is also one of the difficult problems in the transient simulation for active distribution network. In this paper, in view of the fact that the traditional distributed parameter long line decoupling method is not suitable for distribution network, a short line decoupling method based on characteristic line is established to provide support for the transient simulation of active distribution network.

In the transient simulation of large-scale power systems, the waveform relaxation method is used to simplify the modeling of power systems. The transient stability and convergence of the system are preprocessed, and the waveform is predicted. Then, the convergence of the power system and the acceleration of parallel simulation are verified based on the domain decomposition iterative algorithm [1–4]. In terms of network topology parallel simulation, references [5, 6] decouple the known network topology, and establishes the electrical relationship of the parallel network topology through the network segmentation algorithm based on the optimized boundary table through the information transfer interface. Although the complexity of the network is increased to a certain extent, the description of the network relationship is simplified, the acceleration will be improved, and real-time simulation can be achieved. In the transient parallel simulation of power system using graphics processor, algorithm research generally focuses on the preprocessing of coefficient matrix. In references [7–9], the pre-processing algorithm is adopted to reduce the order of coefficient matrix and improve the convergence speed. The preprocessed linear equation is solved by the generalized residual method to realize the parallel calculation of transient simulation. For distributed power generation, it is a highly nonlinear system. If the averaged model is used to model PWM converters and other switching components in distributed photovoltaics, the simulation efficiency can be improved [10, 11].

In this paper, an improved multi-region parallel simulation method for distribution network based on real-time simulator is proposed. This method improves the traditional region segmentation principle, and makes each sub-region port equivalent to ADN-Sim through decoupling the connection line. According to the characteristics of short lines in distribution network, the wave propagation time is extended by increasing the distributed capacitance, and the decoupling method of characteristic line is established, which is suitable for short lines in distribution network and further improves the solution efficiency of parallel computing. Then, the parallel simulation method based on ADN-Sim is given, and the error factors under the transient simulation of distribution network are analyzed. Finally, in an actual 62-node network, a typical distribution network transient scenario is established to verify the effectiveness of the proposed parallel simulation method.

2 Decoupling Equivalent Method for Power Grid Transient Simulation Model

In order to achieve parallel simulation of the active distribution network, the transient simulation model must be decomposed. The traditional segmentation method is mainly divided according to the located area, and the network equivalent is performed at the segmentation location to achieve the purpose of parallel simulation. For the purpose of the simplest network and the fewest breakpoints, the equivalence principle is usually based on electrical logic, and the equivalent method in the form of controlled source is used for decoupling in this paper.

In the distribution network, due to the low voltage level, the resistance is on the same order of magnitude as the inductance, so the existence of the resistance cannot be ignored. Since the lines in the distribution network are not as long as those in the transmission network, the parallel computing method using distributed parameter line decoupling may not converge in the distribution network. The length of the distribution network line cannot meet the minimum requirements of the distributed parameter line. Assuming that the minimum time of traveling wave transmission is Δt, and the traveling wave transmission speed is approximately the light velocity $v = 3 \times 10^8$ m/s, according to the principle of distributed parameter line transmission, the minimum decoupling line length is:

If the simulation is performed with a step size of Ts $= 50\,\mu s$ in the parallel optimization calculation in the active distribution network, a minimum 15 km long distributed parameter line decoupling is required. But in the actual distribution network, especially below 10 kV, the power transmission distance is mostly 6–20 km, of which the line length of 10 km is mostly, less than 15 km, which is called short line decoupling.

Assuming that the propagation speed of the electromagnetic wave on the line is v, and the length of the line is x, the simplified analysis of the distributed parameter line decoupling principle is shown in Fig. 1.

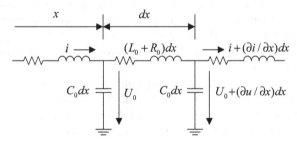

Fig. 1. Equivalent calculation model of short line considering distributed capacitance

The wave equation above is:

$$\begin{aligned}
-\frac{\partial u}{\partial x} &= L_0 \frac{\partial i}{\partial t} + R_0 i \\
-\frac{\partial i}{\partial x} &= C_0 \frac{\partial u}{\partial t} + G_0 u
\end{aligned}$$

(1)

According to the principle of fixed-step parallel computing, there will be a step Δt delay between cores in multi-core parallelism. The following two cores are used as objects to analyze the wave equation. Assuming that the traveling wave propagates forward at the speed v, $f_1(x - v\Delta t)$ is the solution of the forward traveling wave equation in one step propagation time, and similarly $f_2(x + v\Delta t)$ is the solution of the backward traveling wave equation in one step propagation time. The solutions are described by the voltage-current equation as follows:

$$\begin{cases} i(x, t) = f_1(x - v\Delta t) + f_2(x + v\Delta t) \\ u(x, t) = Z_c(f_1(x - v\Delta t) - f_2(x + v\Delta t)) \end{cases} \tag{2}$$

If the distance of the distributed parameter line is set as l, whether the traveling wave propagates in the forward direction or the backward direction, it is regarded as a constant propagation time, that is, $x + v\Delta t$ and $x - v\Delta t$ are constants, then the transmission time from the beginning to the end is as follows:

$$\Delta t = l/v = l \times \sqrt{LC} \tag{3}$$

In parallel computing, there is a step delay between multi-core CPUs. Let the transmission time in the distributed parameter line be Δt, the voltage across the distributed parameter line can be expressed as:

$$u_i(t - \Delta t) + Z_c i_i(t - \Delta t) = u_j(t) - Z_c i_j(t) \tag{4}$$

where $-Z_c i_j(t)$ represents that the traveling wave propagates forward; $t - \Delta t$ means that the CPU operation time lags by one step. During the operation time of this step, the value of the dividing point is the data of the previous step of the kernel connected to it.

$$\begin{cases} i_i(t - \Delta t) = \frac{-u_i(t-\Delta t)}{Z_c} - i_j(t) \\ i_j(t) = \frac{u_j(t)}{Z_c} + i_i(t - \Delta t) \end{cases} \tag{5}$$

The equivalent model of the distribution parameter line is shown in Fig. 2. There is no direct connection between the two ends of this model, but the electrical parameters at both ends are exchanged through the sharing of current data.

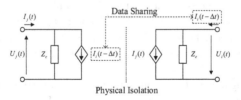

Fig. 2. Controlled source model for short line decoupling

3 Parallel Simulation Method for Distribution Network Based on Real-Time Simulator

3.1 Parallel Partition Simulation Technology in Real-Time Simulator

The parallel partition real-time simulation technology realizes real-time operation in the CPU by compiling the component model and digitizing the electrical relationship. The model is further divided into physical system and control information system. The two systems are connected by technical bus and simulated in CPU respectively. While performing real-time simulation between different arrays in a single CPU, unilateral or multilateral information exchange is also performed to ensure the accuracy of the simulation process. The simulation structure of the CPU formed by the physical system and the information system is shown in Fig. 3.

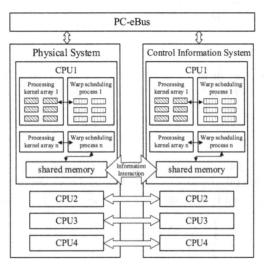

Fig. 3. Active distribution network real-time simulator structure

The ADN-Sim simulation method is constructed from the following aspects:

(1) Split the model and encode it separately. The emulator assigns tasks to different CPUs according to the code names. The electrical information parameter transfer of the physical model between different arrays in a single CPU is realized by controlling the information system to update the number of ports in the array in real time.

(2) After establishing the electrical connection between the subsystems, the subsystem information needs to be downloaded to ADN-Sim. Specifically, the subsystem is compiled through the C compiler, and then the compiled code is downloaded to AND-Sim with a special download tool.

(3) When a subsystem is downloaded to ADN-Sim, its model will be on a different array in a single CPU. There will be unilateral or multilateral information interaction between arrays. Using multi-thread scheduling technology, real-time information exchange between different arrays under a single step size can be realized.

3.2 Parallel Simulation Method Based on Region Segmentation

Through the equivalent decoupling of the distribution network, the variables between the subsystems are the terminal node voltage phasors of each region. The node voltage equation of the i-th subsystem is:

$$
\begin{bmatrix}
Y_{11}^s + \sum_{j=1}^{n_1} y_{1j}^L & Y_{12}^s & \cdots & Y_{1m}^s \\
Y_{21}^s & Y_{22}^s + \sum_{j=1}^{n_2} y_{2j}^L & \cdots & Y_{2m}^s \\
\vdots & \vdots & \ddots & \vdots \\
Y_{m1}^s & Y_{m2}^s & \cdots & Y_{mm}^s + \sum_{j=1}^{n_m} y_{mj}^L
\end{bmatrix} V_{si}^S =
\begin{bmatrix}
-J_1^{ss} - \sum_{j=1}^{n_1} J_{1j}^L \\
-J_2^{ss} - \sum_{j=1}^{n_2} J_{2j}^L \\
\vdots \\
-J_m^{ss} - \sum_{j=1}^{n_m} J_{mj}^L
\end{bmatrix}
\tag{6}
$$

where Y_{ij}^L and J_{ij}^L are the equivalent conductance matrix and current phasor connected to the branch L in the i-th subsystem, respectively. Solving the above Equation can get the subsystem terminal voltage column phasor V_{si}^S.

Calculate the node voltage after the interconnection split according to the injected current of the subsystem, and the injected power is related to the node voltage. In order to update the electrical variables of the subsystem ports, it is necessary to make a convergence judgment. The electrical variable matrix of the ports is established as follows:

$$
V_s = Y^{EQ}
\begin{bmatrix}
J_1^{DG} \\
\vdots \\
J_m^{DG}
\end{bmatrix}
\tag{7}
$$

$$
Y^{EQ} = -\left[Y_{main}^S - Y^{SRGG}
\begin{bmatrix}
Y_1^{GRS} \\
Y_2^{GRS} \\
\vdots \\
Y_2^{GRS}
\end{bmatrix} \right] Y^{SRGG}
\tag{8}
$$

$$
Y^{SRGG} =
\begin{bmatrix}
Y_1^{SRG} \\
Y_2^{SRG} \\
\vdots \\
Y_m^{SRG}
\end{bmatrix}
\begin{bmatrix}
Y_1^{GRG} & & & \\
& Y_2^{GRG} & & \\
& & \ddots & \\
& & & Y_n^{GRG}
\end{bmatrix}
\tag{9}
$$

When the simulation time reaches t:

$$V_i^{DG}(t) = \left(Y_i^{GRG}\right)^{-1}\left(J_{Gi}(t) + J_{Gi}(V_i^{DG}(t)) - Y_i^{GRS}Y^{EQ}\right) \cdot \begin{bmatrix} J_{G1} + J_{G1}V_1^{DG}(t) \\ J_{G2} + J_{G2}V_2^{DG}(t) \\ \vdots \\ J_{Gn} + J_{Gn}V_n^{DG}(t) \end{bmatrix}$$

$$(10)$$

Then there are the following iterations:

$$V_i^{DG(k)}(t) = \left(Y_i^{GRG}\right)^{-1}\left(J_{Gi}(t) + J_{Gi}(V_i^{DG(k-1)}(t)) - Y_i^{GRS}Y^{EQ}\right) \cdot \begin{bmatrix} J_{G1} + J_{G1}V_1^{DG(k)} \\ J_{G2} + J_{G2}V_2^{DG(k)} \\ \vdots \\ J_{Gn} + J_{Gn}V_n^{DG(k)} \end{bmatrix} \quad (11)$$

When $V_i^{DG(k)} - V_i^{DG(k-1)} < \varepsilon$ is satisfied, one step period ends, and then the calculation process of the internal state variables for the next period is carried out.

4 Error Analysis of Parallel Transient Simulation

The active distribution network involves a variety of linear and nonlinear components, and the coupling relationship between them has different effects on the computational efficiency of parallel simulation. The allocation of computing resources for transient processes in parallel simulation will also affect the computing effect. In this section, the relevant factors affecting the efficiency of active distribution networks parallel simulation are mainly analyzed.

Speed-up ratio is an indicator that describes the performance of parallel simulation relative to serial simulation. Assuming that under the same simulation task, the parallel simulation time and the serial simulation time are T_p and T_s respectively, then the speedup ratio can be expressed as:

$$G(N) = \frac{T_s}{T_p(N)} \quad (12)$$

where $G(N)$ is the speedup ratio; $T_p(N)$ is the time-consuming of parallel simulation under N cores, $N = 1, 2 \ldots n$.

Parallel efficiency refers to the average utilization of N CPUs, which can be expressed as:

$$T_p(N) = \frac{G(N)}{N} \tag{13}$$

where $T_p(N)$ represents the parallel efficiency;

The parallel simulation method proposed in this paper can reach the real-time level, and the computing efficiency can be expressed by resource occupancy, as shown below:

$$e(N) = \frac{t_s}{T_{step}} \tag{14}$$

where t_s represents the simulation time under single step; T_{step} represents the simulation step size set by the system.

In parallel real-time simulation, data transmission between cores will experience a step time difference. In this study, the voltage grade is 10 kV, but the substation outlet voltage is 10.5 kV, and the simulation step size is 50 μs, then the voltage difference between peak and trough is:

$$U_{p\_v} = \frac{10.5 \times \sqrt{2} \times 10^3 - (-10.5 \times \sqrt{2} \times 10^3)}{\sqrt{3}} \tag{15}$$

The simulation step size C_{sum} required from the peak to the trough is:

$$C_{sum} = \frac{T}{\Delta t} = \frac{0.01}{50 \times 10^{-6}} \tag{16}$$

Parallel simulation only updates electrical variables (voltage and current) between cores, then the simulation error F_{error} is:

$$F_{error} = \frac{\Delta U}{C_{sum}} = \frac{105\sqrt{2}}{\sqrt{3}} \approx 85.73V \tag{17}$$

In summary, the improved parallel simulation process for active distribution network is shown in Fig. 4.

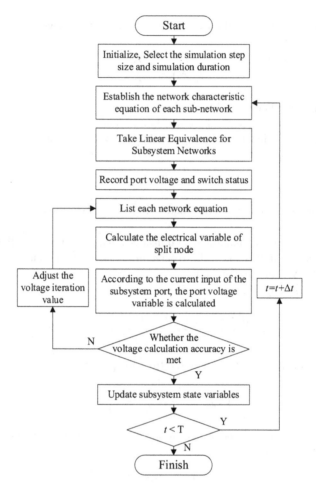

Fig. 4. Region segmentation parallel simulation flow chart

5 Case Study

In order to analyze the application effect of the parallel simulation method for the typical transient events in the distribution network, the real-time simulation system is used to simulate and analyze the active distribution network model in this section. As shown in Fig. 5, in an actual power grid with 62 nodes, the network structure is divided into three main subregions, which are allocated to 5 CPUs for parallel simulation, namely 1#CPU–5#CPU. In the case of sudden power failure at power supply 0, power supply 1 will transfer the load through the tie line connected with bus 37 after 0.2 s.

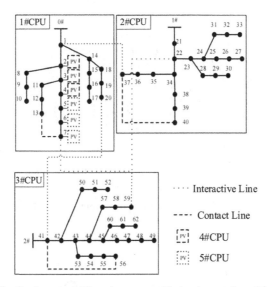

Fig. 5. An actual 62-node power grid structure and partition

In this section, the influence of the closed loop operation after the fault on the output characteristics of the distributed generation is mainly analyzed. The nodes 2–7 are connected to photovoltaics, and the photovoltaic penetration rate is controlled to be kept at the level of 50%, then the loop closing operation is carried out. The transient events are shown in Table 1,

Table 1. Transient event description for serial and parallel simulation mode

Simulation mode	Transient event	Duration	Load shedding time
Parallel	0# failure	1.0 s–1.2 s	1.01 s
Serial	0# failure	0.7 s–0.9 s	1.01 s

Taking the photovoltaic output current at node 4 as an example, the transient process current is shown in Fig. 6. The partial amplification of the distributed photovoltaic output current is shown in Fig. 7

It can be seen from Fig. 6 that a three-phase short-circuit fault occurs at 1 s, and the fault is removed after 0.2 s. During this process, the photovoltaic is not removed. Due to the voltage drop at the access node, the photovoltaic output current will suddenly increase under the constant power control strategy. If the distribution network system does not experience large frequency fluctuation, the photovoltaic current output will gradually return to a stable state. It can be seen that the simulation result accord with the actual situation, and the parallel simulation result are basically consistent with the serial simulation result. It can be seen from Fig. 7 that the error is large when the transient occurs, and the phenomenon of phase deviation will appear.

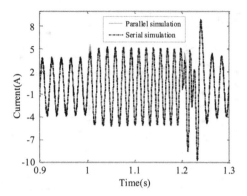

Fig. 6. Distributed photovoltaic output current under transient event

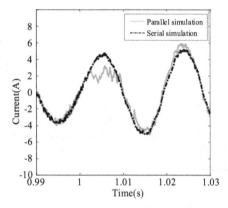

Fig. 7. Local amplification of distributed photovoltaic output current

Figure 8 shows the current error analysis at the Point of Common Coupling (PCC).

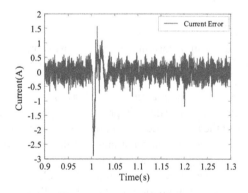

Fig. 8. Current error at PCC

It can be seen from Fig. 8 that the maximum current error is less than 0.5 A in the steady state. When a transient event occurs, the current error will suddenly increase, and after a short fluctuation, the photovoltaic output current tends to be stable. When the fault is removed, the output current will suddenly decrease, and the current error will also have a corresponding fluctuation process.

The series-parallel simulation comparison of the voltage at PCC is shown in Fig. 9, and its partial enlarged detail is shown in Fig. 10.

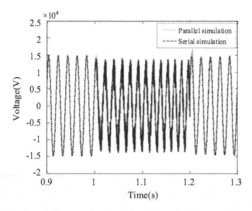

Fig. 9. Voltage variation at PPC under transient event

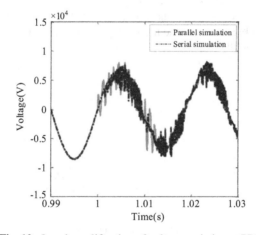

Fig. 10. Local amplification of voltage variation at PPC

It can be seen from Fig. 9 that when a three-phase short-circuit fault occurs, a voltage sag will happen at the PCC, and if the fault is not removed, the voltage will always be at a low level. Combining with Fig. 10, at the moment of the transient event, the voltage error will have an obvious sudden increase, and the voltage error will not be consistent during the transition period when the output current changes from small to large and from large to small.

When the loop closing operation is completed, the PCC can recover to the rated voltage, and the serial-parallel simulation can quickly switch to a stable state. However, in the transient process, the calculation does not converge and the error is too large. The analysis is carried out in conjunction with the error distribution in Fig. 11.

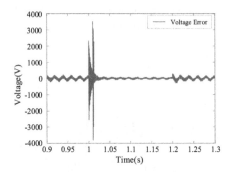

Fig. 11. Voltage error at PCC

In the whole process, the sudden increase of error mainly occurs at the moment of state switching, and the trend of error change is not consistent between the voltage changes from large to small and from small to large. The increase of the error is related to the occupancy rate of computing resources, and the sudden increase of the occupancy rate also occurs at the above two moments. Therefore, in order to express the real-time simulation effect, it is necessary to analyze the computing resource occupancy rate in detail. Figure 12 and Fig. 13 show the transient process computing resource occupancy rate of 3#CPU and 5#CPU respectively.

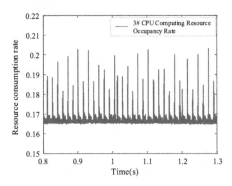

Fig. 12. Local amplification of 3#CPU computing resource occupancy rate

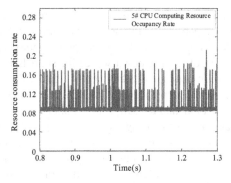

Fig. 13. Local amplification of 5#CPU computing resource occupancy rate

It can be seen from the computing resource occupancy rates in Fig. 12 and Fig. 13 that, on the basis of implementing partitioned parallel simulation, the simulation can reach a real-time level, and when a loop closing transient event occurs, the error is relatively large. Table 2 shows the computing resource usage of each CPU.

Table 2. Computational resource occupancy rate of transient parallel simulation

CPU	Average occupancy rate	Maximum occupancy rate	Timeout times
1#	66.4%	>100%	3
2#	23.2%	>100%	1
3#	17.5%	20.5%	0
4#	17.3%	29.1%	0
5#	13.3%	21%	0

It can be seen from Table 2 that the maximum computing resource occupancy rate can reach 100%, and the time-out phenomenon occurs at the time of event switching. Combining the two transient events can be attributed to the slow convergence speed due to the limitation of the transmission rate.

In addition, the absolute value of the voltage error is 81.56 V (the theoretical error is 85.73 V), and the relative error is 1.49%. From the analysis of the error extreme value, it can be known that the voltage error and the current error are proportional to the permeability. From the perspective of the transient process, when the photovoltaic output current changes from small to large, the error will increase significantly, while the relative error of the photovoltaic output current changes from large to small, and the relative change is not obvious.

6   Conclusion

In order to accurately analyze the complex dynamic process of DG access to the distribution network and improve the speed and efficiency of parallel simulation, a parallel transient simulation calculation method for active distribution network based on node splitting is proposed in this paper. Relying on the real-time simulation system of active distribution network, adopting the short line decoupling method based on characteristic line, the wave equation of up and down waves for short line is proposed, and the relationship of fluctuation impedance is established to reduce the voltage difference. In the aspect of improving parallel simulation, an iterative method of electrical variables at the segmentation port is established to realize the parallel simulation of active distribution network. After that, the error-causing factors in the parallel simulation are briefly analyzed. Taking an actual 62-node network model as a typical scenario, the transient process in the distribution network loop closing operation is simulated and analyzed. The simulation results show that the proposed parallel simulation method can not only be applied to the various transient process of active distribution networks, but also can improve the simulation accuracy of the transient process, and its effect is better than the conventional serial simulation.

References

1. Chen, J., Crow, M.L.: A variable partitioning strategy for the multirate method in power systems. IEEE Trans. Power Syst. **23**(2), 258–266 (2008)
2. Lelarasmee, E., Ruehli, A.E., Sangiovanni-Vincentelli, A.L.: The waveform relaxation method for time-domain analysis of large scale integrated circuits. IEEE Press (2006)
3. Jalilimarandi, V., Dinavahi, V.: SIMD-based large-scale transient stability simulation on the graphics processing unit. IEEE Trans. Power Syst. **25**(3), 1589–1599 (2010)
4. Li, J., Liu, C.C., Schneider, K.P.: Controlled partitioning of a power network considering real and reactive power balance. IEEE Trans. Smart Grid **1**(3), 261–269 (2010)
5. Louri, A., Kodi, A.K.: An optical interconnection network and a modified snooping protocol for the design of large-scale symmetric multiprocessors (SMPs). IEEE Trans. Parallel Distrib. Syst. **15**(12), 1093–1104 (2004)
6. Sheng, H.Z., Wang, C., Li, B.W., et al.: Multi-timescale active distribution network scheduling considering demand response and user comprehensive satisfaction. IEEE Trans. Ind. Appl. **57**(3), 1995–2005 (2021)
7. Aricò, C., Sinagra, M., Tucciarelli, T.: The MAST-edge centred lumped scheme for the flow simulation in variably saturated heterogeneous porous media. J. Comput. Phys. **231**(4), 1387–1425 (2012)
8. Ruan, H., Gao, H., Liu, Y., et al.: Distributed voltage control in active distribution network considering renewable energy: a novel network partitioning method. IEEE Trans. Power Syst. **35**(6), 4220–4231 (2020)
9. Deese, A.S., Nwankpa, C.O., Member, S.: Utilization of FPAA technology for emulation of multiscale power system dynamics in smart grids. IEEE Trans. Smart Grid **2**(4), 606–614 (2012)
10. Chen, Q., Ren, X., Oliver, J.A.: Identifier-based adaptive neural dynamic surface control for uncertain DC-DC buck converter system with input constraint. Commun. Nonlinear Sci. Numer. Simul. **17**(4), 1871–1883 (2012)
11. Ma, Q., Cui, X., Hu, R.: Experimental study on secondary system grounding of uhv fixed series capacitors via EMI measurement on an experimental platform. IEEE Trans. Power Deliv. **27**(4), 2374–2381 (2012)

Numerical Simulation of Acoustic Test Rake Shape Design Optimization

Yehui Chen[✉], Zhibo Zhang, and Anni Wang

AECC Shenyang Engine Design and Research Institute, Shengyang 110015, China
livein2850301@126.com

Abstract. The shape design of the acoustic test rake was optimized by numerical simulation. Firstly, the influence of acoustic test rake on engine inlet flow field was studied. Then combined with the two-dimensional profile optimization design and three-dimensional flow field verification method, the shape design of the acoustic test rake was optimized to reduce the influence of the acoustic test rake on the flow field at engine inlet. The results show that: The test rake wake has a significant influence on the non-uniformity of the inlet flow field at engine inlet. After optimization, total pressure and static pressure non-uniformity of the engine inlet are significantly reduced. The non-uniformity of total pressure is reduced from 5.5% to 2.1%, and that of static pressure is reduced from 0.7% to 0.5%.

Keywords: Aero-engine · Test rake · Total-pressure loss · Numerical simulation · Optimization design

1 Introduction

A continuously rotating rake with radial microphones is developed to measure the inlet and exhaust duct modes on aero-engine [1]. The acoustic test rake is installed in the inlet flow tube of the engine for noise test [2]. During the test, the acoustic test rake can rotate at a certain speed to measure the inlet noise distribution. Placing the acoustic test rake at the inlet flow tube of the engine will have an influence on the flow field of the engine inlet. Therefore, it is necessary to clarify the mechanism of the influence of the test rake on the inlet flow field of the engine, and the design of the test rake shape should be optimized to reduce the influence on the flow field of the engine. A continuously rotating rake with radially distributed microphones was developed for use on the Honeywell TFE731-60 as an acoustic diagnostic tool [3]. Rotating rake technology had been developed and used successfully on model scale1 and a dedicated test bed [4]. Since the study on acoustic test rake at the engine inlet is the first time at home, the study on shape design of the acoustic test rake shape design is very few. Therefore, it is very necessary to analyse the influence of the test rake on the engine inlet flow field, and clarify the optimization design method of the test rake shape [5, 6].

In view of less work related to the shape design of acoustic test rake, the shape optimization design uses the method of blade optimization design for reference [7–10].

W. Fan et al. (Eds.): AsiaSim 2022, CCIS 1713, pp. 375–387, 2022.
https://doi.org/10.1007/978-981-19-9195-0_30

Firstly, the two-dimensional flow field of the original profile is calculated and analyzed, and the profile with low total pressure loss under different inlet Mach numbers is designed. Then, through the three-dimensional stacking, the three-dimensional modeling of the test rake is completed, and the optimization effect is verified through the comparative analysis of the three-dimensional flow field calculation.

In this paper, the numerical simulation method is used to carry out the detailed analysis of the engine inlet flow field when the acoustic test rake is installed, and the law of total pressure non-uniformity variation are studied. The shape design optimization methods of acoustic rake with two-dimensional profile optimization and three-dimensional flow field analysis are proposed.

2   Numerical Model and Calculation Method

2.1   Calculation Mode

The acoustic test rake is placed in the flow tube at the inlet of the engine. Figure 1 shows the relative position relationship between the acoustic test rake and the flow tube at the inlet of the engine and the three-dimensional calculation model of the test rake. Figure 2 shows the two-dimensional profile of the typical section of the test rake.

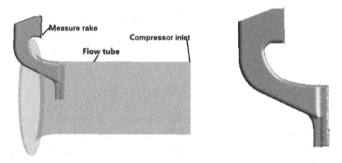

Fig. 1. Geometric model of acoustic test rake

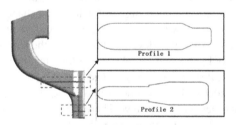

Fig. 2. Typical profiles of acoustic test rake

Fig. 3. Example of 2D grid

2.2 Grid Division

Iceam is used to mesh the computing domain. The two-dimensional grid is a structured grid and the grid diagram is shown in Fig. 3.

Considering the complexity of the acoustic test rake model, the unstructured grid is adopted for the three-dimensional grid. The grid near the wall is densified, and the thickness of the first boundary layer is 0.05 mm. The total number of grids is about 64×10^5, and the quality of the final grid is above 0.3. The schematic diagram of the calculation grid is shown in Fig. 4.

Fig. 4. Schematic diagram of 3D computing grid

2.3 Boundary Conditions and Numerical Calculation Methods

Commercial software CFD ANSYS CFX is used for numerical simulation analysis.

Two-dimensional calculation boundary conditions setting: total temperature at the inlet is 288.15 K, the total pressure is 101325 Pa, and the static pressure is given at the outlet (Table 1).

Three-dimensional calculation boundary conditions setting: the whole calculation domain is set as free flow opening. The flow tube and test rake are set as wall, and the outlet of the flow tube is set as outlet. In order to approach the real test conditions, both models give the total inlet temperature of 288.15 K, the total inlet pressure of 101325 Pa, and the maximum flow at the inlet of the engine.

Considering the convergence and reliability of the calculation results, k- ω model is selected for calculation. The inlet flow is monitored during the simulation, and the simulated medium is ideal air.

Table 1. Boundary conditions for 2D profile calculation

Inlet total temperature (K)	Inlet total pressure (Pa)	Outlet static pressure (Pa)
288.15	101325	90000
		92500
		95000
		97500
		100000

3 Numerical Calculation Results and Analysis

3.1 Influence Analysis of Original Test Rake on Engine Inlet Flow Field

The flow field parameters of the acoustic test rake installed at the inlet were compared and analyzed. Firstly, the analysis section along the axial direction in the flow tube is given. D is the diameter of the flow tube, 0d cross section is located at the inlet of the straight section, where 0.25D cross section is the flow measurement section, and 1.6d cross section is the inlet of the engine (Fig. 5).

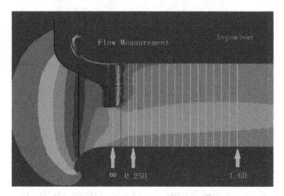

Fig. 5. Axial analysis section in flow tube

Figure 6 shows the change of the total pressure distribution in the flow tube along the flow field when the acoustic test rake is installed. It can be seen from the figure, the changing process of the wake along the flow field can be clearly identified. From the test rake to the engine inlet, the influence range of the test rake wake is gradually increasing.

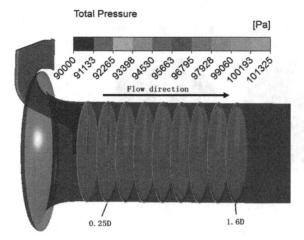

Fig. 6. Distribution of total pressure along flow direction

Figure 7 shows the total pressure non-uniformity of the flow tube in the meridian. It can be seen from the figure that due to the influence of the test rake, the total pressure distribution on the meridional surface is uneven, that the non-uniformity still exists at the outlet of the flow tube.

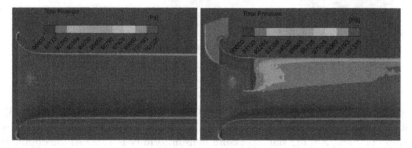

Fig. 7. Effect of test rake on total pressure distribution in meridional plane

Figure 8 and Fig. 9 show the effect of acoustic test rake on the total pressure distribution of flow measurement section and engine inlet section. It can be seen from the figure that for the flow measurement section and engine inlet section, the total pressure of more than 45% blade height is affected. The total pressure loss is bigger near the casing.

380   Y. Chen et al.

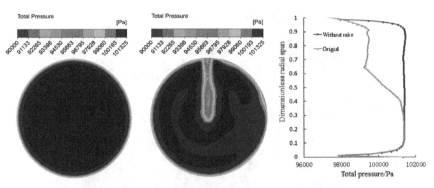

Fig. 8. Total pressure distribution of flow measurement section

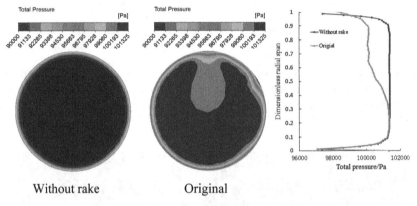

Without rake   Original

Fig. 9. Total pressure distribution at engine inlet

Table 2 shows the influence of the test rake on the flow field non-uniformity of the engine. At the flow measurement section of the flow tube, the total pressure non-uniformity is 6.8%, and the static pressure non-uniformity is 15%. At the engine inlet section, the total pressure non-uniformity is 5.5%, and the static pressure non-uniformity is 0.7%. It can be seen that the test rake has a significant influence on the flow field non-uniformity, so it is necessary to carry out shape design optimization to reduce the influence of the test rake on the flow field.

Table 2. Flow field non-uniformity analysis

Cross section	Total pressure non-uniformity	Static pressure non-uniformity
Flow measurement section	6.8%	15%
Engine inlet section	5.5%	0.7%

3.2   Shape Design Optimization of Acoustic Test Rake

2D Profile Design Optimization. Figure 10 shows the original and optimized geometric comparison of the typical profiles of the acoustic test rake. The blue solid line is the original profile, and the red line is the optimized profile. It can be seen from the figure that compared with the original profile, the axial length of the optimized profile remains unchanged and the thickness decreases.

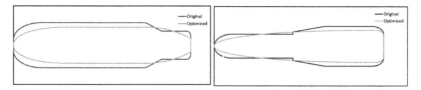

Fig. 10. Geometric comparison of profile optimization

Figure 11 shows the total pressure nephogram and width of wake influence area of the engine inlet between the original and optimized states of profile 1 at the maximum Mach number. It can be seen from the figure that the width of wake influence area of the engine inlet is significantly narrowed after the profile optimization. The width of wake influence area at the engine inlet is reduced from 0.12 m to 0.096 m.

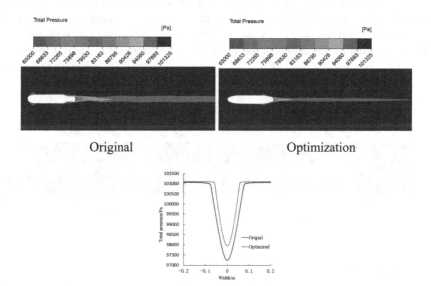

Fig. 11. The influence of profile optimization on total pressure nephogram

Figure 12 shows the change of total pressure recovery coefficient with inlet Mach number after optimization of profile 1. It can be seen from the figure, for the original profile, the total pressure recovery coefficient decreases with the increase of inlet speed.

At the maximum flow rate state, the total pressure recovery coefficient is about 0.982. After optimization, the total pressure recovery coefficient increases from 0.982 to 0.986 at the maximum Mach number state. The total pressure loss is significantly reduced through profile optimization.

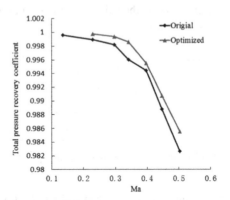

Fig. 12. Total pressure recovery coefficient under different flow conditions

Figure 13 shows the total pressure nephogram and width of wake influence area of the engine inlet between the original and optimized states of profile 2 at the maximum Mach number. It can be seen from the figure that the width of wake influence area of the engine inlet after optimization is narrowed and the low total pressure area at the tail of the profile is reduced. The width of wake influence area of the engine inlet is significantly narrowed after optimization, which the width is reduced from 0.12 m to 0.09 m.

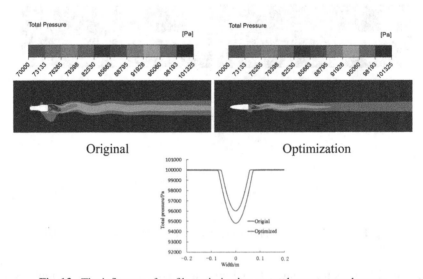

Fig. 13. The influence of profile optimization on total pressure nephogram

Figure 14 shows the change of total pressure recovery coefficient with inlet Mach number after optimization of profile 2. It can be seen from the figure that for the original profile, the total pressure recovery coefficient decreases with the increase of inlet speed. Under the condition of the maximum flow rate state, the total pressure recovery coefficient is about 0.975. After optimization, the total pressure recovery coefficient increases from 0.975 to 0.983, which the total pressure loss is obviously reduced by profile optimization.

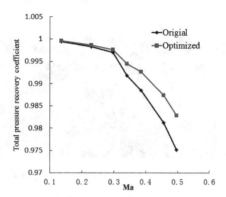

Fig. 14. Total pressure recovery coefficient under different flow conditions

Comparison and Analysis of Three-Dimensional Flow Field. Figure 15 shows the comparison of three-dimensional modeling of the test rake before and after optimization. Figure 16 shows the influence of profile optimization on the distribution of total pressure in the flow tube. It can be seen from the figure that the total pressure loss of the test rake is smaller after optimization and the low-pressure area caused by test rake is smaller. The wake range is smaller and the mixing distance of the wake is significantly shorter.

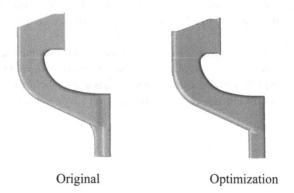

Original   Optimization

Fig. 15. 3D modeling of test rake before and after optimization

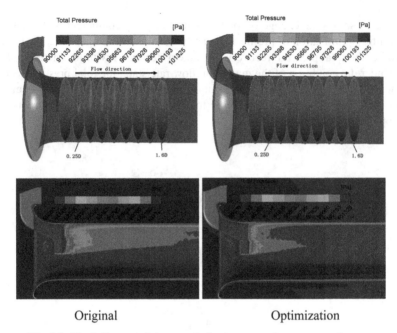

Original   Optimization

Fig. 16. The influence of shape optimization on total pressure nephogram

Figure 17 shows the total pressure distribution of the engine inlet. It can be seen from the figure that the low-pressure area is reduced, and the total pressure loss is reduced. That means the influence of test rake on the total pressure field of engine inlet is reduced after optimization.

Figure 18 shows the comparison of the radial distribution of the total pressure at engine inlet. It can be seen from the figure that, under the condition of original test rake, the total pressure at more than 80% of the radial blade height is significantly affected, and there is a loss for the total pressure at 20% to 60% of the blade height. After optimization, the total pressure near the casing is basically not affected. There is still a total pressure loss at 20%–45% blade height, but the loss is relatively reduced compared with that of original test rake.

Figure 19 shows the static pressure non-uniformity of the flow measurement section and the engine inlet section after optimization. It can be seen from the figure that the static pressure distribution of the flow measurement section is uneven before optimization, and the non-uniformity of the section becomes smaller after shape optimization.

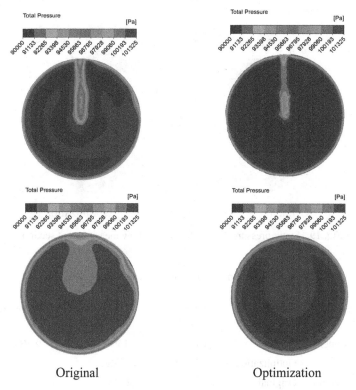

Original   Optimization

Fig. 17. Total pressure distribution of engine inlet

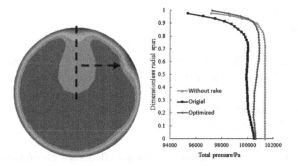

Fig. 18. Total pressure radial distribution at engine inlet

Table 3 shows the total pressure, static pressure non-uniformity and flow deviation at the flow measurement section. For the original shape, the total pressure non-uniformity is 6.8%, the static pressure non-uniformity is 15%, and the flow deviation is 0.8%. After the optimization, the non-uniformity of total pressure is 3.2%, the non-uniformity of static pressure is 3.5%, and flow deviation is 0.3%, which meets the requirements of flow test.

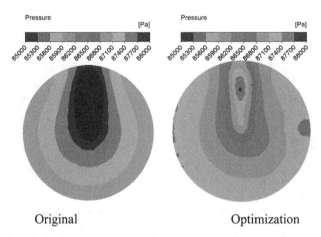

Original   Optimization

Fig. 19. Static pressure non-uniformity of flow measurement section

Table 3. Flow field non-uniformity analysis after optimization

Cross section	Scheme	Total pressure non-uniformity	Static pressure non-uniformity	Flow deviation
Flow measurement section	Original	6.8%	15%	0.8%
	Optimization	3.2%	3.5%	0.3%
Engine inlet section	Original	5.5%	0.7%	0.11%
	Optimization	2.1%	0.3%	0.08%

4 Conclusion

In order to reduce the influence of the acoustic test rake on the inlet flow field of engine, the shape design optimization of the test rake was carried out. From the two-dimensional calculation profile optimization to the three-dimensional calculation verification, the aerodynamic shape optimization of the test rake was realized, and the following conclusions can be drawn:

1. The acoustic test rake installed has a significant effect on the uniformity of the engine inlet flow field. The total pressure non-uniformity is 5.5%, and the static pressure non-uniformity is 0.7% under the condition of original test rake;
2. Through the three-dimensional flow field analysis, the shape optimization significantly reduces the influence of the acoustic test rake on the flow field uniformity. The total pressure non-uniformity at engine inlet is reduced from 5.5% to 2.1%, and the static pressure non-uniformity is reduced from 0.7% to 0.3%;
3. The total pressure non-uniformity at flow measurement section is reduced from 6.8% to 3.2%, and the static pressure non-uniformity is reduced from 15% to 3.5%.

References

1. He, Y., Shi, J., Xing, Y., et al.: Development of a rake for measuring intake case of an aircraft. Gas Turbine Test Res. **21**(3), 59–62 (2008). (in Chinese)
2. Bui, T.T., Oates, D.L.: Design and evaluation of a new boundary-layer measurement rake for flight testing. NASA TM-2000-209014 (2000)
3. Heidelberg, L.J., Hall, D.G.: Inlet acoustic mode measurements using a continuously rotating rake. J. Aircr. **32**, 761–767 (1995)
4. Heidelberg, L.J., Hall, D.G., Bridges, J.E., Nallasamy, M.: A unique ducted fan test bed for active noise control and aero-acoustics research. NASA TM–107213, AIAA 96–1740, May 1996
5. Askari, S., Shojaeefard, M.H.: Numerical simulation of flow over an airfoil with a cross flow fan as a lift generating member in a new aircraft model. Aircr. Eng. Aerosp. Technol. Int. J. **81**(1), 56–64 (2009)
6. Askari, S., Shojaeefard, M.H.: Shape optimization of the airfoil comprising a cross flow fan. Aircr. Eng. Aerosp. Technol. Int. J. **81**(5), 407–415 (2009)
7. Castonguay, P.: Effect of shape parameterization on aerodynamic shape optimization. In: 45th AIAA Aerospace Science Meeting and Exhibit, Reno, Nevada. The American Institution of Aeronautics and Astronautics, Inc. (2007)
8. Hicks, R.M., Henne, P.A.: Wing design by numerical optimization. J. Aircr. **15**(7), 407–412 (1978)
9. Taylor, W., Leylek, J.H., Sommer, R.G.: IC engine intake regine design modification for loss reduction based on CFD methods. SAE981026 (1998)
10. Quinn, R.D., Gong, L.: In-flight boundary-layer measurements on a hollow cylinder at Mach number of 3.0[R]. NASA TP-1980-1764 (1980)

Gain-Scheduling Event-Triggered Load Frequency Control in Power Systems Under Delay Attacks

Yajian Zhang[1], Chen Peng[2(✉)], and Jia Li[1,2]

[1] School of Mechatronic Engineering and Automation and the Shanghai Key Laboratory of Power Station Automation Technology, Shanghai University, Shanghai 200444, China
{zhang_ya_jian,c.peng}@shu.edu.cn
[2] Bozhou Power Supply Company, State Grid Electric Power Co., Ltd., Anhui 236000, China

Abstract. This paper proposes a resilient event-triggered load frequency control scheme based on gain scheduling against the time delay attacks. Based on a series of possible delay intervals representing different attack intensities, the corresponding controller gains are obtained by constructing a Lyapunov functional. Then, under the switching control framework, the controller switching criteria are strictly derived. The control center can switch the most matched event trigger control parameters to the control loop according to the actual transmission delays. Numerical simulation results show that compared with the existing event triggered control schemes, the proposed method reduces the conservatism and improves the control performance with less additional search time consumption.

Keywords: Load frequency control · Delay attack · Adaptive event-triggered control

1 Introduction

Load frequency control (LFC) is a very important application in modern power systems to guarantee real-time power supply-and-demand balance at the rated frequency point [1]. With the large-scale integration of communication networks into the power systems, the threats from malicious cyber attacks to the frequency stability cannot be ignored. Especially, the time-delay attack, as a kind of typical cyber attacks, can inject numerous invalid data packets to seize the limited network bandwidth and then significantly prolong the transmission delays of normal data packets [2]. The control performance of LFC systems may be

Supported by the National Natural Science Foundation of China under Grants 61833011, 62173218, 62103254, and the International Corporation Project of Shanghai Science and Technology Commission under Grant 21190780300.

degenerated or even become unstable under time-delay attacks. Consequently, the event-triggered control (ETC) aiming at reducing the dependence on network has been attracted lots of attention in recent years.

The technical characteristics of ETC is that the data interaction can be triggered only when the control performance has been degenerated to a preset threshold [3]. The analytic relations among the controller gain, triggered threshold and maximal permitted attack duration have been well discussed in [4]. However, the triggered threshold always remained unchanged during the control process. One drawback of such a fixed ETC method is that there still may be frequent packet transmission when the LFC system has been well controlled. Hence, to further reduce the network occupation, some researchers have proposed the adaptive ETC schemes where the triggered threshold can be flexibly adjusted with the current control performance [5,6]. However, there are not a specific upper bound of the event-triggered threshold, which means it will be difficult for the control center to timely monitor the power systems when the LFC systems have been well controlled. In addition, it should also be pointed out that in the existing adaptive ETC schemes, the controller gain still needs to stabilize the maximum attack intensity scenario. Therefore there exists a design conservatism problem and the control performance under low-level attack intensities may be sacrificed.

Actually, the time-delay attacks are stochastic and energy-limited [7]-[8]. It means that the transmission delays of normal data packets randomly change with the attack intensities. Motivated by this, this paper proposes a total adaptive ETC-LFC scheme. First, the ETC parameters corresponding to each preset transmission delay range representing different time-delay attack intensities are well designed by constructing the Lyapunov functional. Second, the switching criteria to select the most matching ETC parameters when attack intensity changes are discussed based on the switching control theory. Compared with the traditional ETC schemes with fixed controller gains, our proposed method can effectively improve the dynamic performance under low-level attack intensities with less design conservatism.

2 Proposed Methodology

2.1 Overall Framework

To reduce the dependence of LFC system on networks, the proposed ETC-LFC scheme contain two modules as follows:

- *ETC parameter set*: In this module, G attack intensities are firstly preset where the transmission delay range under each attack intensity is $(\underline{\tau}_g, \overline{\tau}_g]$. Then the ETC parameters in the set are designed to stabilize the different attack intensities.
- *Switching module*: Too frequent attack intensity changes may cause the state trajectory divergence [9]. Therefore, the function of this module is to determine when to switch which ETC parameters into the loop to stabilize the current attack scenario.

2.2   ETC Parameter Under Arbitrary Attack Intensities

The dynamics of a LFC system can be described by [10]

$$\begin{cases} \dot{x}(t) = Ax(t) + Bu(t) + Ew(t) \\ y(t) = Cx(t) \end{cases} \tag{1}$$

where $x(t) = [x_1^{\mathrm{T}}(t), x_2^{\mathrm{T}}(t), ..., x_N^{\mathrm{T}}(t)]^{\mathrm{T}}$, $x_n = [\Delta f_n, \Delta P_{\text{tien}}, \Delta P_{\text{mn}}, \Delta P_{\text{vn}}, \int ACE_n]^{\mathrm{T}}$, $ACE_n = \beta_n \Delta f_n + \Delta P_{\text{tien}}$ is the area control error; β_n is the frequency bias factor in Area n; Δf_n and ΔP_{tien} are the frequency and tie-line power deviations in Area n, respectively; ΔP_{mn} is the generator's mechanical power output increment; ΔP_{vn} is the valve position difference of the turbine; $u(t) = [u_1(t), u_2(t), ..., u_N(t)]^{\mathrm{T}}$, $A = [A_{nl}]_{N \times N}$, $B = \text{diag}\{B_n\}$, $E = \text{diag}\{E_n\}$, $C = \text{diag}\{C_n\}$,

$$A_{nn} = \begin{bmatrix} -\frac{D_n}{M_n} & -\frac{1}{M_n} & \frac{1}{M_n} & 0 & 0 \\ 2\pi \sum_{l=1,l\neq n}^{N} L_{nl} & 0 & 0 & 0 & 0 \\ 0 & 0 & -\frac{1}{T_{tn}} & \frac{1}{T_{tq}} & 0 \\ -\frac{1}{T_{gn}R_n} & 0 & 0 & -\frac{1}{T_{gn}} & 0 \\ \beta_n & 1 & 0 & 0 & 0 \end{bmatrix},$$

$$A_{nl,l\neq n} = \begin{bmatrix} 0 & 0 & 0 & 0 & 0 \\ -2\pi L_{nl} & 0 & 0 & 0 & 0 \\ 0 & 0 & 0 & 0 & 0 \\ 0 & 0 & 0 & 0 & 0 \\ 0 & 0 & 0 & 0 & 0 \end{bmatrix}, \quad B_n = \begin{bmatrix} 0 \\ 0 \\ 0 \\ \frac{1}{T_{gn}} \\ 0 \end{bmatrix},$$

$$E_n = \begin{bmatrix} -\frac{1}{M_n} \\ 0 \\ 0 \\ 0 \\ 0 \end{bmatrix}, \quad C_n = \begin{bmatrix} \beta_n & 0 \\ 1 & 0 \\ 0 & 0 \\ 0 & 0 \\ 0 & 1 \end{bmatrix}^{\mathrm{T}}.$$

Let K_g denote the controller gain corresponding to the g-th attack intensity with delay range as $(\tau_g, \bar{\tau}_g]$. The event-triggered communication scheme is

$$t_{k+1}h = t_kh + \min_{l\in N}\{lh|e^{\mathrm{T}}(i_lh)\Phi e(i_lh) \geq \sigma(t_kh)x(t_kh)\Phi x(t_kh)\} \tag{2}$$

where h is the sampling period, $e(i_lh) = x(i_lh) - x(t_kh)$ is the error between the system state at the current sampling instant $t = i_lh$ and the previous triggered instant $t = t_kh$, $\Phi_g > 0$ is a symmetric weighting matrix, triggered threshold $\sigma(t_kh)$ is dynamically adjusted with the control performance, i.e.,

$$\sigma(t_{k+1}h) = \begin{cases} \frac{2\alpha}{\pi}\text{atan}\left(\frac{\|y(t_kh)\|}{\|y(t_{k+1}h)\|-\|y(t_kh)\|}\right)\sigma_m + \beta\sigma_m, \|y(t_{k+1}h)\| \geq \|y(t_kh)\| \\ -\frac{2\alpha}{\pi}\text{atan}\left(\frac{\|y(t_{k+1}h)\|-\|y(t_kh)\|}{\|y(t_kh)\|}\right)\sigma_m + \beta\sigma_m, \text{otherwise} \end{cases} \tag{3}$$

where σ_m is the maximum triggered threshold, α and β are the weighting coefficients satisfying $\alpha, \beta > 0$ and $\alpha + \beta = 1$.

Remark 1: When the performance of the LFC system is degenerated, $\sigma(t_k h)$ is approached to the lower bound $\beta\sigma_m$, i.e., the LFC system has a higher event-triggered frequency; Conversely, when the LFC system has been well controlled, $\sigma(t_k h)$ is approached to the upper bound σ_m, i.e., the LFC system has a lower event-triggered frequency. Therefore, in this paper, the control gain corresponding to the g-th attack intensity is required to guarantee asymptotic stability under the upper bound σ_m.

Define $\Omega_g = \cup\Omega_{l,g} = \cup(i_l h + \tau_{i_{l,g}}, i_l h + h + \tau_{i_{l+1,g}}]$, $l = 1, 2, ..., t_{k+1} - t_k - 1$ as the interval between two neighbouring triggered instants, where $\{\tau_{i_{l,g}}, \tau_{i_{l+1,g}}\} \in (\underline{\tau}_g, \overline{\tau}_g]$ are the transmission delays between the sensors and control center at each sampling instant within the two neighbouring triggered instants. In addition, let $\eta_g(t) = t - i_l h \in (\underline{\eta}_g, \overline{\eta}_g] \subset (\underline{\tau}_g, h + \overline{\tau}_g]$. Similar to [11], the control input is given by

$$u(t) = K_g x(t_k h) = K_g(x(t - \eta_g(t)) - e(i_l h)), t \in \Omega_{l,g} \qquad (4)$$

Theorem 1: Give scalars $\lambda, \gamma, \rho, \sigma_m > 0$, the LFC system is exponentially stable with $\gamma - H_\infty$ performance if there exist symmetric matrices $P_g, Q_{1g}, Q_{2g}, W_{1g}, W_{2g}, R_g, X_g, \Phi_g$ satisfying (5)-(7).

$$P_g, Q_{1g}, Q_{2g}, W_{1g}, W_{2g}, R_g, X_g, \Phi_g \geq 0 \qquad (5)$$

$$\begin{bmatrix} \Pi_{11} & \Pi_{12} \\ \Pi_{12}^{\mathrm{T}} & \Pi_{22} \end{bmatrix} \leq 0 \qquad (6)$$

$$\begin{bmatrix} W_{2g} & X_g \\ X_g^{\mathrm{T}} & W_{2g} \end{bmatrix} \geq 0 \qquad (7)$$

where $\underline{\varphi}_g = \dfrac{\lambda\underline{\eta}_g}{e^{\lambda\underline{\eta}_g} - 1}$, $\overline{\varphi}_g = \dfrac{\lambda(\overline{\eta}_g - \underline{\eta}_g)}{e^{\lambda\overline{\eta}_g} - e^{\lambda\underline{\eta}_g}}$, $\Pi_{11}(1,1) = \lambda P_g + A^{\mathrm{T}} P_g + P_g A + Q_{1g} + Q_{2g} - \underline{\varphi}_g W_{1g} - \frac{\pi^2}{4} R_g$, $\Pi_{11}(1,2) = \underline{\varphi}_g W_{1g}$, $\Pi_{11}(1,4) = P_g B K_g + \frac{\pi^2}{4} R_g$, $\Pi_{11}(1,5) = -P_g B K_g$, $\Pi_{11}(1,6) = P_g E$, $\Pi_{11}(2,2) = -e^{-\lambda\underline{\eta}_g} Q_{1g} - \underline{\varphi}_g W_{1g} - \overline{\varphi}_g W_{2g}$, $\Pi_{11}(2,3) = \overline{\varphi}_g X_g^{\mathrm{T}}$, $\Pi_{11}(2,4) = \overline{\varphi}_g(X_g^{\mathrm{T}} - W_{2g})$, $\Pi_{11}(3,3) = -e^{-\lambda\overline{\eta}_g} Q_{2g} - \overline{\varphi}_g W_{2g}$, $\Pi_{11}(3,4) = -\overline{\varphi}_g(-W_{2g} + X_g)$, $\Pi_{11}(4,4) = -\overline{\varphi}_g(2W_{2g} - X_g - X_g^{\mathrm{T}}) - \frac{\pi^2}{4} R_g + \sigma_m \Phi_g$, $\Pi_{11}(5,5) = -\Phi_g$, $\Pi_{11}(6,6) = -\gamma^2 I$,

$$\Pi_{12} = \begin{bmatrix} P_g A & 0 & 0 & P_g B K_g & -P_g B K_g & P_g E \\ C & 0 & 0 & 0 & 0 & 0 \end{bmatrix}^{\mathrm{T}},$$

$$\Pi_{22} = \begin{bmatrix} \rho^2 \left[\overline{\eta}_g^2(W_{1g} + R_g) + (\overline{\eta}_g - \underline{\eta}_g)^2 W_{2g}\right] - 2\rho P_g & 0 \\ 0 & -I \end{bmatrix}.$$

Proof: Choose the following Lyapunov-Krasovskii functional

$$V_g(t) = V_{g,1}(t) + V_{g,2}(t) + V_{g,3}(t) + V_{g,4}(t) + V_{g,5}(t) + V_{g,6}(t) \qquad (8)$$

$$V_{g,1}(t) = x^{\mathrm{T}}(t) P_g x(t) \qquad (9)$$

$$V_{g,2}(t) = \int_{t-\underline{\eta}_g}^{t} x^{\mathrm{T}}(s)e^{\lambda(s-t)}Q_{1g}x(s)\mathrm{d}s \tag{10}$$

$$V_{g,3}(t) = \int_{t-\overline{\eta}_g}^{t} x^{\mathrm{T}}(s)e^{\lambda(s-t)}Q_{2g}x(s)\mathrm{d}s \tag{11}$$

$$V_{g,4}(t) = \underline{\eta}_g \int_{-\underline{\eta}_g}^{0} \int_{t+r}^{t} \dot{x}^{\mathrm{T}}(s)e^{\lambda(s-t)}W_{1g}\dot{x}(s)\mathrm{d}s\mathrm{d}r \tag{12}$$

$$V_{g,5}(t) = (\overline{\eta}_g - \underline{\eta}_g) \int_{-\overline{\eta}_g}^{-\underline{\eta}_g} \int_{t+r}^{t} \dot{x}^{\mathrm{T}}(s)e^{\lambda(s-t)}W_{2g}\dot{x}(s)\mathrm{d}s\mathrm{d}r \tag{13}$$

$$V_{g,6}(t) = \overline{\eta}_g{}^2 \int_{i_l h}^{t} \dot{x}^{\mathrm{T}}(s)e^{\lambda(s-t)}R_g\dot{x}(s)\mathrm{d}s \tag{14}$$

$$-\frac{\pi^2}{4} \int_{i_l h}^{t} [x(s)-x(i_l h)]^{\mathrm{T}} e^{\lambda(s-t)} R_g \left[x(s)-x(i_l h)\right] \mathrm{d}s \tag{15}$$

By adopting reciprocally approach [12], if there exists matrix X_g satisfying (7), we have

$$\begin{aligned}
\dot{V}_g(t) &< V_g(t) + \lambda V_g(t) \\
&\leq \xi^{\mathrm{T}}(t)(\varPi_{11} - \varPi_{12}\varPi_{22}{}^{-1}\varPi_{12}^{\mathrm{T}})\xi(t) \\
&\quad - y^{\mathrm{T}}(t)y(t) + \gamma^2\omega^{\mathrm{T}}(t)\omega(t)
\end{aligned}$$

where $\xi(t) = [x^{\mathrm{T}}(t), x^{\mathrm{T}}(t-\underline{\eta}_g), x^{\mathrm{T}}(t-\overline{\eta}_g), x^{\mathrm{T}}(t-\eta_g(t)), e^{\mathrm{T}}(i_l h), w^{\mathrm{T}}(t)]^{\mathrm{T}}$.

Obviously, if $\varPi_{11} - \varPi_{12}\varPi_{22}{}^{-1}\varPi^{\mathrm{T}}{}_{12} \leq 0$, $\dot{V}_g(t) \leq -y^{\mathrm{T}}(t)y(t) + \gamma^2\omega^{\mathrm{T}}(t)\omega(t)$. Under zero initial condition, we have $\int_0^{+\infty} y^{\mathrm{T}}(t)y(t)\mathrm{d}t \leq \int_0^{+\infty} \gamma^2\omega^{\mathrm{T}}(t)\omega(t)\mathrm{d}t$, i.e., $\|y(t)\| \leq \|\omega(t)\|$. Then under the condition that $\omega(t) = 0$, there exists a scalar $\delta > 0$ making $\dot{V}_g(t, x(t)) \leq \delta e^{-\lambda t}\|x(t)\|$ hold. That means the LFC system is exponentially stable with an H_∞ performance. Finally, by applying Schur complement lemma, we can obtain (6).

∎

Moreover, let $\varLambda_g = P_g{}^{-1}$, $Y_g = K_g\varLambda_g$, $\tilde{Q}_{1g} = \varLambda_g Q_{1g}\varLambda_g$, $\tilde{Q}_{2g} = \varLambda_g Q_{2g}\varLambda_g$, $\tilde{W}_{1g} = \varLambda_g W_{1g}\varLambda_g$, $\tilde{W}_{2g} = \varLambda_g W_{2g}\varLambda_g$, $\tilde{R}_g = \varLambda_g R_g\varLambda_g$, $\tilde{X}_g = \varLambda_g X_g\varLambda_g$, $\tilde{\varPhi}_g = \varLambda_g \varPhi_g\varLambda_g$. Pre- and post-multiplying both sides of (6) and (7) with $diag\{\varLambda_g, \varLambda_g, \varLambda_g, \varLambda_g, \varLambda_g, I, \varLambda_g, I\}$ and $diag\{\varLambda_g, \varLambda_g\}$, respectively. Then if there exist symmetric positive definite matrices $\varLambda_g$, $\tilde{Q}_{1g}$, $\tilde{Q}_{2g}$, $\tilde{W}_{1g}$, $\tilde{W}_{2g}$, $\tilde{R}_g$, $\tilde{\varPhi}_g$ making (16) and (17) hold, then the control gain is give by $K_g = Y_g\varLambda_g^{-1}$.

$$\begin{bmatrix} \tilde{\varPi}_{11} & \tilde{\varPi}_{12} \\ \tilde{\varPi}_{12}^{\mathrm{T}} & \tilde{\varPi}_{22} \end{bmatrix} \leq 0 \tag{16}$$

$$\begin{bmatrix} \tilde{W}_{2g} & \tilde{X}_g \\ \tilde{X}_g^{\mathrm{T}} & \tilde{W}_{2g} \end{bmatrix} \geq 0 \tag{17}$$

where $\tilde{\Pi}_{11}(1,1) = \lambda\Lambda_g + \Lambda_g A^{\mathrm{T}} + A\Lambda_g + \tilde{Q}_{1g} + \tilde{Q}_{2g} - \underline{\varphi}_g \tilde{W}_{1g} - \frac{\pi^2}{4}\tilde{R}_g$, $\tilde{\Pi}_{11}(1,2) = \underline{\varphi}_g \tilde{W}_{1g}$, $\Pi_{11}(1,4) = BY_g + \frac{\pi^2}{4}\tilde{R}_g$, $\tilde{\Pi}_{11}(1,5) = -BY_g$, $\tilde{\Pi}_{11}(1,6) = E$, $\tilde{\Pi}_{11}(2,2) = -e^{-\lambda\eta_g}\tilde{Q}_{1g} - \underline{\varphi}_g \tilde{W}_{1g} - \overline{\varphi}_g \tilde{W}_{2g}$, $\tilde{\Pi}_{11}(2,3) = \overline{\varphi}_g \tilde{X}_g^{\mathrm{T}}$, $\tilde{\Pi}_{11}(2,4) = \overline{\varphi}_g(\tilde{X}_g^{\mathrm{T}} - \tilde{W}_{2g})$, $\tilde{\Pi}_{11}(3,3) = -e^{-\lambda\overline{\eta}_g}\tilde{Q}_{2g} - \overline{\varphi}_g \tilde{W}_{2g}$, $\tilde{\Pi}_{11}(3,4) = -\overline{\varphi}_g(-\tilde{W}_{2g} + \tilde{X}_g)$, $\tilde{\Pi}_{11}(4,4) = -\overline{\varphi}_g(2\tilde{W}_{2g} - \tilde{X}_g - \tilde{X}_g^{\mathrm{T}}) - \frac{\pi^2}{4}\tilde{R}_g + \sigma_m\tilde{\Phi}]$, $\tilde{\Pi}_{11}(5,5) = -\tilde{\Phi}]$, $\tilde{\Pi}_{11}(6,6) = -\gamma^2 I$,

$$\tilde{\Pi}_{12} = \begin{bmatrix} A\Lambda_g & 0 & 0 & BY_g & -BY_g & E \\ C\Lambda_g & 0 & 0 & 0 & 0 & 0 \end{bmatrix}^{\mathrm{T}},$$

$$\tilde{\Pi}_{22} = \begin{bmatrix} \rho^2\left[\underline{\eta}_g{}^2\tilde{W}_{1g} + \overline{\eta}_g{}^2\tilde{R}_g + (\overline{\eta}_g - \underline{\eta}_g)^2\tilde{W}_{2g}\right] - 2\rho\tilde{P}_g & 0 \\ 0 & -I \end{bmatrix}.$$

2.3 Switching Criteria for ETC Parameters

Theorem 2: The LFC system is exponentially stable when the time-delay attack intensities randomly change within $\{g, g'\}$, if the following two conditions hold.

1) There exist matrices $\{\tilde{X}_g, \tilde{Q}_{1g}, \tilde{Q}_{2g}, \tilde{W}_{1g}, \tilde{W}_{2g}, \tilde{R}_g\}$ and $\{\tilde{X}_g', \tilde{Q}_{1g'}, \tilde{Q}_{2g'}, \tilde{W}_{1g'}, \tilde{W}_{2g'}, \tilde{R}_{g'}\}$ satisfying (8) and (9) and scalars $\vartheta_{gg'}, \vartheta_{g'g} \geq 1$ satisfying (18) and (19);

2) The switching frequency between attack intensities g, g' (defined as $f_{gg'}$) is less than $f_{gg'}^{\max} = \dfrac{2\lambda}{\ln(\vartheta_{gg'}\vartheta_{g'g})}$.

$$\{\tilde{Q}_{1g}, \tilde{Q}_{2g}, \tilde{W}_{1g}, \tilde{W}_{2g}, \tilde{R}_g\} \geq \vartheta_{g'g}\{\tilde{Q}_{1g'}, \tilde{Q}_{2g'}, \tilde{W}_{1g'}, \tilde{W}_{2g'}, \tilde{R}_{g'}\} \qquad (18)$$

$$\{\tilde{Q}_{1g'}, \tilde{Q}_{2g'}, \tilde{W}_{1g'}, \tilde{W}_{2g'}, \tilde{R}_{g'}\} \geq \vartheta_{gg'}\{\tilde{Q}_{1g}, \tilde{Q}_{2g}, \tilde{W}_{1g}, \tilde{W}_{2g}, \tilde{R}_g\} \qquad (19)$$

Proof: Assume that time-delay attack intensities changes k times within $\{g, g'\}$ during the time interval $[T_0, t]$. Then the relation between k and switching frequency $f_{gg'}$ satisfies $k = \lfloor f_{gg'}(t - T_0)\rfloor \leq f_{gg'}(t - T_0)$. There exist the two following possible cases.

Case 1: The final attack intensity is as same as the initial one. Suppose that the initial and final attack intensities are g and the attack intensities change at time $t = T_1, T_2, ..., T_k$. According to (10), (11), we have

$$\begin{aligned} V_g(t)|_{t\in(T_k,t)} &< e^{-\lambda(t-T_k)} V_g(T_k{}^+)|_{t\in(T_k,t)} \\ &\leq \delta_{g'g}e^{-\lambda(t-T_k)} V_{g'}(T_k{}^-)|_{t\in(T_{k-1},T_k)} \\ &< \delta_{g'g}e^{-\lambda(T_k-T_{k-1})}e^{-\lambda(t-T_k)} V_{g'}(T_{k-1}{}^+)|_{t\in(T_{k-1},T_k)} \\ &\leq \delta_{gg'}\delta_{g'g}e^{-\lambda(T_k-T_{k-1})}e^{-\lambda(t-T_k)} V_g(T_{k-1}{}^-)|_{t\in(T_{k-2},T_{k-1})} \\ &\vdots \\ &\leq (\delta_{gg'})^{\frac{k}{2}}(\delta_{g'g})^{\frac{k}{2}}e^{-\lambda(t-T_0)} V_g(T_0)|_{t\in(T_0,T_1)} \\ &\leq e^{\frac{f_{gg'}(t-T_0)}{2}\ln(\delta_{gg'}\delta_{g'g})-\lambda(t-T_0)} V_g(T_0)|_{t\in(T_0,T_1)} \end{aligned}$$

If and only if $f_{gg'} \leq \dfrac{2\lambda}{\ln(\delta_{gg'}\delta_{g'g})}$, the LFC system is exponentially stable.

Case 2: The final attack intensity is different from the initial one. Suppose that the initial and final attack intensities are g and g', respectively. Similar to Case 1, we have

$$
\begin{aligned}
V_g(t)|_{t\in(T_k,t)} &< e^{-\lambda(t-T_k)} V_g(T_k^+)|_{t\in(T_k,t)} \\
&\leq \delta_{g'g} e^{-\lambda(t-T_k)} V_{g'}(T_k^-)|_{t\in(T_{k-1},T_k)} \\
&< \delta_{g'g} e^{-\lambda(T_k-T_{k-1})} e^{-\lambda(t-T_k)} V_{g'}(T_{k-1}^+)|_{t\in(T_{k-1},T_k)} \\
&\leq \delta_{gg'}\delta_{g'g} e^{-\lambda(T_k-T_{k-1})} e^{-\lambda(t-T_k)} V_g(T_{k-1}^-)|_{t\in(T_{k-2},T_{k-1})} \\
&\;\;\vdots \\
&\leq (\delta_{gg'})^{\frac{k+1}{2}} (\delta_{g'g})^{\frac{k-1}{2}} e^{-\lambda(t-T_0)} V_g(T_0)|_{t\in(T_0,T_1)} \\
&\leq \sqrt{\frac{\delta_{gg'}}{\delta_{g'g}}} e^{\frac{f_{gg'}(t-T_0)\ln(\delta_{gg'}\delta_{g'g})-\lambda(t-T_0)}{2}} V_g(T_0)|_{t\in(T_0,T_1)}
\end{aligned}
$$

The same conclusion as Case 1 can be obtained. Overall, the maximal allowable switching frequency satisfies $f_{gg'}^{\max} = \dfrac{2\lambda}{\ln(\vartheta_{gg'}\vartheta_{g'g})}$. ∎

Remark 2: Compared with the traditional ETC schemes with fixed controller gain, our proposed method can flexibly adjust the ETC parameter matched with the actual network quality by adding an additional selection mechanism. The corresponding expense is that an extra lookup latency is introduced. However, we can take this latency into consideration in advance during the ETC scheme design process. In addition, such a lookup latency can be ignored since the power control center usually has the sufficient computing resources.

3 Numerical Results and Discussions

To verify the effectiveness of our proposed method, the two-area LFC system as shown in Fig. 1 is simulated and discussed. The simulation parameters is given in Table 1. The ETC scheme with fixed triggered threshold (fixed ETC) [4] and with adaptive triggered threshold (adaptive ETC) [5] are used as the comparisons. The transmission delays are as shown in Fig. 2. Preset three attack intensities with different transmission delay ranges as 1) (0,20 ms]; 2) (20 ms,40 ms]; 3) (40 ms, 60 ms]. Let $\rho = 10, \lambda = 2, \gamma = 5, \sigma_m = 0.01$. The corresponding controller gains and maximal allowable switching frequency are demonstrated in Table 2.

Table 1. Simulation parameters of power subsystems

Areas	T_{ti}/s	T_{gi}/s	R_i/Hz·p.u.$^{-1}$	β_i/p.u.·s	D_i/p.u.·s	M_i/p.u.·s^2
1	0.3	0.03	2.4	0.0045	0.0083	0.1667
2	0.3	0.03	2.4	0.0045	0.0083	0.1667
$L_{12} = L_{21} = 0.545$ p.u.·rad^{-1}						

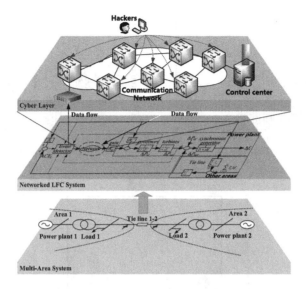

Fig. 1. Transmission delays.

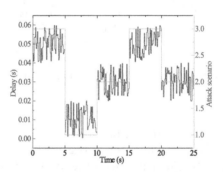

Fig. 2. Transmission delays.

Table 2. Controller gains and allowable attack frequencies

Scenarios	Controller gain K_g
1	$[-2.720, -86.042, -2.783, -0.181, -71.627, 2.481, -87.815, 2.294, 0.275,$ $-65.560; 2.481, -98.682, 2.294, 0.275, -65.508, -2.720, -96.911, -2.782,$ $-0.181, -71.575]$
2	$[-2.755, -14923.825, -2.815, -0.179, -71.819, 2.508, -14925.593, 2.330,$ $0.258, -65.671; 2.508, -14925.629, 2.330, 0.258, -65.671, -2.755,$ $-14923.863, -2.815, -0.179, -71.819]$
3	$[-2.848, -7731.106, -2.887, -0.199, -76.615, 2.564, -7732.847, 2.371, 0.242,$ $-70.279; 2.564, -7732.950, 2.371, 0.242, -70.279, -2.848, -7731.209, -2.887,$ $-0.199, -76.615]$
$f_{gg'}^{\max}$ (Hz)	$f_{12}^{\max} = f_{21}^{\max} = 0.227, f_{13}^{\max} = f_{31}^{\max} = 0.206,\ f_{23}^{\max} = f_{32}^{\max} = 1.242$

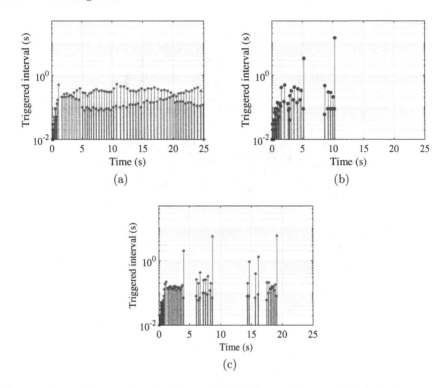

Fig. 3. Triggered interval with (a) fixed ETC [4], (b) adaptive ETC [5], (c) proposed ETC.

Under the 0.1 p.u. and −0.05p.u. load disturbances occurring at $t = 0$s in Areas 1 and 2, the frequency responses and triggered intervals are illustrated in Figs. 3 and 4, respectively. Compared with the fixed ETC method [4], the triggered number with our proposed gain-scheduling ETC method can be reduced by 36.5%, and the integral of absolute error (IAE) of frequency deviations can be reduced by 11.4%. In addition, compared with the adaptive ETC method [5], although the triggered number with our proposed method 26.1% larger, the IAE of frequency deviations can be reduced by 6.85%. The increase of triggered number is because that the LFC system's closed-loop dynamics changes with the time-delay attack intensities. The simulation results show that our proposed method has a better frequency deviation damping performance under the same time-delay attack scenarios. It should also be pointed out that the simulation results show that there exists a certain relation among the control performance, attack intensity division and triggered number, which should be further studied in the future.

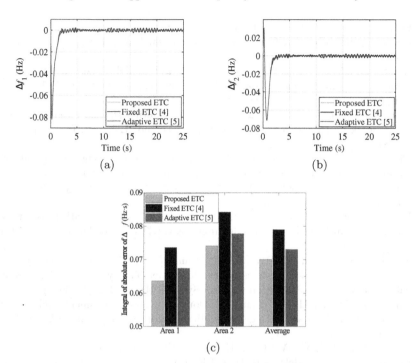

Fig. 4. Frequency responses in (a) Area 1, (b) Area 2 and (c) the corresponding IAEs.

4 Conclusions

This paper proposes a novel resilient event-triggered LFC scheme depending on the time-delay attack intensities. A series of ETC parameters corresponding to the preset transmission delay ranges representing different attack intensities are well designed. Then the switching criteria to determine the most matching ETC parameter in presence of varying attack intensities are strictly deduced under the switching control framework. Numerical results show that the control performance under the low-level attack intensities can be effectively improved. In future, the optimal attack intensity division which compromises the control performance and triggered frequency will be explored in depth.

References

1. Azarbahram, A., Amini, A., Sojoodi, M.: Resilient fixed-order distributed dynamic output feedback load frequency control design for interconnected multi-area power systems. IEEE-CAA J. Autom. **6**(5), 1139–1151 (2019)
2. Zhang, Y., Peng, C., Xie, S., Du, X.: Deterministic network calculus-based H_∞ load frequency control of multiarea power systems under malicious DoS attacks. IEEE Trans. Smart Grid **13**(2), 1139–1151 (2022)

3. Peng, Z., Jiang, Y., Wang, J.: Event-triggered dynamic surface control of an under-actuated autonomous surface vehicle for target enclosing. IEEE Trans. Ind. Electron. **68**(4), 3402–3412 (2021)
4. Liu, S., Luo, W., Wu, L.: Co-design of distributed model-based control and event-triggering scheme for load frequency regulation in smart grids. IEEE Trans. Syst. Man Cybern. Syst. **50**(9), 3311–3319 (2020)
5. Peng, C., Zhang, J., Yan, H.: Adaptive event-triggering H_∞ load frequency control for network-based power systems. IEEE Trans. Ind. Electron. **65**(2), 1685–1694 (2018)
6. Chen, P., Liu, S., Zhang, D., Yu, L.: Adaptive event-triggered decentralized dynamic output feedback control for load frequency regulation of power systems with communication delays. IEEE Trans. Syst. Man Cybern. Syst. https://doi.org/10.1109/TSMC.2021.3129783
7. Wang, C., Zuo, Z., Ding, Z.: Control scheme for LTI systems with Lipschitz non-linearity and unknown time-varying input delay. IET Control Theory Appl. **11**(17), 3191–3195 (2017)
8. Sun, H., Peng, C., Yue, D., Wang, Y.L., Zhang, T.: Resilient load frequency control of cyber-physical power systems under QoS-dependent event-triggered communication. IEEE Trans. Syst. Man Cybern. Syst. **51**(4), 2113–2122 (2021)
9. Cheng, Z., Yue, D., Hu, S., Huang, C., Dou, C., Chen, L.: Resilient load frequency control design: DoS attacks against additional control loop. Int. J. Elect. Power. **115**, 1–12 (2020)
10. Xiahou, K.S., Liu, Y., Wu, Q.H.: Robust load frequency control of power systems against random time-delay attacks. IEEE Trans. Smart Grid **12**(1), 909–911 (2021)
11. Ahmad, I., Ge, X., Han, Q.L.: Decentralized dynamic event-triggered communication and active suspension control of in-wheel motor driven electric vehicles with dynamic damping. IEEE-CAA J. Autom. **8**(5), 971–986 (2021)
12. Park, P., Ko, J.W., Jeong, C.: Reciprocally convex approach to stability of systems with time-varying delays. Automatic **47**(1), 235–238 (2011)

Performance Investigation of Grid-Connected PV System with Dual Current Control Under Unbalanced Voltage Sags

Yu Hu[1]($\boxtimes$), Yue Xia[1], Xiu Liu[2], Peng Zhao[1], Songhuai Du[1], and Juan Su[1]

[1] China Agricultural University, Haidian District, Beijing 100083, China
1149377057@qq.com
[2] Technische Universität Berlin, Straße des 17, 10623 Berlin, Germany

Abstract. Based on the bipolar photovoltaic grid-connected inverter control system, a grid unbalanced fault ride-through control strategy is proposed to support the stable operation of the bipolar photovoltaic grid-connected system. Based on analyzing the operating state of the photovoltaic system when unbalanced voltage sags, a voltage fault ride-through control logic of the bipolar photovoltaic system is proposed to solve the problem of grid-connected current quality during fault. The correctness and usage scenarios of the method are verified by Matlab/Simulink simulation. The results show that for different unbalanced voltage sags, the control method can effectively suppress the grid-connected negative sequence current under the condition of low-voltage sag depth, and keep the grid-connected current within a safe current range.

Keywords: Bipolar photovoltaic system · Fault support · Inverter control

1 Introduction

To achieve its carbon peak and neutrality targets, there is a notable increasing power penetration from PV systems in the global electricity industry during the last decade [1]. Due to its serious power electronics, when the power grid is disturbed or faulty, it is easy to be disconnected from the grid due to a lack of voltage support capability. When a high proportion of photovoltaic systems are connected, a large area of photovoltaic disconnection will cause irreversible damage to the power system. Therefore, the photovoltaic system is required to maintain uninterrupted grid-connected operation during grid failures, and it is necessary to study the failure ride-through technology of photovoltaic grid-connected power generation systems. Improving the voltage support capability of photovoltaic systems during unbalanced voltage sags is of great significance to improving the voltage stability and operational reliability of the entire power system.

When a disturbance or fault occurs in the power grid, formulating a reasonable current reference value can support the stable operation of the photovoltaic system during the voltage fault. In this regard, domestic and foreign scholars have carried out a lot of

© The Author(s), under exclusive license to Springer Nature Singapore Pte Ltd. 2022
W. Fan et al. (Eds.): AsiaSim 2022, CCIS 1713, pp. 399–408, 2022.
https://doi.org/10.1007/978-981-19-9195-0_32

research. The control strategy is divided into two categories according to the coordinate system, which can be divided into two categories: the static coordinate system and the control method under the synchronous rotating coordinate system. In [2–4], the static coordinate system is used for control, the grid-connected current is an alternating current, which is inconvenient to control. It is necessary to use a PR controller with infinite gain at the resonant frequency to track the point currently. In [5–7], the synchronous rotating coordinate system, the grid-connected current is a direct current that is easy to control, which makes the control in the synchronous rotating coordinate system widely used. In addition, some scholars use the method of adding external equipment for support. A method using energy storage batteries is proposed in [8], which stores the excess power of photovoltaic arrays in the form of energy storage. A method using the crowbar circuit is proposed in [9], which uses energy dissipation resistors to dissipate the excess power of the photovoltaic array. A method using a STATCOM device is proposed in [10], which supports the entire system by increasing the grid-side reactive power. The above three methods of adding external devices have yet to be considered in terms of economy and compatibility.

Based on the above analysis of the current situation, this paper takes the quality of grid-connected current as the primary goal, completes phase locking and separation of positive and negative sequence components through DSOGI-PLL, and controls the grid-connected current in a synchronous rotating coordinate system, to ensure that when a voltage fault occurs in the grid, it can be output high-quality grid-connected current. Then through simulation verify the applicable scenarios of deep mining of the method.

The remaining structure of this paper is as follows: In the Sect. 2, the mathematical model of a bipolar grid-connected system under an unbalanced voltage fault is described. In the Sect. 3, the phase-locked loop and control method used is described. In the Sect. 4, the correctness of the control method is verified in Matlab/Simulink. In the Sect. 5, the work done in this paper is summarized.

2 Mathematical Description of the PV Converter Under the Unbalance Condition

The research object in this paper is a bipolar photovoltaic system, and its topology is shown in Fig. 1.

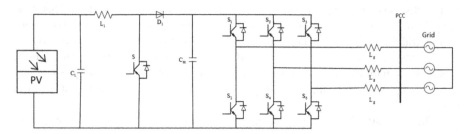

Fig. 1. Bipolar photovoltaic system topology

Among them, the switch tube S implements the MPPT algorithm control in the Boost circuit. The algorithm adopted in this paper is the perturbation and observation method, and the switch tubes S_1-S_6 realize the inverter control for the photovoltaic inverter.

When the power grid is in normal operation, the power emitted by the photovoltaic array is equal to the power emitted by the inverter, and the dc terminal capacitor has no power fluctuation. The power relationship is:

$$P_{pv} = P_e \tag{1}$$

where P_{pv} is the maximum power output by the photovoltaic array, and P_e is the power output by the inverter.

When a voltage sag fault occurs in the power grid, since the current of the grid-side inductance L_g cannot change abruptly, the power absorbed by the grid side decreases. Since the photovoltaic array still outputs the maximum power, the excess power will be accumulated on the dc terminal capacitor, the power relationship is:

$$P_{pv} = P_e + P_{udc} \tag{2}$$

where P_{udc} is the power accumulated by the dc terminal capacitor.

Due to the excess energy accumulated on the dc terminal capacitor, its voltage will rise. The dc terminal capacitor power expression is:

$$P_{udc} = P_{pv} - P_e \tag{3}$$

Multiply Δt both sides of the Eq. (3) to get:

$$(P_{pv} - P_e)\Delta t = \frac{1}{2}C_{dc}\left(U_{dc}^{*2} - U_{dc}^2\right) \tag{4}$$

where C_{dc} is the dc terminal capacitor, U_{dc}^{*2} and U_{dc}^2 are the dc terminal capacitor voltage before and after the fault.

Adjust (4) to get:

$$\begin{cases} U_{dc}^* = \sqrt{\frac{2 \cdot P_{pv} \cdot D \cdot \Delta t}{C} + U_{dc}^2} \\ D = \frac{V_e^* - V_e}{V_e} \end{cases} \tag{5}$$

where D is the voltage drop depth, V_e^* and V_e are the grid phase voltages before and after the grid voltage drop.

It can be concluded that the dc bus voltage will increase with the voltage drop depth and drop time. If it cannot be controlled, the system will collapse.

When an unbalanced voltage fault occurs in the power grid, the voltage at the grid connection point changes abruptly. The symmetric component method is used to decompose it into positive sequence components and negative sequence components. It is worth noting that because the PV inverter adopts the $\Delta - Y$ connection method, the zero-sequence component can be ignored. The grid-connected voltage and grid-connected current can be expressed as:

$$\begin{bmatrix} e_\alpha \\ e_\beta \end{bmatrix} = e^{j\omega t}\begin{bmatrix} e_d^+ \\ e_q^+ \end{bmatrix} + e^{-j\omega t}\begin{bmatrix} e_d^- \\ e_q^- \end{bmatrix} \tag{6}$$

$$\begin{bmatrix} i_\alpha \\ i_\beta \end{bmatrix} = e^{j\omega t}\begin{bmatrix} i_d^+ \\ i_q^+ \end{bmatrix} + e^{-j\omega t}\begin{bmatrix} i_d^- \\ i_q^- \end{bmatrix} \tag{7}$$

where e_α, e_β, i_α, i_β are the voltage and current on the $\alpha - \beta$ axis. e_d^+, e_q^+, e_d^-, e_q^-, i_d^+, i_q^+, i_d^-, i_q^- are the positive and negative sequence components of the voltage and current on the $d - q$ axis.

The expressions for active power and active power can be obtained by applying the instantaneous power theory:

$$\begin{bmatrix} P \\ Q \end{bmatrix} = \begin{bmatrix} e_\alpha & e_\beta \\ e_\beta & -e_\alpha \end{bmatrix}\begin{bmatrix} i_\alpha \\ i_\beta \end{bmatrix} \tag{8}$$

Further arranging formula (8) to obtain:

$$\begin{aligned} P(t) &= P_0 + P_{c2}\cos(2\omega t) + P_{s2}\sin(2\omega t) \\ Q(t) &= Q_0 + Q_{c2}\cos(2\omega t) + Q_{s2}\sin(2\omega t) \end{aligned} \tag{9}$$

where P_0 is the average value of active power, Q_0 is the average value of reactive power, P_{c2} is the peak value of the secondary active power cosine, P_{s2} is the sine peak value of the secondary active power, Q_{c2} is the cosine peak value of secondary reactive power, Q_{s2} is the sine peak value of the secondary reactive power.

Expand the publicity (9) to get the matrix expression of power:

$$\begin{bmatrix} P_0 \\ P_{c2} \\ P_{s2} \\ Q_0 \\ Q_{c2} \\ Q_{s2} \end{bmatrix} = \frac{3}{2}\begin{bmatrix} e_d^+ & e_q^+ & e_d^- & e_q^- \\ e_d^- & e_q^- & e_d^+ & e_q^+ \\ e_q^- & -e_d^- & -e_q^+ & e_d^+ \\ e_q^+ & -e_d^+ & e_q^- & -e_d^- \\ e_q^- & -e_d^- & e_q^+ & -e_d^+ \\ -e_d^- & -e_q^- & e_d^+ & e_q^+ \end{bmatrix}\begin{bmatrix} i_d^+ \\ i_q^+ \\ i_d^- \\ i_q^- \end{bmatrix} \tag{10}$$

When the grid voltage is unbalanced, the grid-connected voltage and current are decomposed into positive sequence and negative sequence components, and the interaction between the two causes the active power and reactive power to have double frequency fluctuations. In addition, the existence of a negative sequence current will increase the grid-connected current, and when it exceeds $1.1 I_N$, the system will be disconnected from the grid.

3 Dual Current Control Scheme Based upon Positive and Negative SRF'S

The traditional phase-locked loop (SRF-PLL) cannot be phase-locked normally when the grid voltage fails, so a new phase-locked loop needs to be used to lock the phase and provide current shunting for the control link.

3.1 DSOGI-Pll

The positive and negative sequence components of the grid-connected voltage are separated by the symmetrical component method, and its expression is:

$$e_{abc}^+ = \begin{bmatrix} 1 & \alpha & \alpha^2 \\ \alpha^2 & 1 & \alpha \\ \alpha & \alpha^2 & 1 \end{bmatrix} e_{abc} \tag{11}$$

$$e_{abc}^- = \begin{bmatrix} 1 & \alpha^2 & \alpha \\ \alpha & 1 & \alpha^2 \\ \alpha^2 & \alpha & 1 \end{bmatrix} e_{abc} \tag{12}$$

where $\alpha = e^{j2\pi/3}$.

The Clark transform is performed on formulas (11) and (12), and its expression in the two-phase stationary coordinate system is:

$$e_{\alpha\beta}^+ = [T_{\alpha\beta}] e_{abc}^+ = \frac{1}{2} \begin{bmatrix} 1 & -q \\ q & 1 \end{bmatrix} e_{abc} \tag{13}$$

$$e_{\alpha\beta}^- = [T_{\alpha\beta}] e_{abc}^- = \frac{1}{2} \begin{bmatrix} 1 & q \\ -q & 1 \end{bmatrix} e_{\alpha\beta} \tag{14}$$

where $q = e^{-j\pi/2}, [T_{\alpha\beta}]$ is the Clark transformation matrix.

This shows that in the two-phase static coordinate system, the positive and negative sequence components of the grid-connected voltage have a 90° phase angle offset from the original signal. In this paper, SOGI-QSG is used to generate quadrature signals. The system structure is shown in Fig. 2.

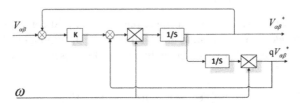

Fig. 2. SOGI-QSG structure diagram

The positive and negative sequence separation structure of the entire DSOGI-PLL system is shown in Fig. 3.

3.2 Dual Current Control

From the analysis of (10), it can be obtained that the degree of freedom of this matrix is 4, and it cannot meet the requirements of the six parameters P_0, P_{c2}, P_{s2} and Q_{s2} control by one of them. This paper mainly discusses the control of suppressing negative sequence current.

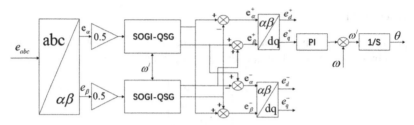

Fig. 3. DSOGI-PLL system structure control diagram

When an unbalanced voltage fault occurs in the power grid, the grid-connected current will exceed the current limit due to the existence of a negative sequence current. The grid-connected current requires that the maximum grid-connected current cannot exceed $1.1 I_N$, and the grid-connected power factor should be maintained as much as possible.

In matrix (10), to eliminate the negative sequence current, we need to make $i_d^- = i_q^- = 0$, then the above matrix can be simplified as:

$$\begin{bmatrix} P_0 \\ Q_0 \end{bmatrix} = \frac{3}{2} \begin{bmatrix} e_d^+ & e_q^+ \\ e_q^+ & -e_d^+ \end{bmatrix} \begin{bmatrix} i_d^+ \\ i_q^+ \end{bmatrix} \tag{15}$$

By inverting the matrix (10), the reference value of the positive sequence current can be obtained:

$$\begin{bmatrix} i_d^{+*} \\ i_q^{+*} \end{bmatrix} = \frac{2}{3\left[\left(e_d^+\right)^2 + \left(e_q^+\right)^2 \right]} \begin{bmatrix} e_d^+ & e_q^+ \\ e_q^+ & -e_d^+ \end{bmatrix} \begin{bmatrix} P_0^* \\ Q_0^* \end{bmatrix} \tag{16}$$

where P_0^* and Q_0^* are the reference values of active power and reactive power dc, respectively.

It is worth noting that to achieve a single power factor grid connection, we need to make $Q_0^* = 0$. In addition, to maintain the stability of the dc terminal voltage, this paper introduces the voltage feedforward decoupling control represent P_0^*, and its expression is:

$$P_0^* = \left[(K_p + K_i/s)(U_{dc}^* - U_{dc}) \right] U_{dc}^* \tag{17}$$

It is worth noting that the matrix in (10) is a 6 * 4 matrix, which is not reversible. Because the degree of freedom of the equation is four, six variables are not controlled at the same time. Therefore, while suppressing negative sequence current, the influence of active power and reactive power fluctuations cannot be eliminated.

In the current inner loop decoupling control, this paper uses the PI regulator to eliminate the static error of the positive and negative sequence current feedback. The inner loop decoupling expression of the positive sequence current and the negative sequence current is:

$$\begin{cases} u_d^+ = (K_p + K_i/S)(i_d^{+*} - i_d^+) + \omega L i_q^+ + e_d^+ \\ u_q^+ = (K_p + K_i/S)(i_q^{+*} - i_q^+) + \omega L i_d^+ + e_q^+ \end{cases} \tag{18}$$

$$\begin{cases} u_d^- = (K_p + K_i/s)(i_d^{-*} - i_d^-) + \omega L i_q^- + e_d^- \\ u_q^- = (K_p + K_i/s)(i_q^{-*} - i_q^-) + \omega L i_d^- + e_q^- \end{cases} \tag{19}$$

From the above control expression, it can be obtained that under the grid voltage unbalance fault, the structure diagram of the control strategy used in this paper is shown in Fig. 4.

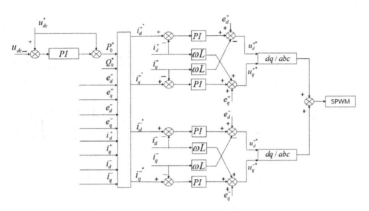

Fig. 4. Structure diagram of unbalanced voltage fault control strategy

4   Case Studies

To further explore the practicality of this control method, this paper simulates different voltage sag depths and then observes the grid-connected current of this method. The simulation model is established in Matlab/Simulink, and its circuit parameters are shown in the following table (Table 1):

Table 1. System parameters for the simulation.

Parameters	Symbol	Value
dc-link voltage reference	V_{dc}^*	700 V
Line-line RMS of PCC voltage	V_{PCC}	380 V
Grid frequency	f	50 Hz
PV panel output maximum power	P_{pv}^*	1600 W
Filter inductor	L_f	0.4 mH

To simulate different degrees of voltage drop in the actual power system, a single-phase voltage drop of 20% is used to represent the low-voltage drop scenario, and a single-phase voltage drop of 80% is used to represent the high-voltage drop scenario.

4.1   Case 1: Single-Phase Voltage Drop by 20%

In this case, the voltage of phase A drops by 20% and the other phases remain unchanged. The method adopted in this paper is used to solve the unbalanced voltage fault. The test results show that when the single-phase voltage drop occurs in the system, the power quality of the grid-connected current output by the proposed control method is good and maintained within $1.1I_N$, which meets the low-voltage ride-through grid-connected standard (Fig. 5).

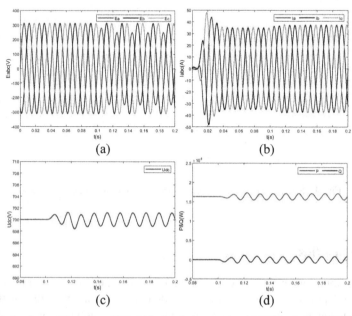

Fig. 5. Single-phase voltage drop 20% fault simulation results; (a) PCC voltage, (b) PCC currents, (c) dc-link voltage, (d) active and reactive power.

4.2   Case 2: Single-Phase Voltage Drop by 80%

In this case, the voltage of phase A drops by 80%, and the other phases remain unchanged. The method adopted in this paper is used to solve the voltage unbalance fault. The test results show that when the single-phase voltage sag fault depth of the system is high, although the proposed control method can ensure the elimination of negative sequence current and obtain good power quality, the grid-connected current exceeds $1.1I_N$, which does not meet the grid-connected requirements.

After many tests, it is concluded that the critical value that the control method proposed in this paper can control is 26% (Fig. 6).

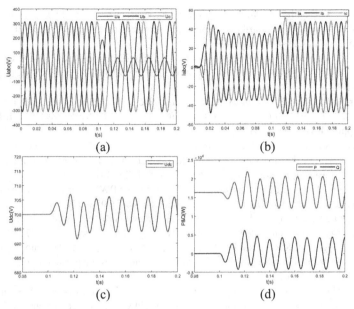

Fig. 6. Single-phase voltage drop 80% fault simulation results; (a) PCC voltage, (b) PCC currents, (c) dc-link voltage, (d) active and reactive power.

5 Conclusion

To improve the dynamic performance of photovoltaic systems under grid unbalanced faults and reduce the risk of off-grid. In this paper, a control algorithm for unbalanced voltage of the bipolar photovoltaic system is used. The control algorithm separates the positive and negative components through DSOGI-PLL, and eliminates the influence of negative sequence current by setting a reasonable current reference value.

In this paper, the practicability of the method used is verified by simulating faults with different voltage sag depths. The results show that when the voltage drop is low, the method can effectively suppress the negative sequence current, avoid overcurrent faults, and output high-quality grid-connected current. However, when the voltage sag depth is large, although the power quality is improved, the grid-connected current exceeds $1.1I\_N$, and the grid-connected state cannot be maintained. The critical value that the control method used in this paper can control is 26%, this method is currently only suitable for fault scenarios with low voltage sag depths.

Acknowledgements. The authors gratefully acknowledge the support of the National Science Foundation of China (52007194).

References

1. Li, Z., et al.: Adaptive power point tracking control of PV system for primary frequency regulation of AC microgrid with high PV integration. IEEE Trans. Power Syst. **36**(4), 3129–3141 (2021)

2. Camacho, A., Castilla, M., Miret, J., Vasquez, J.C., Alarcon-Gallo, E.: Flexible voltage support control for three-phase distributed generation inverters under grid fault. IEEE Trans. Ind. Electron. **6**(4), 1429–1441 (2013)
3. Camacho, A., et al.: Reactive power control for voltage support during type C voltage-sags. In: IECON 2012 - 38th Annual Conference on IEEE Industrial Electronics Society, pp. 3462–3467 (2012)
4. Sochor, P., Tan, N.M.L., Akagi, H.: Low-voltage-ride-through control of a modular multilevel single-delta bridge-cell (SDBC) inverter for utility-scale photovoltaic systems. IEEE Trans. Ind. Appl. **54**(5), 4739–4751 (2018)
5. Camacho, A., Castilla, M., Miret, J., Borrell, A., de Vicuña, L.G.: Active and reactive power strategies with peak current limitation for distributed generation inverters during unbalanced grid faults. IEEE Trans. Ind. Electron. **62**(3), 1515–1525 (2015)
6. Song, H.-S., Nam, K.: Dual current control scheme for PWM converter under unbalanced input voltage conditions. IEEE Trans. Ind. Electron. **46**(5), 953–959 (1999)
7. Xia, Y., Chen, Y., Song, Y., Strunz, K.: Multi-scale modeling and simulation of DFIG-based wind energy conversion system. IEEE Trans. Energy Convers. **35**(1), 560–572 (2020)
8. Talha, M., Raihan, S.R.S., Rahim, N.A., Akhtar, M.N., Butt, O.M., Hussain, M.M.: Multi-functional PV inverter with low voltage ride-through and constant power output. IEEE Access **10**, 29567–29588 (2022)
9. Lin, X., Han, Y., Yang, P., Wang, C., Xiong, J.: Low-voltage ride-through techniques for two-stage photovoltaic system under unbalanced grid voltage sag conditions. In: 2018 IEEE 4th Southern Power Electronics Conference (SPEC), pp. 1–8 (2018)
10. Chen, H., Cheng, P.: A DC bus voltage balancing technique for the cascaded H-bridge STAT-COM with improved reliability under grid faults. IEEE Trans. Ind. Appl. **53**(2), 1263–1270 (2017)

Simulation and Shape Optimization of the Bleeding Pipe in Aero-Engine Internal Air System to Improve Mass Flow Measurement Accuracy

Song Chen-xing[1]([✉]), Wang Hai[1], Zhao Yi-zhen[1], Wang Jun-song[1], and Du Yi-ming[2]([✉]) [iD]

[1] AECC Shenyang Engine Research Institute, Shenyang 110015, People's Republic of China
[2] College of Aerospace Engineering, Shenyang Aerospace University, Shenyang 110137, People's Republic of China
dym9026@163.com

Abstract. Accurate measured mass flow of the bleeding pipe is essential to confirm the behavior of internal air system during design and verification of an aero-engine. However, the measurement accuracy of bleeding pipe mass flow is always insufficient in commonly used mass flow characteristics curve method when the ratio of inlet and outlet pressure is low. To improve the measurement accuracy of mass flow, this paper proposes a convergent-divergent section in bleeding pipe instead of original straight section, and conducts shape design by simulation and surrogate-based optimization. Moreover, a new mass flow characteristics function is constructed by replacing the pressure ratio of inlet and outlet to the pressure ratio of inlet and the throat. In the optimized bleeding pipe, given the same measurement error, the error propagated to the mass flow through the newly constructed mass flow characteristics function is about 1/60 as that of the common approach, when the pressure ratio of inlet and outlet is as low as 1.2.

Keywords: Mass flow · Measurement accuracy · Bleeding pipe · Internal air system

1 Introduction

Internal air system plays the roles of internal engine and accessory unit cooling, air-cooled blade cooling, bearing chamber sealing, axial load control and so on [1], so as to ensure the operation safety and efficiency of the engine [2]. The air to cool the low-pressure turbine is usually from the mainstream of the compressor by bleeding pipe, as illustrated in Fig. 1. Accurate mass flow measured data of the bleeding pipe is essential to confirm the behavior of internal air system during the design and verification of an aero-engine.

However, the measurement accuracy of bleeding pipe mass flows is always insufficient when the ratio of inlet and outlet pressure is low, in the commonly used mass flow characteristics curve method to obtain the tested mass flow of bleeding pipe. Common approach to obtain the bleeding pipe mass flow has two steps: firstly, the mass flow characteristics function(as shown in Fig. 2 in black solid curve) $\overline{G} = f(\pi) = A + Be^{C\pi}$ is fitted based on the tested data(as illustrated in Fig. 2 in blue triangles), where π is the inlet/outlet total pressure ratio, and A, B and C are constant coefficients, and $\overline{G} = G(T_{in})^{1/2}/P_{in}$ is a converted mass flow [3] from original mass flow G, where the indices *in* means the inlet of a pipe, and P and T stand for the total pressure and temperature. Secondly, the converted mass flow $\overline{G}$ is obtained by plugging the measured pressure ratio π into mass flow characteristics function, and then G can be easily obtained. However, the inlet/outlet pressure ratio π of bleeding pipe is usually low, owing that bleeding pipe would not be the throttling unit in internal air system. And corresponding to the large slope region of the mass flow characteristics function (Fig. 2). Therefore, the measurement error of π would be amplified through the mass flow characteristics function, as illustrated in Fig. 2 in red lines.

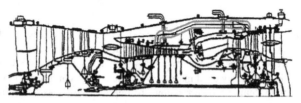

Fig. 1. Profile of an aero-engine [4] with bleeding pipe of internal air system in red. (Color figure online)

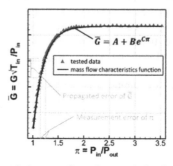

Fig. 2. Error propagation through the mass flow characteristics function ($\pi = 1.2$). (Color figure online)

In this paper, a convergent-divergent section is designed in bleeding pipe instead of the original straight section, and the shaped design of this convergent-divergent section is conducted by simulation and surrogate-based optimization. Moreover, a new mass flow characteristics function is constructed by replacing the pressure ratio from $\pi = P_{in}/P_{out}$ to $\pi' = P_{in}/P_{throat}$, thus to improve the measurement accuracy of mass flow.

2 Methodology

2.1 Bleeding Pipe Simulation

IN this paper, the STAR-CCM + v14.02 is adopted as the computation platform to solve the incompressible Reynolds-Averaged Navier-Stocks (RANS) equations and heat transfer equation in a steady flow. The spatial terms in the governing equation are discreted by second-order upwind scheme. The Spalart-Allmaras (S-A) turbulence model [5] is used. The air is considered as ideal gas and the viscosity coefficient is calculated by Sutherland's law. Other physical parameters like specific heat are treated as functions of temperature.

A set of unstructured grid with totally 4.9×10^5 cells is generated using the STAR-CCM + v14.02, as shown in Fig. 3. To accurately capture the boundary-layer flow, the near-field grid is refined in normal direction from the wall, whose first layer is 2×10^{-4}m height. In addition, the wall function is used at near field to further decease the limitation of y^+ in turbulence model. The mesh convergence curve in Fig. 4 shows that the scale of mesh is adequate.

Fig. 3. The mesh profile of bleeding pipe with original straight section.

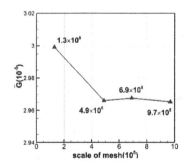

Fig. 4. Mesh convergence curve of the bleeding pipe.

2.2 Convergent-Divergent Section Optimization

TO conduct the design of the convergent-divergent section in bleeding pipe, the spline by 5 poles is employed to describe the outline of the concerned section, as shown in Fig. 5, and the design variables are defined as follows:

design variables A: x coordinate value of the second pole from the left.
design variables B: x coordinate value of the third pole from the left.
design variables C: x coordinate value of the fourth pole from the left.
design variables D: y coordinate value of the third pole from the left.

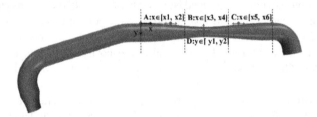

Fig. 5. Definition of design variables.

The optimization objective is maximizing the pressure ratio of total pressure at pipe inlet and static pressure at the throat. The constraints are letting the mass flow of the pipe be not less than a certain value, and letting the maximum Mach number in the whole computational domain be not bigger than 0.8, when the pressure ratio $\pi = P_{in}/P_{out}$ is 1.2.

The optimization problem is defined as follows:

$$
\begin{aligned}
&\text{Max: } \pi' = P_{in}/P_{throat} \\
&\text{s.t. : } \quad \overline{G} \geq 1.78x + y = z \\
&\text{Ma} \leq 0.9
\end{aligned}
\tag{1}
$$

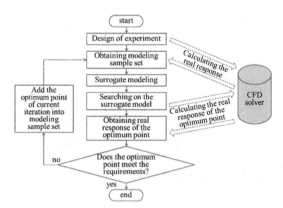

Fig. 6. Workflow of optimization design.

In consideration of saving CFD computation amount, this paper employs surrogate-based optimization [6] method to execute the outline of convergent-divergent section optimization design of bleeding pipe. The workflow of optimization design is illustrated in Fig. 6. In the design of experiment process, the full factor experimental design and

Sobol's sequence [7] is employed. In the surrogate modeling and refinement processes, Kriging surrogate model and minimum punctuating infilling method are used.

2.3 New Mass Flow Characteristics Function Construction

In this part, a new mass flow characteristics function is constructed as $\overline{G} = f'(\pi') = A' + B'e^{C'\pi'}$, where π' is defined as $\pi' = P_{in}/P_{throat}$, i.e., the ratio of total pressure at inlet and static pressure at the throat, and A', B' and C' are constant coefficients.

The mechanism of the advantage of the newly constructed mass flow characteristics function is as follows: at the throat of convergent-divergent section in bleeding pipe, a part of static pressure would be converted to dynamic pressure under the drive of area change of pipe section [8], resulting in a relative low static pressure at the throat. Then the $\pi' = P_{in}/P_{throat}$ is much higher than $\pi = P_{in}/P_{out}$, and corresponding to the small slope region of the constructed mass flow characteristics function. Therefore, the measurement error of π' propagated to the converted mass flow $\overline{G}$ could be restricted.

3 Result

After 37 loops of iteration, the optimal bleeding pipe with convergent-divergent section which meets the requirements is obtained, the corresponding result is shown in Table 1. The propagation from measurement error of π' to the converted mass flow $\overline{G}$ is illustrated in Fig. 7.

Table 1. The results of the optimal bleeding pipe with convergent-divergent section.

Parameters	Bleeding pipe with original straight section	Bleeding pipe with convergent-divergent section
$\pi = P_{in}/P_{out}$	1.2	1.2
$\pi' = P_{in}/P_{throat}$	-	1.756
$\overline{G}$	2.966×10^{-5}	1.785×10^{-5}
Measurement error of π or π'	± 2%	± 2%
Propagated error of $\overline{G}$	6.351%	0.098%

It's shown that the pressure ratio of $\pi' = P_{in}/P_{throat}$ is 1.756 when the pressure ratio $\pi = P_{in}/P_{out}$ is 1.2, for the optimal bleeding pipe with convergent-divergent section. And the newly constructed mass flow characteristics function is obtained based on a series of CFD simulated $\overline{G} - \pi'$ data on different P_{in}/P_{throat} ratios (as illustrated in Fig. 7 in black solid curve and blue triangles). Given the same measurement error of pressure ratio, the error propagated to the converted mass flow $\overline{G}$ through the newly constructed mass flow characteristics function with convergent-divergent section bleeding pipe is 0.098%. And it is about 1/60 as that of the common approach, in which the error propagated to the converted mass flow $\overline{G}$ is 6.351%.

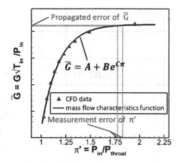

Fig. 7. Error propagation through the newly constructed mass flow characteristics function ($P_{in}/P_{out} = 1.2$, $P_{in}/P_{throat} = 1.756$).

The total pressure and static pressure contours are shown in Fig. 8, which reflect the conversion between static pressure and dynamic pressure in the convergent-divergent section in bleeding pipe, and it is the motivation of the advantage of the developed method in this paper.

Fig. 8. The total pressure contour and static pressure contours of the optimal bleeding pipe.

4 Conclusion

This paper proposes a convergent-divergent section in bleeding pipe instead of original straight section, and conducts shape design by STAR-CCM + v14.02 as CFD solver and surrogate-based optimization.

The conclusions are as follows:

1) A new way is put forward to obtain the measured mass flow of internal air system bleeding pipe.
2) Using the developed method, the accuracy of mass flow measurement can be ensured when the ratio of inlet and outlet pressure is very low.
3) The method and result of this paper are of reference value to support the design and verification of aero-engine internal air system.

References

1. Haiying, L., Yansheng, Y., Ming, W.: Numerical simulation of aero-engine internal air system character. Aeroengine **01**, 1–13 (1997)
2. The Jet Engine. 5th edn. Rolls-Royce PLC, England (1986)

3. Xinyue, W.: Fundamentals of Aerodynamics, 1st edn. Publisher Northwestern Polytechnical University Press, Xi'an (2006)
4. Yufang, L., Huayu, D., GuoWen, H., Yanli, H.: Improvements on engine air system design. Gas Turbine Exp. Res. **14**(3), 48–53 (2001)
5. Yiming, D., Zhenghong, G., Chao, W., Qianhuan, H.: Boundary-layer transition of advanced fighter wings at high-speed cruise conditions. Chin. J. Aeronaut. **32**(4), 799–814 (2019)
6. Liu, J.: Efficient Surrogate-Based Optimization Method and its Application in aerodynamic design, Northwestern Polytechnical University (2015)
7. Joe, S., Kuo, F.Y.: Remark on Algorithm 659: implementing Sobol's Quasirandom sequence generator. ACM Trans. Math. Softw. **29**(1), 49–57 (2003)
8. Jinshan, P., Peng, S.: Fundamentals of Gasdynamics, 1st edn. Publisher National Defense Industry Press, Taiwan (2011)

Application of Modeling/Simulation in Energy Saving/Emission Reduction, Public Safety, Disaster Prevention/Mitigation

Identification of High Emission Mobile Sources Based on Self-supervised Representation Network

Renjun Wang[1,2], Xiushan Xia[3], and Zhenyi Xu[2(✉)]

[1] School of Computer Science and Technology, Anhui University, Hefei 230039, China
e20201122@stu.ahu.edu.cn
[2] Institute of Artificial Intelligence, Hefei Comprehensive National Science Center,
Hefei 230088, China
xuzhenyi@ustc.edu.cn
[3] Institute of Advanced Technology, University of Science and Technology of China,
Hefei 230088, China
xiaxiushan@iat.ustc.edu.cn

Abstract. High-emission mobile sources are the main contributors to road traffic emission pollution, and how to accurately identify high-emission road mobile sources is of great significance to urban traffic pollution control and environmental protection. On-Board Diagnostics (OBD) is a device that records the operating conditions of a vehicle in real time. Usually, for vehicles with excessive tailpipe emissions, OBD monitoring value is compared with the set threshold to identify high emission vehicles. However, it often leads to misjudgment if a vehicle is judged to be high emission only by its excessive emission value. Because this excessive value may originate from external objective factors (such as vehicle idling, uneven road), resulting in pseudo-high emission states. Faced with this challenge, we propose a self-supervised representation network (SRN) for identifying high-emission mobile sources. A self-supervised learning module is integrated to learn general emission representations. Meanwhile, a representation memory module is introduced to make the module retain key emission representations through iterative learning. By reconstructing the time-series characterization of mobile sources, it is achieved for the classification identification of high and normal emissions. Experiments on a real diesel vehicle OBD emission monitoring sequence dataset show that the present method obtains a higher performance for emission source classification compared to other methods, demonstrating the effectiveness of the proposed method.

Keywords: High emissions · OBD · Self-supervised representation

This work was supported in part by the National Natural Science Foundation of China (62103124, 62033012, 61725304), Major Special Science and Technology Project of Anhui, China (201903a07020012, 202003a07020009, 2022107020030), China Postdoctoral Science Foundation (2021M703119).

W. Fan et al. (Eds.): AsiaSim 2022, CCIS 1713, pp. 419–430, 2022.
https://doi.org/10.1007/978-981-19-9195-0_34

1 Introduction

With the improvement of people's living standards, the ownership of motor vehicles has been increasing, resulting in the emission of exhaust gas becoming one of the main threats to urban ecological quality [1]. According to the 2021 China Mobile Source Management Annual Report [2], by the end of 2020, the national motor vehicle fleet exceeded 370 million vehicles, and the total emissions of four pollutants (including CO, HC, NOx, and PM) from motor vehicles reached 15.93 million tons. Therefore, high-emission mobile sources are the main contributors to road traffic emission pollution, and how to accurately identify high-emission road mobile sources is of great significance to urban traffic pollution control and environmental protection [3].

The identification of high emission sources is closely related to the detection devices. Currently, common vehicle emission detection devices include the portable emissions measurement system (PEMS) and the on-road emission remote sensing (OERS).

When a vehicle is equipped with PEMS, it can detect a wide range of different pollutant emission concentration values in a continuous time. In [4], Weiss et al. used PEMS to find that the actual NOx concentrations emitted by diesel vehicles were significantly higher than the established standards, helping regulators to propose new improvement options. In [5], Gallus et al. used PEMS to measure the effects of driving style and road grade on vehicle emissions. CO_2 and NO_x emissions increase linearly with road grade when driving style was kept approximately constant. However, PEMS devices are expensive and complex to install. Also, it can only detect single vehicle, making it difficult to be widely used. OERS can overcome these drawbacks to some extent. OERS has the specialties of short time and large range for detection. Meanwhile, it can identify multiple targets simultaneously without interfering with normal road vehicles. Therefore, this equipment is very suitable for vehicle census detection supervision. Xu et al. [6] proposed a spatiotemporal graph convolution multifusion network to capture the spatial interactions and temporal correlations of high vehicle emissions variation between different road segments. In [7], after statistical analysis for collected 201504 telemetry data, the cut points of CO, HC, and NO were set at 3%, 500 ppm, and 2000 ppm, respectively, and 1373 high emission targets were screened. However, the telemetry value only reflects the transient state of the vehicle and are vulnerable to external factors. Therefore, this statistically based threshold classification often lacks objectivity. Subsequently, several studies have proposed methods to identify high emission sources by considering multiple influencing features. For example, Xu et al. [8] proposed an unsupervised clustering approach to automatically identify high-emission targets using telemetry values of CO, HC, and NO, and combining attributes such as vehicle weight, wind speed, temperature, and pressure. Li et al. [9] also considered multiple factors and propose a high-emission identification model based on a weighted extreme learning machine for the problem of high-emission data imbalance.

Although these methods have been shown to be effective, the precise identification of high emission sources is still not sufficiently studied. PEMS and

OERS are not widely applicable to the accurate identification of high emission sources due to their respective shortcomings. OBD offers advantages they don't, with small size, easy installation, and the ability to monitor in real time. But it causes the problem of excessive emissions under abnormal driving conditions (e.g., idle driving, frequent acceleration and deceleration). It would lead to mis-classification if rely solely on emission concentration values to filter high emission targets (as shown in Fig. 1). Therefore, this study combines a variety of vehicle own attributes in OBD time series data to analyze the vehicle emission status comprehensively and realize the accurate identification of high emission sources.

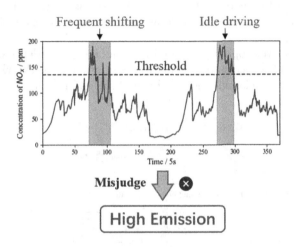

Fig. 1. Excessive concentration of emission due to abnormal driving for vehicle.

The research difficulties in this paper mainly come from two aspects. On the one hand, it is very difficult to manually classify and label the OBD time-series emission sequence data considering the actual scenario. This is mainly because of the high dimensionality and large amount of data. On the other hand, considering the diversity of vehicle driving conditions, it increases the difficulty of the model to learn rich emission representations. Therefore, we propose a self-supervised representation network (SRN) for identifying high-emission mobile sources. Its main contributions can be summarized as follows:

- A self-supervised learning module is designed to help the model learn the general characterization of emission sequences without labels. Self-supervised learning can generate non-target labels for free with the help of the represen-tational properties of the data itself in order to reduce the dependence of the learning task on a large amount of labeled data [10].
- A representation memory module is introduced to record the most represen-tative prototype patterns in emission sequences for multiple different driving states, enhancing the ability of the model to learn the corresponding key representations.

By deeply learning the representations of multivariate OBD timing data, SRN is able to rely on the vehicle's own attributes to comprehensively analyze the emission status and achieve accurate identification of high-emission sources.

2 Methods

The research idea of this paper is based on the method of characterization reconstruction, which uses a limited number of labeled normal emission sequence data for model training to predict unlabeled emission data. Further, during the validation process of the model, when the reconstruction error of the validation set data exceeds a threshold, it is judged as high emission, otherwise it is judged as normal emission.

In this paper, we propose a high emission mobile source identification method SRN based on self-supervised learning representation network, which is used to solve the time series classification problem of high and normal emissions.

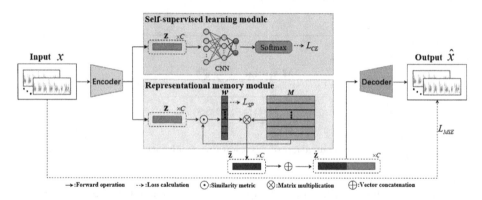

Fig. 2. Framework of Research Methodology for SRN. The notation $\times C$ indicates that one data transformation corresponds to one copy for C times.

SRN belongs to a reconstruction-based approach. Specifically, we employ a convolutional autoencoder (CAE) as the basis of a deep neural network. A autoencoder (AE) is typically contains three parts: an encoder $f_e(\cdot)$, a hidden layer structure $f_h(\cdot)$ and a decoder $f_d(\cdot)$. And they are parameterized by θ_e, θ_h and θ_d, respectively. Firstly, the encoder maps a high-dimensional input x to a potential representation z:

$$z = f_e(x, \theta_e) \tag{1}$$

Then, the potential representation z flows in the hidden layer and is computed as $\hat{z}$:

$$\hat{z} = f_d(z, \theta_d) \tag{2}$$

Finally, the decoder maps $\hat{z}$ back to the input space and outputs the reconstructed features $\hat{x}$ of the original input:

$$\hat{x} = f_d(\hat{z}, \theta_d) \tag{3}$$

SRN contains two main modules, the self-supervised learning module and the representational memory module. The network structure of SRN is shown in Fig. 2. The processing details of each module are shown in the following sections.

2.1 Self-supervised Learning Module

In this section, we introduce the self-supervised learning module, which is used to learn general representations of emission sequences. Since the cost of labeling emission samples is high in real scenarios. Moreover, the proportion of already labeled emission samples is very limited compared to the large amount of unknown and potentially high-emission data. Therefore, we decided to use only known normal emission sequences in the training phase of the model for training. And, to prevent overfit of the high emission detection model trained on the limited normal samples, we use self-supervised learning to improve the generalization ability of the model.

To address the above issues, in combination with self-supervised learning, this paper helps the model learn the general characterization of emission sequences by performing multiple transformations on the raw emission data as a predecessor task without manual annotation. Inspired by [12], we apply six different signal transformations, and they are described as follows (assuming that the raw emission sequence is $T = [t_1, t_2, \ldots, t_n]$):

1) *Noise.* Real-world signals with noise may exist, and adding noise to the signal can help the model improve robustness. Here, we include a common transformation of Gaussian noise.

2) *Reverse.* This transformation reverses the samples along the time dimension to obtain samples in the opposite time direction T_{re}:

$$T_{re} = [t_n, \ldots, t_2, t_1] \tag{4}$$

3) *Permute.* This transformation generates new samples T_{pe} by slicing and swapping different time windows (assuming a slice length of 2) so that the signal is randomly perturbed along the time dimension, enhancing the permutation-invariant nature of the resulting model:

$$T_{pe} = [\ldots, t_1, t_2, \ldots, t_{n-l}, t_{n-l+1}, \ldots] \quad (l < n) \tag{5}$$

4) *Scale.* The size of the signal within a time window is varied by multiplying it by a random scalar. Here, we choose $S = \{0.25, 0.5, 1.5, 2\}$ as the scalar value. The addition of the scaled signal helps the model to learn the scaling invariant pattern, which can be represented as T_{sc}:

$$T_{sc} = [t_1 \times s, t_2 \times s, \ldots, t_n \times s] \quad (s \in S) \tag{6}$$

5) *Mirror*. This conversion is a special type of scalarization conversion. It is scaled by −1 to obtain a mirror sequence T_{ne} of the input signal:

$$T_{ne} = [-t_1, -t_2, \ldots, -t_n] \qquad (7)$$

6) *Smooth*. The exponentially-weighted moving average (EWMA) [13] method is applied to smooth the signal. EWMA is represented as follow:

$$\hat{W}_{t+1} = \alpha W_t + (1 - \alpha)\hat{W}_t, \quad 0 < \alpha \le 1 \qquad (8)$$

in which, $\hat{W}_{t+1}$ denotes EWMA value at moment $t + 1$, $\hat{W}_t$ and W_t indicate EWMA value and the true value at moment t, respectively; α is a weight constant.

The cross-entropy loss function is used to distinguish between different types of transformations, and it is denoted as:

$$L_{CE} = -\sum_{i=1}^{C} y_i \log(p_i) \qquad (9)$$

where, C is the number of data transformation categories (including the original signal and 6 transformed signals, $C = 7$ in this paper). y_i and p_i are the prediction labels and prediction probabilities, respectively.

To facilitate the updating and convergence of the network, the reconstruction error between the input emission sequence x and the output reconstructed sequence $\hat{x}$ is calculated using the mean square error (MSE). It is expressed as:

$$L_{MSE} = \sum_{i=1}^{C} \|\hat{x}_i - x_i\|_2^2 \qquad (10)$$

where, $\|\cdot\|_2^2$ indicate the l_2-norm, C is the number of data transformation categories.

2.2 Representational Memory Module

Traditional autoencoder (AE) is susceptible to noisy information in unknown training data and does not have significant differences in reconstruction effects between normal and abnormal inputs [11]. This can lead to the model not learning key features. Therefore, we introduce a representation memory module to record key patterns in potential representations through a memory matrix, which enhances the ability of the model to distinguish between high and normal emissions.

The representation memory module [7] consists of an encoded latent representation, a memory matrix, and a memory weight vector. Specifically, the memory matrix is initialized as $M \in \mathbb{R}^{T \times F}$, containing T vectors of F-dimension.

Given a potential representation $z \in \mathbb{R}^F$, the memory weight vector $\boldsymbol{w} \in \mathbb{R}^F$ is computed as:

$$\boldsymbol{w} = S(z, \boldsymbol{M}) \tag{11}$$

where, $S(\cdot)$ denotes the similarity function, and its computational expression is shown as:

$$w_i = \frac{\exp(M_i \cdot z)}{\sum_{j=1}^{T} \exp(M_j \cdot z)} \quad (1 \leq i \leq T) \tag{12}$$

where, $M_i \in \mathbb{R}^{1 \times F}$ denotes the i-th row vector of the representation memory matrix $\boldsymbol{M}$.

The memory weight vector $\boldsymbol{w}$ is computed with the representational memory matrix $\boldsymbol{M}$ to obtain the representational memory vector $\tilde{z}$, which is denoted as:

$$\tilde{z} = \boldsymbol{M}^\top \cdot \boldsymbol{w} \tag{13}$$

Further, $\hat{z}$ is obtained by concatenating $\tilde{z}$ and the potential representation z, which in turn is fed to the decoder. $\hat{z}$ is expressed as:

$$\hat{z} = \mathrm{concat}(\tilde{z}, z) \tag{14}$$

To facilitate network update and convergence, the sparse loss value L_{SPAR} of the memory weight vector $\boldsymbol{w}$ needs to be computed as:

$$L_{SP} = -\sum_{i=1}^{T} w_i \log(w_i) \tag{15}$$

2.3   Training and Classification

Model Training. In the training process of the model, by combining the cross-entropy loss L_{CE} in Eq. 9, the reconstruction loss L_{MSE} in Eq. 10, the sparse loss L_{SP} in Eq. 15 and their corresponding weight parameters α_1, α_2, α_3, the optimization objective of SRN can be expressed as:

$$L(\theta) = \alpha_1 L_{CE} + \alpha_2 L_{MSE} + \alpha_3 L_{SP} \tag{16}$$

In this paper, α_1, α_2 and α_3 were set to 1, 1 and 0.002, respectively.

Emission classification. Usually, the reconstruction of the self-encoder-based model is different for different classes of instances. It means that the reconstruction error of high-emission sequences for which SRN is trained on a dataset containing only normal emissions increases significantly. Therefore, we use the reconstruction error in the training phase as a judgment threshold and thus distinguish the high-emission targets in the validation set.

Given a training set $\mathcal{D}_X = \{x_1, \ldots, x_n\}$ that all emission instances are normal, the reconstruction error result $L_{REC} = \{L_{MSE}(x_1), \ldots, L_{MSE}(x_n)\}$ is obtained by SRN. Then, the reconstruction error threshold μ used to judge the validation set is defined as the 96% percentile of L_{REC}. For the true high-emission set $\mathcal{D}_H$ and the normal-emission set $\mathcal{D}_N$ in the validation set, their label classification can be expressed as:

$$d_i = \begin{cases} N, & L_{MSE}(d_i) < \mu \\ H, & L_{MSE}(d_i) \geq \mu \end{cases} \quad (d_i \in \mathcal{D}_H) \tag{17}$$

$$d_j = \begin{cases} N, & L_{MSE}(d_j) < \mu \\ H, & L_{MSE}(d_j) \geq \mu \end{cases} \quad (d_j \in \mathcal{D}_N) \tag{18}$$

in which, H and N denote labels for high and normal emissions, respectively. d_i and d_j indicate any one instance of $\mathcal{D}_H$ and $\mathcal{D}_N$, respectively.

3   Experiments and Discussion

3.1   Datasets

A real OBD real-time emission sequence dataset are used for the experiment, which contains a total of 23 properties, as shown in Table 1.

Table 1. Description of the properties in OBD dataset.

Properties	Unit
Downstream NO_x concentration	ppm
Engine speed	rpm
Percentage of actual torque output	%
Water temperature of engine	°C
Engine oil temperature	°C
Percentage of downstream oxygen	%
Atmospheric pressure	kPa
Temperature of environment	°C
Exhaust gas mass flow rate	km/h
Battery voltage	mV
Longitude	East(°)
Latitude	North(°)
Speed	km/h
Percentage of urea tank level	%
Temperature of urea tank	°C
Openness of the gas pedal	%
Miles per trip	km
Instantaneous fuel injection	L
Instantaneous fuel consumption rate	L/100km
Average fuel consumption rate	L/100km
Fuel consumption per trip	L/100km
Cumulative fuel consumption	L/100km
Cumulative engine runtime	h

After data processing, OBD dataset is divided into training and validation sets. The details about them are presented in Table 2.

Table 2. Illustration of the classification of the OBD dataset.

	Instances	Time interval	Shape	Class
Training set	9960	5 s	83 × 120 × 23	Normal Emission
Validation set	6720	5 s	56 × 120 × 23	Normal Emission
Validation set	5400	5 s	45 × 120 × 23	High Emission

3.2   Baselines

SRN is compared with four classification models, including two reconstruction-based models and two traditional machine learning classifiers. They are presented as follows:

1) *LSTM-AE*: an autoencoder model using *LSTM* on the encoder and decoder [14].
2) *CAE*: an autoencoder model based on convolutional neural network [15].
3) *Random Forest (RF)*: an excellent machine learning algorithm based on integrated learning [16].
4) *SVM*: a traditional machine learning classifier, which is based on the principle of mapping sample points to a feature space and finding a hyperplane to classify different class instances [17].

3.3   Results

In order to verify the effectiveness and superiority of the proposed method, we set up several experiments for comparison in this section. All experiments are run on a Windows 10-64bit operation system with an Intel Core i5-10500 3.10GHz CPU, 16GB RAM and NVIDIA GeForce RTX 3060 GPU.

Our study is a classification task, so the following metrics are used to measure the classification performance of the model: accuracy (Acc), precision (Pre), recall (Rec), and F1. They are calculated as:

$$Acc = \frac{TP + TN}{TP + FP + TN + FN} \tag{19}$$

$$Pre = \frac{TP}{TP + FP} \tag{20}$$

$$Rec = \frac{TP}{TP + FN} \tag{21}$$

$$F1 = \frac{2 \cdot Pre \cdot Rec}{Pre + Rec} \tag{22}$$

where TP denotes true case, FP false positive case, TN true negative case, and FN false negative case. In this paper, high emissions are positive cases and normal emissions are negative cases.

The confusion classification matrixs of each model on the validation set are presented in Fig. 3. Also, their performance scores are recorded in Table 3.

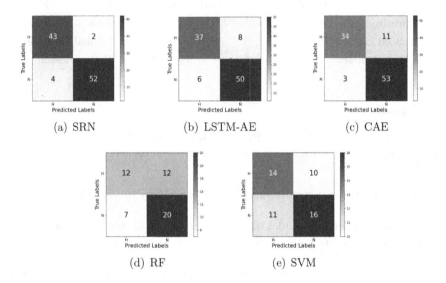

(a) SRN   (b) LSTM-AE   (c) CAE

(d) RF   (e) SVM

Fig. 3. The confusion matrices on four models. (Since the training set in Table 2 contains only normal emission instances, the validation set in Table 2 is divided into the ratio with 1:1 to train and predict for RF and SVM.)

According to Table 3, SRN is best in all metric indicators. This indicates that the reconstruction-based approach to capture the representations of different classes of emission sequences is very meaningful. For LSTM-AE and CAE that are also based on reconstruction, they have only encoder and decoder in their structure, which limits the learning capability for emission representations. For the traditional machine learning algorithms RF and SVM, they are difficult to perform well on temporal datasets compared to deep networks due to the discrepancy in principles.

In addition, we consider three key parameters for sensitivity analysis: the size of potential representation z, the size of memory weight vector w and the percentile of training loss for threshold μ. It is shown as Fig. 4 for the robustness of SRN on three parameters. From Fig. 4, it can be observed that: for the size of potential representation z, the $F1$ value is higher when it is much smaller than the dimension of the input sequence x ($x = 23$); for the size of memory weight vector w, the best value should be taken at 100; for the percentile of the reconstruction error: with the increase of the percentile μ, the gap between Pre and Rec gradually decreases, and in a comprehensive way, the best $F1$ value is obtained at 96.

Table 3. Classification performance results on models.

Model	Acc	Pre	Rec	F1
LSTM-AE	0.861	0.860	0.822	0.841
CAE	0.861	0.919	0.756	0.829
RF	0.627	0.632	0.500	0.558
SVM	0.588	0.560	0.583	0.571
SRN	**0.941**	**0.915**	**0.956**	**0.935**

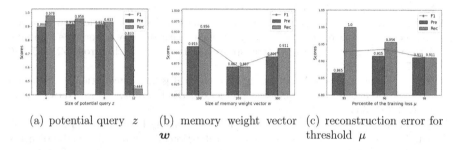

(a) potential query z (b) memory weight vector w (c) reconstruction error for threshold μ

Fig. 4. Parameter sensitivity analysis for SRN.

4 Conclusion

For real-time recording of vehicle emission for OBD datasets, this paper proposes a high emission mobile source identification method (SRN) based on self-supervised learning and characterization memory. Vehicles may generate abnormal excessive emission data under abnormal driving conditions (such as vehicle idling, transient acceleration or rugged working conditions), which leads to the problem of misclassification by the threshold determination method relying on emission concentrations and makes it difficult to achieve accurate identification of high emission sources. Facing this challenge, SRN combines the concentration of emissions with multiple vehicles' own attributes (such as vehicle speed, engine speed, etc.) from OBD and adopts a reconstruction-based approach to delineate the characterization of high-emission and normal-emission sequences. Specifically, considering the difficulty of label annotation in practical scenarios, the self-supervised learning module in SRN uses data type transformation as a manual-free annotation pre-task to assist the model in learning the general representations in emission sequences. Considering the richness of emission sequences, the representation memory module learns and remembers the key representations in emission sequences by the memory structure. The effectiveness of SRN is validated by comparing experiments with several different classification methods. It provides a new reference solution for the relevant authorities to accurately monitor road moving vehicles.

References

1. Franco, V., Kousoulidou, M., Muntean, M., et al.: Road vehicle emission factors development: a review. Atmos. Environ. **70**, 84–97 (2013)
2. China Mobile Source Environmental Management Annual Report in 2021 (Excerpt 1). Environ. Protect. **49**(Z2), 82–88 (2021)
3. Xu, Z., Wang, R., Kang, Y., et al.: A deep transfer NOx emission inversion model of diesel vehicles with multisource external influence. J. Adv. Transp. (2021)
4. Weiss, M., Bonnel, P., Kühlwein, J., et al.: Will Euro 6 reduce the NOx emissions of new diesel cars?-Insights from on-road tests with Portable Emissions Measurement Systems (PEMS). Atmos. Environ. **62**, 657–665 (2012)
5. Gallus, J., Kirchner, U., Vogt, R., et al.: Impact of driving style and road grade on gaseous exhaust emissions of passenger vehicles measured by a Portable Emission Measurement System (PEMS). Transp. Res. Part D Transp. Environ. **52**, 215–226 (2017)
6. Xu, Z., Kang, Y., Cao, Y., et al.: Spatiotemporal graph convolution multi-fusion network for urban vehicle emission prediction. IEEE Trans. Neural Netw. Learn. Syst. **32**(8), 3342–3354 (2020)
7. McClintock, P.M.: 2007 High Emitter Remote Sensing Project (2007)
8. Xu, Z., Wang, R., et al.: Unsupervised identification of high-emitting mobile sources based on multi-feature fusion. In: 2021 China Automation Congress (CAC), pp. 652–657. IEEE (2021)
9. Li, Z., Kang, Y., Lv, W., et al.: High-emitter identification model establishment using weighted extreme learning machine and active sampling. Neurocomputing **441**, 79–91 (2021)
10. Yunyun, W., Guwei, S., Guoxiang, Z., Hui, X.: Unsupervised new set domain adaptation learning based on self-supervised knowledge. J. Softw. **33**(04), 1170–1182 (2022)
11. Gong, D., Liu, L., Le, V., et al.: Memorizing normality to detect anomaly: memory-augmented deep autoencoder for unsupervised anomaly detection. In: Proceedings of the IEEE/CVF International Conference on Computer Vision, pp. 1705–1714 (2019)
12. Zhang, Y., Wang, J., Chen, Y., et al.: Adaptive memory networks with self-supervised learning for unsupervised anomaly detection. IEEE Trans. Knowl. Data Eng. (2022)
13. Lucas, J.M., Saccucci, M.S.: Exponentially weighted moving average control schemes: properties and enhancements. Technometrics **32**(1), 1–12 (1990)
14. Chen, H., Liu, H., Chu, X., et al.: Anomaly detection and critical SCADA parameters identification for wind turbines based on LSTM-AE neural network. Renew. Energy **172**, 829–840 (2021)
15. Wang, G.D., Melly, S.K.: Three-dimensional finite element modeling of drilling CFRP composites using Abaqus/CAE: a review. Int. J. Adv. Manuf. Technol. **94**(1), 599–614 (2018)
16. Xu, Z., Kang, Y., Cao, Y., et al.: Man-machine verification of mouse trajectory based on the random forest model. Front. Inf. Technol. Electron. Eng. **20**(7), 925–929 (2019)
17. Chauhan, V.K., Dahiya, K., Sharma, A.: Problem formulations and solvers in linear SVM: a review. Artif. Intell. Rev. **52**(2), 803–855 (2019)

Belief State Monte Carlo Planning for Multi-agent Visibility-Based Pursuit-Evasion

Xiao Xu[1], Shengming Guo[1], Dong Li[1(✉)], Pinggang Yu[1], Xiaocheng Liu[2], and Yi Ren[3]

[1] Joint Operations College, National Defense University, Beijing, China
donggeat2006@163.com
[2] PLA 31002 Troop, Beijing, China
[3] PLA 32381 Troop, Beijing, China

Abstract. Planning for a team of agents to pursue an evader is a challenging issue. The setting of partial observability in many real-world applications further increases the complexity. This paper presents a belief state based Monte Carlo Tree Search (MCTS) algorithm which is capable of generating online pursuing policy in the multi-agent visibility-based pursuit evasion scenario. The algorithm, Belief State UCT Consider Durations (BS-UCTCD) search, extends the basic MCTS on two main aspects: 1) it forms the tree model considering the uncertainty of forward transitions as well as the ability to reason concurrent and durative moves of involved agents; 2) it employs particle filtering to approximate the belief and combine the tree search to track the belief changes. We evaluate BS-UCTCD in a scenario where the algorithm is used to plan for two patrollers to search and capture one intruder. The comparative results demonstrate that our approach is effective and can perform better than fixed strategies and searching without belief tracking. Some domain specific configurations also enable its use under real-time constraints.

Keywords: Monte carlo tree search · Pursuit-evasion · Limited visibility · Belief state tracking

1 Introduction

Pursuit-Evasion problem is of particular interest for defense or security domains. This paper considers a specific form, *multi-agent visibility-based pursuit-evasion*, where multiple pursuing agents try to detect and capture an evader. In many real-world applications, like environmental monitoring, surveillance/patrol and robot rescue [1], planning effectively for the pursuer-side is strongly required. However, some particular characteristics make the problem quite challenging: 1) the state and action space is often huge due to the combinatorial nature of multiple agents, 2) the optimal pursuing policy changes dynamically with the evader's strategy, and 3) the environment is uncertain as pursuers cannot always observe the evader due to obstacles or limited sensing capabilities.

There have been many approaches for solving the visibility-based pursuit-evasion problem, which can be categorized according to how the problem is formulated, including differential games, graph variants, and geometric variants [2]. In this paper we model the problem as two-player zero-sum game and focus on utilizing game tree search techniques to generate online strategies for a team of pursuing agents. Previous related works mainly use minimax tree search [3] or Monte-Carlo tree search [4] as the backbone to solve the targeted game, configured with some domain specific tricks to tune the search process. There are also efforts incorporating grid partition [5] or Stackelberg Game principles [6] to further improve corresponding performance. However, they often lack a formal description of the problem. The setting of limited visibility makes the environment partially observable and pursuing agents must reason about these uncertainties. There may also be various assumptions about capabilities of pursuers and evaders, e.g. their actions are durative and time should be considered in finding a solution.

To overcome these limitations, this paper presents a general formalism and a game tree search algorithm for the multi-agent visibility-based pursuit evasion problem. our contributions include: (1) A formalism, Partially Observable Multi-Agent Gaming Abstraction (POMAGA), which specifies the multi-agent visibility pursuit evasion problem and incorporates game theoretical elements for tree search techniques. Specifically, it can model environments where agents have concurrent and durative actions. (2) An algorithm, Belief State UCT Considering Durations (BS-UCTCD), which performs Monte Carlo planning within the belief approximation of the real environment. It extends the UCTCD [7] search with particle filtering based belief update process in order to better manage imperfect information reasoning. (3) Experimental results to demonstrate the algorithm's effectiveness. We apply BS-UCTCD to a scenario with some particular configurations. Through a series of comparative analysis, we show the presented approach outperforms searching without belief tracking as well as some fixed strategies. The results also suggest that enhancing opponent modeling is important to enable the algorithm's use in more complex environments.

The rest of this paper is organized as follows: Sect. 2 describes the research background. In Sect. 3 we show details of the presented formalism as well as the algorithm. Section 4 illustrates the evaluated case, experiments and results analysis. Finally conclusions are drawn in Sect. 5.

2   Research Background

Monte Carlo tree search takes random sampling to approximate the optimal strategy. This process contains a number of iterations. Each iteration has four steps [8], i.e. *Selection, Expansion, Rollout,* and *Backpropagation.* The tree model is a directed graph where a node is designed to correspond to a certain domain state and edges are connected with player moves. Starting from the root, the traversing of nodes implies the state's possible future transitions. The key mechanism of MCTS is that it would search deeper on promising branches of the tree, making it effective for large state space problems.

One representative area that MCTS has achieved great success is the board game AI [9]. Recent efforts have also incorporated it to control a team of agents to perform tactical combat in Real Time Strategy (RTS) games [10]. Similar to the multi-agent pursuit-evasion game, the friendly and enemy side entities can move simultaneously rather than alternatively. Thus the vanilla MCTS, which is designed for the turn-based gaming mechanism, cannot be used directly in such settings. A popular approach to address this issue is to incorporate *bookkeeping nodes* [11] to the tree model, where one player chooses a move but does not apply it until the other player finishes move selection. Based on this idea, the UCTCD algorithm is presented in [7], which extends UCT [12], a MCTS implementation, with the capability to reason problems containing concurrent and durative actions. UCTCD with several enhancements has been applied in RTS tactical planning [13]. However, they do not consider the challenge of "fog of war", and assume that the planner has a full grasp of the environment. Necessary extensions must be made for the UCTCD so that it can reason uncertain or hidden information and find proper strategies for agents behaving in partially observable settings.

There are several efforts applying MCTS to imperfect information environments [8]. One popular approach is *determinization*, where a state will be sampled from the indistinguishable information sets, transforming the search to a deterministic problem [14]. The final decision will be chosen by voting among results of those sampled search. Though gained success in some domains [15], the determinization approach is considered to have two main drawbacks [14]: 1) sampled states from the same information set can result in different decisions (**strategy fusion**); 2) a node's value can be affected by not only its subtree but also other regions of the game tree (**non-locality**). The Information Set MCTS (ISMCTS) [16] tackles the first problem by designing the tree node to associate an information set rather than a deterministic state. While ISMCTS assumes each state in the information set appears with equal probability, Belief State MCTS (BS-MCTS) presented in [17] maintains a belief over the search and employs an opponent modeling strategy to learn the belief online. The authors argue that belief-state search with a suitable inference model can weaken the negative impact of non-locality.

MCTS is also making progress in dealing with large Partially Observable Markov Decision Problems (POMDPs). Partially Observable Monte Carlo Planning (POMCP) [18] is considered as the fastest online POMDP planning algorithm. It performs MCTS within the current belief space to seek the optimal strategy, similar to BS-MCTS, and approximates the belief changes using particle filtering. Several of its improvements, like the Determinized sparse partially observable tree algorithm [19] and the Abstract belief tree algorithm [20], all follow this procedure. However, POMCP is originally designed for single-agent problem and requires extension for multi-agent adversarial environments.

Our approach, BS-UCTCD, is similar to BS-MCTS as the search is also performed on the agent's belief state. However, our algorithm is quite different on implementation details. Specifically, it extends the belief state search capable of handling simultaneous and durative actions, thus applicable for POMAGA problems. The other difference is the belief learning strategies. BS-UCTCD mainly learns the probability distribution of individual states in the belief. While our approach draws on POMCP, using an unweighted particle filter to track the belief update under multi-agent settings.

3   Approach

3.1   Partially Observable Multi-agent Gaming Abstraction

We give formal descriptions on the targeted problem. The multi-agent visibility-based pursuit evasion model should contain the two opposing sides of agents as well as their interacting environments. We call it Partially Observable Multi-Agent Gaming Abstraction (POMAGA) here. In the following context of this paper, we represent the model as $M = \{P, S, Z, O, A, Tr, Te, U\}$, where:

- $P = \{MAX, MIN\}$ is the set of force sides. Not losing the generality, we set the MAX as the friendly side and MIN as the enemy.
- $S = \{F, E, T, t\}$ is the set of states where $F = \{u_i^f\}, i \in [1, m]$ and $E = \{u_j^e\}, j \in [1, n]$ are the friendly and the enemy side containing m and n units respectively, T is the terrain, and t denotes the current simulation time.
- Z is the set of observations. Usually the observations for a player reflect only part of the ground state.
- $O(p, s) \rightarrow Z$ is the observation function, returning the observation for player p given the state s. Further, we denote $I(p, z) = \{s \in S | O(p, s) = z\}$ as the information set of player p given observation z, which contains all potential world states that match the observation z from p.
- $A(p, s)$ is the legal action set for player p given the state s.
- $Tr(s, a) : S \times A \rightarrow S$ defines the state transition function, returning the next state given the current state s and the chosen move a. Here a should be the joint actions of corresponding units in s. In the context of game tree search, the joint action issued by a player is also called a 'move'.
- $Te(s) \rightarrow \{true, false\}$ checks whether the simulation can terminate in s.
- $U(s) \rightarrow \mathbb{R}$ returns the utility value of the state s. It is usually defined that higher value means more advantage for the MAX player.

The existence of durative actions mainly concerns the state advancing. Suppose a unit u issued an action a_i with the duration d_i at time t, then it cannot perform a_i again within $(t, t + d_i)$. We denote $u.t_e$ as the nearest expected time that the unit can start execute next actions $(t_e > t)$. The state s will advance to s' with the time $t' = \min\{u_1^f.t_e, \ldots, u_m^f.t_e, u_1^e.t_e, \ldots, u_n^e.t_e\}$.

POMAGA is a special form of Partially Observable Stochastic Game (POSG) [21]. It is a competing environment and set that units in the same side share the same rewarding mechanism. This actually neglects self interests of individual agents and set that they act to maximize the reward of the whole side. Our research concerns the online planning for a side of units. When the side has only one unit, the planner works for the single agent; whereas when there are many agents, the planner acts as the "commander" to guide their actions. The partially observable environment assumes that information of own units is perfect for the planner, while the state and action of the other side units are indistinguishable.

3.2   Belief State UCTCD Algorithm

In the belief state based MCTS, each node is extended to correspond to a belief. The concept of a belief represents the relating agent's subjective understanding of the world. The traversing of nodes relates to belief update. The search process is based on sampled determinized states. That is to say, a *Sampling* step would be firstly performed before a standard MCTS iteration.

Considering the use of MCTS in POMAGA, the main problem is to handle concurrent actions for the forward transition. We continue to use the notations from UCTCD, classifying player moves as three types: the SOLO, the FIRST, and the SECOND. If only one player can issue moves under some node, the move has a SOLO type, which will be applied directly. Otherwise both players can move, and two action selection steps are generated to accomplish the process: take one side player's move as FIRST to generate a book-keeping node, where the move effect will not be applied; then it comes to SECOND moves of the other side player, where both FIRST and SECOND moves will be applied. These extensions ensure states are advanced following the causality. With this idea, we present the BS-UCTCD algorithm as shown in Fig. 1 and Algorithm 1.

Figure 1 describes an example tree model and its search process in BS-UCTCD. We see that in the tree model, nodes with SOLO and SECOND moves are connected with $\langle a, o \rangle$ pairs, while FIRST moves directly lead to a single node. This is due to that the problem is uncertain and even the same move would generate different observations. In the example search, the root node (with observation o_0) selects a_{12} and gets a book-keeping node, denoted as $n_1 \leftarrow root\langle a_{12}\rangle$. After that, the traversing process selects a_{21} and get the observation o_5 with the joint action $a = \{a_{12}, a_{21}\}$ being applied. Here we denote the selected node $n_2 \leftarrow n_1\langle a_{21}, o_5\rangle$. If n_2 is a new node, the Rollout to evaluate n_2 and the Back-propagation for n_1 and *root* will happen, and then the iteration of this time gets ended.

In Algorithm 1, a tree node n is associated with some properties, like $N(n)$ as the number of visits, $N(n, a)$ as the number of selecting a under n, $V(n, a)$ as the value, and $B(n)$ as the belief at n. The search starts from sampling a state of the root's belief, which we denote as $s \sim B(root)$ here. During the Selection step, the action is chosen based on its UCB1 value [12] which is employed by regular UCT search to tackle the exploration and exploitation trade-off. Line 19 is for the MAX player and Line 21 is for the MIN player, where c is referred to as the exploration constant. The available action set under the node n can be different as the sampled state s varies. Here the search chooses actions from $A(\rho(n, s), s)$

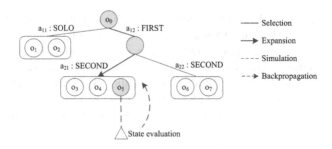

Fig. 1. A conceptual view of BS-UCTCD search. Circles represent tree nodes. Those dark colored are nodes traversed during the example search.

where $\rho(n, s)$ defines which player can move under n and s. In our settings, when both players can move, FIRST moves are for the MAX player and SECOND moves are for the MIN player. We also denote $n\langle a \rangle$ and $n\langle a, o \rangle$ as to return the corresponding child node of n. If it has not existed in the tree yet, a new node will be added for Expansion. Compared to vanilla UCT search, the new node generation will happen if the selected moves produce a new observation. Also this forward transition is only called under SOLO or SECOND moves (Line 39). The following Rollout and Backpropagation steps are the same as the basic UCT, except that $V(n, a)$ means the value of executing a under the belief $B(n)$ rather than under a particular state.

3.3   Belief Update with Particle Filtering

The BS-UCTCD search is used for online planning, where at every decision epoch it is called to produce a decision according to the observation. The root node is created with the planning agent's current belief. Similar to POMCP, we employ an unweighted particle filter to approximate the belief state and track its change. Each particle is a hypothesis about the model's current state, and the particle set collectively form an estimate of the planning agent's belief. We correspond each node with a belief in the tree model, i.e. $B(n) = \{x_i\}, 1 \leqslant i \leqslant K$, where K is the sum of particles. Since nodes are associated with actions of entities, the movement of particles, i.e. the potential state transition, can be naturally simulated with the search.

At the beginning of the environment, $B(root)$ is sampled uniformly from $I(MAX, o)$ where o is the current observation. We set the root player is the MAX side by default. During the search, when the forward transition happens, the resulting child node n' will update its belief (the particle set) with the resulted state s' (Line 42). Note that nodes directed by FIRST moves are not with a belief because it does not lead to belief changes. The algorithm finally returns the action with greatest value for the agent to execute.

At following epochs the algorithm will first evolve the belief. Here we define $root\langle a, z \rangle = \{n_i\}$ where n_i is the root's nearest SOLO or SECOND descendant associated with the move a selected from the root and the observation z at n_i.

Algorithm 1: BS-UCTCD algorithm

Initialize: root $\leftarrow None$, $a^* \leftarrow None$.

1 Function BSUCTCD(Observation o)
2   **if** root *is None* **then**
3 | $\mathcal{L} \leftarrow I(MAX, o)$
4   **else**
5 | $\mathcal{L} \leftarrow \bigcup B(n_i), n_i \in root\langle a^*, o \rangle$
6 | Compensate particles to $\mathcal{L}$
7   **end**
8   root $\leftarrow$ new node with belief $\mathcal{L}$
9   **foreach** $i \leftarrow 1$ *to MaxTraverse* **do**
10 | $s \sim B(\text{root})$
11 | Traverse(root, s)
12   **end**
13 $a^* \leftarrow \arg\max_a Q(\text{root}, a)$
14   **return** a^*
15 end
16 Function SelectAct(Node n, State s)
17 $legal \leftarrow A(\rho(\text{n, s}), \text{s})$
18   **if** $\rho(\text{n, s})$ *is MAX* **then**
19 | $a \leftarrow \arg\max_{b \in legal} \left(V(n, b) + c\sqrt{\frac{\log N(b)}{N(n, b)}} \right)$
20   **else**
21 | $a \leftarrow \arg\min_{b \in legal} \left(V(n, b) - c\sqrt{\frac{\log N(b)}{N(n, b)}} \right)$
22   **end**
23   **return** a
24 end

25 Function Traverse(Node n, State s)
26 $score \leftarrow 0$
27   **if** $N(\text{n}) == 0$ **then**
28 | $score \leftarrow$ Rollout(s)
29   **else**
30 | $a \leftarrow$ SelectAct(n, s)
31 | **if** a.type() *is FIRST* **then**
32 | $n' \leftarrow \text{n}\langle a \rangle$
33 | $s' \leftarrow s$
34 | **else**
35 | $m \leftarrow \{a\}$
36 | **if** a.type() *is SECOND* **then**
37 | $m \leftarrow m \bigcup \{\text{n}.move\}$
38 | **end**
39 | $s' \leftarrow Tr(\text{s}, m)$
40 | $o \leftarrow O(MAX, s')$
41 | $n' \leftarrow \text{n}\langle a, o \rangle$
42 | $B(n') \leftarrow B(n') \bigcup \{s'\}$
43 | **end**
44 | $score \leftarrow$ Traverse(n', s')
45 | $N(\text{n}, a) \leftarrow N(\text{n}, a) + 1$
46 | $V(\text{n}, a) \leftarrow V(\text{n}, a) + \frac{score - V(\text{n}, a)}{N(\text{n}, a)}$
47   **end**
48 $N(\text{n}) \leftarrow N(\text{n}) + 1$
49   **return** $score$
50 end

If a is SOLO, there is only one element; else if a is FIRST, there would be more than one node in the set. Supposing the agent executes a^* and observes o in the real environment, the new belief can be obtained according to the previous root node's belief, i.e. $\mathcal{L} = \bigcup B(n_i), n_i \in root\langle a, o \rangle$. The implied idea is that potential forward transitions have been reasoned in the tree model. The resulted particles will be treated legal only if the corresponding observation matches the real environment. As the exact transition of the world is not able to know fully, we just consider possible enemy actions to happen with equal probability. This ensures computational efficiency and makes the planner general purposed. With the agent continuing to act in the real world, impossible transitions are discarded and the generated particle set is expected to gradually achieve higher accuracy to approximate the exact state.

In practice the particle depletion may happen and $\mathcal{L}$ does not reach a reasonable cardinality. At this point some compensations need be made. Specifically, we randomly perturb some existing particles to nearby states (still consistent to the current observation o) and add them to $\mathcal{L}$. The update iterates until the sum of particles reaches a predefined sum. After the belief updates, a new root with $\mathcal{L}$ will be used for searching.

4   Case Study

4.1   Scenario Description

We test the presented algorithm in a scenario which is shown in Fig. 2. The environment is implemented as a 19×17 grid world, where white tiles are walkable areas and black ones are blocked. There are two patrolling soldiers (blue pentacles) and an intruder (red triangle) in an enclosed area. The intruder's mission is set to perform damages on facilities located at specific positions (denoted as green). The patrolling soldiers need to catch the intruder as soon as possible in order to avoid more loss of facilities. The simulation ends when all facilities get destroyed or the intruder gets caught. There are totally 190 walkable positions in the map, resulting approximately 2×10^8 states ($190^3 \times 2^5$).

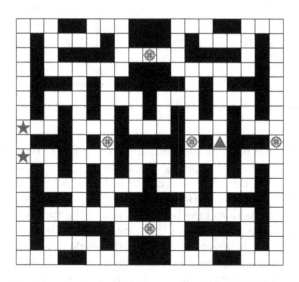

Fig. 2. The visibility-based pursuit-evasion scenario.

For the settings of agent actions, the patrolling soldiers and the intruder can move along four cardinal orientations. They have the same speed. Also, the intruder can perform 'DESTROY' while it is at the position of a facility. Note that destroying a facility will cost more time than a single movement. Therefore, the scenario contains concurrent and durative moves, which describes a typical POMAGA environment. Here we consider that those actions have deterministic effects. The soldier agents cannot get a full view of the environment, though they have prior knowledge about the terrain's shape. They can only percept enemies along the Line of Sight (LoS). What is fortunate is that they can locate the intruder at the time it successfully destroys a facility. The rewarding mechanism sets that the patroller side get a default reward of -1 for every elapsed time. In addition to that, it receives -25 if the intruder successfully destroys a facility, -200 if all facilities are damaged, and +500 if they capture the intruder. Rewards are accumulated with no discount factor.

The intruder has a fixed strategy, with which it will navigate to a random facility and destroy it. If it feels there are patrolling soldiers within certain Manhattan distances (it is not restricted to LoS), it will evade. Also, during the evasion it will try to approach a facility with 50% probability and otherwise choose random routes. The path finding we use is the regular A^* algorithm. How the patrolling soldiers act in this environment is what our experiments target on. As both sides of agents have equal movement speed, it is difficult for patrolling soldiers to catch the enemy unless they cooperate and set traps. The partial observability adds more complexities to the problem. The presented algorithm is used to guide the patrolling soldiers' behavior online, reasoning based on current observations and generating appropriate decisions.

4.2   Experiments and Results

In the presented scenario the main source of uncertainties is that the intruder's position is not known at some time steps. The sum of its possible positions will increase with the time if it continues to be unknown. Supposing the possible position set of the intruder at time t is known as L_t, it will change as $L_{t+1} = \bigcup Adj(l_i)$, where $l_i \in L_t$ and $Adj(l_i)$ are the adjacent coordinates set of l_i. When there is no further information, the possibilities of each position should be the same. The presented BS-UCTCD models the indistinguishable state transition by moving particles, obtaining the information set naturally. This decreases the uncertainty and narrows the searching space when the target is lost. For certain problems, opponent modeling techniques can be employed to make the particle set approximating the exact state better. Here we do not consider this issue but focus on evaluating the effectiveness of our presented algorithm.

MCTS can work with little support of domain knowledge. Whilst it is proved that some specific configurations can be beneficial to get better performance [22]. To improve the search efficiency, we evaluate several domain specific improvements in BS-UCTCD:

- *Move ordering*: The main idea is to priorily expand children nodes with valuable moves, which would enable them to get deeper search in limited time. In our case the search starts with a deterministic state, and the intruder's position is known. We generate the valuable move as the one driving patrolling soldiers moving towards the intruder along the shortest path.
- *Rollout heuristics*: The default MCTS randomly select available actions in the rollout phase. A more pragmatic approach is to simulate probable behaviors of the players, which would obtain more accurate evaluations in a handful

of rollouts. Here we mainly consider the opponent strategy as: 1) randomly choosing a target facility to pursue from the leaf state, and 2) randomly running far away from the nearest patrolling soldier within LoS.

The experiments are set up as three steps to obtain a thorough evaluation on the use of BS-UCTCD in the scenario. We first explore the parameter influence, and then the playing strength compared to other strategies. After that, we make an evaluation on the algorithm's performance with real-time constraints. The computer for running experiments has an Intel(R) Core(TM) i7-6700HQ CPU @ 2.60GHz and 16 GB of RAM available.

Parameter Tuning. While the move ordering and the rollout heuristic are specified, there is a need to optimize parameters of the BS-UCTCD. We mainly consider the effect of exploration constant c (Algorithm 1, Lines 19 and 21) and the rollout depth. They only concerns the regular MCTS iteration process, in which c relates to the Selection step and the rollout depth affects the states evaluation. Therefore, UCTCD is directly used to run parameter tuning experiments, where each iteration will start with the exact state. As BS-UCTCD is sample based UCTCD search, the parameters suitable for UCTCD will also work in BS-UCTCD. The two evaluated parameters are positively related in terms of the algorithm performance. We can make independent evaluations and choose the best values of the two.

The rollout depth is first explored in order to get accurate node estimations. Here the exploration constant is set as $\sqrt{2}$, which is the theoretical value proved in [12]. The actual optimal value should be domain dependent. We will adjust it later. The sum of iterations is set as the fixed 100. Generally more iterations means better performance at the cost of more computation time. This value is chosen for the reason that UCTCD will not achieve convergence with it, thus the results can clearly show the performance difference of variables. For each depth value, 500 independent runs are performed. The results are shown in Fig. 3(a). The symbol ∞ means that the simulation will continue until the termination condition is satisfied. We see that deeper rollout dose not lead to better performance. The score line increases sharply at start and reach a peek when the depth value is 20. After that it steadily decreases. Whether too small or too large is the depth value, a good node estimation cannot be achieved. This is mainly due to that in our rollout configurations the patrolling units move randomly while the intruder has clear purposes. The ∞ depth will mostly cause the simulation terminated as the winning of the intruder. This certainly weakens the estimation accuracy. For our case a proper rollout depth is required.

To further tune the exploration constant, we set the rollout depth as 20. Other settings keep unchanged while c varies. Figure 3(b) shows the exploration constant test results. It can be seen that $c = \sqrt{2}$ obtains a relatively higher score and sum of wins. But overall, the variance of c does not affect the final performance quite much. In the following experiments, we all set the exploration constant as $\sqrt{2}$ and the rollout depth as 20.

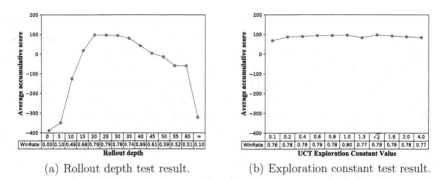

(a) Rollout depth test result.

(b) Exploration constant test result.

Fig. 3. The parameter tuning results. In both figures the lines show the average scores and the tables give the win data (percentage of catching the intruder) with the variance of the rollout depth and the exploration constant respectively.

Playing Strength Evaluation. We show the effectiveness of BS-UCTCD through comparison to other two approaches, i.e. Cheating-UCTCD and Random-UCTCD. Cheating-UCTCD assumes the information is perfect for the patrollers side. Random-UCTCD also performs sample based UCTCD search, but there is no hidden target reasoning. Compared to BS-UCTCD, it assumes that the intruder may be at any of the unseen positions if not within the patrollers' observation or destroying facilities. Expect for those, it is set the same with regard to the tree node expansion, the UCT selection, and the rollout policy.

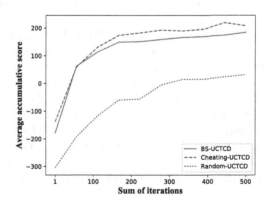

Fig. 4. The performance of three different search algorithms with the variation of iterations. The scores are averaged for every 50 steps from 1 to 500.

In order to show how scores change with the increase of iterations, we perform 100 independent runs for each iteration sum varying from 1 to 500. The obtained results, averaged for every 50 steps, is shown in Fig. 4. It can be seen that the three algorithms all have performance increases as the iteration sum grows. Cheating-UCTCD generates the best scores, running after the BS-UCTCD. Random-UCTCD, however, falls behind quite far. Cheating-UCTCD runs with the perfect information, which is why it outperforms the other two. While Random-UCTCD faces much more uncertainties. It makes the two patrollers move randomly at start. When the intruder comes into sight, they will try to catch it. With the iterations growing, the catching capability also rises, which is why the obtained score increases. However, as the chance of finding the target is still random, the overall score is not high. We can also find that the performance of BS-UCTCD is quite close to Cheating-UCTCD. In our settings, the only indistinguishable information is the intruder's position and it can be fully known under some conditions. The searching space is properly constrained, making BS-UCTCD effective for the evaluated scenario.

It is interesting to note that Cheating-UCTCD represents the performance of searching with the most accurate belief (totally equals to the exact state). While there is no opponent reasoning in Random-UCTCD. It takes all possible states as uniformly distributed. BS-UCTCD employs particle filtering to form the information set based on own and opponents' possible moves. The level of uncertainty is within Cheating-UCTCD and Random-UCTCD. Improving the accuracy of the belief state should be of significant to ensure the performance of BS-UCTCD.

Performance Under Real-Time Constraints. As many applications require the decisions to be generated online, there is a need to test the algorithm's online performance. Similarly, many real-time game AI researches constrain the period of a decision as 40 ms, as the screen refreshes 24 frames per second (41.6 ms). Here we continue to use this budget.

Three algorithms, i.e. Cheating-UCTCD, BS-UCTCD and Random-UCTCD, are evaluated under the real-time constraints. Their settings are the same as previously mentioned but the computation budget changes. Each decision is generated by the 40 ms search from the current observation. As comparison, we also test two fixed strategies. The first is Perfect Information Pursuit (PI-Pursuit), which drives the two patrollers to move to the intruder directly. It works under the assumption that the target's position is known. The other one is Partially Observable Pursuit (PO-Pursuit). It makes the two soldiers patrolling randomly initially and performs PI-Pursuit when they know the exact state of the intruder. Each strategy is performed with 500 independent runs, with results shown in Table 1.

We can get that the two fixed strategies have similar performance, both obtaining quite low wining rates. It would not be easy to catch a target with the same speed if there is no cooperation strategies between pursuers. Though patrollers driven by Random-UCTCD also move randomly at start, they will try to catch the target collaboratively while observing it. That is the reason that its winning percentage approaches to more than 50%. On the other hand, the performance of the three searching algorithms is consistent to the results

Table 1. The performance of different approaches. The search based algorithms are under the 40 *ms* computation budget.

Algorithms	Average score	Winning percentage
Cheating-UCTCD	205.78 ± 201.68	94.0%
BS-UCTCD	166.05 ± 250.62	86.8%
Random-UCTCD	-14.37 ± 317.84	61.2%
PI-Pursuit	-286.93 ± 258.07	17.4%
PO-Pursuit	-245.96 ± 288.23	23.2%

obtained from last section. Cheating-UCTCD wins 94% of the test, which is the best of the three. BS-UCTCD does not fall behind much. It reaches the percentage of 86.8, which is an proper value in terms of real-time performance. Fixing the searching time and fixing the max iteration are different for these algorithms' running. When the searching time is fixed, the sum of iterations performed for each decision should vary with the situations changing. Generally the number should be small at start and rises much at end. This is because the search is easy to access terminal states when the simulation is about to be over. But overall, after a set of domain specific configurations, the main factor to affect the search performance is the level of uncertainty.

4.3   Analysis

The experiments generally show that BS-UCTCD can perform well for planning the two patrollers. There should be two main reasons for the algorithm's proper performance under real-time constraints. Firstly, we incorporate domain independent configurations to improve the search efficiency. The move ordering techniques guide the BS-UCTCD to search more on promising branches. The rollout heuristic is also of great importance to improve the algorithm's anytime performance, as the default random rollout failed to gain good results. Secondly, the presented scenario has a relatively low level of uncertainty. The only indistinguishable state is the intruder's position, and we also set that it can be identified occasionally. That is why BS-UCTCD can get close scores to Cheating-UCTCD in the presented case. The comparative results demonstrate that limiting the information uncertainty should be important to ensure the searching algorithm's performance. If the problem gets more complex (e.g. more intruders are involved), the uncertainty would grow exponentially, and the effectiveness of BS-UCTCD would reduce.

Our current implementation forms the information set by reasoning potential moves of opponents and eliminating those violating the observation history. As each particle in the algorithm corresponds to a determinized search, there is a need to discuss the two problems existed commonly in the determinization approach: strategy fusion and non-locality. For the first problem, BS-UCTCD associates each node to a belief (particle set). Similar to BS-MCTS, it associates the node value $V(n, a)$ as the action utility under a belief rather than individual states. The problem of strategy fusion dose not exist in such settings. The second problem is mainly aroused by ignoring the ability of adversaries. We have not yet addressed this issue. However, opponent modeling is considered to be able to weaken the negative effect of non-locality [17]. Based on these descriptions, exploring opponent modeling techniques to strengthen BS-UCTCD is a reasonable choice to extend its use for more complex scenarios.

5   Conclusion

In this paper we put forward an extension of UCTCD to tackle the sequential decision making of pursuers with limited visibilities catching an evader. The algorithm, BS-UCTCD, performs determinized search based on states sampled form the belief. The detailed implementation employs the unweighted particle filter to approximate belief states as well as their update. It also inherits UCTCD to be capable of reasoning simultaneous and durative moves. We evaluate the algorithm in a typical scenario. Experimental results show that BS-UCTCD is effective for such environments and can perform well under real-time constraints. The comparative analysis also suggests that reducing the level of uncertainty is of significance to improve the algorithm's effectiveness.

As with future work, we intend to incorporate opponent modeling techniques into BS-UCTCD. It helps to approximate the belief with more accuracy, which is the key to obtain satisfying results. On the other hand, heuristics can play important roles to ensure the anytime performance of MCTS. While traditionally they are specified manually, it is worth exploring the use of machine learning techniques to automatically constructing such knowledge.

References

1. Robin, C., Lacroix, S.: Multi-robot target detection and tracking: taxonomy and survey. Auton. Robots **40**(4), 729–760 (2016)
2. Stiffler, N.M., O'Kane, J.M.: Complete and optimal visibility-based pursuit-evasion. Int. J. Robot. Res. **36**(8), 923–946 (2017)
3. Raboin, E., Nau, D.S., Kuter, U., Gupta, S.K., Svec, P.: Strategy generation in multi-agent imperfect-information pursuit games. In: 9th International Conference on Autonomous Agents and Multiagent Systems (AAMAS 2010), pp. 947–954. Toronto, Canada, May 2010
4. Lisý, V., Bosanský, B., Pechoucek, M.: Anytime algorithms for multi-agent visibility-based pursuit-evasion games. In: International Conference on Autonomous Agents and Multiagent Systems, AAMAS 2012, pp. 1301–1302. Valencia, Spain, June 2012
5. Li, A.Q., Fioratto, R., Amigoni, F., Isler, V.: A search-based approach to solve pursuit-evasion games with limited visibility in polygonal environments. In: Proceedings of the 17th International Conference on Autonomous Agents and Multi-Agent Systems, AAMAS 2018, pp. 1693–1701. Stockholm, Sweden, July 2018

6. Karwowski, J., Mandziuk, J.: A monte Carlo tree search approach to finding efficient patrolling schemes on graphs. Eur. J. Oper. Res. **277**(1), 255–268 (2019)

7. Churchill, D., Buro, M.: Portfolio greedy search and simulation for large-scale combat in starcraft. In: 2013 IEEE Conference on Computational Intelligence in Games (CIG), pp. 1–8. Niagara Falls, ON, Canada (2013)

8. Browne, C., et al.: A survey of monte Carlo tree search methods. IEEE Trans. Comput. Intellig. AI Games. **4**(1), 1–43 (2012)

9. Silver, D., et al.: Mastering the game of go with deep neural networks and tree search. Nature **529**(7587), 484–489 (2016)

10. Yang, Z., Ontañón, S.: An experimental survey on methods for integrating scripts into adversarial search for RTS games. IEEE Trans. Games **14**(2), 117–125 (2022)

11. Balla, R., Fern, A.: UCT for tactical assault planning in real-time strategy games. In: 21st International Joint Conference on Artificial Intelligence, pp. 40–45. Pasadena, CA, USA, July 2009

12. Kocsis, L., Szepesvári, C.: Bandit based Monte-Carlo planning. In: Fürnkranz, J., Scheffer, T., Spiliopoulou, M. (eds.) ECML 2006. LNCS (LNAI), vol. 4212, pp. 282–293. Springer, Heidelberg (2006). https://doi.org/10.1007/11871842_29

13. Swiechowski, M., Godlewski, K., Sawicki, B., Mandziuk, J.: Monte Carlo tree search: a review of recent modifications and applications. CoRR abs/2103.04931 (2021). https://arxiv.org/abs/2103.04931

14. Long, J.R., Sturtevant, N.R., Buro, M., Furtak, T.: Understanding the success of perfect information monte Carlo sampling in game tree search. In: Proceedings of the Twenty-Fourth AAAI Conference on Artificial Intelligence, AAAI 2010. Atlanta, Georgia, USA, July 2010

15. Bjarnason, R., Fern, A., Tadepalli, P.: Lower bounding Klondike solitaire with monte-Carlo planning. In: Proceedings of the 19th International Conference on Automated Planning and Scheduling, Thessaloniki, Greece, September 2009

16. Cowling, P.I., Powley, E.J., Whitehouse, D.: Information set monte Carlo tree search. IEEE Trans. Comput. Intell. AI Games **4**(2), 120–143 (2012)

17. Wang, J., Zhu, T., Li, H., Hsueh, C., Wu, I.: Belief-state monte Carlo tree search for phantom go. IEEE Trans. Games **10**(2), 139–154 (2018)

18. Silver, D., Veness, J.: Monte-Carlo planning in large POMDPs. In: 24th Annual Conference on Neural Information Processing Systems 2010, pp. 2164–2172. Vancouver, British Columbia, Canada, December 2010

19. Ye, N., Somani, A., Hsu, D., Lee, W.S.: DESPOT: online POMDP planning with regularization. J. Artif. Intell. Res. **58**, 231–266 (2017)

20. Kurniawati, H., Yadav, V.: An online POMDP solver for uncertainty planning in dynamic environment. In: Inaba, M., Corke, P. (eds.) Robotics Research. STAR, vol. 114, pp. 611–629. Springer, Cham (2016). https://doi.org/10.1007/978-3-319-28872-7_35

21. Bernstein, D.S., Givan, R., Immerman, N., Zilberstein, S.: The complexity of decentralized control of Markov decision processes. Math. Oper. Res. **27**(4), 819–840 (2002)

22. Gelly, S., Silver, D.: Monte-Carlo tree search and rapid action value estimation in computer go. Artif. Intell. **175**(11), 1856–1875 (2011)

Dynamic Modeling and Predictive Control of Heavy-Duty Gas Turbines

Shi Li[1], Shoutai Sun[1], Yali Xue[2(✉)], and Li Sun[1]

[1] School of Energy and Environment, Southeast University, Nanjing 210096, China
[2] Department of Energy and Power Engineering, Tsinghua University, Beijing 100084, China
xueyali@tsinghua.edu.cn

Abstract. The heavy-duty gas turbine system has an excellent energy conversion rate, which can not only significantly improve the power generation efficiency and reduce environmental pollution, but also help to adjust the peak-to-valley difference of the power grid and optimize the energy structure. However, the heavy-duty gas turbine system has the characteristics of nonlinearity and strong coupling. It is a challenge to establish a safe, stable and fast-tracking control system. To this end, this paper establishes a dynamic simulation model for the GE 9FA heavy-duty gas turbine. The state space model with two inputs and two outputs is identified by the subspace identification method. Augmented state space model is used to design predictive control. Under a variety of disturbance conditions, it is proved that the controller designed in this paper has good disturbance rejection ability and robustness compared with the traditional PID controller, and can achieve the expected control effect.

Keywords: Heavy duty gas turbine · Dynamic model · State space · Predictive control

1 Introduction

Energy is the material basis for human survival and one of the decisive factors for national economic development. The goal of carbon peaking and carbon neutrality puts forward more flexible peak regulation requirements for traditional energy sources. Heavy-duty gas turbine is one of the most important components of a combined-cycle gas turbine (CCGT) system. Gas turbine system has extremely high heat-to-power conversion efficiency, low pollutant emissions, and good peak-shaving capacity, which can effectively optimize the energy structure and improve energy utilization [1, 2]. At present, establishing an accurate simulation model for heavy-duty gas turbine units and designing the control system according to the dynamic characteristics is a difficult problem that needs to be solved.

The dynamic characteristics of heavy-duty gas turbines mainly include volume inertia, thermal inertia of components and rotational inertia of shafts. The modular modeling method is concise and clear, which is convenient for assembly, splicing and expansion of system model. At present, this method has become one of the most mainstream and widely used methods among many modeling methods. The models of WI Rowen [3, 4] and GAST [5, 6] are the first classical gas turbine models proposed. They use modular ideas to establish static and dynamic modules of gas turbines and connect these models in series with requirements as a guide. Duan et al. [7] proposed a nonlinear mathematical model of gas turbine system with a regenerator, described the compressor and turbine characteristics with nonlinear analytical expressions, and derived the nonlinear state space equation expression. Hao [8] established a complete gas turbine model,, and studied the dual off-design performance of the gas turbine under three operating conditions of temperature-load, fuel-load and fuel-temperature.

In recent years, there are many researches on the application of advanced control methods in gas turbine control. Iqbal et al. [9] developed a neuro-fuzzy controller using a hybrid learning algorithm, and validated the effectiveness of the controller against load disturbances and setpoint changes in a grid-connected environment. Wiese et al. [10] reduced the order of the nonlinear high-order physical model, and transformed the control problem of the gas turbine system into a multi-input-multi-output (MIMO) optimization problem that can be solved under the nonlinear model predictive control framework. Gengjin Shi al. [11] applied the linear active disturbance rejection controller (LADRC) to the gas turbine shaft speed control system, and compared it with the traditional PID controller. The simulation results show that LADRC has the advantage in suppressing different disturbances.

To summarize, the contributions of this paper are:

1) The dynamic model of the heavy-duty gas turbine is designed based on the mechanism and structure of the gas turbine, and then verified with actual operation data.
2) The MIMO transfer function of the dynamic model is obtained by the subspace identification method.
3) The augmented state space model is designed for predictive control.
4) Through various disturbance tests and comparison with the traditional PID controller, it is proved that the controller designed in this paper has the disturbance rejection ability and robustness and can achieve the expected control effect.

The rest of this paper are organized as follows. The dynamic modeling process of the heavy-duty gas turbine are given in Sect. 2. In Sect. 3, the verification results of the dynamic model using the actual operation data of the heavy-duty gas turbine are presented, and the MIMO transfer function is obtained by the subspace identification method. The design process of the augmented state space model for predictive control are given in Sect. 4. The Sect. 5 is the control experiment, comparing the control effect with the traditional PID controller. Conclusions are discussed in Sect. 6.

2 Modeling of Gas Turbine

The established model conforms to the laws of conservation and thermodynamics. The gas turbine is mainly composed of three basic components: compressor, combustor, and turbine. The model established in this paper also includes modules such as rotating shaft, flow channel, and generator, as shown in Fig. 1. Where the inputs are the inlet air flowrate and fuel flowrate, and the outputs are the turbine exhaust temperature and system power.

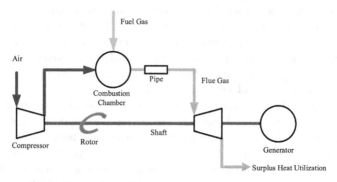

Fig. 1. Dynamic model structure of heavy-duty gas turbine.

2.1 Compressor and Turbine

Compressors and turbines are generally assumed to be quasi-steady-state components. Compressor characteristics are reflected by pressure ratio π_C, corrected rotational speed N_C, corrected flow rate G_C and isentropic efficiency η_C; turbine characteristics are reflected by expansion ratio π_T, corrected rotational speed N_T, corrected flow rate G_T and isentropic efficiency η_T (Fig. 2).

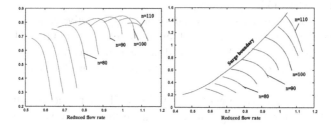

Fig. 2. Compressor characteristic curve.

2.2 Combustion Chamber

According to mass and energy conservation, we have

$$\frac{dM}{dt} = w_{air} + w_{flue} + w_{out} \tag{1}$$

$$\frac{dE}{dt} = w_{air}h_{air} + w_{fuel}h_{fuel} + w_{out}h_{out} + \eta w_{fuel}LHV - \delta S(T_o - T_w) \tag{2}$$

where M is gas mass (kg); E is gas energy (kJ); w is flow rate (kg/s); h is specific enthalpy (kJ/kg); η is combustion efficiency; LHV is low-level calorific value (kJ/kg); δ is heat transfer coefficient (kJ/m²/°C); S is metal wall area (m²); T is temperature (°C).

2.3 Pipe

According to the conservation of momentum, energy, and mass, respectively, we have

$$L/A * \frac{d\omega}{dt} + (P_{out} - P_{in}) + \Delta P = 0 \tag{3}$$

$$A \cdot L \cdot \rho \cdot c_v \cdot \frac{dT}{dt} + w \cdot \Delta h = q \tag{4}$$

$$\frac{dM}{V \cdot dt} = (\frac{d\rho}{dP} \cdot \frac{dP}{dt} + \frac{d\rho}{dT} \cdot \frac{dT}{dt} + \frac{d\rho}{dX} \cdot \frac{dX}{dt}) \tag{5}$$

Where L is length (m); A is cross-sectional area (m²); ΔP is frictional pressure drop; ρ is density; X is gas composition.

2.4 Rotor

$$\frac{d\omega}{dt} = \frac{1}{J\omega} \cdot (P_t - P_c - P_f - P_E) \tag{6}$$

Where ω is angular velocity; J is moment of inertia. P represents different power.

2.5 Dynamic Model Based on Modelica

A dynamic mechanism model system is established for heavy-duty gas turbines, as shown in Fig. 3. In the simulation model, the rotational speed and corrected flow rate of the compressor are obtained by a two-dimensional table interpolation method, to obtain the pressure ratio and adiabatic efficiency of the compressor under different working conditions.

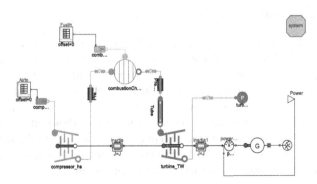

Fig. 3. Heavy duty gas turbine dynamic model.

3   Model Verification and Identification

3.1   Static Operation Condition

The static accuracy of the simulation model is verified by comparing with the rated operating conditions of the GE 9FA gas turbine. The environmental parameters used in the simulation are 101.325kPa, 15°C, and the relative humidity is 60%. The rated speed is 3000 rpm, the inlet air flow rate is 645.02 kg/s, the fuel flow rate is 14.355 kg/s, and the low calorific value of the fuel is 48686.3 kJ/kg.

Table 1 shows the comparison between the calculated value and the design value of the gas turbine model. The calculated results are in good agreement with the design values.

Table 1. Comparison under rated condition

Parameter	Design value	Calculated value	Relative error (%)
Rated speed (rpm)	3000.00	3000	/
Power (MW)	256.50	255.75	0.2924
Cycle thermal efficiency	35.9%	35.86%	0.0111
Inlet air flow (kg/s)	645.02	645.026	≈0
Pressure ratio	15.40	15.40	≈0
Turbine inlet temperature (°C)	1318.00	1319.16	0.0880
Turbine outlet temperature (°C)	600.00	598.81	0.198
Fuel flow (kg/s)	14.355	14.355	/
Low calorific value of fuel (kJ/kg)	48686.3	48686.3	/
Turbine exhaust flow (kg/s)	659.375	659.381	≈0

3.2   Dynamic Operation Condition

The operation data of a GE 9FA unit during the load decrease process and the load increase process for 1 h were selected respectively. The simulation output and operation data are compared under the same inputs. The average relative error and the maximum relative error are used as model accuracy indicators. For the process of load decrease, the comparison of the outputs is shown in Fig. 4.

For the load increase process, the output comparison is shown in Fig. 5.

The results show that when the fuel and inlet air flow are varied, the variation trends of the outputs of the model, that is, the exhaust temperature and power of the turbine, are in good agreement with the actual data. The maximum relative error is about 3%, and the average relative error is not more than 1%.

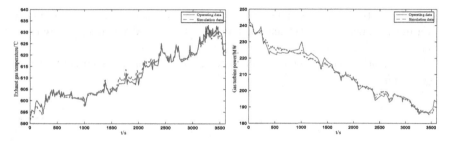

Fig. 4. Model validation for reduced load condition.

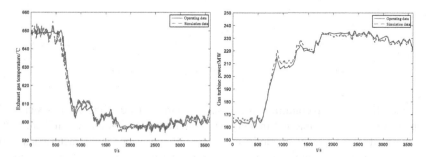

Fig. 5. Model validation for increased load condition.

3.3 Model Identification

When the heavy-duty gas turbine model achieves steady state operation, 600 data points are generated as model inputs by random step variation of the fuel quantity, and the output is obtained by dynamic simulation based on the mechanism model, and the data is de-averaged. The subspace identification method is used to identify the state space model and convert it into the transfer function form. The fuel flow-power transfer function $G_{11}(s)$ and the fuel flow-exhaust temperature transfer function $G_{21}(s)$ are obtained, respectively.

The comparison between the simulation data and the transfer function fitting data is shown in Fig. 6.

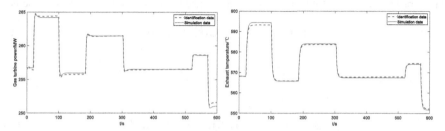

Fig. 6. Comparison of identification under fuel flow step.

Similarly, 600 data points of the air flow is randomly generated to obtain the air mass-power transfer function $G_{12}(s)$, and the fuel mass-turbine exhaust temperature transfer function $G_{22}(s)$, respectively.

The comparison between the simulation data and the transfer function fitting data is shown in Fig. 7.

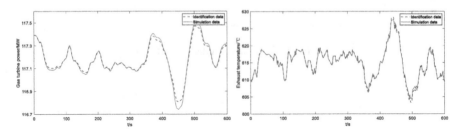

Fig. 7. Comparison of identification under air flow step.

Under the input step condition, the dynamic characteristic changes are basically the same between transfer function model and dynamic simulation model, indicating that the identified transfer function model can accurately describe the characteristics of the heavy-duty gas turbine and can be used as a control model in the subsequent control system:

$$\begin{bmatrix} P \\ T_6 \end{bmatrix} = \begin{bmatrix} \dfrac{3.252s^2 + 8.281s + 1.432}{s^2 + 0.7105s + 0.1133} & \dfrac{0.05366s^2 - 2.78e^{-7}s + 6.95e^{-6}}{s^2 + 1.389e^{-12}s + 0.0001378} \\ \dfrac{7.581s^2 + 15.2s - 0.001833}{s^2 + 0.3502s + 7.273e^{-10}} & \dfrac{-1.311s - 0.001971}{s + 0.001527} \end{bmatrix} \begin{bmatrix} G_f \\ G_a \end{bmatrix} \tag{7}$$

4   Multivariate Predictive Control

4.1   Augmented State Space Model

In the process of predictive control, the state space model that ignores future noise is:

$$x_m(k+1) = A_m x_m(k) + B_m u(k) \tag{8}$$

$$y(k) = C_m x_m(k) \tag{9}$$

The state variable increment is:

$$\Delta x_m(k+1) = A_m \Delta x_m(k) + B_m \Delta u(k) \tag{10}$$

A new state variable $x(k)$ needs to be defined:

$$x(k) = [\Delta x_m(k) \quad y(k)]^T \tag{11}$$

$$y(k+1) - y(k) = C_m A_m \Delta x_m(k) + C_m B_m \Delta u(k) \tag{12}$$

Combining Eqs. (11) and (12), a new augmented state space model is obtained:

$$\begin{bmatrix} \Delta x_m(k+1) \\ y(k+1) \end{bmatrix} = \begin{bmatrix} A_m & o_m \\ C_m A_m & I_{q \times q} \end{bmatrix} \begin{bmatrix} \Delta x_m(k) \\ y(k) \end{bmatrix} + \begin{bmatrix} B_m \\ C_m B_m \end{bmatrix} \Delta u(k) = A \begin{bmatrix} \Delta x_m(k) \\ y(k) \end{bmatrix} + B \Delta u(k) \tag{13}$$

$$y(k) = \begin{bmatrix} o_m & I \end{bmatrix} \begin{bmatrix} \Delta x_m(k) \\ y(k) \end{bmatrix} = C \begin{bmatrix} \Delta x_m(k) \\ y(k) \end{bmatrix} \tag{14}$$

In the formula, $o_m = [0,0,...,0]$ is the n_x-dimensional zero vector; the new A,B,C in the matrix is called the augmented matrix, corresponding to the original matrix:

$$A = \begin{bmatrix} A_m & o_m \\ C_m A_m & I_{q \times q} \end{bmatrix}, B = \begin{bmatrix} B_m \\ C_m B_m \end{bmatrix}, C = \begin{bmatrix} o_m & I \end{bmatrix} \tag{15}$$

The new state variable vector is: $x(k) = [\Delta x_m(k) \; y(k)]^T$, the new augmented state space model is:

$$\begin{aligned} x(k+1) &= Ax(k) + B \Delta u(k) \\ y(k) &= Cx(k) \end{aligned} \tag{16}$$

Assuming that the prediction time domain is N_P and the control time domain is N_C, the future state value can be calculated from (16):

$$x(k+1|k) = Ax(k) + B \Delta u(k)$$

$$x(k+2|k) = Ax(k+1) + B \Delta u(k+1) = A^2 x(k) + AB u(k) + B \Delta u(k+1)$$

$$\vdots$$

$$x(k+N_c|k) = A^{N_c} x(k) + A^{N_c-1} B \Delta u(k) + A^{N_c-2} B \Delta u(k+1) + ... + AB \Delta u(k+N_c-2) + B \Delta u(k+N_c-1)$$

$$\vdots$$

$$x(k+N_p|k) = A^{N_p} x(k) + A^{N_p-1} B \Delta u(k) + A^{N_p-2} B \Delta u(k+1) + ... + A^{N_p-N_c} B \Delta u(k+N_c-1) \tag{17}$$

According to formula (17), the predicted output of the system from time $k+1$ to time $k+p$ is predicted:

$$y(k+1|k) = CAx(k) + CB \Delta u(k)$$

$$y(k+2|k) = CA^2 x(k) + CAB \Delta u(k) + CB \Delta u(k+1)$$

$$y(k+3|k) = CA^3 x(k) + CA^2 B \Delta u(k) + CAB \Delta u(k+1) + CB \Delta u(k+2)$$

$$\vdots$$

$$y(k+N_c|k) = CA^{N_c} x(k) + CA^{N_c-1} B \Delta u(k) + CA^{N_c-2} B \Delta u(k+1) + ... + CAB \Delta u(k+N_c-2) + CB \Delta u(k+N_c-1)$$

$$\vdots$$

$$y(k+N_p|k) = CA^{N_p} x(k) + CA^{N_p-1} B \Delta u(k) + CA^{N_p-2} B \Delta u(k+1) + ... + CB^{N_p-N_c} \Delta u(k+N_c-1) \tag{18}$$

Under the premise that the augmented matrices A,B,C are known, the predicted output is only related to the state variable $x(k)$ and the future control variation $\Delta u(k + i)$, $i = 0,1,2,...,N_C-1$. Converting the above results into matrix form, the structure of the input vector in the control time domain and the form of the output vector in the prediction time domain are expressed as:

$$\Delta U(k) = \begin{bmatrix} \Delta u(k) \\ \Delta u(k+1) \\ \vdots \\ \Delta u(k+N_c-1) \end{bmatrix}_{N_c \times 1}, Y_P(k+1|k) = \begin{bmatrix} y_p(k+1|k) \\ y_p(k+2|k) \\ \vdots \\ y_p(k+N_p|k) \end{bmatrix}_{N_p \times 1} \tag{19}$$

The equation of the predicted output value of the system at the future time is expressed as:

$$Y_P(k+1|k) = \Phi x(k) + \Psi \Delta U(k) \tag{20}$$

Where,

$$\Phi = [CA \quad CA^2 \quad CA^3 \quad \cdots \quad CA^{N_p}]^T \tag{21}$$

$$\Psi = \begin{bmatrix} CB & 0 & 0 & \cdots & 0 \\ CAB & CB & 0 & \cdots & 0 \\ CA^2B & CB & 0 & \cdots & 0 \\ \vdots & \vdots & \vdots & \cdots & \vdots \\ CA^{N_c-1}B & CA^{N_c-2}B & CA^{N_c-3}B & \cdots & CB \\ CA^{N_c}B & CA^{N_c-1}B & CA^{N_c-2}B & \cdots & CAB \\ \vdots & \vdots & \vdots & \cdots & \vdots \\ CA^{N_p-1}B & CA^{N_p-2}B & CA^{N_p-3}B & \cdots & CA^{N_p-N_c}B \end{bmatrix}_{N_p \times N_c} \tag{22}$$

4.2 Feedback Correction

According to formula (20), at the current moment k, an error matrix can be formed by the difference between the predicted output $y_p(k|k-1)$ at the previous moment and the actual output $y(k|k-1)$, and the output Y at the next moment can be fed back and corrected by the correction coefficient, that is:

$$Y(k+1|k) = Y_p(k+1|k) + H\left[y_p(k|k-1) - y(k|k-1)\right] = Y_p(k+1|k) + He(k) \tag{23}$$

$$Y(k+1|k) = \begin{bmatrix} y(k+1|k) \\ y(k+2|k) \\ \vdots \\ y(k+N_p|k) \end{bmatrix}_{N_p-1} \tag{24}$$

$H = [\ h_1,\ h_2,\ ...,\ h_p]$ is the correction coefficient matrix, Eq. (23) is the feedback correction process at the current k time.

4.3 Receding Horizon Optimization

Adopt an objective function that expects the system output value to be as close as possible to the set input value, meanwhile, the control increment is introduced to the objective function:

$$J(k) = \sum_{i=1}^{N_c} \left\| \lambda_{y,i}(y(k+i|k) - r(k+i)) \right\|^2 + \sum_{i=1}^{N_p} \left\| \lambda_{u,i} \Delta u(k+i-1) \right\|^2 \tag{25}$$

Where, λ_y represents the weighting factor for the prediction output error, λ_u is weighting factor acts on the control increment.

Therefore, the optimization problem of state-space model predictive control can be regarded as the problem of minimizing the prediction difference and the control action:

$$\min_{\Delta U(k)} J\left(x(k), \Delta U(k), N_c, N_p\right) \tag{26}$$

$$J\left(x(k), \Delta U(k), N_c, N_p\right) = \sum_{i=1}^{N_c} \left\| \lambda_y(y(k+i|k) - r(k+i)) \right\|^2 + \sum_{i=1}^{N_p} \left\| \lambda_u \Delta u(k+i-1) \right\|^2 \tag{27}$$

The matrix-vector form of the above formula is:

$$J\left(x(k), \Delta U(k), N_c, N_p\right) = \left\| \Gamma_y(Y(k+1|k) - R(k+1)) \right\|^2 + \left\| \Gamma_u \Delta U \right\|^2 \tag{28}$$

where, Γ_y represents the predictive control output error weighting matrix, and Γ_u represents the control increment weighting matrix:

$$\begin{cases} \Gamma_y = diag\left(\lambda_{y,1}, \lambda_{y,2}, ..., \lambda_{y,p}\right) \\ \Gamma_u = diag\left(\lambda_{u,1}, \lambda_{u,2}, ..., \lambda_{u,m}\right) \end{cases} \tag{29}$$

Derivation to solve extreme value problems:

$$\frac{\partial J\left(x(k), \Delta U(k), N_c, N_p\right)}{\partial \Delta U} = 0 \tag{30}$$

The optimal control sequence imposed at the current k moment is:

$$\begin{aligned}\Delta U^*(k) = (\Psi^T \Gamma_y^T \Gamma_y \Psi + \Gamma_u^T \Gamma_u)^{-1} \\ \cdot \Psi^T \Gamma_y^T \Gamma_y (R(k+1) - \Phi x(k))\end{aligned} \tag{31}$$

The control increment acting on the system in real time is:

$$\begin{aligned}\Delta u(k) = [I_{n_u \times n_u} \quad 0 \quad \dots \quad 0]_{1 \times N_c} \Delta U^*(k) \\ = [I_{n_u \times n_u} \quad 0 \quad \dots \quad 0]_{1 \times N_c} (\Psi^T \Gamma_y^T \Gamma_y \Psi + \Gamma_u^T \Gamma_u)^{-1} \Psi^T \Gamma_y^T \Gamma_y (R(k+1) - \Phi x(k) + He(k)) \\ = K_{mpc}(R(k+1) - \Phi x(k) + He(k))\end{aligned} \tag{32}$$

K_{mpc} expressed as the gain of state-space model predictive control.

5   Control Simulation Experiment

As a comparative experiment, a decentralized PID control system is built in MATLAB, which are the power control loop and the exhaust temperature control loop. The parameters of prediction control are set as: sampling period is T = 0.01s; prediction time domain NP = 10; control time domain NC = 6; softening coefficient α = 0,75; output deviation weight matrix Γ_y = $diag[0.02, 0.02]$; control increment weight matrix Γ_u = $diag[0.1, 0.1]$.

5.1   Setpoint Tracking

After the system runs to the steady state, the setpoint value of power changes by + 10% at 100s, and then by --0% at 300s; at 500s, the given value of exhaust temperature changes by + 10%; the setpoint value of exhaust temperature changes by + 10% at 500s and by --10% at 700s, as shown in Fig. 8. The comparison results between the predictive controller and the PID controller are shown in Fig. 9.

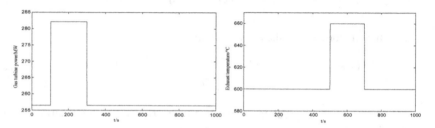

Fig. 8. Setpoint disturbance.

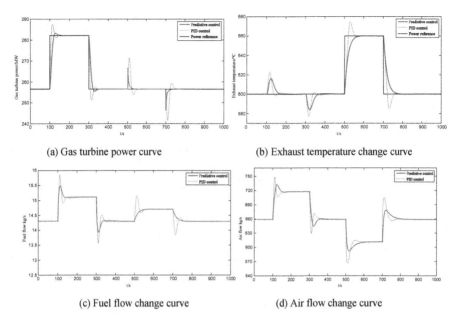

(a) Gas turbine power curve   (b) Exhaust temperature change curve

(c) Fuel flow change curve   (d) Air flow change curve

Fig. 9. Controlled quantity and control strategy effect comparison.

It can be seen from Fig. 9 that in the case of setpoint value changes, the predictive controller can track setpoint more quickly, and the overshoot is smaller, and the change of the control variable is smaller and faster. Its control effect is obviously better than that of the traditional PID controller, which ensures the stability and robustness of the unit operation.

5.2 External Disturbance

When the power step disturbance occurs, the output power suddenly increases by 10% from the stable operating point. At this time, the control results of the predictive controller are compared with the traditional PID control results, and the results are shown in Fig. 10.

When the external disturbance output power increases, the power setpoint value remains unchanged. Under the action of the predictive controller, the fuel flow rate and air flow rate decrease rapidly, so that the power is adjusted to the original setpoint value again. Predictive controllers adjust fuel flow rate and air flow rate with smaller and more rapid changes. The designed predictive controller can meet the anti-external disturbance requirements of the heavy-duty gas turbine system.

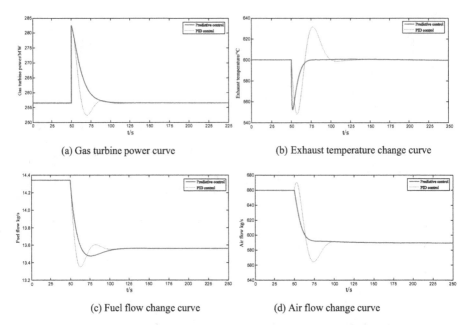

(a) Gas turbine power curve   (b) Exhaust temperature change curve

(c) Fuel flow change curve   (d) Air flow change curve

Fig. 10. Comparison between MPC and PID under external disturbance.

5.3 Internal Disturbance

When the step disturbance of fuel flow occurs, the fuel flow suddenly increases by 10% from the stable operating point. At this time, the control results of the predictive controller are compared with the traditional PID control results, and the results are in Fig. 11.

Fuel flow rate is increased, resulting in an increase in power output and exhaust temperature, as well as an increase in air flow rate. Since the setpoint value of the power and exhaust temperature do not change, under the action of the controller, the fuel flow rate decreases rapidly, and the fuel flow rate and air flow rate rapidly return to the setpoint value. The predictive controller returns to the setpoint value faster and more smoothly, with less overshoot of the controlled variable, and changes more quickly and agilely. Therefore, the predictive controller can meet the anti-internal disturbance requirements for the normal operation of the heavy-duty gas turbine system.

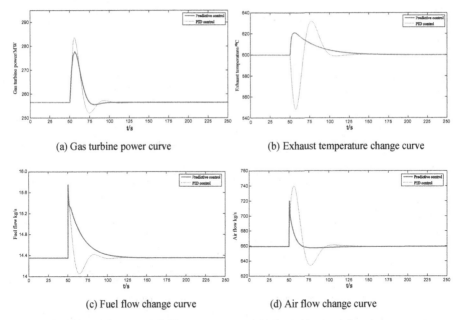

(a) Gas turbine power curve

(b) Exhaust temperature change curve

(c) Fuel flow change curve

(d) Air flow change curve

Fig. 11. Comparison between MPC and PID under internal disturbance.

5.4 Parameter Perturbation

In the actual operation, the perturbation of the model parameters may occur. The transfer function model parameters are randomly fluctuated by 10%, and the new model can be obtained as follows:

$$
\begin{bmatrix} P \\ T_6 \end{bmatrix} = \begin{bmatrix} \dfrac{3.2s^2 + 8.4s + 1.432}{1.1s^2 + 0.7105s + 0.1133} & \dfrac{0.05066s^2 - 2.74e^{-7}s + 6.95e^{-6}}{0.96s^2 + 1.5e^{-12}s + 0.0001378} \\ \dfrac{7.521s^2 + 15.2s - 0.001833}{0.99s^2 + 0.3502s + 7.273e^{-10}} & \dfrac{-1.111s - 0.001971}{1.08s + 0.001527} \end{bmatrix} \begin{bmatrix} G_f \\ G_a \end{bmatrix}
$$

$$(33)$$

On this basis, the setpoint value of power is increased by 10%, the dynamic response curve is shown in Fig. 12.

When the model parameters are perturbed, because the prediction model adopts the state space form, the structure of the controlled model does not change too much. Even if the model parameters have a certain mismatch, it can better meet the control requirements, and the overshoot is still small. Therefore, the predictive controller designed in this paper can meet the anti-parameter perturbation disturbance requirements for the normal operation of the heavy-duty gas turbine system.

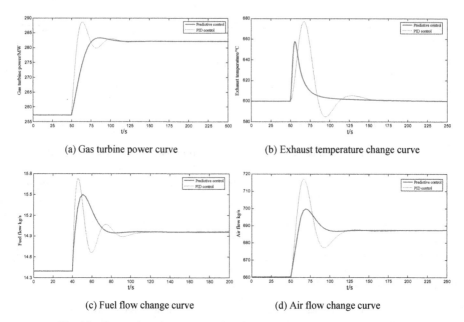

(a) Gas turbine power curve

(b) Exhaust temperature change curve

(c) Fuel flow change curve

(d) Air flow change curve

Fig. 12. Comparison between MPC and PID under parameter perturbation.

6 Conclusion

In this paper, a dynamic model of a heavy-duty gas turbine is established based on Modelica. The model structure is derived from the conservation law and the thermodynamic theorem, and the accuracy of the model is ensured by comparison with the actual operating data. In addition, the transfer function model is obtained by the subspace identification method, which is convenient and accurate. A multivariable model predictive controller is established, and the augmented state space prediction model replaces the conventional state space prediction model, which can reduce the complexity of the optimal control law solution process, simplify the calculation process of rolling optimization and feedback correction, and improve the control speed. The comparison with the PID controller shows that the controller designed in this paper can meet the control requirements of heavy-duty gas turbines under various disturbance conditions, and has good robustness.

Acknowledgment. This work is supported by the National Science and Technology Major Project of China (2017-I-0002–0002).

References

1. Oh, H.T., Lee, W.S., Ju, Y., Lee, C.H.: Performance evaluation and carbon assessment of IGCC power plant with coal quality. Energy **188**(1), 116063 (2019)
2. Xing, F., et al.: Flameless combustion with liquid fuel: a review focusing on fundamentals and gas turbine application. Appl. Energy **193**(1), 28–51 (2017)

3. Rowen, W.I.: Simplified mathematical representations of single shaft gas turbines in mechanical drive service. Turbomach. Int. **33**(5) (1992)
4. Rowen, W.I.: Simplified mathematical representations of heavy-duty gas turbines. J. Eng. Power **105**(4) (1983)
5. Nagpal, M., Moshref, A., Morison, G.K., Kundur, P.: Experience with testing and modeling of gas turbines. In: 2001 IEEE Power Engineering Society Winter Meeting, Conference Proceedings (Cat. No.01CH37194) (2001)
6. Mahat, P., Zhe, C., Bak-Jensen, B.: Control strategies for gas turbine generators for grid connected and islanding operations. Transm. Distrib. Conf. Expos. (2010)
7. Duan, J., Sun, L., Wang, G., Wu, F.: Nonlinear modeling of regenerative cycle micro gas turbine. Energy **91**, 168–175 (2015)
8. Xuedi, H., Lei, S., Jinling, C., Shijie, Z.: Off-design performance of 9F gas turbine based on gPROMs and BP neural network model. J. Therm. Sci. **31**(1), 261–272 (2022)
9. Iqbal, M.M.M., Xavier, R.J., Kanakaraj, J.: A neuro-fuzzy controller for grid-connected heavy-duty gas turbine power plants. Turk J. Electr. Eng. Co. **25**(3), 2375–2387, (2017)
10. Wiese, A.P., Blom, M.J., Manzie, C., Brear, M.J., Kitchener, A.: Model reduction and MIMO model predictive control of gas turbine systems. Control. Eng. Pract. **45**, 194–206 (2015)
11. Shi, G., Wu, Z., He, T., et al.: Shaft speed control of the gas turbine based on active disturbance rejection control. IFAC-PapersOnLine **53**(2), 12523–12529 (2020)

Research on Modeling of a Fully Loaded Ultra-SuperCritical Coal-Fired Units Coordination System

Liu Qian[1,2(✉)], Lv Jianhong[2], Zhang Qiusheng[1], and Zhuo Hua[1]

[1] CHN Energy New Energy Technology Research Institute Co., Ltd, Beijing 102206, China
liuqian0614@126.com
[2] School of Energy and Environment, , Southeast University, Nanjing 210000, China

Abstract. Through the rational simplification and mechanism analysis of the boiler system and the steam-water system of the ultra-supercritical coal-fired unit under deep peak regulation, considering the influence of the different characteristics of the unit under dry and wet operation on the coordinated control system, the segmented lumped parameter method is adopted. A nonlinear model of the coordination system in the full load section of the unit is established. Combined with the operating data of a 1000MW ultra-supercritical unit, the model parameters of the unit working within the full load range were obtained by means of particle swarm algorithm identification. The model is verified and analyzed by the actual operation data of the unit. The results show that the model can better reflect the operation characteristics of the unit and has a certain accuracy. It can be used to design the coordinated control of ultra-supercritical coal-fired units under deep peak regulation.

Keywords: Ultra-supercritical unit · Coordinated control · Full load model · System modeling

1 Introduction

In recent years, with the rise and rapid development of various clean energy power generation such as wind power, photovoltaic, nuclear power, geothermal, biomass, etc. in China, the proportion of new energy in the power grid has shown a trend of rapid growth year by year. Due to the objective issues such as the regionality, timing and stability of natural resources such as wind and light, their development has entered the stage of large-scale incremental replacement and regional stock replacement from the previous incremental supplementation. It has brought a great impact on the stable operation of the power system, and the problem of large-scale clean energy consumption is very serious, which restricts the construction and development of new power systems. In order to consume more clean energy power, the power grid urgently requires thermal power units to undertake the task of peak shaving, and thermal power units must improve their operational flexibility to meet the grid scheduling requirements [1]. At this stage, thermal power units are still the main power source in my country. Improving the flexibility of thermal power units and tapping their peak shaving potential is not only an effective way to solve the current dilemma of new energy consumption, but also to extend the life cycle of thermal power enterprises and realize green power. The inevitable choice for transformation [2, 3].

The load variation range of deep peak shaving is wide, and most of the current modeling research focuses on the dry state. It is necessary to establish a full-load nonlinear model that can correctly reflect the changes of various equipment parameters in the unit and the dynamic response of the unit [4, 5]. This paper takes a 1000MW ultra-supercritical unit in China as the research object, aiming at the complex steam-water system in the unit's boiler and steam turbine, and the difficulty of using moving boundary modeling for the working fluid [6]. The segmented lumped parameter method is used to establish a nonlinear model [7, 8]. The parameters in the model are calculated with the learning algorithm, and the validity of the established model is proved by the simulation verification with the actual operation data in the field.

2 Full Load Segment Nonlinear Model

2.1 Model Structure and Assumptions

The coordination system of the ultra-supercritical unit mainly includes three parts: the combustion system of the boiler, the steam-water system and the steam turbine system. The combustion system is that coal is pulverized by the coal mill and then enters the boiler for combustion. The steam-water system includes an economizer, a water-cooled wall, a superheater and a reheater. The separated steam enters the superheater for further heating, the generated high temperature and high pressure steam enters the high pressure cylinder of the steam turbine to do work, and the exhaust steam from the high pressure cylinder is heated by the reheater and enters the medium and low pressure cylinder to do work [9].

For the coordinated system modeling of the ultra-supercritical unit, the general input quantities are the feed water flow, the coal feed amount and the opening of the steam turbine valve. It is necessary to solve the intermediate point temperature (the steam temperature at the outlet of the steam-water separator), the main steam pressure, the main steam temperature and the unit [10]. Power these 4 parameters. The usual dry state model is applied to the wet state model under all working conditions, and there is a large error [11]. The source of the error is: (1) After the unit enters the low load section, the outlet of the steam-water separator carries water, the unit enters the wet state, and the steam-water separator A part of the flow is recirculated back to the water wall, and the mass balance equation in the dry state will be inaccurate; (2) The ratio of the heat absorption of the water wall and the superheater varies due to the different endothermic characteristics, resulting in the outlet parameters of the steam-water separator. (3) When the low load enters the wet state, the water level in the water storage tank of the steam-water separator should also be used as the output. Therefore, it is necessary to establish a full-load segment model suitable for dry and wet states. Firstly, this paper simplifies the real system appropriately based on the literature [12]. The boiler side model regards the economizer and the water wall as a uniformly heated pipe, and the heat absorption is

Q_1, and the superheater is regarded as a single pipe, and the heat absorption is Q_2. The simplified boiler model structure is shown in Fig. 1.

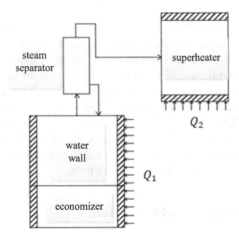

Fig. 1. Simplified model of boiler section

The following assumptions are made for the simplified boiler model:

(a) It is assumed that the metal temperature of the economizer-water wall and superheater is the same as the temperature of the working fluid, and changes synchronously with the temperature of the working fluid;
(b) Treat the steam separator and water storage tank as a cylindrical container;
(c) The influence of the working fluid cyclone in the separator on the energy is ignored;
(d) Ignore the effect of separator heat dissipation on energy loss;
(e) It is assumed that the distribution of the working medium inside the separator is uniform and the separation efficiency is 1.

The simplified coordination system can be divided into three processes: boiler combustion process, economizer-water wall heat absorption process, superheater heat absorption process, and an intermediate link of steam-water separator. The outlet parameters are used as lumped parameters [13, 14].

2.2 Full Load Segment Model Establishment

Boiler Section. (a) Boiler combustion process. The combustion process of the boiler adopts the pulverizing process and combustion heat release of the model established in the literature [5]:

$$u_B = \frac{e^{-\tau s}}{t_0 s + 1} D_B \tag{1}$$

$$Q_b = \mu_b u_B \tag{2}$$

Where D_B is the coal mass flow (kg/s), u_B is the pulverized coal mass flow (kg/s), s is the complex parameter, t_0 is the inertia time, τ is the pure delay time, Q_b is the total heat release of the boiler, kJ/s; μ_b is the calorific value of coal combustion, kJ/kg.

(b) Economizer-Water Wall Endothermic Process. To further simplify the model, it is assumed that the dynamic process on the flue gas side is ignored; the heat transfer from the flue gas side to the pipe side is forced heat flow; the axial heat transfer in the direction of the pipe length is ignored. Considering the flow change caused by the density change of the working medium in the tube and the heat storage of the metal, the mass balance equation and energy balance equation of the working medium are:

$$V_{lb}\frac{d\rho_{lb}}{dt} = D_{fw} - D'_{sp} + D_{re} \tag{3}$$

$$V_{lb}\frac{d(\rho_{lb} \cdot h_{lb})}{dt} = D_{fw}h_{fw} - D'_{sp}h_{lb} + D_{re}h_{re} + Q_j \tag{4}$$

where V_{lb} is the inner volume of the economizer and the water wall (m^3), ρ_{lb} is the outlet working medium density (kg/m^3), D_{fw} is the feed water mass flow (kg/s), D'_{sp} is the steam-water separator inlet working medium flow (kg/s), and D_{re} is the recirculating water flow (kg/s). h_{lb}, h_{fw} and h_{re} are the outlet enthalpy value of the working fluid (kJ/kg), the enthalpy value of the feed water and the enthalpy value of the recirculating water (kJ/kg), and Q_j are the metal heat absorption of the economizer and the water wall.

Part of the heat absorbed by the metal of the economizer-water wall is transferred to the working medium in the tube, and the other part is used for its own heat storage, so there is an energy balance equation:

$$c_j m_j \frac{dT_j}{dt} = Q_{lb} - Q_j \tag{5}$$

$$Q_j = kD^n(T_j - T_{lb}) \tag{6}$$

where k is the heat transfer coefficient, k_{lb} is the ratio of the heat absorption of the water wall Q_{lb} to the total heat release Q_b of the boiler, that is, the total heat release of the boiler can be calculated from the total coal amount and the low calorific value of the coal, and T_j is the metal of the economizer-water wall part The wall temperature c_j is the metal specific heat capacity, and m_j is the metal mass of the economizer-water cooling wall, generally 0.8.

Since the outlet enthalpy value of the water wall is the inlet enthalpy value of the steam-water separator h_{sp}(kJ/kg), which is regarded as a state variable, the following Eqs. (3) and (4) can be obtained:

$$(V_{lb}\frac{\partial\rho_{lb}}{\partial h_{sp}})\frac{dh_{sp}}{dt} = D_{fw} - D'_{sp} + D_{re} \tag{7}$$

$$V_{lb}h_{lb}\frac{\partial\rho_{lb}}{\partial h_{sp}}\frac{dh_{lb}}{dt} + V_{lb}\rho_{lb}\frac{dh_{lb}}{dt} = D_{fw}h_{fw} - D'_{sp}h_{lb} + D_{re}h_{re} + Q_j \tag{8}$$

Let $\alpha_1 = 2V_{lb}(\partial \rho_{lb}/\partial h_{sp})$, $\alpha_1 = 2V_{lb}\rho_{lb}$, substitute Eq. (7) into Eqs. (8) and (6) into Eq. (5) to obtain the partial model of the water wall:

$$\begin{cases} \dfrac{dh_{lb}}{dt} = \dfrac{D_{fw}h_{fw}+D_{re}h_{re}-(D_{fw}+D_{re})h_{lb}+kD_{sp}'^{n}(T_j-T_{sp})}{a_1 h_{lb}+a_2} \\ c_j m_j \dfrac{dT_j}{dt} = Q_{lb} - kD_{sp}'^{n}(T_j - T_{sp}) \end{cases} \tag{9}$$

Among them, the outlet density of the water wall $\rho_{lb}(kg/m^3)$ is equal to the density of the working medium at the inlet of the steam-water separator $\rho_{sp}(kg/m^3)$, α_1 is the dynamic parameter to be identified.

(c) Steam Water Separator. The modeling of the steam-water separator is mainly to obtain the two parameters of the recirculating water flow rate and the separator liquid level when the unit enters the wet state, so the mass and energy balance equations are listed for the steam-water separator on the basis of the previous assumptions.

$$V_{sp}\frac{d\rho_{sp}}{dt} = D_{sp}' - D_{sp}'' - D_{re} \tag{10}$$

where V_{sp} is the volume of the working medium in the steam-water separator, and D_{sp}'' is the flow rate of steam flowing out of the steam-water separator (kg/s).

At the same time, the liquid level of the separator and the cross-sectional area F of the separator can be obtained:

$$V_{sp} = V_w + V_i = H \cdot F + V_i \tag{11}$$

where V_w and V_i is the volume of water and steam in the steam-water separator, referring to formula (8), the water level variation equation of formula (12) can be obtained by separately listing some mass balance equations for:

$$\frac{dH}{dt} = \frac{(1-x')D_{sp}' - D_{re}}{\rho_w F} \tag{12}$$

where x' is the dryness of the inlet working fluid of the steam-water separator.

The energy balance equation is:

$$\frac{d(V_w \rho_w h_w + V_i \rho_i h_i + M_j c_j T_j)}{dt} = D_{sp}' h_{sp}' - D_{sp}'' h_{sp}'' - D_{re} h_{re} \tag{13}$$

Formula (13) takes the partial derivative of the outlet pressure of the separator p_{sp}(Mpa) and substitutes it into the water level variation Eq. (12) to obtain:

$$\frac{dp_{sp}}{dt} = \frac{(D_{sp}' h_{sp}' - D_{sp}'' h_{sp}'' - D_{re} h_{re}) + \frac{(\rho_i h_i - \rho_w h_w)}{\rho_w}[D_{sp}' - D_{sp}'' - D_{re}]}{V_{sp}\frac{\partial \rho_i h_i}{\partial p_{sp}} + HF(\frac{\partial \rho_w h_w}{\partial p_{sp}} - \frac{\partial \rho_i h_i}{\partial p_{sp}}) + m_j c_j \frac{\partial T_j}{\partial p_{sp}}} \tag{14}$$

Let $b_1 = \partial \rho_i h_i/\partial p_{sp}$, $b_2 = \partial \rho_w h_w/\partial p_{sp}$, $b_3 = \partial T_j/\partial p_{sp}$ be the dynamic parameters to be identified. In the formula, the saturated water density ρ_w, the saturated steam density ρ_i, the saturated steam enthalpy h_i and the saturated water enthalpy h_w can all be obtained from the pressure of the steam-water separator p_{sp}, so let $g = (\rho_i h_i - \rho_w h_w)/\rho_w) = f(p_{sp}, h_{sp})$, the model of the steam-water separator is:

$$
\begin{cases}
\dfrac{dp_{sp}}{dt} = \dfrac{(D'_{sp} h'_{sp} - D''_{sp} h''_{sp} - D_{re} h_{re}) + W}{V_{sp} b_1 + HF(b_2 - b_1) + m_j c_j b_3} \\[2ex]
\dfrac{dH}{dt} = \dfrac{(1 - x')D'_{sp} - D_{re}}{\rho_w F}
\end{cases}
\tag{15}
$$
$$
W = g[D'_{sp} - D''_{sp} - D_{re}]
$$

(d) Superheater Endothermic Process. The modeling of the coordination system is mainly studied, so the influence of superheated desuperheating water is not considered, the steam pressure at the outlet of the separator can be obtained from the steam-water separator model, and the main steam pressure can be obtained according to the momentum Eq. (14):

$$
\lambda_{st} \frac{D_{st}^2}{\rho_{st}} = p_{sp} - p_{st}
\tag{16}
$$

where D_{st} is the main steam flow, p_{st} is the main steam pressure, λ_{st} is the flow resistance coefficient, and formula (16) can be simplified as:

$$
D_{st} = k_3(p_{sp} - p_{st})
\tag{17}
$$

where: k_3 is the parameter to be identified. The mass balance equation and energy balance equation are similar to the water wall model. The superheater outlet parameters are used as lumped parameters, the steam flow D''_{sp} into the superheater is the outlet flow of the steam-water separator, and the flow into the steam turbine is the main steam flow D_{st}. The mass of the superheater section can be obtained. Equilibrium equation:

$$
V_{gr} \frac{d\rho_{st}}{dt} = D''_{sp} - D_{st}
\tag{18}
$$

where V_{gr} is the inner volume of the superheater section, and the heat absorption of the superheater section can be obtained from the ratio of the heat absorption $Q_g = (1 - k_{lb})Q_b$, so the capacity balance equation of the superheater section is:

$$
\frac{d(V_{gr} \cdot \rho_{st} \cdot h_{st} + m_g c_j T_g)}{dt} = D''_{sp} h''_{sp} - D_{st} h_{st} + Q_g
\tag{19}
$$

where m_g is the metal mass of the superheater section and T_g is the metal wall temperature of the superheater section. The partial model of the superheater can be obtained by referring to the partial model of the economizer-water wall:

$$
\begin{cases}
\dfrac{dh_{st}}{dt} = \dfrac{D''_{sp} h''_{sp} - D_{st} h_{st} + k D_{st}^n (T_g - T_{st})}{c_2 h_{st} + c_1} \\[2ex]
c_j m_g \dfrac{dT_g}{dt} = Q_g - k D_{st}^n (T_g - T_{st})
\end{cases}
\tag{20}
$$

where T_{st} is the main steam temperature, °C.

Steam Turbine. For the steam turbine part, the relationship between the steam flow at the steam turbine inlet and the pressure and temperature is expressed as:

$$D_{st} = g(u_t)f(p_{st}, T_{st}) \tag{21}$$

Among them, u_t is the opening degree of the steam turbine valve, $g(\cdot)$, $f(\cdot)$ respectively represent a functional relationship. Reference [64] proposed that the functional relationship can be expressed as:

$$D_{st} = k(u_t)\frac{p_{st}}{\sqrt{T_{st}}} \frac{1}{T_{tur}s + 1} \tag{22}$$

where: T_{tur} is the inertia time (s). In this paper, the function $k(u_t)$ is identified on the basis of Eq. (20), and $k(u_t)$ is assumed to be in the form of a polynomial:

$$k(u_t) = \alpha u_t^2 + \beta u_t + c \tag{23}$$

The generating power of the unit can be calculated from the energy balance:

$$N_e = k_2 D_{st}(h_{st} - h_{fw}) \tag{24}$$

where k_2 is the equivalent factor considering the generating efficiency of the unit.

3   Model Parameter Calculation

3.1   Static Parameter Calculation

From the above modeling results, it can be seen that the parameters to be identified are $\alpha_1, b_1, b_2, b_3, c_1, c_2, k_{lb}, k_1, k_2, k_3$, and the sum can be statically identified according to the data in Table 1. In addition, the time constants t_0 and τ of the milling process can refer to the value of the same type of mill [5].

Table 1. Model static parameters

$D_{st}/$ kg·s-1	k_2	$k_{lb}/\%$
254.1667	0.000174	63.2
367.2222	0.000174	61.1
512.5	0.000173	56.8
748.0556	0.000172	54.3
805.8333	0.0001719	53.5
833.3333	0.0001718	52.4

The steam flow rate and the heat absorption ratio of the water cooling wall in Table 1 are obtained from the historical operation data of a 1000MW unit in China. The quality of some metals is similar, so it can be approximated that the heat absorption ratio of the water wall is the ratio of the heat absorption of the working medium in the water wall to the total enthalpy of the working medium [4]. It can be seen from Table 1 that there is little change with the change of steam flow, so it is taken as 0.00173. After testing, rational function approximation is the best, and the rational function can be written as:

$$f(x) = \frac{P(x)}{Q(x)}, Q(x) \neq 0 \tag{25}$$

where $P(x)$ and $Q(x)$ are both polynomials of $(ax - b)^n$, a and b are the coefficients to be fitted, which are obtained by fitting a first-order rational function using the MATLAB fitting toolbox:

$$k_{lb} = (9.63D_{st} + 195200)/(0.278D_{st} + 2665) \tag{26}$$

3.2 Dynamic Parameter Calculation

The dynamic parameter part uses the actual operation data of the unit to identify through the teaching and learning algorithm (TLBO). Teaching and learning algorithm, particle swarm algorithm and genetic algorithm are equivalent to a swarm intelligence optimization algorithm. Different from other swarm intelligence algorithms, the teaching and learning algorithm does not need to set the calculation parameters of the algorithm itself, and it is simple and easy to implement. Since there are basically no parameters to be set in the teaching and learning algorithm itself, it is only necessary to set the search range for the identification parameters. There are 8 parameters $\alpha_1, b_1, b_2, b_3, c_1, c_2, k_1$ and k_3, which need to be identified. The search range is shown in Table 2.

Table 2. Search range of parameters to be identified

Identification parameters	α_1	b1	b2	b3	c1	c2	k1	k3
Upper limit	10	20	200	150	15	50	2500	2000
Lower limit	1	1	50	50	1	10	500	500

The algorithm fitness function is:

$$f(x) = \sum_{t=0}^{t_1} \left(\left| \frac{\Delta N_e}{N_{e0}} \right| + \left| \frac{\Delta p_{st}}{p_{st0}} \right| + \left| \frac{\Delta T_{st}}{T_{st0}} \right| + \left| \frac{\Delta T_{sp}}{T_{sp0}} \right| + \left| \frac{\Delta H}{H_0} \right| \right) \tag{27}$$

where t is the simulation time, t_1 is the end time of the simulation, Δ is the difference between the calculated value of the model and the actual value of the field, the subscript 0 is the actual value of the field, the smaller $f(x)$ is, the closer the model is to the actual process, when the algorithm reaches the upper limit of iteration or has been Convergence stops the iteration to get the final model parameters.

Set the population number of TLBO algorithm to 50 and the upper limit of iteration to 50 times. Using the field operation data of a 1000MW unit in China, input the coal feed, water feed, recirculating water flow and turbine valve opening. The calculation results are shown in Table 3.

Table 3. Calculation results of parameters to be identified

Identification parameters	Calculation results	Identification parameters	Calculation results
α_1	0.1259	c_1	115.7
b_1	12.25	c_2	0.1625
b_2	192.6	k_1	1090.0
b_3	74.85	k_3	512.2

4   Model Validation

The model is verified by using the data of the process of dry state to wet state when the on-site unit is reduced in load. At 0s, the actual power of the unit is 338MW, the coal feed volume is 120t/h, the feed water volume entering the economizer is 1066t/h, and the main steam pressure is 9.8Mpa. When the actual power is 319MW, the unit starts to enter the wet state from the dry state. Input the coal feed, water feed, recirculating water flow and turbine valve opening to the model, and get the model comparison and validation curve in Fig. 2.

Figure 2 compares the model output and the actual running power, main steam pressure, water level of the water storage tank and outlet enthalpy of the steam-water separator during the process of the unit from dry state to wet state. The simulation is more accurate. In terms of the water level of the water storage tank, the error of the dry state process is small, while the water level of the wet state process starts to rise significantly at 250s, and then drops to a stable level at 610s. This is because the increase in the recirculating water flow makes the model The output water level changes greatly. The water level change rate calculated by the model is faster and the change range is larger. This is due to the deviation between the volume calculation of the water storage tank and the actual volume, and the difference in the rate of water level change due to different inertial volumes. In terms of the outlet enthalpy value of the steam-water separator, the initial process model and the actual calculation error in the dry state and the wet state are small, and there is a certain deviation after 430s. The reason may be that the model calculation is simplified on the one hand, and the energy such as heat dissipation is ignored. Losses, on the other hand, may be the calculation error or uneven distribution of the dryness of the steam-water separator, and the rapid action of the water supply during the power change process causes the calculated value to deviate from the actual value in the wet state. But in general, it can correctly reflect the changing state of the unit and meet the requirements of simulation.

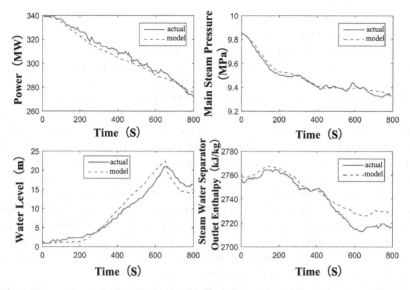

Fig. 2. Comparison of 540 MW superheated steam temperature model and actual output

Figure 3 shows the model to simulate the process of the unit from wet state to dry state. At 0s, the actual power generation of the unit is 388 MW, the coal feed volume is 176t/h, the feed water volume entering the economizer is 1100/h, and the main steam pressure is 11.46 MPa, which is verified by model comparison. The curve is shown in the figure below.

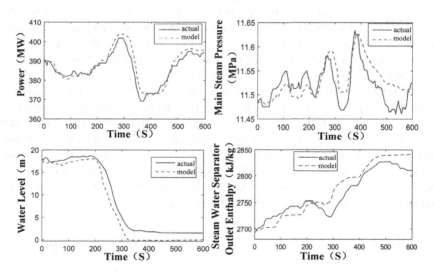

Fig. 3. Comparison of 540 MW superheated steam temperature model and actual output

It can be seen from Fig. 3 that the change of coal amount in the process of changing from wet state to dry state is larger than that when changing from dry state to wet state, which causes the main steam pressure and power to oscillate. The maximum deviation of main steam pressure is 0.06 MPa, and the power is the largest The deviation is 5.4MW, and the model output is basically consistent with the actual situation during the whole oscillation process, and the model output is relatively accurate in general. In the water level diagram of the storage tank, the actual trend is consistent with the model calculation results, but the water level rate of the model changes rapidly. The theoretical analysis of the water level in the dry state is zero water level, but the actual water level still has a little water level in the dry state. Because the water level is already below the lower limit of measurement. In the outlet enthalpy diagram of the steam-water separator, in the process of changing from wet state to dry state, the actual outlet enthalpy decreased to a certain extent compared with the model outlet enthalpy and then increased. The reason is that the actual water action is faster than the coal action in the actual adjustment process, and Due to the deviation of the heat capacity of the actual steam-water separator from the model calculation, the outlet enthalpy of the separator decreases to a certain extent during the process of changing from wet state to dry state.

Combining Figs. 2 and 3, it can be seen that the model is relatively accurate in the dry state process, and there is a certain error in the wet state process, but the overall trend is consistent, which can reflect the operating state of the unit and meet the simulation requirements.

5   Conclusion

With the increasing demand for deep peak shaving of ultra-supercritical units, in the study of ultra-supercritical models, the usual dry state model applied to the wet state model under all working conditions has a large error, due to the separation of steam and water. There is water at the outlet of the device, so that the dry state calculation formula is no longer applicable to the wet state, and the model needs to be remodeled. The coordinated system model of all working conditions proposed in this paper uses the mass and energy balance to establish the nonlinear model of the economizer water wall, steam-water separator, superheater and steam turbine. For the parameters that cannot be determined in the model, the theory and experience are combined with each other. Method, based on historical data, using teaching and learning heuristics to determine. Finally, the simulation experiment of the process of dry state to wet state and wet state to dry state is carried out. The simulation results show that the model can better reflect the characteristics of the unit in dry state and wet state. The dry state process is more accurate, and the wet state process has certain characteristics error.

References

1. Tan, Y., Kashem, M., Ciufo, P., et al.: Enhanced frequency regulation using multilevel energy storage in remote area power supply systems. IEEE Trans. Power Syst. 34(1), 163–170 (2019)
2. IRENA (2020). Renewable capacity statistics 2020. AbuDhabi: International Renewable Energy Agency (IRENA) (2020)s

3. Zeng, J., Wang, Q.A., Liu, J.F., et al.: A potential game approach to distributed operational optimization for microgrid energy management with renewable energy and demand response. IEEE Trans. Industr. Electron. **66**(6), 4479–4489 (2019)
4. Tan, W., Fang, F., Tian, L., et al.: Linear control of a boiler-turbine unit : analysis and design. ISA Trans. **47**(2), 189–197 (2008)
5. Fan, H., Zhang, Y.F., Su, Z.G., et al.: A dynamic mathematical model of an ultra-supercritical coal fired once-through boiler-turbine unit. Appl. Energy **189**, 654–666 (2017)
6. Terzi, E., Fagiano, L., Farina, M., et al.: Learning-based predictive control for linear systems: a unitary approach. Automatica **108**, 108473 (2019)
7. Moeini, A., Kamwa, I., Brunelle, P., et al.: Open data IEEE test systems implemented in SimPowerSystems for education and research in power grid dynamics and control. In: 2015 50th International Universities Power Engineering Conference (UPEC), Stoke on Trent, UK, pp. 1–6, 1–4 September 2015
8. Åström, K.J., Bell, R.D.: Drum-boiler dynamics. Automatica **36**(3), 363–378 (2000)
9. Leva, A., Maffezzoni, C., Benelli, G.: Validation of drum boiler models through complete dynamic tests. Control Eng. Pract. **7**(1), 11–26 (1999)
10. Chaibakhsh, A., Ghaffari, A., Moosavian, S.A.A.: A simulated model for a once-through boiler by parameter adjustment based on genetic algorithms. Simul. Model. Pract. Theory **15**(9), 1029–1051 (2007)
11. Liu, J.Z., Yan, S., Zeng, D.L., et al.: A dynamic model used for controller design of a coal fired once-through boiler-turbine unit. Energy **93**, 2069–2078 (2015)
12. Zhang, C., Chen, H.Y., Ngan, H., et al.: Solution of reactive power optimization including interval uncertainty using genetic algorithm. IET Gener. Transm. Distrib. **11**(15), 3657–3664 (2017)
13. Oh, S.K., Lee, J.M.: Interative learning model predictive control for constrained multivariable control of batch processes. Comput. Chem. Eng. **93**, 284–292 (2018)
14. Dindarloo, S.R., Hower, J.C.: Prediction of the unburned carbon content of fly ash in coal-fired power plants. Coal Combust. Gasification Prod. **7**, 19–29 (2015)

A Study on Accuracy of Flow Measurement of Intake Flow Tube Used in Fan Forward Acoustic Tester

Li-ping Zhu$^{(\boxtimes)}$, Bo-bo Jia, and Zhi-bo Zhang

AECC Shenyang Engine Design and Research Institute, Shengyang 110015, China
zhulp2602@163.com

Abstract. Fan forward acoustic tester is immersed in the anechoic room. Under many working conditions, the Mach number in the intake flow tube is low, which affects the accuracy of flow measurement. Therefore, it is necessary to study its accuracy of flow measurement. Taking the intake flow tube of a fan forward acoustic tester as the research object, the variations of flow coefficient and flow measurement error with Mach number ranging from 0.09 to 0.51 are quantitatively analyzed by the numerical method, and their affecting factors are revealed, aiming to support the accuracy of flow measurement. The numerical model adopting SST turbulence model is reliable after the test data verification. The flow coefficient increases with Mach number, and it ranges from 0.985 to 0.989. The flow coefficient is mainly affected by the boundary layer and the static pressure distribution, both of which tend to decrease with the increase of Mach number. The main factor is boundary layer, which accounts for about 71%~78%. If the uneven static pressure distribution is ignored, a maximum flow measurement error of about 0.4% will be brought.

Keywords: Flow tube · Flow coefficient · Static pressure distribution · Boundary layer · Flow measurement error

1 Introduction

Inlet flow is an important parameter of compressor general characteristics [1], and it is the main measurement physical parameter of compressor test. According to the principle of mass conservation, the measured section is generally selected at the intake flow tube of the tester to measure the physical flow. In order to ensure the flow measurement accuracy, the Ma number in the flow tube is generally controlled between 0.2 and 0.6 to meet the 0.5% flow measurement error, which is specified in the literature [2] (Hereinafter referred to as HB7115). The intake flow tube of the fan forward acoustic tester is installed at the central height of the anechoic room as a cantilever, without convergence section, pressure stabilization box and other components. At this time, the flow tube is connected with the test article directly, and the diameter of the flow tube is consistent with the test article, generally 1.0 m~1.4 m. The range of operating Ma number is wide, therefore, the Ma number in the flow tube is low under many working conditions. The applicability requirements for the flow tube of the same size under a wide Ma number range are put forward.

In view of the research on the flow measurement of the engine intake flow tube, Xiang Honghui et al. [3] studied the variation law of the flow coefficient with Mach number and the size of the flow tube, and found that the influence degree of viscous blockage of the boundary layer of different flow tubes is similar. Zhang Zhihong et al. [4] pointed out that the flow coefficient not only represents the influence brought by the boundary layer, but also covers the measurement error of the whole flow measurement system. Beale et al. [5] conducted a detailed study on the relationship between the flow coefficient and the measurement position, Reynolds number and Mach number. The results showed that under high Mach number condition, the boundary layer conversion has a great impact on the static pressure measurement at low Reynolds number. Wang Hong et al. [6] found that the symmetry of the total pressure and static pressure map at the engine inlet is related to the rotation direction of the rotor, and the flow distribution is non-uniform. Shi Jianbang et al. [7] calculated the air flow by using the method of combining area partition and boundary layer correction. After flight test, the measurement results of this method are relatively accurate. Then Li Bing et al. [8] found that the measurement error of engine inlet air flow and boundary layer can be significantly reduced by using pressure combination measuring rake and differential pressure sensor. Zhuang Huan, Xiao Min, et al. [9, 10] studied the inconsistency of air flow measurement between the ground platform and the high altitude platform of the core aircraft, and the results showed that the problem was caused by the uneven static pressure distribution. Li Yajin, Zhong Huagui, Liu Rongrong et al. [11–13] carried out relevant research on the calibration of flow measurement of flow tube to provide guarantee for the improvement of flow measurement accuracy.

To sum up, the boundary layer and static pressure distribution in the flow tube are important factors, that affect the accuracy of compressor flow measurement. At present, there are many domestic studies on the flow coefficient considering the boundary layer correction, but there is a lack of research on the impact of static pressure distribution on flow measurement quantitatively, especially for the intake flow tube designed by non navigation mark design methods such as fan forward acoustic tester, it is necessary to study the accuracy of flow measurement. In this paper, the numerical simulation method is mainly used to study the pressure, boundary layer distribution, and flow coefficient under different Mach number, and their influence on the accuracy of flow measurement is considered. The results can provide support for the accuracy evaluation of flow measurement of flow tube in some specific engineering applications.

2   Research Object and Calculation Method

2.1   Geometric Mode

The research object is the intake flow tube of the fan forward acoustic tester in the anechoic room, including the bell mouth and the test section. Its structure and installation position are shown in Fig. 1.

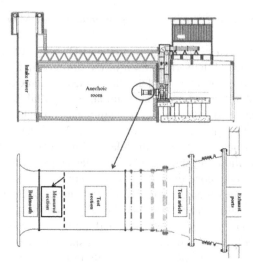

Fig. 1. Structure diagram of the tester

2.2 Calculation Method

Calculation Method of Flow. In the test study, the flow calculation adopts the formula specified in HB7115, namely formula (1):

$$Q = K_W mA \frac{P_0^*}{\sqrt{T_0}} \left(\frac{\gamma + 1}{2}\right)^{\frac{1}{\gamma - 1}} \left(\frac{\gamma + 1}{\gamma - 1}\right)^{\frac{1}{2}} \left[\left(\frac{P_0}{P_0^*}\right)^{\frac{2}{\gamma}} - \left(\frac{P_0}{P_0^*}\right)^{\frac{\gamma+1}{\gamma}}\right]^{\frac{1}{2}} \tag{1}$$

where: K_W is the flow coefficient; A is the cross-sectional area for the flow tube, m^2; T_0 is the atmospheric temperature, K; P_0^* is the total pressure of the measured section for the flow tube, Pa; P_0 is the static pressure of the measured section for the flow tube, Pa; m is the coefficient, calculated according to formula (2), (kg · K)$^{0.5}$/J$^{0.5}$.

$$m = \sqrt{\frac{\gamma}{R}\left(\frac{2}{\gamma + 1}\right)^{\frac{\gamma+1}{\gamma-1}}} \tag{2}$$

where, $\gamma = 1.4$, R $= 287.06$J/ (kg · K), then $m = 0.0404$, formula (1) can be changed to:

$$Q = 0.1561 K_W A \frac{P_0^*}{\sqrt{T_0}} \left[\left(\frac{P_0}{P_0^*}\right)^{\frac{2}{\gamma}} - \left(\frac{P_0}{P_0^*}\right)^{\frac{\gamma+1}{\gamma}}\right]^{\frac{1}{2}} \tag{3}$$

Calculation Method of Flow Coefficient. In Eq. (3), the flow coefficient is one of the important parameters for flow calculation. HB7115 stipulates that the velocity coefficient λ is calculated by the total pressure radial distribution and the wall static pressure, and 0.99 times of the mainstream area λ is taken as the starting point of the boundary layer to determine the thickness of the boundary layer. Taking 0.125 times of the boundary layer thickness as the boundary layer displacement thickness, the area of the flow measured section minus the boundary layer displacement thickness torus area is the effective flow area. The ratio of the effective flow area to the measured section area is the flow coefficient [2]. The theoretical basis of this method is that the static pressure gradient in the mainstream and boundary layer is zero[14], which satisfies Eq. (4).

$$\frac{\partial p}{\partial r} = 0 \tag{4}$$

In the above formula, the determination of the flow coefficient is based on the assumption that the static pressure is uniformly distributed along the radial direction, that is, the flow loss is mainly caused by the boundary layer. However, in the actual test process, the static pressure radial distribution of the measured section may be uneven, resulting in the uneven distribution of velocity, which will lead to the flow calculation error.

When the wall static pressure is used to calculate the flow, the flow calculation method under the condition of uneven static pressure distribution is explained: If the average flow velocity calculated by the wall static pressure and the incoming total pressure is taken as the ideal flow velocity, that is, the red straight line in Fig. 2. The black solid line are the actual velocity distribution obtained by the simulation, it can be seen that the flow loss includes two parts: the one caused by the boundary layer ΔQ_1 and the other one caused by the uneven static pressure distribution ΔQ_2. The calculation method specified in HB7115 ignores the flow loss ΔQ_2. When the static pressure distribution is uneven extremely, the calculation error of flow is large, and the ΔQ_2 should be taken into account in the calculation of flow coefficient.

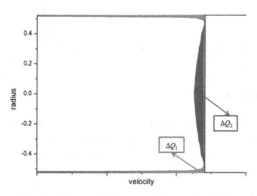

Fig. 2. Schematic diagram of flow calculation.

Compared with HB7115, the accuracy of flow calculation depends on the impact of uneven static pressure distribution. When it is within the acceptable range, the method specified in HB7115 can be used to calculate the flow coefficient. If it is large, the impact of this part should be considered.

In engineering application, when the Ma number is different, the same flow coefficient is used for flow calculation to facilitate data processing. When there is great difference in flow coefficients at different Mach numbers, this method is bound to cause a large error. In order to study the impact of uneven static pressure distribution on the flow coefficient, the following two methods are used to calculate the flow coefficient of this type of flow tube at different Mach numbers.

Method 1. The ratio of the actual flow Q to the flow calculated by the wall static pressure Q_{PS} (where $K_W = 1$) is defined as the flow coefficient K_{W1}, and the calculation formula is as follows:

$$K_{W1} = \frac{Q}{Q_{PS}} \tag{5}$$

Method 2. Based on the provisions of HB7115 and the simulation results, take 0.99 times of the velocity in the mainstream as the starting point of the boundary layer. The velocity coefficient is integrated along the radius to calculate the boundary layer displacement thickness, which are calculated by the total pressure and wall static pressure. The ratio of the effective flow area A_{eff} to the geometric area A of the flow tube is K_{W2}. This method only considers the influence of the boundary layer, and the formula is as follows:

$$K_{W2} = \frac{A_{eff}}{A} \tag{6}$$

Calculation Method of Flow Measurement Error. In order to quantitatively analyze the influencing factors of flow error, the error E is calculated by Eq. (7), the error E_1 caused by boundary layer is calculated by Eq. (8), and the error E_2 caused by static pressure distribution is calculated by Eq. (9).

$$E = 1 - K_{W1} \tag{7}$$

$$E_1 = 1 - K_{W2} \tag{8}$$

$$E_2 = K_{W2} - K_{W1} \tag{9}$$

3   Research Scheme

In this paper, the flow field in the flow tube is obtained by simulation to study the variation of two types of flow coefficient and measurement errors with Ma number, which provides a basis for the calculation of flow coefficients in practical engineering.

3.1   Simulation Model

Figure 3 shows the geometric model and main dimensional parameters of the flow tube of a certain compressor. In the modeling process, the wall thickness of the flow tube is ignored and replaced by a rotating surface. A straight section with a length of about 3D is reserved behind the bell mouth, where D is the diameter of the measured section of the flow tube, that is, the components behind the bell mouth are uniformly simplified to a straight section with a diameter of D. The cylindrical surface with a radius of about 6.1D is taken as the boundary of the calculation domain, and its axis is same as the flow tube.

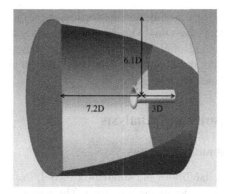

Fig. 3. Schematic diagram of geometric model.

3.2 Grid Division

In this paper, ANAYS CFX is used for simulation calculation, and ICEM is used for grid topology division in the computational domain. The grid type is hexahedron structured grid. The symmetry of the flow tube and the method of "O net over O net" are used to divide the grid, which can quickly and easily get a good quality mesh. In order to reduce the calculation error, mesh refinement is carried out at the wall. The height of the first layer of grid on the wall is 0.01 mm, the wall YPLUS < 3.3, and the number of grids is 3.52 million. The schematic diagram of flow tube grid is shown in Fig. 4.

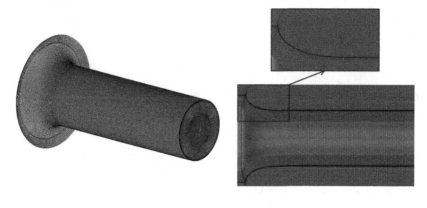

(a) Flow tube mesh   (b) Flow tube local enlarged mesh

Fig. 4. Schematic diagram of flow tube grid

3.3 Boundary Conditions and Convergence Judgment

Assuming that the flow in the tube is steady, the setting of the computational domain is as follows: The medium is ideal air. The far-field fluid temperature is 15°C. The model is total energy. The wall of the flow tube and the section where the outlet is located are set as no slip wall. The outlet is mass flow. Other calculation fields are set as opening and the pressure is 1atm. The high resolution scheme is used for solution, and the convergence residual is set to 10^{-5}. As the flow in the boundary layer is an important part of the calculation, SST turbulence model is selected.

4   Calculation Results and Analysis

4.1   Verification of Simulation Results

In order to verify the reliability of the simulation model, this paper obtained the test data of the real physical model of the flow tube. The measured section is 0.3D from the bell outlet, as shown in Fig. 5(a). The layout of probes of measured section are shown in Fig. 5(b), including two 4-point total pressure probes (PT probe), and two 10-point boundary layer probes (Boundary layer probe). The distribution of wall static pressure measuring points (PS point) is shown in Fig. 5(c), including 8 points uniformly distributed in the circumferential direction.

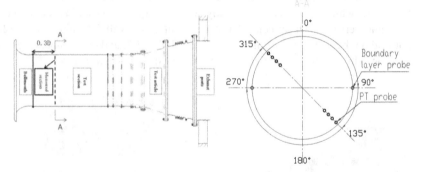

(a) Schematic diagram of measured section position

(b) Diagram of probes layout of measured section

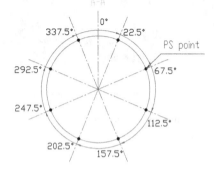

(c) Layout diagram of wall static pressure measuring points

Fig. 5. Test layout diagram

The test performance range of a certain fan is Ma = 0.15~Ma = 0.51. In order to understand the variation law of flow coefficient with Mach number more accurately, this paper extends the test range to Ma = 0.1~Ma = 0.52.

The simulation model is checked based on the boundary layer results under Ma = 0.3 condition. The dimensionless total pressure distribution of the boundary layer is shown in Fig. 6. It can be seen that the pressure gradient is similar, and the simulation results can calculate the boundary layer distribution better. Compared with the test results, the total pressure of the simulation results is lower. The difference comes from the difference between the simplified model and the actual physical model. The flow field in the simplified model is uniform. In the actual engineering, the flow tube is in an asymmetric intake anechoic room, and the flow field in the flow tube is circumferential non-uniformly.

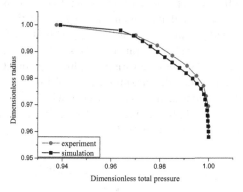

Fig. 6. Verification of the simulation results of the dimensionless total pressure radial distribution in boundary layer

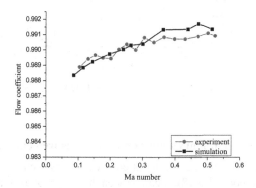

Fig. 7. Comparison of simulation result and test results of variation law of flow coefficient with Ma number

For the boundary layer test data, the flow coefficient K_{W2} is calculated by using method 2 in the previous section, and compared with the simulation results. The results are shown in Fig. 7. It can be seen that the simulation results are basically consistent with the test results, and the flow coefficient increases with Ma number. The results show that the model is reliable, and the results can provide data support for flow measurement.

4.2 Comparison of Calculation Results of the Two Flow Coefficients

Figure 8 shows the comparison between the flow coefficient K_{W1} and the flow coefficient K_{W2} calculated based on the simulation results. It can be seen that both of them increase with Mach number, and the change trend is consistent with the law described in literature [3]. At the same Mach number, $K_{W1} < K_{W2}$, the law conforms to the theoretical analysis in Sect. 2.2 Under Ma = 0.09 condition, the difference between them is the largest, about 0.004, that is, using K_{W2} of substitution K_{W1} will bring about 0.4% of the flow calculation error, and the flow error specified in HB7115 is not greater than 0.5%. Therefore, the correction of flow coefficient is particularly important to reduce the test error.

For the convenience of calculation in engineering application, the flow coefficient is generally taken as a certain value, such as the average flow coefficient. It can be seen from Fig. 8, when Ma number is in the range of 0.09~0.51, the variation range of flow coefficient K_{W1} is 0.985~0.989, and the variation value is about 0.004. Therefore, if the average flow coefficient is 0.987, the maximum error of flow measurement is 0.2%.

In conclusion, only considering the boundary layer will bring about a maximum flow calculation error of about 0.4%; If the average value of flow coefficient is selected, the maximum flow calculation error will be 0.2%.

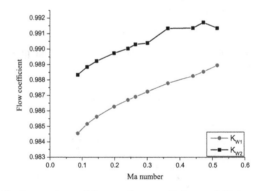

Fig. 8. Comparison of the two flow coefficients with Ma number

4.3 The Influence of Boundary Layer on Flow Measurement Results

Figure 9 shows that the flow measurement error caused by the boundary layer, and the error calculation method is Eq. (8). It can be seen that the error decreases gradually with the increase of Mach number for the same flow tube, which varies from 1.17% to 0.86%.

Figures 10 and 11 show the velocity distribution on the meridional plane of the flow tube at Ma = 0.09 and Ma = 0.51. It can be seen that the thickness of the boundary layer decreases with the increase of Mach number. Therefore, the error caused by the thickness of the boundary layer decreases gradually.

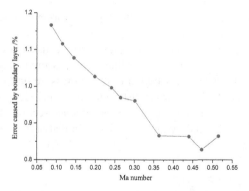

Fig. 9. Variation of flow measurement error caused by boundary layer with Ma number

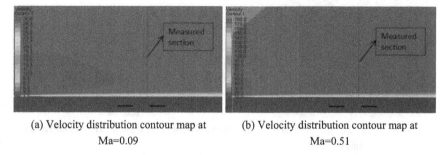

| (a) Velocity distribution contour map at Ma=0.09 | (b) Velocity distribution contour map at Ma=0.51 |

Fig. 10. Comparison of velocity contour map at different Mach numbers

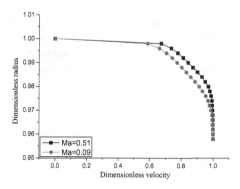

Fig. 11. Comparison of velocity distributions at different Mach numbers

4.4 The Influence of Static Pressure Distribution on Flow Measurement Results

As mentioned above, the Ma number in the flow tube of the fan forward acoustic tester is low under many working conditions. Therefore, the meridional static pressure distribution of the flow tube at Ma = 0.26 is taken as an example, to analyze the impact of uneven static pressure distribution on the flow measurement results. The results are shown in Fig. 12.

Figure 12(a) shows the contour map of static pressure distribution on the meridional plane of the flow tube, and the black straight line is the measured section. The black curve in Fig. 12(b) shows the radial distribution of static pressure on the measured section. It can be seen that the static pressure is distributed unevenly along the radial direction. This is because the air flow accelerates at the convergence section of the bell mouth, but the acceleration near the boundary layer is faster than that at the center, so the pressure near the boundary layer is lower.

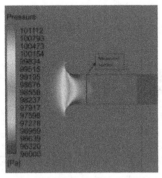 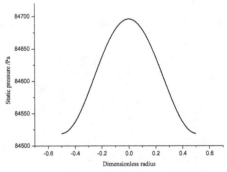

(a) Static pressure distribution con-   (b)Radial distribution of static pressure in measured
              tour map   section

Fig. 12. Static pressure distribution on the meridional plane of flow tube when Ma = 0.26

Figure 13 shows the flow measurement error caused by uneven static pressure distribution. The calculation method is formula (9), which corresponds to the difference between the two in Fig. 8. It can be seen that in the same measured section, the error decreases basically with the increase of Mach number, from 0.38% to 0.24%.

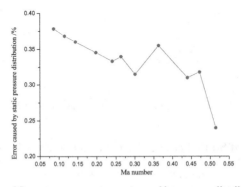

Fig. 13. Variation law of flow measurement error caused by uneven distribution of static pressure with Ma number

4.5 Comprehensive Influence Results of Static Pressure Distribution and Boundary Layer

The error caused by uneven static pressure distribution and boundary layer thickness is calculated by formula (7), and it is shown in Fig. 14, corresponding to the sum of errors in Figs. 9 and 13. It can be seen that with the increase of Mach number, the error decreases gradually, so the flow coefficient increases gradually.

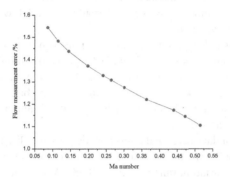

Fig. 14. Variation of flow measurement error with Ma number

Figure 15 shows the respective proportions of flow measurement errors caused by uneven static pressure and boundary layer distribution. It can be seen that the error caused by the boundary layer is dominant, accounting for about 71%~78%, and the error caused by uneven static pressure distribution accounts for about 22%~29%. The former is about 3 times of the latter.

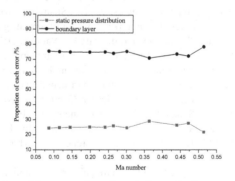

Fig. 15. Variation of the proportion of two influencing factors with Ma number

5 Conclusion

In this paper, the flow field characteristics in the flow tube of the fan forward acoustic tester are studied by simulation. The effects of static pressure and boundary layer distribution on the flow measurement error and flow coefficient are analyzed in the range of $Ma = 0.09$ to $Ma = 0.51$. The following conclusions are obtained:

(1) A numerical model is established based on the inlet flow tube of a fan forward acoustic tester. The model adopts SST Turbulence model, which has been verified by experiments to be reliable.

(2) Based on the physical meaning of flow coefficient, a quantitative analysis method for the influence of static pressure and boundary layer distribution is formed. The results show that the flow measurement errors caused by uneven static pressure distribution and boundary layer thickness decrease with Mach number. The flow measurement error is mainly caused by the boundary layer, which is about 71%~78% of the sum of the above two errors.

(3) The effect of boundary layer and uneven static pressure distribution are considered comprehensively. The result shows that the flow coefficient increases with the Mach number, and the variation range is 0.985~0.989. When the Mach number is 0.09, the difference between the flow coefficient K_{W1} and the flow coefficient K_{W2} is the largest, and the former is about 0.004 lower, that is, the maximum flow error is about 0.4% if only considering the boundary layer.

References

1. Gui, X., Teng, J., Liu, B., et al.: Compressor Aerothermodynamics and its Applications in Aircraft Engines. Shanghai Jiaotong University Press, Shanghai (2014)
2. Test method for compressor aerodynamic performance of aircraft gas turbine engine. State Administration of Science\Technology and Industry for National Defence, Beijing (2020)
3. Xiang, H., Hou, M., Ge, N., et al.: Aerodynamic design and application validation of flow tube used in compressor experiment. Gas Turbine Technol. **28**(4), 28–34 (2015)
4. Zhang, Z., Wu, F., Ma, H., et al.: Numerical and experimental study on the air flow coefficient determination for small aeroengine in ATF. Measur. Control Technol. **33**(8), 23–26 (2014)
5. Beale, D.K., Hand, T.L., Sebourn, C.L.: Development of a bellmouth airflow measurement technique for turbine engine ground test facilities. In: Proceedings of the 37th AIAA/ASME/SAE/ASEE Joint Propulsion Conference and Exhibit, pp. 1–20. AIAA, Salt Lake City (2013)
6. Wang, H., Ma, M.: Study on the factors affecting airflow calculation of an aero-engine. Gas Turbine Exp. Res. **23**(3), 22–26 (2010)
7. Shi, J., Shen, S., Gao, Y., et al.: Research on the aero-engine airflow measurement and calculation method. Eng. Test **51**(4), 15–18, 41 (2011)
8. Li, B., Hao, X., Shen, S.: Measuring methods of aero-engine inlet airflow. Gas Turbine Exp. Res. **26**(4), 54–57 (2013)
9. Zhuang, H., Guo, X., Ma, Q., et al.: Numerical and experimental study of the air flow determination of a core engine with bypass duct. Gas Turbine Exp. Res. **21**(2), 11–14, 52 (2008)

10. Xiao, M., Zhuang, H., Guo, X.: Numerical and experimental study of air flow field of a core engine with bypass duct. J. Aerosp. Power **24**(4), 836–842 (2009)
11. Li, Y., Wang, Y., Wang, H.: A scaling model test facility for turbo engine airflow calibration. Metrol. Meas. Technol. **37**(4), 27–30 (2017)
12. Zhong, H.: An evaluation method on calibration and measurement uncertainty of an airflow rate measuring tube. Gas Turbine Exp. Res. **24**(2), 1–4 (2011)
13. Liu, R.: Design of air flow tube calibrated device on LabVIEW. Tech. Autom. Appl. **36**(1), 88–90 (2017)
14. Ding, Z.: Fluid Dynamics, 2nd edn. Higher Education Press, Beijing (2013)

Modeling/Simulation Applications
in the Military Field

A Review of Cyberspace Operations Modeling and Simulation Research and Suggestions

Bingtong Liu, Anchao Cheng$^{(\boxtimes)}$, and Yihui Zhou

School of Information and Communication, National University of Defense Technology, Wuhan 430000, China
chinachao@189.cn

Abstract. Cyberspace Operations have become a regular part of future warfare. However, due to its significant strategic value, numerous elements involved, strong operational concealment and prominent chain reaction, directly conducting relevant research in real space will have incalculable consequences. At present, modeling and simulation methods have become a common practice and an important scientific method to develop and test network countermeasures equipment, assist in the design of cyberspace operations concepts and tactical research, and enhance cyberspace operations capabilities at home and abroad. By analyzing the current situation and future trends of modeling and simulation research in the field of cyberspace operations in countries around the world, and propose "five accelerated" development suggestions for cyberspace operations modeling and simulation in China.

Keywords: Cyberspace operations · Modeling and simulation · Research status · Development trends · Policy recommendations

Cyberspace is a new field that is independent in form from the four traditional physical spaces of land, sea, air and space, but depends on them in connotation. The Internet of Everything is an ideal effect of the network. It has become a new symbol of the development and progress of human society, and it is also the basic form of existence in which various elements of the modern battlefield can play a "system of system" role. Cyberspace can provide important support for the smooth implementation of other military operations, making joint operations more and more dependent on cyberspace. How to carry out combat operations in cyberspace and how to use cyberspace to carry out combat operations has become a new pattern of operations that countries around the world are competing to develop.

In recent years, the military powers have invested a lot of human, material and financial resources to study new equipment, new forces, new mechanisms and new efficiencies for Cyberspace Operations(CO); among them, adopting Modeling and Simulation(M&S) methods and using simulation platforms such as Cyber Range and Cyber Proving Ground have become a common practice and an important scientific method for developing and testing network countermeasures equipment, assisting CO concept design and tactical research, and improving CO capabilities at home and abroad.

© The Author(s), under exclusive license to Springer Nature Singapore Pte Ltd. 2022
W. Fan et al. (Eds.): AsiaSim 2022, CCIS 1713, pp. 491–502, 2022.
https://doi.org/10.1007/978-981-19-9195-0_39

1 Research Status of Cyberspace Operations Modeling and Simulation

In recent years, countries around the world have continuously deepened their understanding of CO by improving the relevant M&S theory and methods in the field of CO, breaking through key technologies of M&S, and developing M&S support platforms, and have continued to improve their CO capabilities steadily. Among them, as a country with highly developed information technology, the US is also leading the world in the field of CO M&S.

1.1 Analysis of the Research Status of the Theoretical Method of Cyberspace Operations Modeling and Simulation

Starting from the unified top-level design in the field of cyberspace, countries around the world have actively explored and innovated a series of theoretical methods suitable for CO M&S, and have made important progress in modeling framework, simulation standards, breakthroughs in key and difficult problems, and model verification.

Continuously Improve the Cyberspace Modeling Framework and Simulation Standards. In the joint operation environment without network and "system of system", CO simulation has become a key field of military modeling and simulation around the world, and CO training has become an essential training subject for improving joint operations capabilities. The team of Si Guangya of National Defense University proposed a new war system modeling framework EBNI [1] (Entity Behavior Network-Interaction) reflecting the characteristics of cyberspace under the background of joint operations, which can describe the information interaction process and network interaction characteristics in cyberspace, and reflect the "system of system" confrontation characteristics of physical domain, information domain and cognitive domain. In order to realize the joint training of various services around the world, the US takes the LVC (Live-Virtual-Construction) architecture as the traction, and constantly explores the LVC training mode in the field of cyberspace. In 2016, the US Army led the development of the "Persistent Cyber Training Environment" (PCTE) project, which effectively promoted distributed joint network training among different services in different regions. The US military also proposes a cyberspace simulation system based on recognized standards and an open architecture, which can integrate new technologies, continuously upgrade and improve, and achieve the fastest iterative development. In 2018, the Simulation Interoperability Standards Organization (SISO) held a Winter Simulation Innovation Seminar, which put CO into the key areas of modeling and simulation applications, and took CO simulation standards as one of the focuses of SISO standard formulation [2].

Continue to Break Through the Simulation Modeling Theory of Important and Difficult Problems. In recent years, from the M&S of a single network attack and defensive weapon to small virtualized cyber ranges to a ubiquitous network, a large-scale virtual-real combined CO simulation systems [3], M&S theories such as complex information environments, complex adaptive behaviors, and network interaction effects have been continuously promoted at home and abroad, the background of "system of system" confrontation is becoming stronger and stronger, the modeling elements is becoming more and more complete, and the simulation scenarios is becoming more and more realistic. Modeling methods based on complex network and super-net theory for solving large-scale, heterogeneous and dynamically evolving information environment have gradually matured, and Zhang Qingjun of the National Defense University proposed a modeling method combining multi-Agent, multi-resolution and complex networks [4] to reflect the characteristics of adaptive, multi-level and cascading information networks under "system of system" confrontation conditions. In the M&S of complex adaptive behaviors such as information behavior and cognitive behavior in cyberspace, the action modeling method based on complexity theory proposed by the US Army and the rules-oriented adaptive behavior modeling in China can better describe complex cyberspace operations behaviors including cognitive behavior [5–10]. In describing the interaction mechanism and cascading effects of CO, it is a hot topic at home and abroad in recent years [11–19]. The I2Sim developed by the Complex Systems Integration Center of the University of British Columbia in Canada can model interconnected critical infrastructure and its interdependencies, and its simulation process can also reveal some hidden interdependencies [13] and has been widely used [14–16]; China, has also explored a new way to understand the phenomenon of network cascade failure in the Agent modeling method [17–19].

Explore New Ways to Validate Simulation Models for Complex Systems. Cyberspace is a complex giant system with highly uncertainty characteristics, huge M&S projects, and the higher the complexity of the simulation model, the more critical the verification and credibility assessment of the model is, which is related to the application value of the simulation model. However, the unique characteristics of CO, such as network virus model, network attack behavior model and related models spanning physical domain, information domain, and cognitive domain, are programs of programs, codes of codes, and the virtuality of virtuality, with real-time, evolutionary and uninterpretable, the existing model verification theories and methods are not enough to fully support the analysis, verification and credibility verification of such models. In addition, there is no single general assessment method for the credibility assessment of models at home and abroad, and the verification results are mostly inaccurate. Generally, the model credibility is indirectly verified through simulation experiments on data and result analysis. Therefore, the credibility assessment, Verification, Validation and Accreditation (VV&A) technology of complex simulation system is both a research hotspot and a difficulty at home and abroad. There are all committed to finding a common standardized assessment method and index system for models to support the whole process verification for the whole life cycle of current simulation models. NASA has developed the VV&A's 3 phases process and 4 confidence assessment criteria to rigorously assess the credibility of simulation model predictions, but there are still some problems in the application of model plausibility verification in the field of cyberspace [20]. Professor Zhang Lin of Beihang University also mentioned many difficult problems in quantitative analysis and credible assessment of complex models in modeling and simulation-based "system of system" engineering (MSBS2E) [21]. However, the wide application of emerging technologies such as AI and big data has brought new ways to solve the problem of credibility assessment of complex systems simulation model.

1.2   Analysis of the Research Status of Key Technologies in Cyberspace Operations Modeling and Simulation

The ultimate goal of M&S is to build a virtual model that is as similar as possible to the real object. Under the support of scientific M&S theory, M&S technology is the key to helping the theory to be implemented and realize the virtualization of modeling objects. At home and abroad, the M&S basic technologies such as network information environment, combat entities, combat behavior and combat effects have been continuously exerted efforts, and emerging technologies have been deeply integrated to help M&S difficulties make major breakthroughs.

Cyberspace Operations Information Environment Modeling and Simulation Technology. At present, the information environment combining virtual and real is the main trend of M&S, mainly including large-scale heterogeneous network simulation technology and rapid deployment technology. In order to highly fit with the real information environment, in terms of large-scale heterogeneous network simulation, network simulation based on virtualization technology has become the main research direction of countries around the world; in terms of rapid network deployment, AI-enabled simulation technology is the mainstream direction to solve the rapid construction of large-scale network scenarios. Elex Cybersecurity Inc. has developed a twin cyber range that integrates the information domain and the physical domain [22]. And this full-spectrum cyber security range is based on digital twin and virtualization technology to digitally simulate the information domain and physical domain of the target system. The US PCTE uses network virtualization, cloud services, software definition, AI and other technologies to realize the organic integration of physical space and virtual space. It can flexibly build and quickly deploy various training scenarios, and automatically plan and design the training content, and meet the needs of the unified network training environment of the network combat units scattered in various places and services. It can carry out large-scale network training in a virtual environment that is highly close to the real, and supports the playback of the training process and the reuse of training data and scenarios, realizing standardization, simplification, and automation of the training process [23]. The US Department of Defense's Joint Cyberspace Operations Range (JCOR), which is used for learning, training, and military exercises, also uses AI, digital twins, and other technologies to achieve automatic and rapid construction of large-scale cyber training environments.

Cyberspace Operations Entity Modeling and Simulation Technology. CO entities are all objects that can produce combat effects in the information environment, including physical entities such as operational forces, weapons and equipment systems, as well as virtual entities such as network information, network viruses and vulnerability patches. It has the characteristics of combining virtual and real, soft and hard. At present, CO entities simulation mainly uses virtualization technology to perform multi-granular and multi-level entity simulation according to mission requirements. In terms of physical entity modeling, the US Air Force's SIMATEX simulator uses AI, digital twins and other technologies to provide the simulation of critical information infrastructure and industrial control systems for the JCOR. In terms of virtual entity modeling, Sichuan University proposed a knowledge graph-based method for building a cyber range arsenal [24]. This method designs and constructs a schema layer ontology structure of the cyber range arsenal for formalizing and standardizing network attack tools, and modelers only need to obtain attack tools, malicious code and related vulnerabilities, infrastructure and other description features from the Open Source Intelligence database, and instantiate the data according to the schema layer ontology structure to realize the instantiation and visualization of the virtual entity model.

Cyberspace Operations Behavior Modeling and Simulation Technology. The diversity of CO entities determines that CO behavior have a high degree of complexity and obvious adaptability. At present, CO behavior M&S are mainly realized by using emerging technologies to extract the characteristics of combat behavior, process restoration and playback. BAE Systems developed an automated cyber defense tool "Cyber Hunting at Scale" (CHASE) project for DARPA [25], which combines advanced machine learning and cyberattack modeling techniques to enable the automatic detection of undetectable advanced cyber threats, improving cyber resilience and cyber security [24]. The SIM-TEX simulator realizes the ability to simulate attack events, large-scale users, and attack management in real traffic through computer simulation technology. In 2013, MITRE introduced the ATT &CK framework [26], which has been updated to its 10th version. It is a knowledge base and model for describing attack behavior from the attacker's perspective, and provides a common language and framework. Users describe and classify the adversarial behavior based on actual observational data firstly, and then use the framework to develop attack plans and simulate specific threat attacks. At present, it has been widely used in various scenarios such as attack behavior simulation, red-blue confrontation, threat intelligence analysis, and assessment of defense capabilities. In addition, Strike Pack, the UK's cybersecurity and malicious code attack simulation system, can simulates 4,500 attacks and more than 28,000 malicious code attacks [27].

Cyberspace Operations Effects Modeling and Simulation Technology. Cyberspace is a kind of artificial space that spans the physical domain, information domain and cognitive domain. CO can produce independent combat effects or be integrated with other military operations to complete operational tasks collaboratively and produce a "system of system" effect, which is the essence of CO and the difficulty of M&S. At present, CO effects M&S mainly uses the method of collecting simulation operation data and extracting the assessment index. However, compared with the combat effects in the traditional field, the generative mechanism of CO effects is more complex. For example, whether a combat message is stolen or not directly affects the operations in various combat domains such as land, sea, air, and space, and even changes the tactical warfare of the whole battlefield, which will be not reflected in the simulation operation data, so it is difficult to fully describe its inherent characteristics by using mechanism analysis methods. At home and abroad, most of them only put forward ideas and methods in studying the modeling and simulation of CO "system of system" effects. The team of

Si Guangya of National Defense University proposed the modeling idea of multi-scale network interaction effect [28], which includes three levels: unit level, system level and "system of system" level. On the "system of system"-level effect modeling, the inter-network dependency relationship based on flow balance theory is used as an effective attempt to solve the problem of inter-network effect propagation. In the future, the M&S of CO effects should pay attention to the overall effect from the perspective of the "system of system", and establish a non-mechanistic model of nonlinear, time-varying, and uncertain CO effects by exploring non-mechanical modeling techniques based on neural networks and fuzzy regression.

1.3 Analysis of the Research Status of Cyberspace Operations Modeling and Simulation Support Platform

In recent years, countries around the world have increased their investment in the field of CO M&S, and accelerated the construction of the CO M&S support platform with the cyber range as the core. The cyber range has become an important symbol that reflects the capability of CO M&S in various countries. In terms of cyber ranges construction, the major military powers have continuously improved the comprehensive performance of the simulation support platform by expanding construction scale, enhancing the diversity of functions, and increasing the practicality of the cyber range.

Multi-party Cooperation to Expand the Scale of Cyber Range Construction. At present, the construction of cyber range around the world mainly aims at the integration of domestic construction and deployment and the support of multi-party joint cyber exercises. The British national "Federal Cyber Range" (FCR) [29] uses the existing cyber range to form a cyber experiment platform by means of federal integration, which is used to simulate large and complex networks and support the British military, government and academic institutions to carry out relevant cyber drills. The test site will also be interconnected with the US Army's "Cyberspace Solution Center" and other cyber laboratories around the world to enhance cyber simulation capabilities and conduct cyber offense and defense tests worldwide. The US National Cyber Range (NCR) is led by DAPRA and jointly built by more than 60 companies, academic institutions, and commercial entities. In 2016, to meet the rapidly evolving demand for cybersecurity test and assessment and the growing training and certification efforts of the US Department of Defense's Cyber Mission Force (CMF), the US Test Resource Management Center (TRMC) implemented a program called National Cyberspace Range Complex (NCRC), which expands NCR's network capacity and connectivity by incorporating new facilities in different locations as "new sites". The Joint Information Operations Range (JIOR) is a larger and more functional joint test range formed by interconnecting the network test environments of US military bases around the world. Its application has expanded from the US military to government agencies, security alliances, National Guard and other fields. And with the progress of construction and the expansion of application scope the number of nodes is also increasing [23]. Peng Cheng Cyber Laboratory, the largest and fully functional in China, is a national cyber range covering the Internet + Internet of Things + Industrial Control Network for the major national needs, and more than 20 universities, research institutes and related enterprises have participated in the construction of sub-ranges [30].

Develop Diversified Cyber Range Service Capabilities. Driven by M&S theories and technologies in the field of cyberspace, the functions of cyber ranges in various countries have developed from initially focus on cyber security testing to training and drill, equipment research and development, theoretical verification, and technical testing. Among them, the US, as the country with the highest level of informatization, has carried out the construction of various types of cyber ranges at the national and professional levels, and services and arms has also established its own cyber range with its own demands [31]. In terms of functions, these cyber ranges not only support the research and development of network weapons and equipment, and the research on CO tactics, but also train cyber professionals and test and assess new cyber security technologies. They also provide experimental platforms for new operational concepts to play a role, greatly improving the US military's CO capability. Under this influence, the United Kingdom, Japan, Sweden and other countries have also started the construction of cyber range. The FCR integrates four functions of assessment, testing, research and development and training, and plays an important role in the cyber threaten research, critical infrastructure survivability assessment and other aspects. The Cyber Range and Training Environment (CRATE) established by the Swedish Defense Research Agency [32] is mainly to provide cyberspace security testing services for the Swedish government and military, as well as military exercises and cyber weapons testing. The domestic Guiyang National Big Data Security Range has played an important role in national offense and defense live drills, product testing, technological innovation and other functions. At present, it is stepping up cooperation with domestic universities and well-known enterprises in the industry to establish an "Industry-University-Research Joint Innovation Laboratory", carrying out key technology research, and developing towards a comprehensive range and digital twin range [33].

Widely Used in Virous Military Exercises and Training Activities. The cyber range is vividly called "virtual training ground", which has attracted more and more countries' attention. The PCTE first joined the "Cyber Flag 20-2" mission in June 2020, followed by the "Cyber Flag 21-2" and the National Guard's "Cyber Yankee" exercise. The fundamental Cyber Operations "Plan X" [34], launched by DARPA in 2013, is also an important part of annual joint cyber exercises such as "Cyber Guard" and "Cyber Flag". In 2013, The JCOR integrated with the NCR, and provided infrastructure support and service support in a series of exercises such as "Cyber Storm", "Cyber Flag", and "Cyber Lightning". The JIOR and the NCR have also provided platform support for a number of information warfare and cyber operations, including "Cyber Flag", "Cyber Guard", "Cyber Knight" and many other exercises. In addition, the world's largest cybersecurity drill NATO "Lock Shield" live exercise [35], has been participated by more than 2,000 experts from nearly 30 countries by 2021, involving about 5,000 virtualization systems. Every year, "Lock Shield" has clear goals and specific scenarios, and the participating teams play different network offense and defense roles. The cyber range built for the drill can provide a real and safe training environment for cyber security experts from various countries participating in the drill.

2 Future Trends in Cyberspace Operations Modeling Simulation

At present, the construction of cyberspace operational forces is in the stage of rapid development, and M&S technology provides strong support for relevant experiments. In-depth analysis of the research status of CO M&S at home and abroad, starting from a unified top-level architecture design, with the development direction of integrating emerging technologies to solve the difficult problems in modeling, and aiming at the comprehensive integration of multiple simulation systems, CO M&S will develop in the direction of standardization, intelligence and integration.

2.1 The Theoretical Methods of Modeling and Simulation Are Developing in the Direction of Systematization and Standardization

With the continuous maturity of M&S technology, countries around the world have unified the model development process by continuously optimizing the top-level architecture of CO M&S, continuing to promote the construction of CO simulation standards, and actively exploring universal assessment standards for model credibility, the reusability and interoperability of systems related to CO continue to improve. Aiming at the "chimney" problem of simulation platform in various services and arms, the integration of existing systems and the development of new distributed systems are continuously promoted through unified technical standards. The US military attaches great importance to the construction of a multi-service distributed CO simulation system [36] in order to achieve interconnection between various CO simulation platforms and other combat systems, which is of fundamental significance for realizing cross-regional and cross-service joint simulation drills.

2.2 Modeling and Simulation Technology is Developing in the Direction of Intelligent, the Combination of Virtual and Real

In information warfare with intelligent characteristics, the elements and correlations of CO are becoming more and more complex, with a high degree of uncertainty. Traditional M&S techniques are difficult to effectively simulate various elements of CO. Machine learning, AI, digital twin, virtual reality and other emerging technologies integrate M&S technology to provide new solutions to the difficult problems under new combat modes, new combat spaces, and new combat rules [37], and intelligent modeling and simulation technology combining virtual and real has become the mainstream direction, promoting the development of future CO M&S platforms in the direction of intelligence and digitalization.

2.3 Modeling and Simulation Support the Development of the Platform in the Direction of Integration and Actual Combat

With the gradual deepening of research on CO in various countries, new application requirements for CO simulation systems continue to emerge, and the construction mode has developed from independent development and decentralized construction to integrated, collaborative, and actual combat. The national-level cyber simulation systems, puts more emphasis on the integration of the characteristics in multiple fields, taking into account different test purposes, and is developing towards the comprehensive integration with other combat systems, showing the characteristics of high comprehensiveness, high integration and strong versatility; The military-level cyber simulation systems, more emphasis is placed on the simulation of CO in the context of joint operations, requiring combat training and experiments that can support different mission requirements; The professional-level cyber simulation systems, more attention is paid to coordination, professional division of labor, and functional complementarity, and the promotion method of technology integration and collaborative co-construction is generally adopted.

3 Countermeasures and Suggestions for Promoting Cyberspace Operations Modeling and Simulation

Comparing and analyzing domestic and foreign research results in the field of CO M&S, to promote the supporting role of M&S methods in China's CO concept verification, equipment research and development, personal training, security testing, etc., puts forward five suggestions from the theoretical research, standard and norms, key technologies, personnel training and platform construction.

3.1 Accelerate the Theoretical Research on Modeling and Simulation of Cyberspace Operations

China's research on M&S theory of complex systems has been supported by a relatively mature theoretical system, but there is no consensus on M&S theory specifically for the field of CO. It is commended to incorporate CO M&S into the key areas of simulation applications, extract relevant forces to set up special organizations, in-depth analysis of the requirements and characteristics of CO M&S, sort out and summarize the relevant modeling theories that have been formed and highly applicability, to form a set of theoretical systems for guiding and standardizing M&S in this field.

3.2 Accelerate the Development of Standard and Norms for Cyberspace Operations Modeling and Simulation

Compared with foreign countries, in terms of strategic guidance documents for CO and the normative documents for CO M&S in China, the background of joint operations is not enough, and the pertinence and normalization need to be strengthened. On the one hand, accelerate the formulation and improvement of strategic documents such as relevant operational rules, military scenarios and technical standards in the field of CO, provide standardized and normative guidance and support for M&S; On the other hand, vigorously promote a unified and standardized top-level architecture for CO M&S to form an integrated and standardized standard specification for CO M&S.

3.3 Accelerate Breakthroughs in Key Technologies of Cyberspace Operations Modeling and Simulation

The difficulties of CO M&S are the modeling and simulation of virtual cyber entities, cognitive combat behavior, large-scale complex heterogeneous network information environment, and operations effects with nonlinear, time-varying, and uncertainties. Cyberspace is an abstract space spanning physical, information and cognitive domains, which greatly increases the difficulty of M&S. However, the traditional M&S technology that focuses on the physical domain and is led by the concept of mechanism modeling has been difficult to fully describe and characterize the operation mechanism of CO, and cannot effectively meet the needs of CO M&S. Therefore, it is urgent to increase breakthroughs in M&S methods and core key technologies. It is recommended to fully learn from the advantages of emerging technologies such as big data, AI, digital twins, and virtualization, and combine them with the needs of CO M&S, and then put forward a new technical route to crack CO M&S.

3.4 Accelerate the Cultivation of Cyberspace Operations Modeling and Simulation Professionals

At present, there is a shortage of network professionals in China, the team construction is in short supply, the overall structure is out of proportion. In particular, there is a lack of comprehensive talents who understand both tactics and techniques [38]. Therefore, it is necessary to vigorously strengthen the cultivation of CO professionals and M&S professionals. Shorten the growth cycle of command-technology integrated talents who understand CO and master M&S technology by making up for the shortcomings of knowledge in related fields, promoting professional cross-integration and in-depth participation in project practice and other pragmatic measures to better meet the urgent demand of cyberspace modeling and simulation professionals in future joint operations.

3.5 Accelerate the Construction of a Cyberspace Operations Modeling and Simulation Platform

The CO M&S support platform represented by the cyber range is a scientific testing ground for maintaining national cyberspace security and a virtual military exercise ground for training elite soldiers in CO. In recent years, China has emerged a new generation of cyber ranges represented by the "Fire Sky Cyber Realm" [39] and the Sai Ning Cyber Security Digital Range [40], but due to the late start in China, there are gaps with foreign countries in terms of construction scale, functional positioning and application background. On the one hand, it is necessary to highlight key points. Focus on key projects, key engineering and other national strategic projects, and strengthen the construction of cyberspace test ground that are highly matched with mission needs; On the other hand, it is necessary to make overall planning, through scientific planning of the construction cycle, continuous improvement of the construction mechanism, comprehensive coordination of the construction force, and form a simulation support platform construction mode of industry-university-research integration and theory-technology-application coordination; In addition, it is also necessary to do well in comprehensive integration, integrate the existing scattered cyber ranges and build a comprehensive and integrated distributed cyber test grounds to enhance the functional diversity and coordinate task capabilities.

References

1. Zhang, Y., Si, G., Wang, Y.: Research on cyberspace operations based on EBNI model frame. Journal 29(9), 1886–1894, 1906 (2017)
2. China Aerospace Science and Industry Corporation Second Research Institute 208, Beijing Simulation Center. Development report on military modeling and simulation (2018). 1st edn. National Defense Industry Press, Beijing, pp. 19–31,114–123, 319 (2019)
3. Fang, B., Jia, Y., Li, A.: Research on cyberspace range technology. Journal 1(3), 1–9 (2016)
4. Zhang, Q., Zhang, M., Zhang, Q., Wu, X.: Research on modeling of spatial information SoS based on complex network theory. Journal 29(9), 1907–1920 (2017)
5. Li, C., Si, G., Wang, Y., Zhang, M.: Research on DDos defense model based on MAS. Journal 2(S1), 157–166 (2013)
6. Wang, X., Ku, T., Jin, G., et al.: Adaptive behavior modeling method based on cognitive system. Journal 16(25), 111–115 (2016)
7. Gong, Y., Liu, Z.: A review on the application of agent-based modeling and simulation technology in the study of consumer behavior. Journal 22(3), 185–189 (2022)
8. Si, G., Hu, X., Wang, Y.: Practice and experience of novel combat space modeling and simulation. Journal 28(4), 5–10 (2014)
9. Li, Y., Zhang, G.: Research on behavior modeling of CGF based on agent. In: Seventh Symposium on Novel Photoelectronic Detection Technology and Applications, pp. 1176333.1–1176333.6 (2020)
10. Mei, S., Zarrabi, N., Lees, M., Sloot, P.M.A.: Complex agent networks: an emerging approach for modeling complex systems. Appl. Soft Comput. 37, 311–321 (2015)
11. Potts, M.W., Sartor, P.A., Johnson, A., Bullock, S.: A network perspective on assessing system architectures: robustness to cascading failure. Syst. Eng. 23(5), 597–616 (2020)
12. Lopez, C., Martí, J.R., Sarkaria, S.: Distributed reinforcement learning in emergency response simulation. IEEE Access 6, 67261–67276 (2018)
13. Martí, J.R.: The I2Sim simulator for disaster response coordination in interdependent infrastructure systems (2006)
14. Yang, Z., Martí, J.R.: Resilience of electrical distribution systems with critical load prioritization. In: International Conference on Critical Information Infrastructure Security, Lucca, Italy (2017)
15. Juarez-Garcia, H.: 'Multi-hazard risk assessment: an interdisciplinary approach. PhD thesis, University of British Columbia, Vancouver, Canada (2010)
16. Wang, J.: Vulnerability and risk analysis of the Guadeloupe Island for disaster scenarios using the modified i2SIM toolbox. MASc thesis, The University of British Columbia, Vancouver, Canada (2013)
17. Zhang, M., Xiao, B., He, X., et al.: Proceedings of the 8th National Annual Conference on Simulators, Conference 2013, pp. 297–300 (2013)
18. Xiao, F., Li, J., Wei, B.: Cascading failure analysis and critical node identification in complex networks. Stat. Mech. Appl. 596, 127117 (2022)
19. Gao, X., Li, X., Yang, X.: Robustness assessment of the cyber-physical system against cascading failure in a virtual power plant based on complex network theory. Int. Trans. Electr. Energy Syst. 31(11), e1309 (2021)

20. Yang, X., Xu, Z., Zhang, X., et al.: Review and difficulty analysis of credibility assessment of simulation models. Journal **46**(S1), 23–29 (2019)
21. Zhang, L., Wang, K., Lai, L., Ren, L.: System engineering based on modeling simulation. Journal **34**(02), 179–190 (2022)
22. Zheng, Y., Hu, F., Wang, L., Jiang, Z.: Proceedings of the "Network Industry Development Forum" of the 2021 National Cyber Security Publicity Week, pp. 100–107 (2021)
23. Li, F., Wang, Q.: Research on network range and its key technologies. Journal **58**(5), 12–22 (2022)
24. Ye, Y., Wang, Y., He, J.: Construction of cyber range weapon library based on knowledge graph. Journal **05**, 19–22 (2022)
25. NetEase News. https://www.163.com/dy/article/H3I7V5350552MROB.html. Accessed 17 Jun 2022
26. CSDN. https://blog.csdn.net/CoreNote/article/details/123508440. Accessed 17 Jun 2022
27. Wang, H., Song, L., Zhang, G.: Research status and key technology analysis of network range. Journal (09), 46–51 (2020)
28. Guangya, S.I., Yanzheng, W.A.N.G.: Cyberspace Operations Modeling and Simulation, 2nd edn. Science Press, Beijing (2020)
29. Li, Q., Hao, W., Li, C., Xu, L.: Status quo and enlightenment of foreign network range technology. Journal **09**, 63–68 (2014)
30. Doc Baba. https://www.doc88.com/p-66916078249843.html. Accessed 26 Jun 2022
31. Li, L., Wang, X., Hao, Z.: A review of the construction of cyberspace ranges in the United States. Journal **06**, 53–60 (2021)
32. Tencent Cloud. https://cloud.tencent.com/developer/article/1555473. Accessed 26 Jun 2022
33. Zhao, Q., Li, Y., Jiang, H., Li, Y.: Construction status, basic characteristics and development ideas of network range. Journal **06**, 62–64 (2020)
34. Wang, Y.: Research on the development and construction ideas of the US military X-plan network warfare system. Journal **48**(5), 367–371 (2018)
35. LiREEBUF: NATO 2022 "Lock Shield" network exercise held in Estonia, 2,000 security experts jointly trained troops. https://www.freebuf.com/news/330139.html. Accessed 21 Jun 2022
36. Chen, Q., Ma, T., Yin, J., Chen, Y.: A review on the development of simulation system for joint training of the US military. Journal **09**, 37–40 (2022)
37. Tong, J., Yang, D., Li, T., Li, H.: Proceedings of the 33rd China Simulation Conference (2021)
38. Journal of Information Security and Communication Confidentiality. https://mp.weixin.qq.com/s/ESbe6lnq_yGyJEBuh5N3QA. Accessed 20 Jun 2022
39. IT168. https://net.it168.com/a2022/0529/6739/000006739052.shtml. Accessed 20 Jun 2022
40. Cyberspace. https://baijiahao.baidu.com/s?id=1733970518126176003. Accessed 20 Jun 2022

Conceptual Modeling and Simulation of Multi-unmanned Cluster Air-to-Ground Battle for Battalion-Level Units in Mountainous Areas

Boyu Jia, Weidong Bao, and Yanfeng Wang[✉]

National University of Defense Technology, Changsha 43000, Hunan, China
1531927486@qq.com

Abstract. In order to solve the problems of battlefield environment restriction and heavy casualties in Synthetic Battalion, this paper takes a U.S. military battle in the valley of Hindu Kush in 2016 as an example, innovates the military demand generation process, and puts forward the operational concept of "multi-unmanned cluster air-ground integrated battle". The specific research process is to take the total annihilation of the terrorist battalion-level formed enemy as the combat target, and use the modeling method of operational concept process, joint mission thread modeling method and QFD (quality function deployment) equipment index optimization method to study and put forward the requirements for the establishment and equipment construction of mountain unmanned combat battalion-level units. Finally, Mozi Warfare Deduction System is used to verify the validity and rationality of the new operational concept. The innovation of this paper lies in the establishment suggestion of mountain unmanned combat battalion through operational concept modeling, and exploring the warhead ratio of patrol missile with the best killing effect by simulation deduction, so as to provide tactics, organizational reference and equipment construction suggestions for mountain unmanned combat battalion.

Keywords: Mountain warfare · Unmanned equipment cluster · Conceptual simulation

1 Introduction

The U.S. military has been carrying out security and anti-terrorism wars in mountainous areas of Afghanistan all the year round. In large-scale encirclement and suppression operations, several mountain synthetic battalions are often used to attack the terrorist jihad brigade (one battalion), which has accumulated rich combat experience. It also exposes some problems such as heavy casualties of traditional mountain combat troops, high combat costs, and strong domestic anti-war thoughts endangering political stability. "Fighting for the enemy", the U.S. military urgently needs a new operational concept to guide battalion-level mountain operations to reduce casualties and save money. At the same time, our country has similar battalion-level combat needs to those of the U.S.

Army in the western mountainous areas. This paper studies the combat concept of the U.S. Army's battalion-level units in mountainous areas, which has certain reference significance for our army construction.

This paper advances according to the idea of operational concept-equipment requirements-experimental simulation. [1] Absorb ideas such as mosaic warfare and global assault [2], designed the combat concept of "multi-unmanned cluster air-ground integrated battle", which supplemented the theory of unmanned combat in mountain areas of battalion-level units.

2 Basic Conceptual Modeling

This case is based on the conflict between the 2nd Battalion, 3rd Brigade, 10th Division of the U.S. Mountainous Army in 2016 and the Taliban armed forces in Afghanistan. It is envisaged that the U.S. Army (M) will use the unmanned army barracks to carry out precision strike and fire annihilation operations against the established terrorists (K). The specific scenario is as follows: 10: 00 on May 10, 202X, K continuously carried out special reconnaissance, information surveillance and sneak attack on M stationed in the base, and attempted to kill the personnel of M by means of light weapons shooting, light artillery harassment and roadside improvised explosive device ambush, thus distracting the military morale of M and turning the domestic public opinion of M against the war, thus achieving the purpose of dismantling the base of M. At the command of the superior, M patrol missile battalion, unmanned aircraft company, unmanned vehicle platoon, electronic jamming platoon, infantry assault platoon and logistics unit assembled in the designated area, in order to encircle the effective forces of K in one fell swoop, smash the combat attempt of K and eliminate the bloody conflict crisis. Table 1 the mission scenario.

Table 1. Operational mission background scenario table.

Serial number	Assumed condition	Content
One	Time	10: 00 on May X, 202X
Two	Site	M, k square Hindu Kush mountain area
Three	K-square firepower allocation	K harassment squadron (20 armed trucks, 4 command vehicles, 1 anti-aircraft sqaud, some light weapons)
Four	M-square firepower allocation	M Army Unmanned Combat Unit, Special Battalion, Patrol Missile Battalion, Air Defense Squad

(continued)

Table 1. (continued)

Serial number	Assumed condition	Content
Five	M combat phase	There are five stages: reconnaissance, wolf-hunting, beheading, striking and evaluation
Six	M combat mission	Wolf encirclement, precision strike, panic in
Seven	Combat operations	Advance reconnaissance operation, detailed reconnaissance operation, deployment control operation, beheading operation, fire cover operation and clean-up operation

2.1 Operation Stage Division

In this battle, M is divided into six combat systems, namely: reconnaissance system, fire strike system, combat communication system, air defense system, service support system and target guidance system.

Step 1: Reconnaissance Stage. The reconnaissance stage is divided into two small stages, including pre-war reconnaissance and on-site reconnaissance before firing. The main purpose of reconnaissance is to find out the general military deployment of the enemy, find out the initial situation of the enemy, and form a basic judgment on the next action of the enemy, mainly using spies, satellites and high-altitude silent drones for exploration. Pre-fire reconnaissance is responsible for obtaining the formation and deployment of enemy forces before the start of the war, further arranging enemy bunkers, trenches, secret posts and hidden heavy firepower by unmanned vehicles and reconnaissance platoons, paying special attention to the movements of high-value targets, and the communication support forces should monitor the enemy in the whole process. Before firing, supplement the war report according to the reconnaissance results.

Step 2: "WOLf" Stage. The main purpose of the "wolf" stage is twofold. The first is to destroy the enemy's mobile forces, attack the enemy's frontier targets, block the passage of the follow-up teams, prevent the enemy's formation from unfolding, and concentrate the enemy's forces on the frontier attack positions to facilitate the subsequent annihilation. The second is to feint, so that the enemy's fighting center of gravity (such as leaders and confidential documents) can be shifted, so that we can quickly identify, quickly advance the battle to the beheading stage, break the enemy's center of gravity, and build our hub situation.

Step 3: Beheading Stage. "The way to use soldiers is to attack the heart and attack the city." The beheading stage has only one goal, that is, to eliminate the enemy's center of gravity, destroy the enemy's combat hub, cooperate with the dissemination of leaflets, and paralyze the enemy's command hub, so as to ensure that the attack scene has a strong psychological deterrent effect (for example, in 2021, the U.S. military used a blade Hellfire missile to "kill" the terrorist armed leader, or used a military dog to hunt baghdadi), thus destroying the enemy's campaign objectives and causing great psychological shock to the enemy personnel.

Step 4: Strike Stage. The attack stage is also divided into two small stages. The first stage is the lag stage, which is responsible for delaying the enemy's reinforcements, making use of aerial firepower such as unmanned aerial vehicles (UAVs) and hovering patrol missiles, and ambush offensive unmanned vehicles, and carrying out a high-intensity attack on the support (rescue) reinforcements that move along the road. The idea of fighting is to "besiege the state of Wei to save the state of Zhao", "lure him with profit, and wait for him with a pawn", and ambush him on the road that the enemy must save, further killing the effective forces of the enemy. The second stage is the clearance stage. After the end of the main battlefield, commandos, CCT teams, military dogs, unmanned vehicles and unmanned aerial vehicles are dispatched to jointly advance, and the remaining enemy bunkers and sporadic troops are found to be destroyed, so as to "encircle" and "wipe out" the enemy's regional combat capability completely.

Step 5: Evaluation Stage. The purpose of the evaluation stage is to finish the post-war period, clean the battlefield, collect intelligence and sort out the combat materials. The two key points are to do a good job in humanitarian work, provide timely assistance to the wounded of both sides M and K, and start a war based on the materials collected on the battlefield, which is beneficial to M's public opinion and psychological warfare, deter the remaining terrorists in the region, deal a heavy blow to the enemy's politics and morale, and smash K's strategic goal of making M withdraw its troops by relying on M's domestic public opinion. It is a further attack on K's psychological domain and network domain, and a continuation of the hot war.

2.2 Operation Action Division

Six operations are designed in this campaign, among which the early reconnaissance and detailed reconnaissance correspond to the reconnaissance stage, the deployment control corresponds to the wolf stage, the decapitation corresponds to the beheading stage, the fire cover corresponds to the first half of the strike stage, and the clearance corresponds to the second half of the strike stage and the assessment. The author will elaborate on each stage of the operation.

Step 1: Advance Reconnaissance Operation. Make use of satellites (one or two) assigned by superiors to take photos and conduct covert reconnaissance on the combat area. The communication force monitors the K square array electronically to further acquire the enemy's deployment, and the UAV team dispatches a high-altitude silent UAV for further detailed reconnaissance to acquire the enemy's deployment and search for high-value targets. Note that this stage should usually be conducted in secret.

Step 2: Detailed Reconnaissance Operations. Three groups of two low-altitude UAVs and two groups of two unmanned vehicles were sent to approach and reconnaissance more than one day before the operation, and it was required to further discover the enemy's deployment, deliberately expose the unmanned platform to the square's field of vision, and let it relax its vigilance. At the same time, the air defense capability of K was tested, and this battle was regarded as an ordinary reconnaissance operation. Pay attention to the fact that do not fire at this time. After two waves, four UAVs and four unmanned vehicles were completely destroyed by the enemy and completed the designated reconnaissance mission.

Step 3: Deployment Control Action. After the command department planned the cruise missile route, it dispatched unmanned vehicle groups (4 vehicles in groups of 5), attack unmanned vehicle groups (5 vehicles in groups of 6) and Wolf patrol missiles (4) to carry out an early strike, damaging the enemy motor vehicle groups, blocking the effective forces of Party K in their own positions and motor roads, and placing drones and unmanned vehicles on the periphery to stand by, blocking the enemy in the "pocket array". After the lure, three reconnaissance patrol missiles were quickly launched to attract the enemy's anti-aircraft firepower and lock in the enemy's high-value targets. The target high-altitude reconnaissance drone was upgraded and opened, ready to be beheaded.

Step 4: Decapitation. After quickly identifying the head target, a group of attack drones (5) and a group of attack unmanned vehicles (4) quickly entered the field and launched an attack. Three reconnaissance patrol missiles bombarded to support the head force of the K party, attracting the attention of enemy infantry, making the enemy relax their defense against high-altitude weapons, paying attention to attacking armored targets such as tanks with armor-piercing bullets, and attacking accusation node targets such as communication vehicles with graphite bombs or jamming bombs. After identifying the target of the leader, quickly use the blade Hellfire missile to attack the leader of Party K, complete the beheading deterrent task, and use the third patrol missile to distribute leaflets and launch a psychological offensive.

Step 5: Fire Coverage Operation. The patrol group of missiles (15), the remaining 5 groups of 30 drones and 16 unmanned vehicles carried out long-range strikes, further killing the enemy's effective forces, striving to eliminate more than 70% of the effective forces of the K side, paralyzing the combat system of the K side, and creating a relatively safe environment for subsequent manual intervention. After the unmanned platform is destroyed, it can quickly use swarm intelligence algorithm and quickly reorganize into a new combat formation again. At the same time, we should explore the blind spots of early reconnaissance and further expand our field of vision.

Step 6: Clean-up Operation. The patrol missile group (9), the special battle platoon and the unmanned vehicle unmanned aerial vehicle (UAV) combined cluster further explored the dead corner, and carried out operations in the order of the unmanned platform breaking into the open space, the infantry approaching the ground for reconnaissance, and the patrol missile clearing at a fixed point, so as to further destroy the remaining forces of the K side and wipe out more than 90% of the enemy.

According to the division of combat stages and operations, combined with the opinions of military experts, the operations are further refined into operational processes. See Table 2 for the operational process table and Fig. 1 for the operational schematic diagram.

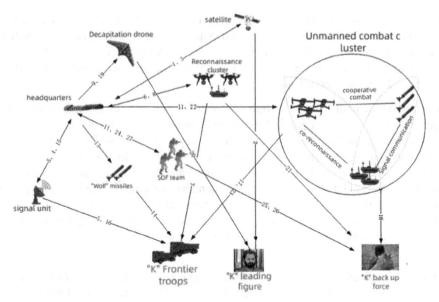

Fig. 1. Operational schematic diagram

Table 2. Operational sequence table

Activity serial number	Content
1	Coordinate the superior two satellites and one high-altitude UAV for early reconnaissance
2	2 satellites and 1 high-altitude UAV for early reconnaissance
3	Reconnaissance information return command post
4	Dispatch communication troops from the command post to implement electronic monitoring
5	Communication troops implement electronic monitoring and transmit back in real time
6	The headquarters dispatched 6 low-speed drones and 4 unmanned vehicles to conduct approach reconnaissance
7	6 drones and 4 unmanned vehicles approached reconnaissance
8	Reconnaissance information return command post
9	Command ordered to check and hit one drone to take off
10	Check one drone to take off
11	The headquarters dispatched unmanned clusters and special battle platoons to maneuver in the controlled areas

(continued)

Table 2. (*continued*)

Activity serial number	Content
12	Unmanned, special combat platoon maneuvers to the controlled area
13	The command post assigned the wolf patrol missile group to strike first
14	The wolf missile group made an strike
15	Dispatching an electronic detachment from the headquarters to interfere with enemy communication
16	Electronic detachment interferes with enemy communication
17	Unmanned cluster attack assessment
18	Unmanned cluster searches for high-value targets and feints
19	Command ordered the UAV to fire
20	Check one drone and fire
21	The missile patrol group distributed psychological leaflets
22	Command the unmanned cluster to strike
23	Unmanned cluster long-range strike
24	Command special operations platoon to rush in and finish up
25	The special platoon broke into the tail
26	Strike effect evaluation
27	Report to the headquarters to carry out propaganda war
28	Carry out post-war propaganda

3   Modeling the Guidance System of Joint Mission Thread Method

In this chapter, the joint mission thread method is used to analyze the operational conceptual structure from three aspects: the mission of the troops undertaking the mission, the connotation of the mission of the troops and the requirements of the equipment construction. Finally, the equipment tactical indicators needed by the troops completing the mission are obtained, which can provide simulation indicators for the follow-up simulation experiments.

Step 1: Mission Level Decomposition. Using 5w1h analysis method, explain clearly the events, people, time, place, reasons and methods involved in military operations [3]. Complete the basic description of the combat mission. Through mission description, combined with expert opinions, we can get the required effect to achieve the combat goal. Further, refer to the U.S. Joint Operations Regulations, and match the mission attribute with the demand effect. It should be noted that an effect corresponds to at least one behavior. Further, by consulting the list of joint operations tasks [4] and the list of services and arms, corresponding mission attributes and mission indicators to complete mission decomposition.

Step 2: Task Level Decomposition. Task-level decomposition aims to obtain quantifiable task indicators, so that pre-war simulation and post-war re-evaluation can quantitatively evaluate the task completion quality. The specific decomposition idea is shown in Fig. 2, in which the mission description is obtained from the mission decomposition in Sect. 2.1, and the combat nodes are obtained from the operational schematic diagram. According to the operational sequence table, combined with the opinions of military experts, specific combat tasks are further obtained, and then the corresponding relationship between combat tasks and combat task attributes is obtained by referring to the US joint combat task list. By consulting the existing conceptual modeling cases such as crossing the sea and landing on the island, joint air defense and anti-missile, emergency rescue, etc., the relationship between combat mission attributes and combat mission indicators is obtained. Further, by consulting the combat mission lists, joint combat mission lists, and special combat mission lists of various US military services, the relationship between combat mission and combat mission indicators is obtained, and the evaluation basis of military tasks is obtained, and task-level decomposition is completed.

Step 3: Equipment Level Decomposition. This section starts with the combat system, and further clarifies the relationship between combat nodes and system functions, system functions and combat activities, system functions and system attributes, and system attributes and tactical indicators in the combat concept of "multi-unmanned cluster air-ground integrated combat" through matrix, which lays the data and logic foundation for the next simulation deduction [5].

System function is the function of combat system to ensure the operation of combat nodes and the completion of combat tasks. System attribute is a qualitative and quantitative characteristic that describes a system and reflects the essence of the system. System index is the measurement value that the system should achieve when it exerts its established ability. Usually, we think that the system index of weapons and equipment is the tactical index set at the time of appearance. By consulting the list of US joint operations tasks and combining the opinions of military experts, 37 key tactical indicators related to this operation are obtained [6].

4   Index Optimization and Simulation Experiment

4.1   Ration Concept Index Based on JMT

In last chapter, based on the joint mission thread method, 37 tactical indicators needed to complete the task are obtained [7]. If they are verified by simulation one by one, a huge amount of scheme space will be generated. However, most of them have little impact on the combat task [8]. In order to save resources and speed up the decision-making time, this chapter uses quality function deployment(QFD) to make reasonable filtering [9]. Experts are invited to assign weights to each task process index, and through quantitative calculation, key combat activities can be obtained, and their own combat center can be found. Then, according to the guiding ideology of "save yourself and complete the task" and key combat task indexes, equipment tactical indexes can be optimized, and based on the key indexes, compressed scheme space can be generated, and equipment tactical indexes meeting the task needs can be designed [10]. Finally, through Mozi's war chess deduction system, the paper compares the combat situation between traditional mountain infantry battalion and unmanned combat battalion, and verifies the rationality of key tactical indicators and the advanced operational concept.

4.2 Equipment Index Optimization

This section starts with mission objectives, uses QFD method to deconstruct mission objectives-mission demand effects-combat tasks-combat nodes-combat activities layer by layer from top to bottom, and quantitatively finds out the key combat activities that affect mission objectives by combining the mission-level indicators and expert scoring weight table obtained in Sect. 2. In the end, the three most critical combat activities were selected, namely, the command post assigned UAV/patrol missile tasks, the command post ordered unmanned cluster to feint, and unmanned cluster and patrol missile cluster cooperated to strike [11]. The corresponding three simulation indicators are: whether there is a wolf patrol missile, whether the patrol missile groups cooperate in combat, and whether the patrol missile carries different warhead proportions when striking. Through simulation experiments, the execution mode of key combat activities is verified [12].

4.3 Simulation Experiment of Key Combat Activities

According to the data, it is known that the best cost-effectiveness ratio can be achieved when the battalion-level unit is engaged in combat and the patrol missile has a detection distance of 35 km and a range of 25 km. When the size of the cluster is 4 pieces, it has the best cooperative efficiency. By consulting the relevant equipment parameters of the US military, it can be found that MQ-8 UAV meets the requirements of relevant technical indicators and can be used as beheading UAV. According to the basic assumptions in Sect. 1, select the above equipment and cluster mode to verify the key combat activities [13].

Wolf patrol missile: using Mozi to carry out simulation, according to the desired area, at 10:00 a.m., without target assignment in advance, it will strike itself, and carry out 25 simulations. On average, 25.0 targets were destroyed when there were Wolf patrol missiles, and 21.5 targets were destroyed when there were headless Wolf patrol missiles. Therefore, Wolf patrol missiles were used in the follow-up experiments.

Whether there is coordination between groups of patrol missiles: Mozi is used to carry out simulation. At 10:00 a.m., 40 missiles are used to attack 30 ground armored targets guarded by anti-aircraft missiles. With coordinated patrol missiles, only each group of combat areas are designated, and no targets are assigned in advance. No coordinated patrol missiles are assigned to strike targets in advance, and 25 simulations are carried out. On average, 15.5 targets were destroyed by cooperative patrol missiles, and 10.5 targets were destroyed by non-cooperative patrol missiles. Therefore, the cooperative mode was adopted in the follow-up experiments [14].

Different warhead ratios: The explicit target deduction can be seen by both sides. The M-side strike mission is assigned in advance, and the K-side starts the automatic battle. The force of party M is 40 attack patrol missiles. The initial position of patrol missiles is 70.00° east longitude and 29.90° north latitude. Patrol missiles are LAM type. Every four missiles are grouped and equipped with armor-piercing warheads: 1:5, 1:3, 1:2, 1:1, 1:2, 1:3 and 1:5 respectively. Party K's strength is a Mobu battalion, with a total of 35 sets of equipment: 25 infantry fighting vehicles, 5 command vehicles, 5 sets of anti-aircraft missiles (missile vehicles used in simulation experiments), two guard companies (120 people) and an elite class (10 people). Initial position: N 29.15, E 70.75, the ammunition carried by infantry fighting vehicle is PG-15, and the ammunition carried by missile vehicle is Arrow 10 missile. Based on Mozi platform, the simulation deduction is carried out, and 25 simulation experiments are carried out for each warhead ratio, and the average value is taken. The specific results are shown in Fig. 3.

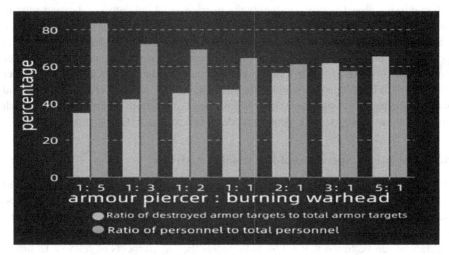

Fig. 3. Influence of warheads with different ratios on combat

For ease of presentation, the ordinate in Fig. 3 is the percentage of targets destroyed or killed.

Percentange = The total number of goals/The number of targets lost in combat   (1)

It can be seen that after the ratio of armor-piercing projectiles to incendiary projectiles is greater than 3, increasing the ratio of armor-piercing projectiles has little effect on the ratio of destroying armored targets. Considering that the follow-up rotary-wing unmanned aerial vehicle and unmanned aerial vehicle can effectively attack the enemy's effective force, and meet the requirement that the ratio of initial personnel killing and armor killing is greater than 55%, the ratio of armor-piercing projectiles to incendiary projectiles is 3:1 for follow-up experiments.

Considering that the warhead ratio is 3:1, the killing effect is obvious. Through expert discussion and literature reading, two factors are considered:

When the K-Party fought, each infantry unit was equipped with three combat vehicles, and the ratio of vehicles to formed units was 3:1. Therefore, simulation experiments were carried out to verify that 42 missiles attacked an armored company, and the armored vehicles were modified: formed units were 2:1, 3:1, 4:1, and 6:1, respectively. There were 210 enemy troops in total, and each ratio was used to carry out armor-piercing bombs: incendiary bombs were 1:1. Using Mozi system, 25 simulation experiments were conducted on the warhead ratio of each armor system. The average ratio of killing armor targets is shown in Fig. 4, and the average ratio of killing living targets is shown in Fig. 5, in which different colors represent different warhead ratios, and the abscissa shows different armor ratios. It can be seen that only when the armor target: organizational system = armor-piercing projectile: incendiary projectile, the condition that the ratio of the first killing live target to the armor target in the field task list is greater than 55% can be met.

The second guess is that the plateau mountainous terrain is suitable for 3:1 armor-piercing projectile: incendiary bomb. The battle sites were exchanged to islands (Zhanxian County, Hainan Province), plains (Harbin, Heilongjiang Province) and cold mountainous areas (Baku, Georgia). It was found that terrain had little effect on warhead ratio, and the killing ratio was basically consistent with Fig. 4. No conjecture 2.

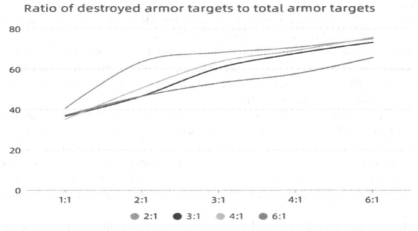

Fig. 4. Proportion diagram of armored targets destroyed by different warhead ratios

4.4 Basic Scenario Simulation Verification

Time: 10:00 am, May, 20XX.

Venue: A valley floor in the Hindu Kush Mountains, 71.50° east longitude, 29.10° north latitude.

Weather conditions: sunny weather, temperature 30.0° Celsius, humidity 30%

Troop strength of unmanned battalion M: initial position: 70.00° east longitude, 29.90° north latitude, 40 attack patrol missiles are LAM type, every 4 missiles are grouped, and the whole unit is equipped with armor-piercing warheads: the ratio of burning warheads is 3:1, and 10 reconnaissance patrol missiles are Quicklook type, every 2 missiles are grouped. 15 rotor reconnaissance drones, Coyote, 30 attack drones, Harpy, MQ-5A, 30 unmanned aerial vehicles, Ripsaw M5, of which 10 are equipped

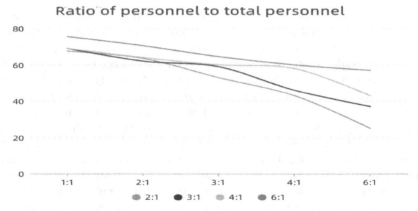

Fig. 5. Proportion diagram of different warhead ratios destroying living targets.

with reconnaissance modules and 20 with attack modules, one special combat platoon is replaced by Delta combat platoon simulation, one electronic jamming platoon, one air defense squad and two air defense missile vehicles.

The M-side strength of the traditional infantry battalion: the initial position: 70.00° east longitude, 29.90° north latitude, one helicopter company, one anti-armor company, one medical platoon, one sniper reconnaissance platoon and two anti-aircraft missile vehicles, which are simulated by Mozi.

Party K's strength: initial position: 70.75° east longitude, 29.15° north latitude, 4 aircraft platoons, 5 armored vehicles in each platoon, model T-72, one reconnaissance unit, 3 armored vehicles, model T-80, 40 armed men, 2 anti-aircraft missile vehicles, model Tiangong, 4 anti-tank vehicles, model IT-1, infantry. Except that the head target is calculated as three armored targets and 30 living targets, the rest of the vehicles are calculated as one armored target, and Party K has a total of 40 armored targets and 400 living targets.

Mozi simulation deduction is adopted to simulate the M traditional infantry battalion vs. K party, and the M unmanned combat battalion vs. K party, respectively. The simulation deduction is made 25 times, and the average data is taken. One M soldier killed in battle is calculated according to the cost of $600,000, the injured soldier is calculated according to twice of the killed soldier, and the compensation amount is calculated according to $100,000 [15]. The rest of the equipment damage is calculated according to the numerical value given by Mozi deduction system, and the combat time is 3 h. In the traditional M Mountain Battalion, 38.2 people were killed, 76.4 people were injured, and 5.2 chariots were destroyed. Ammunition and other maintenance costs were calculated at USD 4 million. This operation cost USD 46.24 million on average, destroying 29.2 enemy armored targets and killing 323 living targets on average. One simulation failed to destroy the leader target. M unmanned combat battalion killed 2.0 people and injured 4.0 people, destroyed 3.2 unmanned combat vehicles and 5.2 unmanned aerial vehicles. The cost of 40 patrol missiles was USD 4 million, and the cost of two beheaded missiles was USD 450,000. Ammunition and other maintenance costs were calculated at USD 4 million. The average cost of this operation was USD 24.53 million, with an average of

36.0 enemy armored targets destroyed and 375.2 living targets killed. Each simulation was destroyed. See Figs. 6 and 7 for detailed data relationship. Only the equipment consumption is taken into account when the compensation for the fallen soldiers is dropped. The traditional mountain camp combat cost and unmanned combat camp cost are shown in Fig. 8. From the perspective of the U.S. military, according to the amount of military compensation in other countries and regions, one South Korean soldier was killed, with compensation of 103,000 dollars, another wounded soldier with compensation of 20,800 dollars, and the number of wounded soldiers was twice that of killed soldiers, with compensation of 27,000 dollars for one Vietnamese soldier, 0.5 million yuan for wounded soldiers, 165,000 dollars for dead Taiwan soldiers, and 31,500 dollars for injured soldiers [16]. The rest of the equipment damage is calculated according to the numerical value given by Mozi deduction system, and the combat time is 3 h. In the traditional M mountain camp, 38.2 people were killed, 76.4 people were injured and 5.2 chariots were destroyed. Ammunition and other maintenance costs were calculated at 4 million dollars, and the average cost of this battle was 46.24 million dollars. According to the traditional mountain camp, South Korea, Vietnam and Taiwan spent 19.524 million dollars and 15.413 million dollars respectively. According to the establishment of unmanned combat battalion, South Korea, Vietnam and Taiwan spent USD 23.219 million, USD 23.004 million and USD 23.386 million, respectively. See Fig. 9 for the specific quantity relationship. It can be seen that, judging from the economic point of view, different armies adapt to different combat modes.

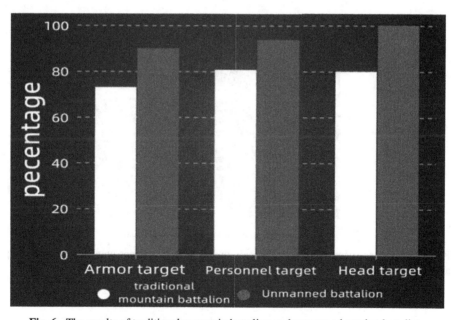

Fig. 6. The results of traditional mountain battalion and unmanned combat battalion.

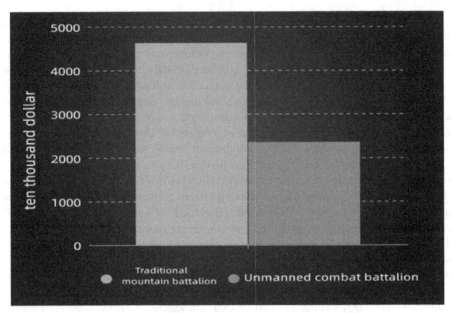

Fig. 7. Comparison of spending between traditional mountain camp and unmanned combat camp

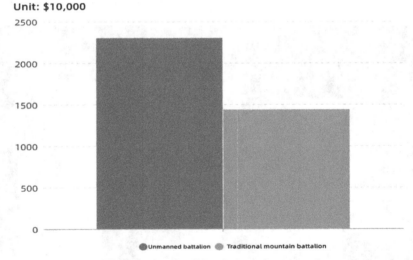

Fig. 8. Comparison of spending between traditional mountain camp and unmanned combat camp after removing labor cost

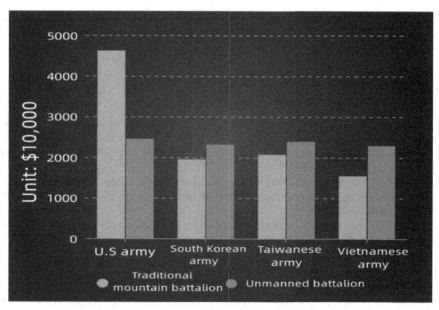

Fig. 9. Comparison of spending between traditional mountain battalions and unmanned combat battalions of different armies

5   Conclusion

At present, by reasonably transforming and combining the existing weapons and equipment of the U.S. Army, a new type of battalion-level combat unit is set up, and a close-range multi-unmanned cluster air-ground battle is carried out against the battalion-level unit of the K side under mountain conditions, which can greatly reduce the political cost of combat, reduce negative domestic public opinion, improve the task completion, test the feasibility and rationality of the concept of global assault operations, form a certain degree of campaign deterrence to the K side, and eliminate potential risks for subsequent operations. It is worth popularizing this new type of combat concept to the battalion-level unit of close-range plateau mountain operations.

At the same time, it should be noted that the economic consumption caused by using different combat methods in mountainous areas is different, and unmanned combat usually costs a lot. At the same time, due to the uncertainty of mechanical and electronic systems, the commander should flexibly choose the combat style when carrying out specific combat tasks, so as to avoid armchair strategist and rigid command.

This paper is a supplement to the operational theory of the plateau synthetic battalion, explores the system establishment of the future intelligent troops, clarifies the operational mission and key evaluation indicators of the battalion-level units, and designs a series of tactical parameters of mountain combat equipment, which provides a certain reference for the unmanned combat construction of the plateau battalion.

At the same time, through technical means, this paper explores the tactical problem of warhead carrying ratio of unmanned combat synthetic battalion in plateau, which provides technical reference for subsequent troops to carry out combat operations, further explains the necessity of timely tactical deployment and in-depth deployment, and provides a relatively perfect process and method for breaking the combat system of terrorist brigade.

References

1. Wang, Y., Lin, M., Li, X., Wang, W., Zhu, Y.: Research on modeling and measuring methods of operational concepts based on JMT and ABM. In: 2020 Proceedings of China High-level Forum on System Simulation and Virtual Reality Technology, pp. 209–214 (2020). https://doi.org/10.26914/C.CNKIHY.200
2. Jiang, Z., Zhang, Y., Wu, J., Peng, H.: The development and enlightenment of the concept of distributed killing operations in the U.S. Navy. Air Missile (01), 83–85+96 (2020). 10.16338/J. ISSN: 1009-1009119109
3. Almeida, D., Machado, D., Andrade, J.C., Mendo, S., Gomes, A.M., Freitas, A.C.: Evolving trends in next-generation probiotics: a 5W1H perspective. Crit. Rev. Food Sci. Nutr. **60**(11), 1783–1796 (2020)
4. Lei, X., Yang, Y.: Evaluation of unit combat action plan based on warchess deduction. Firepower Command Control **46**(88–92), 98 (2021)
5. Li, L.: Research on the US special operations Doctrine. Command Control Simul. **42**(6), 135–140 (2020)
6. FM 5–100–15 19950606 Corps Engineer Operations, pp. 78–83, 126–127
7. Billaud, S., Daclin, N., Chapurlat, V.: Interoperability as a key concept for the control and evolution of the system of systems (SoS). In: Proceedings of the 6th International IFIP Working Conference on Enterprise Interoperability, pp. 53–63. Nmes, France (2015)
8. Joint Staff J6: Joint Capabilities Integration and Development System (JCIDS). CJCSI 3170.01I, January 2015
9. The Chief of Army Operations: Army Doctrine Publication 3-05: Army Special Operations. Department of the Army, Georgia (2019)
10. Joint Chiefs of Staff: Joint Publication 3-05: Special Operations. The Joint Staff, Washington DC (2014)
11. The Chief of Air Force Operations: Air Force Doctrine Document 3-05: Special Operations. Air Force Command, Virginia (2013)
12. The Chief of Naval Operations: Naval Warfare Publication 3-05: Naval Special Warfare. Naval Warfare Development Command, Norfolk (2013)
13. Zhou, H., Li, H.: Joint operation command system and command and control system of U.S. Army in Jiechong. Command Inf. Syst. Technol. **7**(5), 11–18 (2016)
14. Karaman, S., Walter, M.R., Perez, A., et al.: Anytime motion-planning using the RRT. In: Proceedings of the International Conference on Robotics and Automation (2011)
15. Chong, C.Y., Kumar, S.P.: Sensor networks: evolution, opportunities and challenges. In: Proceedings of the Sensor Networks and Applications, pp. 1247–1256. IEEE Press, Washington (2003)
16. Xiaofeng, H., Jingyu, Y.: Si Guangya's Simulation Analysis and Experiment of Complex System of War. National Defense University Press, Beijing (2009)

Research on Intelligent Algorithm for Target Allocation of Coordinated Attack

Lin Zhang[✉], Manguo Liu, Shixun Liu, and Xiang Zhang

Xi'an Institute of Modern Control Technology, Xi'an 710000, Shanxi, China
15991787558@163.com

Abstract. The coordinated attack target allocation problem is essentially a combinatorial optimization problem. Firstly, the relevant knowledge of the weapon target allocation problem is introduced, and several commonly used intelligent algorithms for solving combinatorial optimization problems are introduced and compared by simulation. Secondly, the genetic algorithm is introduced in detail, and it is improved. An improved genetic algorithm based on greedy initialization, bucket sorting selection and adaptive operator is proposed, and the traveling salesman problem is used to analyze the algorithm before and after improvement. By comparison, the superiority of the improved algorithm is verified.

Keywords: Coordinated attack · Target allocation · Combinatorial optimization problem · Genetic algorithm

1 Introduction

In modern warfare, formation warfare has become the most important combat method. However, the weapon resources of the offensive formation are limited. How to reasonably allocate weapons to attack which or multiple targets, and obtain the maximum attack benefit under the limited resources, is the current army. A very important problem in combat decision-making is the problem of coordinated attack target allocation [1].

For the problem of attack target allocation, it is mainly divided into two parts. One is the establishment of the attack effect evaluation model, which is used to evaluate the pros and cons of the allocation results. Through this model, the objective function of the allocation can be designed and the effectiveness matrix of the allocation can be calculated, which is Mathematical basis for target allocation; the second is the selection of allocation algorithm, that is, the optimization algorithm for combinatorial problems, which is the main tool to solve the target allocation model, and a suitable optimization algorithm based on the efficiency matrix can calculate the allocation plan under the maximum benefit.

In the early target allocation, traditional algorithms such as dynamic programming method [2], maximum element method [3] and Hungarian method [4] are fast and effective, which have good allocation effect. However, with the increase of the scale of confrontation, the phenomenon of combinatorial explosion is prone to occur in the calculation, and the allocation scheme cannot be given in a short time. Compared with

W. Fan et al. (Eds.): AsiaSim 2022, CCIS 1713, pp. 519–527, 2022.
https://doi.org/10.1007/978-981-19-9195-0_41

the traditional algorithm, the intelligent algorithm has stronger search ability and faster convergence speed. It has a very good effect in solving large-scale resource allocation problems and is simple to implement [5]. Commonly used intelligent algorithms include genetic algorithm [6], ant colony algorithm [7], particle swarm optimization algorithm [8], etc.

2 Coordinated Attack Target Allocation Problem

When solving the target allocation problem, it is assumed that our weapon resources are $R_i, i \in \{1, 2, \cdots, N\}$ and the enemy targets are $T_j, j \in \{1, 2, \cdots, M\}$. Construct the attack benefit matrix B is constructed, which is used to quantitatively evaluate the attack effect. The elements in the matrix represent the income obtained by attacking the target n with the m-*th* resource. The problem of attack target allocation can be summarized as finding the best resource allocation scheme according to the specific resource allocation constraints under the known R, T and B, so as to obtain the maximum attack income S.

$$S = \max \sum_{i=1}^{N} \sum_{j=1}^{M} d_{ij} \cdot b_{ij} \tag{1}$$

d_{ij} Indicates whether the weapon resource R_i attacks the target T_j, when $d_{ij} = 1$, it means to execute the attack, when $d_{ij} = 0$, it means not to attack. b_{ij} represents the income obtained by using the weapon R_i resource to attack the target T_j.

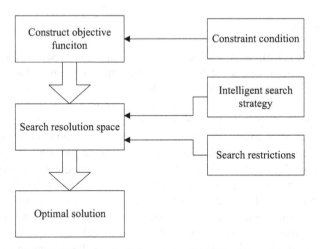

Fig. 1. The framework for solving the allocation problem

According to the above definition, the coordinated attack target allocation problem is a typical combinatorial optimization problem, which belongs to the category of optimization problems. The solution method is shown in Fig. 1.

As can be seen from Fig. 1, for solving the target allocation problem, the key point is the choice of intelligent search strategy under the determination of the attack payoff matrix and specific constraints. As a combinatorial optimization problem, the simplest method is to traverse all the combinations to find the combination scheme with the greatest benefit, which belongs to the category of N-P problems. For the solution of N-P problems, various intelligent algorithms are often used to search the solution space, and the heuristic probability search based on intelligent algorithms can find a suitable suboptimal solution or even an optimal solution in a limited time.

In order to verify the effectiveness of each intelligent algorithm, the traveling salesman problem (TSP) is selected for algorithm comparison. In the sense of graph theory, the TSP is the minimum Hamiltonian cycle problem, which is widely used in many fields and is often used as a symbol for comparing the performance of optimization algorithms. The TSP is one of the well-known N-P hard problems in modern combinatorial optimization problems [9]. The importance of studying it is that all N-P problems are mathematically equivalent to TSP. The TSP can be described as: under knowing the distances between n cities, a traveling salesman starts from a certain city and visits each city once and only once, and finally returns to the starting city. The problem is how to arrange it so that taking the shortest route.

3 Intelligent Algorithm

Intelligent algorithm refers to an algorithm that is inspired by natural laws and simulates its principles to solve problems, which is a heuristic optimization algorithm. Because intelligent algorithms have some characteristics similar to biological intelligence, they belong to abstract machine learning. Compared with traditional algorithms, they have the advantages of being fast and efficient in dealing with some complex problems. Commonly used intelligent algorithms include particle swarm algorithm, ant colony algorithm, simulated annealing algorithm, etc.

3.1 Ant Swarm Algorithm

The basic idea of ant swarm (AS) algorithm is to simulate the collective foraging behavior of ants in nature. When ants go out to look for food, they often act collectively. When ants find food on a path, they will leave pheromones here, and their companions will according to the pheromone. The higher the pheromone concentration, the greater the tendency to choose the current path. The specific algorithm principle is as follows:

In the initial state, the pheromone on all paths is consistent, and the ants $k(k = 1, 2, \cdots m)$ follow the pheromone concentration on each path to determine the transfer direction at the next moment. At time t, the probability $p_{ij}^k(t)$ that the ant turns from the path point i to the path point j is as follows

$$
p_{ij}^k(t) = \begin{cases} \dfrac{[\tau_{ij}(t)]^\alpha [\eta_{ij}(t)]^\beta}{\sum_{s \in J_k(i)} [\tau_{is}(t)]^\alpha [\eta_{is}(t)]^\beta} & j \in J_k(i) \\ 0 & others \end{cases} \tag{2}
$$

$J_k(i)$ represents the path point that the ants are allowed to transfer at the next moment, $\tau_{ij}(t)$ represents the pheromone concentration between the path point i and the path point j at the current moment, $\eta_{ij}(t) = 1/d_{ij}$, d_{ij} represents the distance from the path point i to the path point j, α and β are constants. After all ants walk the path once, the pheromone on the path needs to be updated, and the update formula is as follows

$$\tau_{ij}(t+n) = (1-\rho)\tau_{ij}(t) + \Delta\tau_{ij}(t) \tag{3}$$

$$\Delta\tau_{ij}(t) = \sum_{k=1}^{m} \Delta\tau_{ij}^{k} \tag{4}$$

ρ represents the information volatilization coefficient. $\Delta\tau_{ij}(t)$ represents the increase degree of pheromone on the path at the current moment. $\Delta\tau_{ij}^{k}$ represents the amount of pheromone released by the ant k on the current path.

3.2 Simulated Annealing Algorithm

Simulated Annealing (SA) Algorithm [10] is derived from the principle of solid annealing. The solid is heated to a certain temperature. As the temperature increases, the molecules in the solid will intensify, changing from a steady state to a disordered state. After the heating is stopped, the temperature will gradually decreases over time, and the molecules recombine and eventually return to an ordered state. The starting point is based on the similarity between the annealing process of solid matter in physics and general combinatorial optimization problems.

From the perspective of solid annealing, according to the Metropolis criterion, the probability that the particles tend to equilibrium at temperature $\exp(-\Delta E/(kT))$ T is, where is E the internal energy ΔE at temperature,T is the change in internal energy, and k is the Boltzmann constant. The Metropolis criterion is commonly expressed as formula (5):

$$p = \begin{cases} 1 & \text{if } E(x_{new}) < E(x_{old}) \\ \exp(-\frac{E(x_{new})-E(x_{old})}{kT}) & \text{if } E(x_{new}) \geq E(x_{old}) \end{cases} \tag{5}$$

3.3 Particle Swarm Optimization

Particle Swarm Optimization (PSO) simulates the predation behavior of a flock of birds, which is to solve the target optimization problem based on the nearby search principle of bird flocks. In the algorithm, each bird is regarded as a feasible solution in the solution space, which is called a particle. Each particle has its own fitness value according to the objective function, and determines the movement direction and distance of the particle according to a certain speed. The particles will select the optimal particle in the current space to move closer, and finally all particles reach the optimal state.

The position and velocity update formulas of particles in particle swarm optimization are as follows

$$V_i^{k+1} = \omega V_i^k + c_1 rand(0,1)(P_i^k - X_i^k) + c_2 rand(0,1)(P_g^k - X_i^k) \tag{6}$$

$$X_i^{k+1} = X_i^k + V_i^{k+1} \tag{7}$$

V_i^k represents the velocity of the i-th particle at the k-th iteration. X_i^k represents the position of the i-th particle at the k-th iteration. c_1 and c_2 are learning factors. P_i and P_g are respectively the historical optimal position and the global optimal position of the particle at the k-th iteration.ω is the inertia weight of the particle.

3.4 Genetic Algorithms

The genetic algorithm (GA) was first proposed by a student of Professor Holland in his paper in 1962, and some genetic operators were introduced in the paper. As a random search method, genetic algorithm can effectively use the existing information to search for an excellent solution to the problem, instead of blindly conducting random search. The basic genetic algorithm is to directly operate on the individuals in the solution space set of the problem, and the mathematical model of the basic genetic algorithm can be expressed by the formula

$$SGA = (C, E, P_0, M, \Phi, \Gamma, \psi, T) \tag{8}$$

In the formula: C refers to the coding method adopted by the chromosome. E refers to the fitness function that evaluates the individual's strengths and weaknesses.P_0 refers to the initial population.M refers to the population size.Φ refers to the selection operator.Γ refers to the crossover operator.ψ refers to the mutation operator. T refers to the termination condition of the algorithm running.

4   Improved Genetic Algorithm

The basic genetic algorithm has some disadvantages. It is easy to fall into the local optimal problem, which makes the algorithm unable to achieve the global optimal convergence. It belongs to the random search method, which has the characteristics of randomness and no guidance. Because of these disadvantages, some improvements are made to the algorithm, and an improved genetic algorithm (IGA) is proposed.

1 Improvement of Population Initialization

When initializing the population, using random generation is easy to cause uneven distribution of the population in the basic genetic algorithm. A population initialization method based on greedy algorithm [11] is proposed.

Taking the TSP problem as an example, the specific greedy initialization steps are as follows:

1) Randomly select a city A as the starting point, and find the n cities closest to city A.
2) Randomly select a city from the n cities as the next city B to arrive at, and record the city. It is forbidden to pass through it twice
3) Taking city B as the starting point, go back to step 1) until all cities are passed.

2 Improvement of Selection Operator

The optimal individual preservation strategy [12] and the idea of bucket sort are adopted to improve the roulette selection method.

1) Calculate the fitness value of each individual, retain the individual with the highest fitness value.
2) Turn the wheel N/2 times when an individual is selected, and count the number of times each individual is selected in the interval. Select the individual with the highest number of times.
3) Replace the individual with the lowest fitness with the optimal individual.

3 Introducing an Adaptive Operator

When the crossover probability P_m is too large, the crossover frequency between individuals in the population will be too high, which will easily destroy the goodness of the population. If it is too low, the generation of new individuals will not be utilized, making the evolution of the population slow. For the mutation probability P_m, if the value is too large, the number of mutant individuals will increase, and the search method of the algorithm is similar to random search. If the value is too small, it is difficult for the algorithm to jump out of the local optimum. In order to solve this problem, an adaptive operator is introduced [13], and the expression is as follows

$$P_c = \begin{cases} K_c \times e^{\frac{f'-f_{avg}}{f_{avg}-f_{min}}} & f' \leq f_{avg} \\ K_c & f' > f_{avg} \end{cases} \tag{9}$$

$$P_m = \begin{cases} K_m \times e^{\frac{f-f_{avg}}{f_{avg}-f_{min}}} & f \leq f_{avg} \\ K_m & f > f_{avg} \end{cases} \tag{10}$$

$K_c \in (0, 1)$, $K_m \in (0, 1)$, f_{min} is the minimum fitness value in the population, f_{avg} is the average fitness value of the population, f' is the smaller fitness value of the two crossed individuals, and f is the fitness value of the mutated individual.

5   Simulation

In this paper, the TSP is selected to carry out the simulation comparison test of the IGA, and the comparison algorithms include the AS, PSO, GA and SA.

31 cities are selected for simulation comparison, and the coordinate is as follows:
C = [1304 2312;3639 1315;4177 2244;3712 1399;3488 1535;3326 1556;
3238 1229; 4196 1004; 4312 790; 4386 570; 3007 1970; 2562 1756;
2788 1491;2381 1676;1332 695;3715 1678;3918 2179;4061 2370;
3780 2212;3676 2578;4029 2838;4263 2931;3429 1908;3507 2367;
3394 2643; 3439 3201; 2935 3240; 3140 3550; 545 2357; 2778 2826;
2370 2975];

There are five algorithms in one simulation experiment, and the results are shown in Table 1 and Fig. 2.

According to Table. 1, in terms of convergence speed, the PSO has the fastest convergence speed, and the GA has the slowest convergence speed. The IGA has a faster convergence speed, especially compared with the GA. In terms of operation time, the AS takes the longest, reaching 29 s, the SA takes the shortest time, and the other three algorithms take a little longer, between 5-8 s. From the point of view of the convergent

Table 1. Comparison of simulation results of various algorithms

	Convergent algebra	Convergent solution	Operation time
PSO	30	20010	8.02s
GA	149	16800	4.79s
SA	137	17150	3.44s
AS	72	15602	29.11s
IGA	74	15378	4.88s

solution, the result of the PSO is the worst, falling into a local optimum. The IGA finds the optimum solution after 50 iterations. Although the convergent solutions of the other three algorithms are also locally optimum.

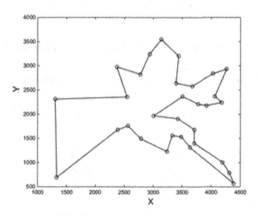

Fig. 2. Shortest path diagram

The shortest path is shown in Fig. 2. The shortest path length is 15378.

According to the convergence curves of various algorithms in Fig. 3, the convergence curve of the AS is the smoothest. It can find a shorter path in the early stage of the algorithm, and the convergence speed is also fast. The quality of the solution is better. The PSO has the fastest convergence speed, but falls into a local optimum prematurely. The convergence curve of the GA in the early stage of the algorithm declines rapidly, indicating that the evolution of the population in the early stage is fast, while the evolution speed of the population in the later stage is slow, and it gradually falls into a local optimum. The curve of the IGA is more volatile, which means that the diversity of the population is guaranteed during the search process, and the global optimum is finally achieved, which reflects the advantages of the improved selection operation and the introduction of self-adaptive operators. The randomness in the optimization process improves the stability and convergence speed of the algorithm.

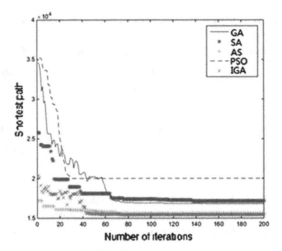

Fig. 3. Fitness curve graph

Table 2. Comparison of simulation results of various algorithms

	Convergent optimal solution	Worst Convergent Solution	Average solution	Average operation time
PSO	19061	24585	21380	7.96s
GA	15378	18551	16468	4.82s
SA	15378	17990	16444	3.38s
AS	15602	15602	15602	28.96s
IGA	15378	16390	15634	4.95s

In order to avoid the randomness of algorithm optimization in one simulation experiment, 100 simulation experiments were carried out for each of the five optimization algorithms, and the final optimization results are counted, as shown in Table.2

Table 2 shows the results of 100 repeated experiments on the five algorithms. Comparative analysis shows that the PSO does not reach the global optimum once in 100 experiments, and the quality of the solution is very poor. The stability of the AS is very good. In the process of 100 iterations, the convergent solution remains unchanged, but it is not the global optimum, and the operation time is too long, reaching 29s, so the operation efficiency of the algorithm is very low. The running time of the SA is the shortest, but the stability is poor, and the convergence between the optimal solution and the worst solution is quite different. The performance of the GA is relatively general in terms of running speed and convergent solution. Although the IGA will still fall into the local optimal situation, its stability has been improved.

6   Conclusion

In this paper, the algorithm of problem of coordinated attack target allocation is studied, and the related concepts of combinatorial optimization. Several common intelligent algorithm for solving combinatorial optimization problems are introduced. An improved genetic algorithm based on greedy initialization, bucket sorting selection and adaptive operator is proposed. Finally, the effectiveness and superiority of the improved algorithm is verified by comparing several common algorithms and the improved algorithm through the TSP.

References

1. Shi, R., Liu, J.: Application of intelligent optimization methods in jamming resource allocation: a review. Electron. Opt. Control. **26**(10), 54–61 (2019)
2. Mahmoudimehr, J., Loghmani, L.: Optimal management of a solar power plant equipped with a thermal energy storage system by using dynamic programming method. Proceedings of the Institution of Mechanical Engineers **230**(2), 219–333 (2016)
3. Ren, C.L., Jiang, L.Q., et al.: Prediction for allocation of enemy air strike weapon based on maximum element method. Ship Electronic Eng. **30**(04), 53–55 (2010)
4. Rabbani, Q., Khan, A., Quddoos, A.: Modified Hungarian method for unbalanced assignment problem with multiple jobs. Appl. Math. Comput. **361**, 493–498 (2019)
5. ElSoud, M.A., Anter, A.M.: Computation intelligence optimization algorithm based on meta-heuristic social-spider: case study on CT liver tumor diagnosis. Int. J. Adv. Comput. Sci. Appl. **1**(7), 466–475 (2016)
6. Nikravesh, A.Y., Ajila, S.A., Lung, C.-H.: Using genetic algorithms to find optimal solution in a search space for a cloud predictive cost-driven decision maker. J. Cloud Computing **7**(1), 1–21 (2018)
7. Bahar, K., Mehran, Y.: A new optimized thresholding method using ant colony algorithm for mr brain image segmentation. J. Digit. Imaging **32**(1), 162–174 (2018)
8. Shahraki, H., Zahiri, S.-H.: Fuzzy decision function estimation using fuzzified particle swarm optimization. Int. J. Mach. Learn. Cybern. **8**(6), 1827–1838 (2017)
9. Zhang, L.Y., Gao, Y., Fei, T.: Firefly genetic algorithm for traveling salesman problem. Computer Engineering Design **40**(07), 1939-1944 (2019)
10. Kim, J., Lee, S.: A simulated annealing algorithm for the creation of synthetic population in activity-based travel demand model. KSCE J. Civ. Eng. **20**(6), 2513–2523 (2016)
11. Chen, C.H., Liu, T.K., et al.: Optimization of teacher volunteer transferring problems using greedy genetic algorithms. **42**(1), 668–678 (2015)
12. Wang, L., Luo, X.H., Yu, M., et al.: Genetic algorithm used in functional verification based on elite strategy. J. East University of Science and Technology (Natural Science Edition) **42**(05), 676–681 (2016)
13. Jafar-Zanjani, S., Inampudi, S., Mosallaei, H.: Adaptive genetic algorithm for optical metasurfaces design. Scientific Reports **8**(1), 116 (2018)

A Review of Beyond Landing Modeling and Simulation Research

Q. I. Haoliang[1(✉)] and A. N. Jing[1,2]

[1] Graduate School of China People's Liberation Army, National Defense University,
Beijing 100000, China
1550154752@qq.com
[2] Joint Service College of China People's Liberation Army, National Defense University,
Beijing 100000, China

Abstract. This paper first introduces the military significance of beyond landing and summarizes the current status of beyond landing military theory, operational and experimental research. Then it focuses on the analysis and summary of domestic and foreign research on modeling and simulation framework of beyond landing, operational planning and operation modeling. Finally further analysis on the problems and challenges of beyond landing modeling and simulation is presented which is based on the problems and challenges of system countermeasure level.

Keywords: Beyond Landing · Modeling and simulation framework · Operational planning · Operation modeling

1 Introduction

Beyond landing refers to the three-dimensional landing style that relies on helicopters, ground-effect vehicles and other high-throughput landing equipments in landing operations to cross the enemy's forward inter-water beachhead obstacles and achieve a combination of wide frontal and shallow entry depth. At present, most countries are focusing on reinforcing beyond landing capability, strengthening the construction of equipment and forces, studying the innovative use of warfare, and organizing the operational trials. Modeling and simulation are effective ways of operational analysis and research, and evaluation of the warfare case innovation, which can test the technical performance of beyond landing equipments, optimize the operational plan, and guide the development direction of beyond landing. This paper reviews the military theory and practice of beyond landing, as well as the domestic and foreign researches on beyond landing modeling and simulation, and analyzes the problems of modeling and simulation.

2 Current Situation of Beyond Landing Military Theory and Practice

2.1 The Current Situation of Beyond Landing Military Theory Research

Beyond landing theory is originated from the U.S. Army "beyond the horizon landing operations" concept in the 1980s, and is constantly developed and improved. Now it

W. Fan et al. (Eds.): AsiaSim 2022, CCIS 1713, pp. 528–536, 2022.
https://doi.org/10.1007/978-981-19-9195-0_42

is generally defined as: "A operation of using a variety of sea-skimming and vertical landing vehicles, coming across the enemy inter-water beachhead obstacles, conveying landing soldiers to the enemy front and deep vantage point (landing or docking) to implement landing operations" [1].With the development of military equipment and guidance changes in modern warfare, it is difficult for traditional sea level landings to support the rapid requirements of joint landing operations, and the importance of beyond landings are gradually increasing in operational design and guidance.

The U.S. Army has clarified the concept of "beyond-the-horizon landing assault" and "the first landing echelon with 2/3 of the force transported to the enemy's landing defense depth, and the rest landed from the sea [2]" in the "Marine Corps Operations Concepts (2016)" and "JP3–02 Amphibious Operations Manual (2019)". In May 2020, Russia firstly launched the construction of amphibious assault ships, which represented a shift in the direction of its landing tactics and equipments construction [3].

2.2 The Current Situation of Beyond Landing Operations and Its Experiments

Beyond landing operations have been tested and evolved in actual operational. In the Grenada operation, the U.S. military quickly seized key airfields through transporting helicopters across the sea, and then used airborne operations, transport and attack helicopters to cooperate to end the war within three days [4]; the Gulf War quickly changed the battlefield posture by investing more than 300 helicopters at one time [5]; in the Libya operation, Britain, France, and the United States used amphibious assault ships with shipborne attack aircraft, marines, and medical facilities to participate in the Libyan military operations and hostage rescues [6]. In recent years, countries have strengthened the construction of beyond landing forces and focused on the research and development and application of new system security equipments.

Beyond landing experiments are effective ways to validate the concept of beyond landing operations and to test tactics and the project design. The related research of U.S. Army started earlier. It has successively established the "Joint Operational Center", "Federal Operations Laboratory", "Joint C4ISR Operations Center" and other joint force level test institutions, the U.S. Marine Corps Warfighting Laboratory (USMCWL) and other warfighting laboratories, relying on Amplified Mobility Platform (AMP), Naval Simulation System (NSS), Research, Evaluation, and System Analysis (RESA), Joint Flow and Analysis System for Transportation (JFAST), Joint Theater Level Simulation (JTLS), etc., to validate operational concepts, deduce design options, and lead equipment construction. In 2019, U.S. Marine Corps Commandant, David Berger, focused on leading major changes, urged the organization of multiple levels and rounds of push-back, culminated in a new Marine Corps structure plan approved by the Department of Defense.

3 Current Situation of Beyond Landing Modeling and Simulations

3.1 The Current Situation of Framework Research on Modeling and Simulation

Currently, the main frameworks on beyond landing modeling and simulation include the modeling framework of ABM/MAS [7, 8], the modeling framework of complex network theory [9, 10], and the integrated modeling framework with C4ISER [11] modeling as the core.

Agent's Based Modeling (ABM) and Multi-Agent System (MAS) frameworks: This class of frameworks describes the pair of beyond landing entities and entity local behaviors from the micro level, and realizes the overall system modeling through entity interaction modeling. Literature [12] analyzed the operational entity model types, entity relationship models, interaction models, decision models, and adaptive models, modeled the system for operational warfare simulation experiments based on MAS framework, studied the complexity of operational systems in a way of bottom-down, individual-to-whole, micro-to-macro. The ABM/MAS framework focuses on the emergence of the system level by reflecting the adaptive nature of entities, with the disadvantage of not considering the hierarchical and cross-domain nature of operationalants and combat operations, and the high resolution of the model, which is not applicable to the Joint Operational level.

The modeling framework of complex network theory: This type of framework abstracts the interacting subsystems in the beyond landing system as network nodes, abstracts the subsystem interrelationships as the edges of the network, and conducts a comprehensive study on the dynamical behavior of the internal components within the system and the interactions between the components. The current research on beyond landing mostly analyzes the system network model and reveals the nature of the network model, for example, literature [13] designed the future amphibious operational architecture and its construction algorithm based on complex network theory and mosaic warfare concept, and confirmed through simulation and deduction that the constructed future amphibious operational architecture has higher integrated operational effectiveness compared with the traditional or current architecture. The disadvantage of the complex network theory modeling framework is that it cannot reflect the adaptive and intelligent characteristics of network nodes, which is not conducive to reflecting the adversarial process. Now a better solution is to combine ABM framework with complex network theory [14], and make up for the lack of adaptiveness and intelligence of complex network nodes by ABM, and describe the network node interrelationship by complex network and provide analysis tools and representation for agent interaction emergence.

An integrated modeling framework based on C4ISR: New operational concepts and new quality operational forces require modeling and simulation to be closer to actual operational, and the integrated modeling and simulation framework with C4ISR as the core is gradually becoming the frontier direction of countermeasure simulation system modeling, which has inspirational and guiding significance for integrating beyond landing modeling into the whole system operational modeling. The NSS developed by the U.S. Navy is a typical simulation system based on this framework, which can effectively evaluate the impact of C4ISR on weapon effectiveness and operational processes. The SEAS developed by RAND also focuses on C4ISR, with special emphasis on the impact and role of space and C4ISR systems in the modeling framework, providing a robust Agent-based modeling simulation environment to support joint warfighting imagination analysis at different scales, which can support different system design, system architecture, and operational concept of operations exploration [11].

3.2   The Current Situation of Beyond Landing Operational Planning Research

The purpose of operational planning is to optimize the allocation of various operational resources and coordinate the action plans of various types of forces and weapons as a whole. It can be dynamically adjusted in real time according to changes in the battlefield situation to ensure that the operational objectives are achieved in an optimal or near-optimal way [15]. Domestic and foreign beyond landing planning technologies mostly focus on applied research, and mainly address the problems of resource domain, path domain, and target domain.

Resource domain aspect: The allocation and integration of beyond landing operational resource facilitate the improvement of utilization efficiency, also adapt to mission requirements and overall effectiveness. At present, the main research is in landing force formation, balanced loading of landing equipments [16, 17] and product allocation optimization [18, 19], and the related methods are Pareto multi-objective planning method [20], genetic algorithm [21, 22], Ripple consensus mechanism planning method [23] and planning method based on multi-priority task chain [24], etc. The Pareto-based multi-objective planning method is easy to realize the local optimal solution under the specific preference of the commander, the genetic algorithm-based planning method can better reflect the reusability and the temporal logic of using beyond landing operational resources, the Ripple consensus mechanism-based planning method is easy to realize the data consistency and ensure the allocation efficiency to a certain extent when distributing operational resources among distributed nodes, and the multi-priority task chain-based planning method constructs parallel optimal sets for ensuring the priority of core important tasks, while introducing a negative feedback adjustment mechanism for the excessive redundancy of operational resources, which optimizes the planning of operational resources and makes the scheduling process more balanced [25].

In terms of path domain, domestic and international research has focused on beyond-landing path planning geared toward improving path search capability and obstacle avoidance, which can be divided into local path planning and global path planning according to the level and granularity of research. Local path planning includes artificial potential field method, velocity barrier method and vector field histogram method [26–28], and this type of algorithm has accurate solution results in a limited range, but it is mostly applicable to tactical level and laboratory model validation. Global path planning includes Dijkstra's algorithm, A* algorithm and improved algorithms, ant colony algorithm and genetic algorithm [29–31]. Dijkstra algorithm is simple in principle and high in accuracy, and has been maturely used in large simulation systems such as JTLS, with the problem of large computational effort. A* algorithm and its improved IDA* algorithm, LPA* algorithm, Bidirectional A* algorithm [32], etc. introduce heuristic functions to estimate the cost of the current node to the target point to achieve the purpose of reducing the search range and improving the search efficiency, which is more efficient in finding the shortest path in static networks, with the problem of relying on the heuristic function, and the planning results not in line with the actual performance of landing equipments; ant colony algorithm and genetic algorithm and other heuristic algorithms have been widely used in the complex path solving in recent years, and have high accuracy. But in the applications of battle level simulation, the problems such as premature caused by many control parameters need to be further resolved.

Target domain: The present research mostly focuses on the preferential ranking of landing points, landing fields or landing territories and provides relevant basis for the commander's decision, using methods such as triangular fuzzy function, hierarchical analysis, gray ideal scheme and multi-attribute decision method [33–36]. Due to the problems of simple calculation logic, subjective indicator weights, distorted gray correlation, and low evaluation resolution in the above single methods, the existing solution is to use composite methods for modeling. In the literature [34], a gray correlation analysis evaluation method based on the gray ideal scheme was proposed by combining the gray correlation analysis theory, entropy method and the technique of approximating to the ideal. Through data normalization processing, entropy value method to determine the weights, gray ideal scheme evaluation model and algorithm, an example simulation analysis of alternative landing points was carried out. The literature [36] establishes an evaluation model of landing point scheme with multi-attribute decision method, uses hierarchical analysis method to calculate the weights of each element of the landing point evaluation index, then conducts multi-attribute evaluation based on fuzzy comprehensive evaluation for the landing point selection scheme, and finally judges the merits of the scheme based on the ranking results of fuzzy comprehensive evaluation, and selects the optimal scheme to provide operational basis for command decision makers.

3.3 Current Situation of Beyond Landing Operation Modeling Research

Beyond Landing Environment Modeling
Beyond landing is influenced by the battlefield environment and the natural environment. In terms of battlefield environment modeling research, existing studies focus on the impact level of a warfare technology performance on a single piece of equipment, and the research elements involve battlefield smoke and fumes, radar electromagnetic environment, communication electromagnetic environment and obstacle minefield [37–39], while the research on the impact effect of beyond landing cluster landing is relatively weak. In terms of natural environment modeling, domestic and foreign scholars distinguish between sea and air battlefields, analyze and study the geographic environment, climate environment, marine environment, chemical environment and other environmental factors related to beyond landing, design various types of natural environment models and evaluation systems at each level, and argue the impact on the landing link of beyond landing parts [40–42].

Beyond Landing Equipments Entity Modeling
In terms of beyond landing operation equipment entity modeling, the landing equipment and its key systems are now mainly modeled in terms of geometric modeling and physical characteristics [43–45]. In terms of extracting the tactical technical indicators needed to complete the overlanding mission for physical modeling, Xu Zifang et al. used the leaf element theory and flight dynamics model to calculate the power required for the rotor and tail rotor of the helicopter in the full leveling attitude, and established the engine power altitude temperature characteristic model and the fuel consumption rate altitude temperature characteristic model to calculate the helicopter flight performance [46]. Shen Li et al. constructed a helicopter transport capability index model from flight

speed, flight altitude, engine fuel consumption characteristics, effective commercial load to fuel quantity ratio, single maximum load, maximum transport distance, and maximum transport speed [47].

Beyond Landing Key Operation Modeling

Beyond landing operations can be divided into landing force loading, navigation (ferrying), wave-formation landing and force unloading and assembly. The problem of operational resource for loading and unloading and the problem of path-finding for navigation (ferrying) have been discussed in the previous paper. As for the modeling of wave braiding landing operations, the process node decomposition scheme [48] and the system dynamics (SD) modeling approach [49] have better solutions due to the parallelism of wave braiding and the closed-loop nature of the whole process. For relay service assurance, no research results directly oriented to beyond landing are identified. Chen Jinbo studied the electric vehicle path problem containing time windows in the battery replacement mode, and used the framework of mixed integer planning model to construct an EVRPTW-BSS model based on time windows and switching stations, which has some reference significance [50].

4 Problems and Challenges of Modeling and Simulation

4.1 There is a Gap Between Model Rules and Reality Operations

Domestic and foreign scholars have conducted a lot of research around beyond landing modeling and simulation, but the disconnect between some model rules and operational reality leads to the low scientificity of modeling and the low usefulness of the results. It should analyze and set up beyond landing operation entity rules, operation rules, and adjudication (interaction) rules from the operational reality, so as to lay the foundation for modeling and simulation research.

4.2 Complex Warfare Environment Modeling is Relatively Weak

Beyond landing is obviously constrained by the complex warfare environment. For the war environment modeling data of beyond landing, foreign countries are basically in a state of secrecy, so valuable information is less. Domestic research is mainly for a specific type of equipment or a single link, and for the research on the modeling of whole process impact effect, the environmental settings are relatively simple or too ideal, which are unable to objectively body to Beyond landing of the whole process of combat operations impact.

4.3 Battle Level Modeling and Simulation Are Needed to Be Strengthened

The current modeling and simulation of beyond landing mostly focus on the experimental platform level and tactical level, which have support and auxiliary significance to verify the technical performance of equipment warfare and simple background tactical operations. Facing the contradictory aspects of modeling accuracy and battle-level simulation derivation calculation complexity, there is still a lack of norms and unified guidance, which needs to be further strengthened.

5   Conclusions

Beyond landing theory meets the requirements of modern landing operations, and is of great significance to solve realistic problems. This paper summarizes the development frontiers knowledge of beyond landing theory and practice, examines the current situation of modeling and simulation researches from three aspects: modeling and simulation framework, operational planning and combat operations modeling, and analyzes their existing problems and challenges.

References

1. Yang, S.-X., Chen, S.-H.: Introduction to Landing Operations. Hai Chao Publishing House, Beijing, 6 (2002)
2. JP3–02, Amphibious Operations. Joint Chiefs of Staff (2019)
3. Jiang, Y.W.: Exploration of the development of Russian Navy amphibious assault ships (below). Arms Knowledge **01**, 58–61 (2020)
4. Rao, Y., Du, Y.: The evolution of amphibious landing operation concept form revelation. Military Digest (1):6 (2021)
5. Yu, L., Zhang, Z.F.: Research on the development trend of U.S. amphibious assault fleet under the new situation in Asia-Pacific. Defense Science and Technol. **41**(3), 5 (2020)
6. Zheng, J., Huang, Y., Wang, Y.: Analysis of the development of U.S. Navy amphibious assault ships. Command Control and Simulation **37**(03), 85–89+93 (2015)
7. Committee on Network Science for Future Anny Applications. network Science [RJ. National Academies Press, Washington DC, USA (2005)
8. Hu, X., Si, G., Luo, B., Yang, J., Zhang, W., Wu, L.: Research on complex systems of war and war simulation. Journal of Systems Simulation (11), 2769–2774
9. Xiaofeng, H.: Overview of warfare complex network research. Complex Systems and Complexity Science **7**(2–3), 24–28 (2010)
10. Dekker, A.H.: Network topology and military performance. In: 2005 International Congress on Modeling and Simulation, Modeling and Simulation Australia, New Zealand:Modeling and Simulation Society of Australia and New Zealand, pp. 2174-2180 (2005)
11. Cao, Z., Tao, S., Hu, X., et al.: Advances in foreign military chess deduction and systems. J. Systems Simulation **33**(9), 7 (2021)
12. Li, G., Chen, J., Yin, W.: Experimental system modeling of multi-agent-based amphibious operational warfare simulation. In: System Simulation and Applications (Vol. 16), pp. 251–255 (2015)
13. Xu, S., Peng, L., Wu, K., Zhou, X.: Research on system construction for future amphibious operations. In: Proceedings of the Ninth China Command and Control Conference, pp. 168–178 (2021)
14. Ma, L., Zhang, W.: Research progress on complex system modeling of warfare based on complex networks. J. Systems Simulation **27**(02), 217–225+245 (2015)
15. Maorong, C., Wei, L., Jinliang, L., et al.: Research on air operational mission planning under information-based conditions. Flying missiles **9**, 61–64 (2015)
16. Wu, B.: Research on amphibious landing force balanced loading and delivery planning technology. Nanjing University of Science and Technology (2018)
17. Gaosheng, Y., Yanyan, H.: Balanced loading of amphibious forces based on loading constraints. Military Automation **40**(9), 7 (2021)
18. Li, W., Wang, X.-L., Wang, S.: Design of loading scheme for transport helicopters based on improved layer construction algorithm. Firepower and Command and Control **46**(02) (2021)

19. Tian, Y., Tian, W., Li, W.-L.: A differential evolutionary algorithm based personnel loading problem for transport helicopters. Firepower and Command Control **46**(05) (2021)
20. Klein, G., Orasanu, J., Calderwood, R.: Decision Making in Action : Models and Methods, pp. 17–84. Ablex Publishing Corporation, Norwood, NJ (1993)
21. Klein, G.: Sources of Power: How People Make Decisions. The MIT Press, Cambridge, MA, pp. 6–23 (1988)
22. Xu, Q.H., Song, J.: Optimization of ship loading scheme based on genetic algorithm. Science and Technology Herald **38**(21), 110–117 (2020)
23. Klein, G.: Naturalistic Decision Making. Human Factors **50**(3), 456–460 (2008)
24. Ancker, C.J., Flynn, M.: Field manual 5–0: exercising commands and control in an era of persistent conflict. Mil. Rev. **41**(5), 13–19 (2010)
25. Long, H.J., Liu, J.A.M.: A review of marine battlefield operational resource planning methods. Ship Electronics Eng. **41**(09), 16–19 (2021)
26. Li, S.: Research on path planning algorithm of quadrotor UAV based on A* and artificial potential field method. Harbin Institute of Technology (2017)
27. Douthwaite, J.A., Zhao, S., Mihaylova, L.S.: Velocity obstacle approaches for multi-agent collision avoidance. Unmanned Systems **7**(1), 5564 (2019)
28. Babinec, A., Duchoň, F., Dekan, M., et al.: VFH* TDT (VFH* with Time Dependent Tree): a new laser rangefinder based obstacle avoidance method designed for environment with non-static obstacles. Robotics and Autonomous Systerms **62**(8), 1098–1115 (2014)
29. Li, G.J.: Research on Path Following Control Method for Full Cushion Lift Hovercraft. Harbin Engineering University (2019)
30. Shi, L., Xu, J.: Research on terrain-based helicopter path planning algorithm. Helicopter Technol. (02), 38-42 (2021)
31. Wang, J., Song, W., Shang, S., Lan, T., Sheng, S.: Unmanned helicopter flight path planning based on improved particle swarm algorithm. Mechanics and Electronics **39**(06), 66–69 (2021)
32. Nannicini, G., Delling, D., Schultes, D., et al.: Bidirectional A* search ontime - dependent road networks. Networks **59**(2), 240251 (2012)
33. Huang, H., Chen, S.-H., Xiao, L.-H., Jia, Z.-Y.: A landing area selection model based on hierarchical analysis and gray correlation. Sichuan Journal of Military Engineering (05), 138–139+145 (2008)
34. Tong, H.P., Jin, T.Y., Zhu, Q.: Preferred landing point of hovercraft based on gray ideal scheme. Firepower and Command Control **37**(12), 95–98 (2012)
35. Shuangxi, C., Jianglong, G.: Optimization model of landing point selection for air cushion landing craft based on triangular fuzzy function. Ship Electronics Eng. **35**(02), 41–45 (2015)
36. Zhao, Y., Huang, Y.: A landing point selection model for amphibious operations based on multi-attribute decision making method. In: Proceedings of the 5th China Command and Control Conference, pp. 470–475 (2017)
37. Yang, C.: Research on three-dimensional modeling and drawing method of electromagnetic environment in virtual battlefield. National University of Defense Technology (2010)
38. Jun, Z., Jun, Y., Da Feng, Z., Zhijun, C.: Analysis of the blind area of helicopter-borne fire control radar and the impact of complex battlefield environment. Fire Control Radar Technology **49**(04), 10–13 (2020)
39. Deng, J.: Study on the mine dispersion model of self-propelled mines and its simulation application. Ship Electronics Eng. **33**(09) (2013)
40. Li, S., Yang, S., Sun, D.: Military aircraft environmental adaptability evaluation model. Journal of Aeronautics **30**(06) (2009)
41. Xie, K.: Modeling and simulation data modeling of the natural environment of the naval battlefield. Computers and Modernization (11) (2011)

42. Wang, Q.: Research on modeling and visualization technology of marine environmental data field. National University of Defense Technology (2014)
43. Yang, Y.C.: Research on modeling and simulation of helicopter suspension system based on multi-body system. Nanjing University of Aeronautics and Astronautics (2020)
44. Fan, X.: Dynamics modeling and control system synthesis of composite co-axial twin-propeller unmanned helicopter. Northern Polytechnic University (2020)
45. Zhang, Z., Xu, S.J., Liu, Y.: Numerical analysis of hovercraft bow ternary flexible apron indentation resistance based on CATIA. Ship **31**(05) (2020)
46. Zifang, X., Yijun, H., Qinghua, Z.: Research on the calculation method of helicopter flight performance in plateau environment. Helicopter Technology **03**, 1–5 (2016)
47. Shen, L.: Mission effectiveness analysis and optimization of transport helicopters based on test flights. Aviation Science and Technology **28**(05) (2017)
48. Hao, T.: Analysis of amphibious assault ship helicopter sortie recovery process. Ship Engineering **42**(05) (2020)
49. Luan, T.: Study on the capability of shipborne aircraft departure process model and evaluation method. Harbin Engineering University (2017)
50. Chen, J.: Research on the path optimization problem of electric vehicles with time windows based on switching stations. Tsinghua University (2016)

Research on Simulation Sample Generation Technology Based on Multiple Variable Points

Fangyue Chen[1](✉), Yue Sun[2], Xiaokai Xia[1], and Luo Xu[2]

[1] North China Institute of Computing Technology, Beijing 100083, China
samelechen@outlook.com

[2] Artificial Intelligence Institute of China Electronics Technology Group Corporation,
Beijing 100041, China

Abstract. In the combat simulation field, how to construct numerous corresponding simulation samples based on the simulation deduction scene, which are used to optimize the combat plan or choose a better one, is an urgent problem to be solved. Current sample generation methods, such as the generative adversarial network, cannot solve the problem of simulation sample generation very well, so a new approach is needed to generate simulation samples. A simulation sample generation technology based on multiple variable points is proposed to solve this problem. This technology can automatically create simulation samples and their corresponding scripts through rule-based means according to a given series of variable points. To verify the effectiveness of the technology, the system effectiveness was compared between the baseline version and the simulation samples generated based on variable points through the AHP, and the combat effectiveness of the generated samples improved by nearly 10%. The simulation sample generation technology based on multi-variable points solves the problem of constructing numerous simulation samples based on the simulation deduction scene and dramatically reduces labor consumption.

Keywords: Multi-variable points · Combat simulation · Sample generation · System analysis · Anti-missile interception

1 Introduction

With the development of the times, the demand for digitalization continues to increase. As a result, the importance of simulation verification also increases due to the high cost of trial and error. For example, in combat simulation, to verify the feasibility of the combat plan in different environments, it is usually necessary to construct numerous simulation samples based on the simulation scene to locate the problems in the combat plan quickly.

At present, common sample generation methods include statistical-based sample generation methods and Artificial Intelligence methods. The sample generation methods based on statistical laws include the Bayesian and Bootstrap methods. Zhang et al. [1] researched the normal distribution sample data generation using the Bootstrap method.

They used the Bootstrap method to fit the normal distribution sample data to expand the maximum number of samples from 500 to 50,000. Zhang et al. [2] conducted a comparative study on the Bootstrap method and the Bayes Bootstrap method. Finally, they verified the accuracy of the Bayes Bootstrap method in parameter and interval estimation. Yang et al. [3] proposed a virtual sample generation method based on Gaussian distribution, which uses smoothness to enhance rationality and adaptability. Although statistics-based sample generation methods have advantages in sample generation, they require a certain number of initial samples for expansion. In our situation, simulation scene-based simulation sample generation usually has no basic sample or only one basic sample. Therefore, the statistical-based sample generation method is not suitable for this scenario.

Artificial Intelligence-based sample generation is mainly a generative adversarial network method. Goodfellow et al. proposed Generative Adversarial Networks in 2014 [4]. By building a generator and a discriminator for confrontation, it gradually achieves the generation of hard-to-distinguish image samples [5]. Wang et al. [6] proposed a single-sample-based image enhancement method, which can enhance the color of food images based on a single-sample generative adversarial network. These methods have significant advantages for image sample generation but require intelligent algorithms to learn the samples, which need to be supported by primary samples, and are not suitable for generating combat simulation samples that only have text descriptions. Zhou et al. [7] designed a training sample generation framework for Artificial Intelligence, referring to generative adversarial networks. Enough training samples are generated for subsequent Artificial Intelligence training through intelligent algorithms and rule-based adversary confrontation. The method of Zhou et al. can generate enough simulation samples for training. Still, the simulation samples generated based on Artificial Intelligence are usually not interpretable, so they are unsuitable for simulation verification of combat scenarios.

The purpose of generating simulation samples based on combat scenarios is to create a significant number of simulation samples to verify the effect of the same combat scheme in different environments. Since the simulation samples do not have statistical laws and need interpretability, the existing methods cannot perfectly solve the simulation sample generation based on combat scenarios. We propose a multi-variable point-based combat simulation sample generation method to solve this problem. Depending on the selected simulation scenario, a series of variable points can be designed that influence combat effectiveness. Through simulation verification, variable points with a more significant impact can be found, and improvements and optimizations to the combat plan can be made based on these points. This paper studies the simulation script generation technology based on multi-variable points and verifies the effectiveness of the technology through Analytic Hierarchy Process (AHP). We selected four variable points for verification. Compared with the baseline samples, the combat effectiveness of each variable point sample increased by a minimum of 8% and a maximum of 27%.

2   Simulation Experiment Process Based on Variable Points

Since the current work cannot solve the simulation sample generation based on combat scenarios, this paper proposes a simulation sample generation and running framework based on variable points. The framework includes scene design, sample generation, script execution, evaluation and analysis, and other links. The scenario design link will give basic combat scenarios, including essential information such as the geographical location of the battlefield, the combat objectives of both sides, and the deployment of troops on both sides. Based on scene design, simulation samples must be generated according to the corresponding information. After the samples are constructed, the simulation scripts corresponding to these samples need to be executed. During the execution process, it is necessary to collect the necessary data and perform corresponding calculations on the data after the execution to obtain the final evaluation result. The overall research method flow is shown in Fig. 1.

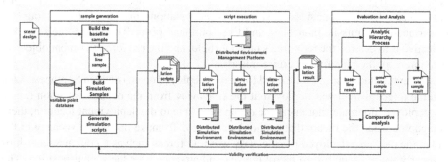

Fig. 1. Overall research method flow chart

Simulation scene design is usually provided by domain experts with corresponding information, which is not the focus of this paper. In the generation stage of the simulation samples, it is necessary to construct a baseline sample as a control group and then modify it according to the simulation requirements to build a series of simulation samples. This paper studies the simulation sample generation technology based on multi-variable points summarizes and classifies the points that can be modified in the system, and generates simulation samples according to the rules.

After building the simulation sample, it is necessary to find an efficient means to execute the script corresponding to the sample. Since the number of simulation scripts is usually large, the simulation scripts are executed in a distributed running manner. Multiple simulation scripts can be run simultaneously by deploying several virtual machines on the server, configuring the simulation environment on the virtual machine, and uploading the simulation script to the virtual machine. After the operation, it is necessary to evaluate the combat effectiveness based on the data recorded in the process. Evaluating the combat system allows those who design simulations to identify problems. In this paper, by comparing the efficiency difference between the generated simulation samples and the primary samples, it is proved that the modification of variable points can impact combat effectiveness. Furthermore, it is proven that the simulation sample generation technology based on multi-variable points is effective.

This paper will describe the work of the paper in the following structure: Sect. 3 is the selection of multi-variable points and simulation scene design, Sect. 4 is the effectiveness evaluation method, Sect. 5 is the experimental results and analysis, and the last section is the conclusion.

3   Simulation Scene Design and Selection of Multi-variable Points

3.1   Simulation Scene Design

To better conduct simulation experiments, it is necessary to set up a basic simulation scene. After preliminary investigation and analysis, we selected a scenario of a coordinated anti-cruise missile with multiple sources of information. In this scenario, we set the blue team as the attacker and the red team as the defender. Blue forces include a destroyer carrying two cruise missiles. The red forces include two AWACS aircraft that have taken off, two reconnaissance ships, four radar stations, one target airfield, four sets of surface-to-air missile launchers, and five command posts. The blue force's combat objective is to destroy the red force's target airfield, and the red force's combat objective is to intercept the blue force's cruise missiles.

In the simulation scenario, the red force was deployed on an island in Southeast Asia, and the target airfield was deployed at the side away from the ocean. Detection radars are deployed along the shore near the ocean side. Due to the deployment location, there is a gap between the radars that cannot be covered. To make up for the vacancy in the radar, the red force deployed mobile early warning forces such as reconnaissance ships and early warning aircraft to patrol the sea and airspace near the vacant position. The destroyer of the blue troops is arranged in the ocean, close to the red force's deployment island.

3.2   Variable Point Selection Method

After the simulation scene is determined, it is necessary to decide on the large categories of variable points according to the characteristics of the simulation scene to extract the corresponding variable points from the variable point library for subsequent simulation sample generation. This part mainly expounds on the analysis process for the simulation scene and the method of constructing the simulation script.

In the actual combat and simulation process, it is usually necessary to consider the impact of different aspects on completing the combat mission. The first is the influence of the surrounding environment on weapons and equipment. These effects include meteorological changes in weapon flight and detector performance due to environmental changes. When shooting, it is necessary to add the calculation of weather effects to achieve more accurate shooting accuracy. When conducting reconnaissance, the influence of the environment on the sensor should also be considered. For the above reasons, the meteorological environment should be considered.

Second, the initial operational planning can have an impact on the outcome of the operation. The operational planning includes the invested troops' formation, deployment position, and the following maneuvering route. The number of troops dispatched, the ratio of various types of troops, and the deployment positions and maneuvering routes of each combat unit during the combat process may influence the actual strike effect, so it is necessary to consider combat planning.

A third aspect to consider is the influence of the mode of operation. Combat methods include command and control, communication links, and fire strikes. Changes in command-and-control will result in different response speeds to the battlefield, changes in communication links will lead to changes in information transmission efficiency, and changes in fire strike methods will affect strike effects. Therefore, the variable points related to the combat mode must also be considered.

The fourth aspect is the influence on equipment performance. Equipment performance includes the performance of several types of equipment such as detection equipment, communication equipment, and weaponry. The performance of the detection equipment will affect the timing of detecting the enemy, thereby affecting the combat effectiveness. The performance of communication equipment affects the efficiency of intelligence transmission and impacts battlefield situational awareness. Finally, the performance of weapons and equipment directly affects the striking effect, affecting the completion of combat missions. Therefore, the variable point of equipment performance also needs to be considered.

Based on the above considerations, there are four categories of variable points in the simulation scenario of anti-missile interception: meteorological environment, battle planning, combat mode, and equipment performance. Environmental variables mainly include global meteorological changes and the impact of the meteorological environment on the detection range of various detectors. Operational planning consists of the formation and formation of combat forces, the initial deployment positions, and the setting of the maneuvering routes of each unit. Equipment performance includes the performance of various types of equipment such as detection equipment, communication equipment, and weaponry. Finally, the combat method consists of the command-and-control process, the communication method, and the choice of the means of firepower. Then we extracted the corresponding variable points from the variable point library and built simulation scripts based on the platform. We chose SSG as the simulation platform to save performance and cover the variable point class. After selecting the simulation platform, it is necessary to map the variable points to the parameters of the simulation platform.

Table 1 shows the variable points under each category and the variable parameters mapped to the SSG simulation platform. The first column in the table is the large category of variable points, and the second is the subdivision category of variable points. The third column shows the specific variable points under each category, which are particular variable points that should be considered in various aspects of the previous discussion. Finally, the fourth column shows that each variable point corresponds to the variable parameters in the SSG. Due to the limitation of the SSG platform, some variable points are mapped to the same variable parameter, while others are mapped to multiple variable parameters.

Table 1. Variable points and their mapping variable parameters

Category of variable points	Subcategory	Variable Points	Map to variadic parameters in SSG
Meteorological Environment	Weather Changes	temperature	air temperature
		humidity	humidity
		wind speed	wind speed
	Detection range	environment influence on radar detection	radar detection range
		environment influence on optical detection	visual detection range
Battle Planning	Grouping	number of troops	force configuration
		equipment ratio	
	Deployment location	radar station deployment location	radar station deployment location
		anti-aircraft missile deployment location	anti-aircraft missile deployment location
		the initial position of early warning aircraft	the initial position of early warning aircraft
		the initial position of the scout ship	the initial position of the scout ship
	maneuvering route	early warning aircraft movement trajectory	early warning aircraft movement trajectory
		reconnaissance ship movement track	reconnaissance ship movement track
Combat Mode	command and control	decision node	initial tactics settings
		object of instruction	
	communication link	intelligence report	
		situation distribution	
	Fire strike	weapon selection	
		interception method	
Equipment Performance	Detection equipment	radar transmits power	radar transmits power
		radar detection gain	radar detection gain

(continued)

Table 1. (*continued*)

Category of variable points	Subcategory	Variable Points	Map to variadic parameters in SSG
	communication device	communication equipment reception performance	receive frequency
		communication equipment launch performance	bandwidth
			transmit power
			antenna gain
	Weaponry	weapon range	maximum range
			effective range
		weapon maneuverability	maximum speed
			minimum speed
			angular rate
		weapon killing performance	hit rate
			kill radius

4 Effectiveness Evaluation Method

After the variable point is selected, it is necessary to verify the influence of the variable point on the combat effect. This paper argues that the impact of variable points can be measured by assessing the overall effectiveness of the combat system. If a particular variable point has a relatively noticeable effect on the combat effectiveness, the system efficiency will undergo significant changes. Therefore, it is imperative to construct a suitable evaluation index system. This chapter will describe the construction of the evaluation scheme and the corresponding calculation method.

For the same simulation scenario, constructing the same evaluation scheme will make the evaluation fairer and more credible. Since the focus of this paper is not to build a brand-new evaluation scheme, it will draw on the existing evaluation index system and refer to expert opinions to construct an experimental index evaluation system. Analytic Hierarchy Process (AHP) [8] is a standard evaluation method that decomposes the object to be evaluated so that the evaluation of the object can be calculated according to a series of quantitative indicators. When assessing the overall efficiency, the overall efficiency is usually decomposed according to the system's capability, and the capability score calculates the overall efficiency score. The system's capabilities can be refined to specific indicators, which could calculate the corresponding scores.

The traditional AHP method requires the scheme to be assessed as a layer in the assessment system. The assessment system needs to be modified when adding a plan to be evaluated. Therefore, the AHP assessment method needs to be adjusted. In our scenario, many schemes are to be assessed, and there is no need to choose among multiple plans. Therefore, we remove the scheme layer from the evaluation system and use the specific indicators as the bottom layer. The particular values of each scheme obtained are used as input, and the total score of the top layer is used as the final output.

In the scenario, cruise missile interception must consider early warning and detection capabilities, information reporting capabilities, threat judgment capabilities, command decision-making capabilities, and firepower strike capabilities. In the interception process, the interception object must be found before the interception can begin, so the early warning detection capability must be considered. After the target is found, it is necessary to report the relevant information, such as the target location, to the decision-makers to make subsequent judgments. Therefore, the information reporting capability needs to be considered.

After the decision-makers receive the target information, they need to make a threat judgment on the target, so the threat judgment ability must also be considered. After judging the threat level of the target, it is necessary to decide what means of counterattack to use and issue the combat instruction to the executor. Hence, the command and decision-making capability also needs to be considered. Finally, the interception mission's success depends on the fire strike's effectiveness, so the fire strike capability also needs to be considered.

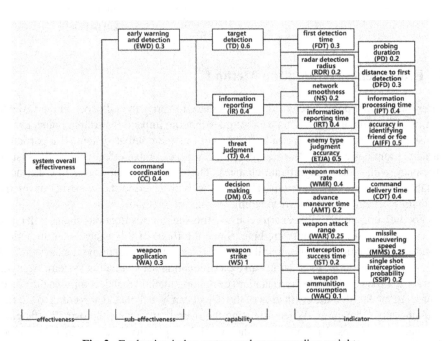

Fig. 2. Evaluation index system and corresponding weights

Based on the above analysis and careful consideration of the relevant literature on the performance evaluation of the anti-missile interception system in the past, we have constructed an evaluation system as shown in Fig. 2. Three sub-effectiveness support the system's overall effectiveness: early warning and detection (EWD), command coordination (CC), and weapon application (WA) effectiveness. The target detection (TD) and information reporting (IR) capabilities support the early warning detection efficiency. The threat judgment (TJ) and command decision-making (CDM) capabilities support the command coordination efficiency, and the weapon strike (WS) capability supports the weapon application efficiency. There are corresponding evaluation indicators under each capability to support it.

When selecting indicators, ensure each is related to the capability and can support it. In the performance calculation, the weight of each indicator needs to be given. The weight this paper used is directly provided by considering the opinions of various experts and previous related work [9–14]. Each indicator has an abbreviation, which shows in the brackets. The number under the indicator name represents the weight of the indicator.

5 Experimental Results and Analysis

Based on the descriptions in Sects. 2 and 3, this paper describes the multi-variable point-based simulation sample generation method and its verification method. Due to the length limitation of the article, it is impossible to verify and analyze each variable point in detail, so this section only selects the specific variable points from the four categories of variable points for discussion.

For example, in the variable point of meteorological environment, we selected the influence of the environment on radar detection; in the variable point of combat planning, we selected the number of troops; in the variable point of combat mode, we selected the decision node; Among the variable points of equipment performance, we chose the range of weapons. The selection of these four variable points can be used in all aspects of the combat process and can cover various variable point categories. For example, the number of troops dispatched involves pre-war planning, the impact of the environment on the radar detection range involves early warning and detection, the selection of decision-making nodes involves command decision-making, and the range of weapons and equipment affects the final fire interception link.

In the Environmental Impact Change sample (EIC), we downgraded the environmental impact on radar detection, increasing the radar detection radius from 94 km to 103 km. In the Troops' Number Change sample (TNC), we added two reconnaissance ships and two early warning aircraft to patrol the sea and airspace. In the Decision Node Change sample (DNC), we moved the node that issued the order from the highest command post to the frontline command post, reducing the information reporting time and information processing time from 10 s to 1 s. Finally, in the Weapons Shooting Range sample (WSR), we increased the shooting range of the surface-to-air missile by 10%, and the original range of 30 km was expanded to 33 km. The original simulation data are shown in Table 2.

Table 2. Original simulation data for each simulation sample

simulation sample	FDT	PD	RDR	DFD	NS	IPT	IRT	AIFF	ETJA
baseline	282	250	94	131.3	0.8	10	10	1	1
EIC	190	331	103	158.5	0.8	10	10	1	1
TNC	45	470	94	200.7	0.8	10	10	1	1
DNC	277	263	94	133.7	0.8	1	1	1	1
WSR	276	240	94	133.8	0.8	10	10	1	1
	WMR	CDT	AMT	WAR	MMS	IST	SSIP	WAC	
baseline	1	10	60	30000	764	532	0.33	6	
EIC	1	10	60	30000	764	521	0.33	6	
TNC	1	10	60	30000	764	515	0.5	4	
DNC	1	1	60	30000	764	540	0.29	7	
WSR	1	10	60	33000	764	517	0.5	4	

We evaluated each simulation sample's performance based on the collected simulation data. Table 3 shows each capability, sub-effectiveness, and overall effectiveness calculation result. The last row in the table is the overall score of the anti-missile interception system's effectiveness, with a total score of 1. The score of the baseline version is 0.72. All other samples have improved based on the baseline version.

The most significant improvement is the change in the number of troops. Due to the addition of reconnaissance ships and early warning aircraft, the two indicators of the first detection time and the first detection distance have significantly improved. At the same time, due to the pre-detection of the enemy, the air defense troops have more time to occupy better. In addition, the shooting position reduces ammunition consumption, thereby improving the probability of a single shot and the successful interception time. According to the data shown in Table 3, it is not difficult to find that the change of variable points impacts the combat effectiveness of the simulated samples, which shows that the selection of variable points is effective.

Table 3. System effectiveness calculates result

	Baseline	EIC	TNC	DNC	WSR
TD	0.3	0.54	0.89	0.32	0.3
IR	0.46	0.46	0.46	0.91	0.46
TJ	1	1	1	1	1

(continued)

Table 3. (*continued*)

	Baseline	EIC	TNC	DNC	WSR
DM	0.48	0.48	0.48	0.48	0.48
WS	0.46	0.47	0.56	0.42	0.6
EWD	0.36	0.51	0.72	0.55	0.36
CC	0.69	0.69	0.69	0.69	0.69
WA	0.46	0.47	0.56	0.42	0.6
System overall effectiveness	0.52	0.57	0.66	0.57	0.56

6 Conclusion

To solve the problem of simulation sample generation based on the simulation scene, this paper proposes a simulation sample generation technology based on multi-variable points. This technology can construct a corresponding variable point set based on a given simulation scene and use a rule-based generation method to generate samples and their related simulation scripts. To verify the technology's feasibility, this paper designed a combat scenario, and simulation experiments are carried out based on this scenario.

A baseline version is first constructed based on the simulation scenario, and then four variable point samples covering different operational processes are generated based on the baseline version. After the samples are generated, the overall efficiency of the constructed simulation samples is analyzed by the AHP. After analysis, it is found that by making improvements on four specific variable points, the system performance has been improved to different degrees, with a minimum increase of 8% and a maximum increase of more than 25%. Therefore, this paper believes that the sample generation technology based on variable points is feasible.

In future research, there is still room to improve the work of this paper. The first is that the category of variable points can be further expanded. In the face of different combat scenarios, the variable points of concern will also change accordingly. This technology can be adapted to different types of combat scenarios in the future by expanding the variable point library.

The second is to improve the generation process. The current generation method is based on the characteristics of variable points, using a rule-based approach to assist generation, and the design of rules still requires manual operation. Artificial Intelligence technology allows the computer to automatically generate corresponding simulation samples according to the combat scene, reducing human resource consumption.

References

1. Zhang, S.Y., Feng, S.W.: Research on normal distribution sample data generation based on bootstrap method. J. Equipment Command Technol. College **2**(20), 97–100 (2009)
2. Zhang, J.B.: Research on evaluation of equipment's combat effectiveness under the condition of small sample. Computer & Digital Eng. **49**(8), 1516–1519 (2021)

3. Yang, J., Yu, X., Xie, Z.Q., et al.: A novel virtual sample generation method based on Gaussian distribution. Knowl.-Based Syst. **24**(6), 740–748 (2010)
4. Goodfellow, I.J., Pouget-Abadie, J., Mirza, M., et al.: Generative adversarial nets. In: Proceedings of the 27th International Conference on Neural Information Processing Systems - Volume 2 (NIPS'14), pp. 2672–2680. MIT Press, Cambridge, MA, USA (2014)
5. Creswell, A., White, T., Dumoulin, V., et al.: Generative adversarial networks: an overview. In: IEEE Signal Processing Magazine, **35**(1), pp. 53–65 (2018)
6. Wang, S.D., Sun, L., Dong, W.M., et al.: Food photo enhancer of one sample generative adversarial network. In: Proceedings of the ACM Multimedia Asia (MMAsia'19), Article 51, 1–8. Association for Computing Machinery, New York, NY, USA (2019)
7. Zhou, F., Ding, R., Mao, S.J., et al.: The training sample generation framework based on confrontation deduction. Fire Control & Command Control **47**(4), 145–149 (2022)
8. Saaty, T.L.: The Analytic Hierarchy Process. Agricultural Economics Review 70. McGraw Hill, New York (1980)
9. Duan, J.K.: Antimissile operational effectiveness evaluation of naval ships based on analysis network process. J. Naval Aeronautical Astronautical Univ. **31**(4), 489–494 (2016)
10. Zhang, G.F., Wu, L.: Effectiveness evaluation of surface ship air defense and antimissile combat in complex electromagnetic environment. J. System Simulation **34**(3), 640–650 (2022)
11. Ji, J.L., Wang, M.L., Han, H.H., et al.: Application of AHP method for top end of anti-missile system positions air defense. Fire Control & Command Control **44**(4), 126–130 (2019)
12. Shi, W.K., Liu, Q., Li, S.L.: Evaluation method research based on short range antimissile weapon system operation capability of shipborne platform. Computer Measurement & Control. **30**(1), 282–287 (2022)
13. Wang, Y.K., Ren, X.W.: Effectiveness evaluation of formation cooperative anti-missile based on AHP and fuzzy comprehensive evaluation. Ship Electronic Eng. **32**(11), 21–22 (2012)
14. Zheng, Y.J., Tian, K.S., Chen, G., et al.: Operational efficiency evaluation for anti-missile early warning radar based on grey AHP. J. Equipment Academy **27**(1), 111–115 (2016)

Simulation Verification of Cruise Missile Route Planning Based on Swarm Intelligence Algorithm

Yang He$^{(\boxtimes)}$, Kai Qu, and Xiaokai Xia

North China Institute of Computing Technology, Beijing 100083, China
Heyang147@126.com

Abstract. Cruise missile route planning aims to improve the penetration capability and survivability of cruise missiles, and ensure the accuracy of hits, which plays a significant part in ensuring the effective completion of combat missions. The essence of the cruise missile route planning is to determine a flight route under the given constraints, so that the cruise missile can reach the target position safely and meet the condition of satisfying the maneuvering characteristics of the cruise missile to the maximum extent. In this paper, a cruise missile route planning space model is established, and three swarm intelligence algorithms, including grey wolf optimizer (GWO), firefly algorithm and particle swarm optimization (PSO), are applied to the cruise missile route planning problem. To verify the effectiveness and compare the performance of these swarm intelligence algorithms in this problem, relevant simulation experiments is designed and implemented.

Keywords: Cruise missile · Route planning · Swarm intelligence algorithm

1 Introduction

A cruise missile is an unmanned air-breathing aircraft with a large combat range, strong penetration capability and autonomous navigation capability, which can be launched from land, sea or air, fly at ultra-low altitude, and attack important targets located within enemy depth [1]. The high precision and high survivability of cruise missiles mainly depends on the ability of cruise route planning, which makes it fly autonomously along a preset flight route. The main purpose of route planning is to provide cruise missiles with a flight route which maximizes the penetration capability of cruise missiles and possibility of destroying predetermined enemy targets [2].

The current mainstream route planning methods are mainly consist of global route planning methods and local route planning methods. The global route planning methods contain the Steepest Descent Method, the Genetic Algorithm and the Voronoi Diagram Method [3]. The local route planning methods mostly include the Artificial Potential Field Method and some other method. These methods have their own advantages and disadvantages, which mainly depend on the optimality of the solution and the computational complexity of the method. In this paper, several typical swarm intelligence

W. Fan et al. (Eds.): AsiaSim 2022, CCIS 1713, pp. 549–560, 2022.
https://doi.org/10.1007/978-981-19-9195-0_44

algorithms are applied to the cruise missile route planning problem, and a simulation verification experiment is designed to compare the effectiveness and convergence speed of these algorithms.

2 Mathematical Model of Cruise Missile Route Planning

As a key technology for cruise missiles to carry out precise strikes, ultra-low-altitude penetration and improve actual combat effectiveness, cruise missile route planning effectively makes up for the shortcomings of cruise missiles with long flight time, slow speed and weak ultra-low-altitude penetration capability. The theoretical cruise missile route planning space is a continuous space. In this space, the optimal route random search will have extremely low efficiency or even be failed because of the exponential expansion. By dividing the route planning space, the scale of the route planning space and difficulty of route planning can be effectively reduced and the planning efficiency can be improved [4].

2.1 Route Planning Space Modeling

The cruise missile route planning space model is shown in Fig. 1. Points S and T are respectively defined as the beginning point and end point of the cruise missile route and some dark circular areas are defined as dangerous areas, such as radar detection areas, anti-aircraft missile kill areas, and anti-aircraft artillery kill areas. When part of the cruise missile's route falls within the danger zone, the cruise missile will face the threat of being intercepted when passing through this part of the route. The task of cruise missile route planning is to find an optimal route from point S to point T so that the cruise missile is under as few threats as possible while satisfying the maneuvering characteristics of the cruise missile.

In order to further quantify this problem, connect the points S and T, divide line segment ST into $D + 1$ equal parts, and take a discrete point on each vertical line Lk to form a discrete point set as follows.

$$C = \{S, (x_1, y_1), (x_2, y_2), \ldots, (x_D, y_D), T\} \tag{1}$$

Then, the cruise missile route planning problem is transformed into an optimization problem of a D-dimensional function. Convert the danger zone information from the original coordinate system XOY to the rotating coordinate system X'SY'.

Assuming that the coordinates of point S are (x_1, y_1) and the coordinates of point T are (x_2, y_2), the formula to transform the coordinates (x, y) in the original coordinate system to the coordinates (x', y') in the rotating coordinate system is as follows.

$$\theta = arcsin\frac{y_2 - y_1}{|\overrightarrow{AB}|} \tag{2}$$

$$\begin{bmatrix} x \\ y \end{bmatrix} = \begin{bmatrix} cos\theta & sin\theta \\ -sin\theta & cos\theta \end{bmatrix} \begin{bmatrix} x' \\ y' \end{bmatrix} + \begin{bmatrix} x_1 \\ y_1 \end{bmatrix} \tag{3}$$

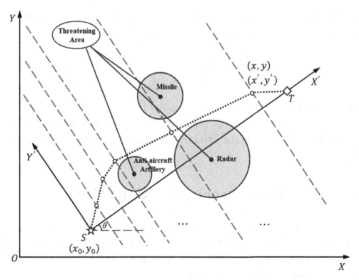

Fig. 1. Modeling diagrams of cruise missile route planning space

2.2 Fitness Function

When applying the swarm intelligence algorithm to cruise missile route planning problem, the fitness function, which is used to estimate the performance of the generated route and is also the basis of the algorithm population's iterative evolution, determine the efficiency and quality of the algorithm execution. The main performance indicators of generating routes include danger zone threat cost $J_{i,t}$ and route cost $J_{i,f}$, and the total cost is expressed as follows.

$$J_{total} = \sum_{i=1}^{D+1} (1 - \varphi) \cdot J_{i,t} + \sum_{i=1}^{D+1} \varphi \cdot J_{i,f} \tag{4}$$

where φ is a constant between 0 and 1. As a weighting parameter, φ expresses the importance of danger zone threat cost or the route cost. When φ approaches 1, more attention is paid to route cost, and when φ is closer to 0, more attention is paid to the danger zone threat cost. In this paper, the value of φ is set to 0.3. The formulas to calculate danger zone threat cost and route cost are as follows.

$$J_{i,t} = \int_0^{L_i} \omega_{i,t} dl \tag{5}$$

$$J_{i,f} = \int_0^{L_i} \omega_{i,f} dl \tag{6}$$

where $\omega_{i,t}$ represents the weight parameter of the danger zone threat cost, which changes with the route segment, $\omega_{i,f}$ represents the route cost weight parameter, which is set to 1 in this paper.

To simplify the calculation, if the route segment numbered i is within the threat range, the danger zone threat cost is expressed as follows.

$$\omega_{i,t} = \frac{L_i}{4} \cdot \sum_{k=1}^{N_t} \left(\frac{R_k}{d_{0.1,i,k}^4} + \frac{R_k}{d_{0.3,i,k}^4} + \frac{R_k}{d_{0.5,i,k}^4} + \frac{R_k}{d_{0.7,i,k}^4} \right) \quad (7)$$

where N_t is the number of threat areas, L_i is the length of route segment numbered i, $d^4_{0.1,i,k}$ represents the distance from 1/8 of the route segment numbered i to the threat area numbered k, R_k represents the level of the threat area numbered k, which is all set to 3 in this paper.

3 Cruise Missile Route Planning Based on GWO

The GWO [5] is inspired by the prey activity of grey wolves and developed a swarm intelligence optimization search algorithm. Due to its strong convergence performance, few preset parameters, and easy implementation, this algorithm has received extensive attention from scholars in recent years [6].

3.1 Rationale of GWO

In the gray wolf population, the population is generally divided into four levels. The first level is the wolf α, who leads the individual of the entire population. The second level is the wolf β, who mainly gives assistance to the alpha wolf to make decisions. The third level is the wolf δ, who executes tasks such as reconnaissance and sentry. The fourth level is the wolf ω, which is the lowest gray wolf and obeys the command of the upper-level wolf. The GWO is a mathematical simulation of the hunting activity of the gray wolf population according to the hierarchical system.

In the process of hunting prey in gray wolf populations, their predation behaviors are defined as follows.

$$D = \left| C \cdot X_p(t) - X(t) \right| \quad (8)$$

$$X(t+1) = X_p(t) - A \cdot D \quad (9)$$

where D represents the distance between the individual wolf and the quarry, $X_p(t)$ represents the position of the individual in the t-th generation wolf pack tribe, A and C are the parameter, which is calculated by follows.

$$A = 2a \cdot r_1 - a \quad (10)$$

$$C = 2r_2 \quad (11)$$

where A is the convergence factor which decreases linearly from 2 to 0 with the number of iterations, r_1 and r_2 are random numbers between 0 and 1.

When gray wolves capture their prey, wolf α at tier 1 lead other tiers of wolves to surround the prey. Among the gray wolves, wolves α, wolf β and wolf δ are the closest

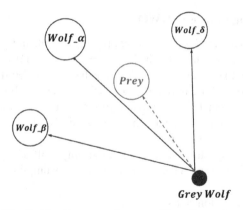

Fig. 2. The diagrams of grey wolf location update

and most sensitive groups to the presence of their prey. The location updates of other gray wolf individuals were determined based on the location of these three wolves. The diagrams of gray wolves hunting for prey is shown in Fig. 2, and the mathematical model is as follows.

$$D^j_{i,\alpha}(t) = \left| C \cdot X^j_\alpha(t) - X_i(t) \right| \tag{12}$$

$$D^j_{i,\beta}(t) = \left| C \cdot X^j_\beta(t) - X_i(t) \right| \tag{13}$$

$$D^j_{i,\delta}(t) = \left| C \cdot X^j_\delta(t) - X_i(t) \right| \tag{14}$$

$$X^j_{i,\alpha}(t) = X^j_\alpha - A \cdot D^j_{i,\alpha}(t) \tag{15}$$

$$X^j_{i,\beta}(t) = X^j_\beta - A \cdot D^j_{i,\beta}(t) \tag{16}$$

$$X^j_{i,\delta}(t) = X^j_\delta - A \cdot D^j_{i,\delta}(t) \tag{17}$$

$$X^j_i(t + 1) = [X^j_{i,\alpha}(t) + X^j_{i,\beta}(t) + X^j_{i,\delta}(t)]/3 \tag{18}$$

where $D^j_{i,\alpha}(t)$ represents the distance between the individual of the t-th generation of wolves and the individual of the wolf α, $D_j{}^{i,\beta}(t)$ represents the distance between the individual of the t-th generation of wolves and the individual of the wolf β, $D_j{}^{i,\delta}(t)$ represents the distance between the t-th generation wolf pack individual and the wolf δ. Equation (15–17) respectively represents the step size and direction of ω wolf moving to wolf α, wolf β and wolf δ. Equation (20) represents the position of the new generation of gray wolf individuals after the position update.

3.2   Route Planning Process of GWO

The process of applying the GWO to cruise missile route planning are as follows.

Step 1: initialize gray wolf individuals X_i (i = 1, 2, 3,..., D), a, A, C, N, D, M.

Step 2: according to Eq. (2) and Eq. (3), rotate the original coordinate system into the new coordinate system whose horizontal axis is the connection line from the beginning point to the end point, transform the threat information of the circular danger zone into the new coordinate system, and divide the line connecting the beginning point and the ending point into D + 1 equal parts.

Step 3: calculate the total cost of each initial gray wolf according to Eq. (4), and record the top three wolves with the lowest total cost consumption, namely wolf α, wolf β and wolf δ.

Step 4: the loop iteration starts.

Step 5: update a, A, and C.

Step 6: Update the position of each individual gray wolf according to Eq. (18).

Step 7: Calculate the total cost consumption of each new gray wolf individual according to Eq. (4), and update wolf α, wolf β and wolf δ.

Step 8: go back to step 5 until the number of iterations is satisfied.

Step 9: output the optimal value of wolf α and the corresponding position information.

Step 10: transform the coordinates of the final optimal route back to the origin coordinate system and output.

4   Cruise Missile Route Planning Based on Firefly Algorithm

The firefly algorithm [7], a swarm intelligent algorithm, simulates the light-emitting mechanism of fireflies in nature, that is, fireflies with strong light attract fireflies with weak light, and use this method to communicate with each other, communicate, courtship or prey and other behaviors. With its advantages of less parameters, simple operation and pretty stability, the firefly algorithm has received extensive attention and research [8].

4.1   Rationale of Firefly Algorithm

The inspiration of the firefly algorithm comes from the flashing behavior of fireflies which is regarded as a special signal, through which information exchanges such as foraging, courtship, and alertness among fireflies are carried out. The core concept of the firefly algorithm is that within the firefly's perception range, the firefly is attracted to the brighter firefly than it, and thus moves to a new position and that the brightest fireflies in the population moved randomly. The mathematical model of the standard firefly algorithm includes the following key equations.

The equation for the relative brightness of fireflies is shown as follows.

$$I_{ij} = I_i \times e^{-\gamma r_{ij}} \tag{19}$$

where I_{ij} is the relative brightness between the ith firefly and the jth firefly, I_i is the absolute brightness of the ith firefly, γ is the light absorption parameter which is generally a constant, r_{ij} is the distance between the ith firefly and the jth firefly.

The equation for the attraction of fireflies is as follows.

$$\beta = \beta_0 \times e^{-\gamma r_{ij}} \tag{20}$$

where β_0 is the maximum attraction factor, which is generally a constant.

The firefly's movement equation is shown as follows.

$$x_i = x_i + \beta \times (x_j - x_i) + \alpha \times \left(rand - \tfrac{1}{2}\right) \tag{21}$$

where x_i and x_j respectively represent the spatial positions of the i-th firefly and the j-th firefly, α is the step factor which is generally a constant, rand is a random number between 0 and 1.

4.2 Route Planning Process of Firefly Algorithm

The process steps of applying firefly algorithm to cruise missile route planning are as follows.

Step 1: initialize firefly individuals X_i ($i = 1, 2, 3, \ldots, D$), α, γ, N, D, M.

Step 2: according to Eq. (2) and Eq. (3), rotate the original coordinate system into the new coordinate system whose horizontal axis is the connection line from the beginning point to the end point, transform the threat information of the circular danger zone into the new coordinate system, and divide the line connecting the beginning point and the ending point into $D + 1$ equal parts.

Step 3: calculate the maximum fluorescence brightness I_i of each firefly according to Eq. (4).

Step 4: the loop iteration starts.

Step 5: calculate the relative fluorescence brightness and attraction between each pair of fireflies according to Eq. (19) and Eq. (20).

Step 6: update the position of each individual firefly according to Eq. (21).

Step 7: calculate the maximum fluorescence brightness I_i of each new firefly according to Eq. (4).

Step 8, go back to step 5 until the number of iterations is satisfied.

Step 9: output the brightness value and position information of the firefly with the highest maximum fluorescence brightness value.

Step 10: transform the coordinates of the final optimal route back to the initial coordinate system and output it.

5 Cruise Missile Route Planning Based on PSO

PSO is a swarm intelligence algorithm inspired by the foraging behavior of birds. In the process of searching for the largest food source, the birds pass their position information to each other, so that other birds know the location of the food source and all birds can finally gather around the food source, which means the optimal solution to the problem is found. With strong versatility, easy-to-implement algorithm principles and pretty global optimality, PSO has become one of the most classic intelligent algorithms.

5.1 Rationale of PSO

PSO simulates individuals in natural bird flocks by using points in the search space, and compares the process of their foraging behavior to the process of optimization iterations of feasible solution changes. The search process basically uses little external information and only uses the fitness function as the evolutionary basis, and each individual approaches the optimal solution on the basis of the part extreme value and the global extreme value.

There is a particle population of size M with an N-dimensional space search area, which denoted as $X = [x_1, x_2, ..., x_i, ..., x_M]^T$, and the position of the i-th particle is represented as $xi = [x_{i1}, x_{i2}, ..., x_{iN}]^T$. The distance the particle moves in each iteration is expressed as $vi = [v_{i1}, v_{i2}, ..., v_{iN}]^T$, and the position passed by the i-th particle with the best fitness, which means the individual extreme value of the i-th particle, is expressed as $Pi = [p_{i1}, p_{i2}, ..., p_{iN}]^T$. The global extreme value which means the fitness value of the entire population is the highest is indicated as $P_g = [p_{g1}, p_{g2}, ..., p_{gN}]^T$. The feasible solution of the optimization problem is the position of the optimized particle. The particle updates its position and velocity at each iteration according to the following equations.

$$v_{id}^{k+1} = \omega v_{id}^k + c_1 \varepsilon \left(p_{id}^k - x_{id}^k \right) + c_2 \eta \left(p_{gd}^k - x_{gd}^k \right) \tag{22}$$

$$x_{id}^{k+1} = x_{id}^k + v_{id}^{k+1} \tag{23}$$

where v_i^k is the velocity of the k-th iteration of particle i, v_{id}^k is the d-th dimension component of v_i^k. X_i^k is the position of particle i in the k-th iteration, x_{id}^k is the d-th dimension of x_i^k component, p_i^k is the individual extreme value of particle i in the k-th iteration, p_{id}^k is the d-dimensional component of p_i^k, p_g^k is the global extreme value of the particle swarm in the k-th iteration, p_{gd}^k is the d-dimensional component of p_g^k, ω is the inertia weight, c_1 and c_2 are learning factors which respectively adjust the particles to their own individual extreme value and global extreme value, ε and η are uniform distribution of random numbers between 0 and 1.

5.2 Route Planning Process of PSO

The process steps of applying PSO to cruise missile route planning are as follows.

Step 1, initialize particle population individual Xi (i = 1, 2, 3,..., D), ω, c_1, c_2.

Step 2, according to Eq. (2) and Eq. (3), rotate the original coordinate system into the new coordinate system whose horizontal axis is the connection line from the beginning point to the end point, transform the threat information of the circular danger zone into the new coordinate system, and divide the line connecting the beginning point and the ending point into D + 1 equal parts.

Step 3, calculate the total cost of each particle according to Eq. (4), and record the position information of the particle with the lowest cost.

Step 4, the loop iteration starts.

Step 5, calculate the next moving speed $v_{id}^{(k+1)}$ of each particle according to Eq. (22), and record the speed.

Step 6, calculate the position of each new particle according to Eq. (23).

Step 7, calculate the total cost f each particle according to Eq. (4), update the optimal position of each particle, and record the position information of the particle with the lowest cost.

Step 8, go back to step 5 until the number of iterations is satisfied.

Step 9, output the total cost of the particle with the lowest total cost consumption and its position information.

Step 10: transform the coordinates of the final optimal route back to the initial coordinate system and output.

6 Simulation Verification and Analysis

To validate the feasibility and effectiveness of the swarm intelligence algorithm applied to cruise missile route planning, set the coordinates of the launch point of the missile to be (1600, 900) and the coordinates of the attack target point to be (200, 200), and the information of the 10 circular danger zones is as follows (Table 1).

Table 1. Information of the circular danger zone

Number of danger zones	Center coordinate of danger zone	Radius of danger zone
0	(551, 119)	1180
1	(979, 217)	672
2	(372, 167)	1239
3	(200, 160)	400
4	(573, 93)	695
5	(988, 40)	1648
6	(115, 195)	1725
7	(824, 136)	1421
8	(347, 72)	104
9	(777, 60)	911

For the three swarm intelligence algorithms, we uniformly set the initial number N of particles to be 300 and the number M of iterations to 100. The firefly algorithm additionally sets the light absorption coefficient γ to be 0.5, the step factor α to be 0.2, and the PSO additionally sets Fixed inertia weight ω to be 0.1, learning factors c_1 to be 0.5 and learning factors c_2 to be 1.5. Figure 3 and Fig. 4 respectively show the optimal planning route diagrams and algorithm convergence curves of the three swarm intelligence algorithm under a certain simulation.

As can be seen from Fig. 3 and Fig. 4, the three swarm intelligence algorithms applied to the cruise missile route planning problem can all search for a feasible route while avoiding obstacles, so that the cruise missile can safely reach the end point from the

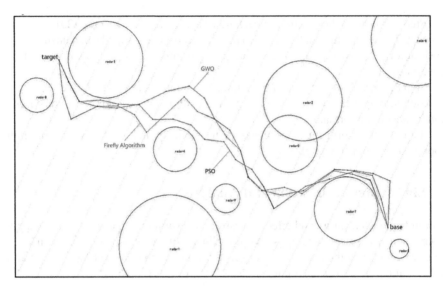

Fig. 3. Planning route diagram of three population intelligence algorithms

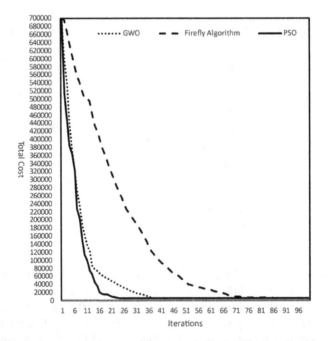

Fig. 4. Convergence curves of three population intelligence algorithms

beginning point. But the total cost of searching for the optimal route and the convergence speed of these algorithms are different. It can be seen in Table 2 that the average total cost values of the GWO and the PSO are 5610 and 5603, which are evidently better than the average cost value of the firefly algorithm. In addition, the average number of iterations of the particle swarm algorithm is 23, which is the best among the three swarm algorithms, while the average number of iterations of the firefly algorithm is 76, which is far inferior to the other two algorithms.

Table 2. Comparison of simulation results of three population intelligence algorithms

Algorithms	Best Total Cost	Worst Total Cost	Average Total Cost	Average Number of Iterations
GWO	5402	5903	5610	37
Firefly Algorithm	5626	7147	5697	76
PSO	5517	5764	5603	23

7 Conclusions

In this paper, three swarm intelligence algorithms including GWO, firefly algorithm and PSO, are applied to the cruise missile route planning problem, and relevant simulation experiments are designed and implemented to verify the effectiveness of the three algorithms in this problem. PSO has the best performance in the cruise missile route planning problem due to its fast convergence speed and low total cost. GWO also has the same fast convergence speed and low total cost consumption, which shows good applicability in the cruise missile route planning problem. The convergence of the firefly algorithm is the worst among the three swarm intelligence algorithms, and occasionally shows a high total cost during the simulation process, which is needed to be improved to be better applicable to the cruise missile route planning problem.

References

1. Wang, M., Fang, M.: Research on the intelligent operations of cruise missile. Tactical Missile Technology **03**, 18–22 (2013). https://doi.org/10.16358/j.issn.1009-1300.2013.03.013
2. Fang, M.Y., Wang, M.L., Bi, Y.M.: Zhang, B.: Research on intelligent autonomous online route planning of cruise missile in local area. Tactical Missile Technology (05), 60–64+75 (2013). https://doi.org/10.16358/j.issn.1009-1300.2013.05.024
3. Xie, C.S., Liu, Z.Y.: Research on path planning of land cruise missile based on improved V-ACO algorithm. Tactical Missile Technology (05), 122–131 (2021). https://doi.org/10.16358/j.issn.1009-1300.2021.5.146
4. Zhang, S., Zhou, Y.Q., Li, Z.M., Pan, W.: Grey wolf optimizer for unmanned combat aerial vehicle path planning. Adv. Eng. Softw. **99**, 121–136 (2016)
5. Mirjalili, S., Mirjalili, S.M., Lewis, A.: Grey wolf optimizer. Adv. Eng. Softw. **69**, 46–61 (2014)
6. Liu, N.N., Wang, H.W.: Path planning of mobile robot based on the improved grey wolf optimization algorithm. Electrical Measurement and Instrumentation **57**(01), 76–83+98 (2020). https://doi.org/10.19753/j.issn1001-1390.2020.001.010

7. Ding, J.W., Qu, S.C.: On the path planning of mobile robots based on improved firefly algorithm. J. Hunan Univ. Technol. **31**(01), 64–68 (2017)
8. Liu, J., Huang, Y.: Research on path planning of unmanned surface vehicles based on improved chaotic firefly algorithm. Control Engineering of China **28**(11), 2209–2214 (2021). https://doi.org/10.14107/j.cnki.kzgc.20210062
9. Tian, Y.L., Mi, Z.C., Zhou, Y.L., Wang, H., Lu, F.X.: Charging path planning method of UAV based on the improved hybrid particle swarm algorithm. Radio Communications Technology, 1–7 (2022). https://kns-cnki-net-s.nudtproxy.yitlink.com:443/kcms/detail/13.1099.TN.20220615.1442.002.html
10. Liang, J.Q., Zhou, Z.C., Liu, X.Y.: Robot path planning based on particle swarm optimization improved grey wolf algorithm. Software Guide **21**(05), 96–100 (2022)

A Joint Operation Simulation Environment for Reinforcement Learning

Dong Li$^{(\boxtimes)}$, Xiao Xu, and Lin Wu

National Defense University, Beijing 100091, China
donggeat2006@163.com

Abstract. Reinforcement learning has received broad attention from multiple areas due to its remarkable successes nowadays. And the intelligence about decision making is becoming the new frontier of artificial intelligence. Among various real-world scenarios need accurate decision making, military decisions, however, been studied by few people. This paper describes a reinforcement learning environment powered by a war game, which is considered as a high level simulation for military operation. We define the observation and action space for this environment, along with a system designed for programmatic access. We also provide a series of mini-games for evaluation.

Keywords: Military decision · Joint operation simulation · Reinforcement learning

1 Introduction

Reinforcement learning (RL) has shown great potentials to deal with complex decision making recently, highlighted by super-human performance in the classical video game of Atari [9], board game of GO [15], Chess and Shogi [12], and complex real-time strategy (RTS) video game Dota [11] and StarCraft II [19]. Especially in RTS games, RL agents face challenges in real-world, which represent multiple participation, imperfect information due to partially observation, huge observation involving a broad map, large action space due to lots of units, and delayed reward requiring long-term considerations [20].

The success of RL has pushed the frontier of artificial intelligence (AI) to the area of decision making. As a special real-world scenario need accurate and rapid decision making, military decision is less studied due to several reasons, among which the lack of universal environment and benchmark might prevent more professional effort coming into this field. While this field is becoming vital today, as reported in [4], military decision in complex battle field featured by autonomous systems could even shape the future warfare.

In this paper, we describe a reinforcement learning environment powered by a war game which is traditionally studied as a modeling and simulation tool for training military staffs, and most of the research are outside of AI. We summarize a few unique characteristics of this environment compared with board cards and video games, and provide an analysis of difficulties about applying RL on it. Based on which, we define the observation and action space along with a system designed consideration for programmatic access. In order to evaluate RL algorithms on this environment, we provide a series of mini-games to let agents could finish simple tasks, which may lay the foundation for further research. The main contribution can be summarized as three aspects:

- We describe a joint operation simulation environment from the learning perspective, and provide an analysis of features and challenges about applying RL on it.
- We design the observation and action space with a programmatic access consideration, which makes a complete learning environment.
- We design a series of mini-games on this warfare simulation and report the initial baselines.

1.1   Background

Before we dive into the details of learning environment, here we briefly review the fundamentals of Markov Decision Processes (MDPs) [17], the mathematical form of problem setting in RL. A MDP is normally defined by the 5-tuple $(\mathcal{S}, \mathcal{A}, \mathcal{P}, \mathcal{R}, \gamma)$, which means state space, action space, transition probability, rewards and discount factor on rewards. At each step t, the agent receives a state s_t from $\mathcal{S}$, and choose an action from $\mathcal{A}$ following the policy $\pi(a_t \mid s_t)$ which is mapping from state s_t to action a_t, receives a scalar reward r_{t+1} given by the reward function $\mathcal{R}_s^a$, and goes to the next state s_{t+1} following the transition probability $\mathcal{P}[s_{t+1} \mid s_t, a_t]$. The reward is normally discounted and we expect the overall return in T steps to be maximized,

$$G_T = r_{t+1} + \gamma r_{t+2} + \gamma^2 r_{t+3} + \ldots = \sum_{k=0}^{T} \gamma^k r_{t+k+1}. \tag{1}$$

An RL agent learns the policy to maximize the overall return by interacting with the environment, or by collecting the samples $s_t, a_t, r_{t+1}, \ldots$ first. Kinds of algorithms have been developed to achieve this goal, and we can categorize these algorithms into value-based or policy based RL by what to learn, since directly maximizing the overall return (1) is hard. The value-based RL instead tries to optimize value function (or state value function) of the current state s w.r.t. the policy π,

$$\begin{aligned} V_\pi(s) &= \mathbb{E}[G_t \mid s_t = s], \\ Q_\pi(s, a) &= \mathbb{E}[G_t \mid s_t = s, a_t = a]. \end{aligned} \tag{2}$$

In contrast, the policy-based RL tries to optimize the policy directly, by maximizing an objective w.r.t. the parameterized policy π_θ.

2   Related Work

2.1   Learning Environments

Environments and benchmarks have been used to evaluate RL algorithms and measure the progress of decision intelligence in a quantitative way. The classical video games Atari 2600 have been benchmarks for RL as the Arcade Learning Environment (ALE) [1] was released, later gained broader interests as a interface along with OpenAI Gym [3], a more famous and standard de facto environment for RL studies. Here we highlight several special examples related to our topic.

As mentioned in Sect. 1, RTS games share several common features with warfare simulation. As the design of RTS games takes inspiration from human activities including competition and warfare, characterized by imperfect information caused by fog of war, multiple players sequential interacting, huge observation space caused by high-resolution map, large action space for dealing with hundreds of units, and delayed reward requiring long-term consideration.

In order to push the RL research to more difficult problems, DeepMind and Blizzard jointly release the StarCraft II Learning Environment (SC2LE) [20], which contains a StarCraft II game engine, the StarCraft II API and PySC2, an opensource python wrapper for accessing the game and optimizing the agent. SC2LE becomes a foundation of studying RL for complex decision making on RTS games, and attracts researches from a wide range of institutes. Among which DeepMind achieves the milestone, AlphaStar [19], which was the first agent to defeat top human players in the full game of StarCraft II.

Multi-player Online Battle Arena (MOBA) game is a special and up-to-date form of RTS, and also receives attention form RL community. Compared with traditional RTS games like StarCraft, MOBA emphasizes on collaboration and complex control, as each player takes in charge of his own role. On the popular e-sport game Dota 2, OpenAI Five [2] integrates various tricks on a large RL agent which eventually defeat the world champions. On another popular MOBA 1v1 game among Chinese community *Honor of Kings*, Ye et al. [21] developed the RL agent based on knowledge from experienced human player, which also defeat top human players.

2.2   RL Research on Military Decision

Compared with games, military decision is less studied in AI community due to various factors. Conventionally, military decision made by human commanders could be partly evaluated by warfare simulation, which model the procedures of wars on tactical, operational, or strategic level. The warfare simulation is also called wargaming [16], serving as basic tool to train military staffs even from ancient times. As decision intelligence is becoming more important in real-world applications other than games, the bridge between computer-based military simulation and RL studies could also play an important role in this field.

The tactical simulation has been used as for evaluating intelligent decision making for long time, and there exists numerous examples of conventional knowledge-based and machine learning methods applied on computer generated forces (CGFs) [22]. Moy et al. [10] considers applying a AlphaZero deep RL algorithm on a turn-based wargame, Coral Sea [18], and the agent could outperform heuristics that trained it. Goodman et al. [5] conducts an extensive research on discussing the relationship between AI and wargaming, pointing out several potential directions of applying AI techniques on wargaming.

3   A Joint Operation Simulation Environment

3.1   Elements of a Joint Operation Simulation

The modern warming system aims to restore the campaign process by leveraging the computer modeling & simulation (MS) techniques. The joint operation simulation is generally carried out according to the needs of analysis or training of commanders and command institutions. And different drill seats are set up according to the established plan, mainly including coordinators, players and technical support. Among them, the players are the training objects of the drill, generally placed in the hypothetical situation, together with the computer simulation forces form a human-in-loop drill environment.

The elements of a joint operation simulation include entities, behavior and interaction [6], which can be further featured by diverse entities, heterogeneous behavior and complex interactions. The joint operation simulation system generally simulates the activities of ground, air, sea, and other support. In order to simulate relatively complete progress of joint operations, it should cover the following basic functions in Table 1.

Table 1. Functions of joint operation simulation.

Combat field	Operations
Ground	Move, attack, defense, artillery fire, fire damage
Air	Defense, interception, anti-submarine, ground attack, patrol
Sea	Transport, ship guidance, mines, submarine, amphibious
Logistics	Intelligence sharing, supply, pipeline operations, repairing

From Table 1 we can see that different actions vary greatly in the time and space dimensions. When all activities are carried out under the same world view, there must exist differences in the effect of action. For example, in the same hour, ground forces may move only 40 km, while aircrafts can make several trips over a vast area of several thousand kilometers. While Marine units may move even closer. Most ground or marine units may keep stationary while a few moving. The impact of intelligence or logistics on battlefield situations may be delayed as well. These differences in the effects of actions are caused by the simulation of the objective world, which in turn affects decision-making under such a world view.

In terms of the interaction, a joint operation simulation system involves a large number of participating forces, and the interaction process is extremely complex which is generally approximated by discrete event simulation (DES) mechanism. The DES normally use an event queue to maintain the life cycle of events, and all events take effect in sequence and output the results in turn. When the event processing service is lower than the event generation, queuing occurs. Generally, a large number of events will cause the delay of the interaction effect. In addition, the semi-theoretical and semi-empirical method used in the calculation of interaction effects bring certain randomness and delay to the inter-action results. And this delay is further coupled with the partially observation of the battle simulation.

3.2 Procedures of Joint Operation Simulation Drill

In a typical medium-scale battle, each side has several hundred units to carry out independent combat missions, with a simulation time of several hours, usually involving the main operations in Table 1. Before the direct combat with offensive and defensive actions, players usually go through a longer period of planning, including the analysis and assessment of battlefield situation, the organization of operational meetings, and the preparation of operational plans, so as to select the final plan. This stage is also known as pre-war planning. Since the main activities is making plans, which is difficult to directly illustrate the process interacting between the two sides, the basic tool for decision is computation in operational research. But the battle plan is subject to the change of battlefield situation, as the direct combat begins, a large number of actions are concentrated at a short period, with occasionally coming up throughout the rest of the time.

In a typical battle, except for a large number of interactions in the stage of direct combat, most of the events are driven by indirect contact. During which there exist operations carried out according to the prepared plan, as well as oper-ations emergently arranged subject to the change of battlefield situation. At the same time, there exist not only events that change the overall battlefield, but also a large number of events that maintain local or individual state. Furthermore, the actions range from those that cause immediate damage or reconnaissance effects to those lasting for hours and then triggered randomly. Although the joint operation simulation follows causality in individual modeling, it can also be regarded as a black box as a whole due to the complex interactions between a large number of entities.

3.3 Unique Features of Joint Operation Simulation

From the above description we can see that war game is similar to RTS games in terms of combating in huge and partially observed state and large action space. But there exits several key differences between as well, along with the lack of universal environment and benchmark, preventing the effort of applying RL directly to military decision making on the war game. Here we discuss several important ones brought by the war game and decision making as below.

Since the computation of situation changing with or without damaging effects is accomplished by randomness and approximation schemes, which brings uncertainty in state transition and reward emotion, two basic functions of model in RL. Furthermore, the reward signal can be shifting as the campaign goes on, coupled with credit assignment problem (CAP) caused by a combination of several operations required by complex tasks.

In addition to the uncertainty, the state transition is non-Markovian due to the hidden driving factors and the tasks dependency as in [8]. Therefore, the decision of lasting and randomly triggered operations are hard to learned in a series of subtasks. Finally, the heterogeneity of decision-making frequency also brings a difficulty for RL: since the situation is changing by default in most time of the episode, the RL agent is barely motivated.

4 Components of Joint Operation Simulation Environment

The overall structure of joint operation simulation environment follows the MDP design in Sect. 1.1, with special consideration of observation, action and reward from this simulation, as shown in Fig. 1. We provide the description of each component in details in this section, followed by an interface design to let agent access to this environment.

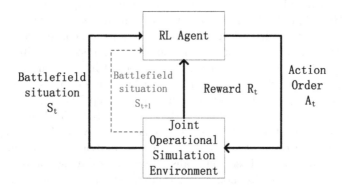

Fig. 1. The overall structure of agent interacts with the joint operation simulation

4.1 Observations

Generally speaking, the global battlefield situation of a joint operation simulation drill is too large. We take a simplified way to focus on tasks related scenes, which selects and encodes certain battlefield situation (observation) elements to maintain a reasonable state space for the RL agent.

First of all, two levels of selection are carried out: unit and unit specific situation. For example, we only keep missile artillery for the shore-distance mission in ground forces when the drill involves missile artillery in ground. And the remaining missile artillery is screened again to make sure only the current state, position and available ammunition, which are strongly related to the task, have been retained. For naval units, 6 elements such as current state, position, heading and speed are reserved at present. For air missions, 12 elements such as current oil, status, flying altitude and speed are retained.

Secondly, the battlefield situation is encoded for further processing. However, the encoding may face the problem of uncertainty from both players. For example, units may be reduced by being destroyed or sunk, and air formations may be increased by new missions on one side. Missions can increase or decrease on the other side. Furthermore, the same position in the vector-based representation may change subject to the situation change, which would increase the difficulty of training. To solve these two problems, we design a situation encoding method based on fixed entity code and reserved buffer bits. The whole situation was divided into sides and corresponding positions are filled in according to the name of the unit. For the incremental situation that might occur, certain vacancies are reserved in each type of code bits as buffers. The overall situation selection and encoding are shown in Fig. 2, resulting in a vector-based representation with 1749 elements.

Fig. 2. The encoding on observation in joint operation simulation environment.

4.2 Actions

The overall idea of action design is to provide certain type of specific tasks after simplified control, and encode the action vector output by the task agent. The process of action encoding is similar to the process of observation encoding. It carries out bit encoding according to the unit performing the action. If the individual unit does not perform the action at current moment, the corresponding position is set to be 0. For a large number of control parameters in the output action, we only keep several key parameters such as the target, moving position and whether to strike, and fixing the rest parameters to reduce the size of action space. The overall encoding process of the action is shown as Fig. 3.

Fig. 3. The encoding on action in joint operation simulation environment.

4.3   Reward and Mini-Games

Reward signals determine what agent learns and how agent learns partly. But reward is also related to specific scenario and mission tasks. In order to evaluate RL algorithms on the joint operation simulation environment, as in SC2LE [20], we provide a series of mini-games involving two sides battle as follows.

Mini-game 1. Side A plans to form a team to go to front, and side B sends a flying formation to patrol in the air to prevent the side A formation from assembling, as Table 2 shows. The reward for side A means units have reached the assembly area. In order to make the reward signal denser, agent get positive incentives proportional to the reciprocal of the distance between unit α_i and assembly point x.

Table 2. Mini-game 1.

	Side A	Side B
Units	unit $\alpha \times 6$, unit $\beta \times 2$	Fighter $\chi \times 6$
Mission	Reach to assembly area	Prevent side A
Stopping criteria simulation lasts for one hour		

Mini-game 2. Side A units have arrived at certain point, planning to launch the operation L. Side B is waiting with an fire artillery, as Table 3 shows. The reward signal for Side A is to go point D and hold on for at least 20 min.

Table 3. Mini-game 2.

	Side A	Side B
Units	Unit $\alpha \times 6$, fighter $\zeta \times 2$	Fire artillery $\mu \times 6$
Mission	Control the point	Prevent side A
Stopping criteria simulation lasts for one hour		

Mini-game 3. Side B intends to send units ω to ambush Side A landing formations, while Side A use anti-aircrafts to find and destroy ω, as Table 4 shows. The reward signal for Side B is the number of destroyed units from Side A.

Table 4. Mini-game 3.

	Side A	Side B
Units	Unit $\alpha \times 6$, unit $\beta \times 2$	Unit $\omega \times 2$
Mission	Find and destroy ω	Ambush side A
Stopping criteria simulation lasts for one hour		

Mini-game 4. Side A planning to launch the operation L and the units have arrived certain area. Side B needs to prevent side A by sending the air attack mission, as Table 5 shows. The reward signal of side B is the number of destroyed units from Side A.

Table 5. Mini-game 4.

	Side A	Side B
Units	Unit $\alpha \times 6$, aircraft $\kappa \times 2$	Fighter $\chi \times 6$
Mission	To the L point	Attack side A
Stopping criteria simulation lasts for one hour		

4.4   Application Programming Interface

We provides a packaged application programming interface (API), mainly supported by Python. The interface uses client/server architecture to implement the communication between the interface client (agent) and the interface server (simulation service), using TCP protocol. We adopts the cross-platform cross-language open source efficient communication library ZeroMQ for data transition. The interface server distributes the situation data and parse the orders sent by the client, acting as the information hub between the simulation engine and the agent. Interface client support four types of programming specifications:

- Observation interface: we provide situation awareness, various information such as units, geography and weather. It is worth noting that in order to simulate the real cases as much as possible, one can only partially observe the situation. And this is the same for human players looking at the board or for agents accessing through the API.
- Action interface: we provide an interface for the agent to control units and complete actions through the instructions allowed by the simulation system.
- Simulation control interface: we provide interface to let agent start, stop, accelerate or decelerate the simulation engine, which could be helpful for training.
- Auxiliary interface: we provide connection, login, logout, error reporting and other functions.

The learning environment packages a simplified version of simulation engine, support service, API and mini-games in a Docker image. The whole system is used as in Fig. 4 shows.

5   RL Baseline

In order to apply RL algorithms to the joint operation simulation environment described in Sect. 3, we conduct the baseline experiments using the opensource

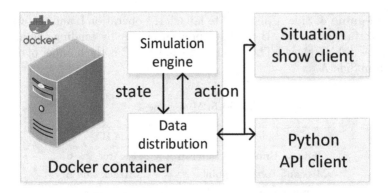

Fig. 4. The system design of joint operation simulation environment.

Ray RLlib [7], which provides a user-friendly support for production-level, highly distributed RL workloads. We mainly take the state-of-the-art on-policy algorithm proximal policy optimization (PPO) [14] for example, which is efficient for continuous state and discrete action space. PPO aims to solve the following objective (3) as in Trust Region Policy Optimization (TRPO) [13],

$$\max_{\theta} \mathbb{E}_{s,a\sim\pi_{\theta_{\text{old}}}} \left[\frac{\pi_\theta(a \mid s)}{\pi_{\theta_{\text{old}}}(a \mid s)} A_{\pi_\theta}(s, a) \right] \text{ s.t. } \mathbb{E}[KL[\pi_{\theta_{\text{old}}}(\cdot \mid s), \theta(\cdot \mid s)]] \le \delta, \quad (3)$$

where θ and θ_{old} are the parameters of policy after and before updating, $A_{\pi_\theta}(s, a) = Q_\pi(s, a) - V_\pi(s)$ is the advantage function. And the KL-divergence avoid too large update of the policy, which is implemented by clipping in PPO, i.e. removing the changing outside a certain interval. Using PPO or other RL algorithms in Ray RLlib is straightforward, we just need to specify several configs for the algorithm and turn the rest to the system.

Take the mini-game 4 (see Table 5) for example, side A is driven by a predefined rule where aircrafts are covering units to conduct operation L, and side B is an RL agent with the policy $\pi_\theta(a \mid s)$ represented by the default multilayer perceptron (MLP). We evaluate the performance by the convergence of reward, and the sampling steps needed to reach the stable performance, as Fig. 5 shows.

We can see that in such scenario, the PPO could achieve expected performance: as the training goes on, the mean of reward (Fig. 5a) increases to show the agent learns how to attack. At the same time, the KL-divergence of policy (Fig. 5b) decreases steadily which means the policy goes stable as the training steps. Both criteria shows that the convergence achieved at 100k sampling steps, a relatively low sample complexity compared with large RTS game.

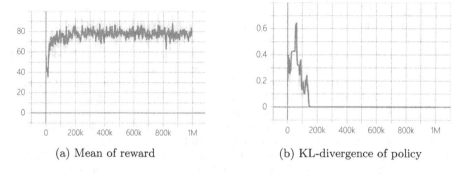

(a) Mean of reward   (b) KL-divergence of policy

Fig. 5. Performance of PPO on mini-game 1.

6   Conclusion

This paper introduces a joint operation simulation environment for RL. We describe the joint operation simulation environment in details, including an analysis from the learning perspective. We conduct a system design of components in this learning environment, as well as the API to access it. Finally we provide the preliminarily results of applying basic RL algorithms on mini-games in this environment.

The joint operation simulation environment we introduce is challenging for RL because of the unique features of this environment. Based on the initial work, we hope more efforts can be done to accomplish intelligent decision making in this field.

Acknowledgement. This research was supported by the National Natural Science Foundation of China under Grant No. 62006235.

References

1. Bellemare, M.G., Naddaf, Y., Veness, J., Bowling, M.: The arcade learning environment: an evaluation platform for general agents. J. Artif. Intell. Res. **47**, 253–279 (2013)
2. Berner, C., et al.: Dota 2 with large scale deep reinforcement learning CoRR abs/1912.06680 (2019). http://arxiv.org/abs/1912.06680
3. Brockman, G., et al.: Openai gym. arXiv preprint arXiv:1606.01540 (2016)
4. Clark, B., Patt, D., Schramm, H.: Mosaic Warfare: Exploiting Artificial Intelligence and Autonomous Systems to Implement Decision-Centric Operations. Center for Strategic and Budgetary Assessments (2020)
5. Goodman, J., Risi, S., Lucas, S.: AI and wargaming. CoRR abs/2009.08922 (2020). https://arxiv.org/abs/2009.08922
6. Xiaofeng, H., Guangya, S.: War Gaming & Simulation Principle and System, W.L. (2009)

7. Liang, E., et al.: RLlib: abstractions for distributed reinforcement learning. In: Dy, J.G., Krause, A. (eds.) Proceedings of the 35th International Conference on Machine Learning, ICML 2018, Stockholmsmässan, Stockholm, Sweden, 10–15 July 2018. Proceedings of Machine Learning Research, vol. 80, pp. 3059–3068. PMLR (2018). http://proceedings.mlr.press/v80/liang18b.html

8. Mao, H., et al.: SEIHAI: a sample-efficient hierarchical AI for the MineRL competition. In: DAI 2021. LNCS (LNAI), vol. 13170, pp. 38–51. Springer, Cham (2022). https://doi.org/10.1007/978-3-030-94662-3_3

9. Mnih, V., et al.: Human-level control through deep reinforcement learning. Nature **518**(7540), 529–533 (2015). https://doi.org/10.1038/nature14236

10. Moy, G., Shekh, S.: The application of AlphaZero to wargaming. In: Liu, J., Bailey, J. (eds.) AI 2019. LNCS (LNAI), vol. 11919, pp. 3–14. Springer, Cham (2019). https://doi.org/10.1007/978-3-030-35288-2_1

11. OpenAI: OpenAI Five. https://openai.com/blog/openai-five/

12. Schrittwieser, J., et al.: Mastering Atari, go, chess and shogi by planning with a learned model. Nature **588**(7839), 604–609 (2020)

13. Schulman, J., Levine, S., Abbeel, P., Jordan, M.I., Moritz, P.: Trust region policy optimization. In: Bach, F.R., Blei, D.M. (eds.) Proceedings of the 32nd International Conference on Machine Learning, ICML 2015, Lille, France, 6–11 July 2015. JMLR Workshop and Conference Proceedings, vol. 37, pp. 1889–1897. JMLR.org (2015). http://proceedings.mlr.press/v37/schulman15.html

14. Schulman, J., Wolski, F., Dhariwal, P., Radford, A., Klimov, O.: Proximal policy optimization algorithms. CoRR abs/1707.06347 (2017). http://arxiv.org/abs/1707.06347

15. Silver, D., et al.: Mastering the game of Go without human knowledge. Nature **550**(7676), 354–359 (2017). https://doi.org/10.1038/nature24270

16. Surdu, J.B.: Wargaming: past, present and future. In: Rajaei, H., Wainer, G.A., Chinni, M.J. (eds.) Proceedings of the 2008 Spring Simulation Multiconference, SpringSim 2008, Ottawa, Canada, 14–17 April 2008, p. 619. SCS/ACM (2008). http://dl.acm.org/citation.cfm?id=1400549.1400647

17. Sutton, R.S., Barto, A.G.: Reinforcement Learning: An Introduction. MIT Press, Cambridge (2018)

18. Tregenza, M.: Coral sea 2042: Rules for the maritime/air analytical wargame (2018)

19. Vinyals, O., et al.: Grandmaster level in starcraft ii using multi-agent reinforcement learning. Nature **575**(7782), 350–354 (2019)

20. Vinyals, O., et al.: StarCraft II: A New Challenge for Reinforcement Learning (2017). https://deepmind.com/documents/110/sc2le.pdf, http://arxiv.org/abs/1708.04782

21. Ye, D., et al.: Mastering complex control in MOBA games with deep reinforcement learning. In: AAAI 2020: The Thirty-Fourth AAAI Conference on Artificial Intelligence (2020). https://academic.microsoft.com/paper/2996896271

22. Zhang, Q.: Research on Learning Behavior Modeling Methods for Decision Making of Computer Generated Forces (CGFs). National University of Defense Technology, Thesis (2018)

Time Optimal Control of Mini-submarine Missile Based on Deep Reinforcement Learning

Canhui Tao[✉], Zhiping Song, and Baoshou Wang

China Ship Scientific Research Center, Wuxi 214082, China
taocanhui@cssrc.com.cn

Abstract. In this paper, the time optimal control problem of a submarine launched missile is discussed and analyzed. Vertical separation and horizontal launch has become one of the key technologies of mini-submarine missile. The key problem is how to reach the vertical as quickly as possible. Deep reinforcement learning is a very popular algorithm recently, and it is considered as a promising method to solve sequential decision problems, especially when the model structure is poorly understood. We apply proximate policy optimization algorithm to control the missile. Through mathematical simulation experiments, the proportional differential controller and the deep reinforcement learning algorithm are compared, and the effectiveness of the deep reinforcement learning algorithm is shown.

Keywords: Missile · Deep reinforcement learning · Optimal control

1 Introduction

The miniature submarine-launched missile is a new type of underwater test vehicle launched by a certain submarine [1]. It consists of a large number of sensors that can acquire data when the water comes out vertically. The propeller, strapdown inertial measurement unit (SIMU) and control swing nozzle are mounted on the hull, so the vehicle has great maneuverability. Typically, the experimental process is completed in a manual test tank. Considering the limited space in the experimental tank, the main task is to quickly pull the missile to a vertical position when it is launched horizontally underwater.

In recent years, reinforcement learning method has received extensive attention [2], which is mainly used to solve sequential decision problems. Inspired by the trial-and-error method in animal learning, reinforcement learning takes the reward value obtained by the interaction between the agent and the environment as a feedback signal to train the agent. Deep learning has a strong perceptual ability, which has even exceeded the human perceptual level in some application scenarios [3]. Deep reinforcement learning (DRL) is based on the strong perceptive ability of deep learning to deal with complex and high-dimensional environmental characteristics, and combines the idea of reinforcement learning to interact with the environment to complete the decision-making process. It is considered as a promising method to solve sequential decision problems, especially when the model structure is poorly understood [4].

© The Author(s), under exclusive license to Springer Nature Singapore Pte Ltd. 2022
W. Fan et al. (Eds.): AsiaSim 2022, CCIS 1713, pp. 573–579, 2022.
https://doi.org/10.1007/978-981-19-9195-0_46

2 Problem Formulation

Figure 1 shows the coordinate relationship between the missile and the tank. During launch, the coordinates $ox_by_bz_b$ of the missile are consistent with the coordinates $ox_py_pz_p$ of the pool. In order to avoid collision with the wall of the experimental tank, the missile needs to be vertical as soon as possible, which means that the pitch angle must be controlled to 90° as soon as possible. Therefore, the control problem can be regarded as a time optimal regulation problem.

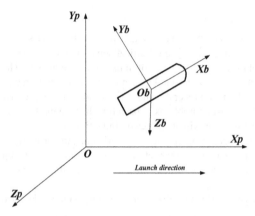

Fig. 1. Coordinate relationship

In this paper, only the longitudinal plane is considered, and the yaw and roll motions on the horizontal plane are not considered. Then, the kinematic model of the missile can be obtained as follows:

$$\begin{cases} \dot{\phi} = r \\ \dot{X} = u\cos\phi - v\sin\phi \\ \dot{Y} = u\sin\phi + v\cos\phi \\ V_T = \sqrt{u^2 + v^2} \\ \alpha = \arctan(-v/u) \end{cases} \tag{1}$$

where X, Y denote position in a longitudinal plane, ϕ denotes pitch angle, V_T denotes velocity of the missile, r and (u, v) denote angle velocity and velocity of the missile's $Ox_by_bz_b$, and α denotes attack angle.

The dynamic equations of the model can be simplified as:

$$\begin{cases}
(m + \lambda_{11})\dot{u} - (m + \lambda_{22})vr - \lambda_{26}r^2 = F_x \\
(m + \lambda_{22})\dot{v} + \lambda_{26}\dot{r} + mur = F_y \\
(J_{zz} + \lambda_{66})\dot{r} + \lambda_{26}\dot{v} = N_z \\
F_x = -mg\sin\phi + F_B\sin\phi + 0.5\rho V_T^2 S \cdot C_x + T\cos\delta_e \\
F_y = -mg\cos\phi\cos\gamma + F_B\cos\phi\cos\gamma + 0.5\rho V_T^2 S \cdot C_y + T\sin\delta_e \\
N_z = F_B(x_B\cos\phi\cos\gamma - y_B\sin\phi) + 0.5\rho V_T^2 SL \cdot C_N - T\sin\delta_e L_T \\
C_x = C_x \\
C_y = C_y^\alpha \alpha + C_y^r(rL/V_T) \\
C_N = C_N^\alpha \alpha + C_N^r(rL/V_T)
\end{cases} \qquad (2)$$

where m denotes mass, $(\lambda_{11}, \lambda_{22}, \lambda_{26}, \lambda_{66})$ denote added mass, (F_x, F_y) denote dynamic force, N_z denotes moment, J_{zz} denotes inertia, F_B denotes the buoyance, (x_B, y_B) denote the distance from the buoyant center to the barycenter, L_T denotes the arm of force, T denotes thrust, L and S denotes the length and the cross-sectional area of the missile respectively. $(C_x, C_y, C_N, C_y^\alpha, C_y^r, C_N^\alpha, C_N^r)$ denote hydrodynamic force coefficients. δ_e denotes thrust vector deflection angle, which is the actuator.

3 Deep Reinforcement Learning

DRL algorithms are mainly divided into two categories: value function algorithm and strategy gradient algorithm. The value function algorithm indirectly obtains the agent's strategy by iteratively updating the value function. When the value function iteration reaches the optimal value, the agent's optimal strategy is obtained by the optimal value function. The strategy gradient algorithm directly uses the method of function approximation to establish the strategy network, and obtains the reward value by selecting the action of the strategy network, and optimizes the parameters of the strategy network along the gradient direction to obtain the optimized strategy maximization reward value. In the application scenario of the algorithm, the value function algorithm needs to sample actions, so it can only deal with discrete actions. The strategy gradient algorithm directly uses the strategy network to search actions, and can be used to deal with continuous actions.

The main control issue of this paper concentrated on the Proximal Policy Optimization (PPO). PPO uses the strategy function and value function approximator constructed by multi-layer artificial neural network to parameterize and optimize a large number of decision-making and state spaces [5]. PPO is an on policy algorithm with good performance. It is a general and reliable reinforcement learning algorithm, and it is also a kind of policy gradient algorithm.

PPO uses two neural networks. By inputting the current state of the intelligent body into the neural network, the corresponding action and reward will finally be obtained. Then, the state of the intelligent body will be updated according to the action. According to the objective function containing reward and action, the weight parameters in the neural network will be updated by gradient rise, so as to obtain the action judgment that makes the overall reward value larger.

Take the states as $s = [\ u\ v\ r\ X\ Y\ \phi\ ]$, and action as $a = \delta_e\ (-8° \le \delta_e \le 8°)$, then the state is updated according to the dynamic Eq. (2).

Define the reward as

$$R = -(1 * r^2 + 10 * (\phi - \frac{\pi}{2})^2)$$ (3)

So The update of the environment can be described as follows:

$$s', R, done = env.step(a)$$ (4)

4 Mathematical Simulation

By taking the simulation of missile launching process control, the effect of the proposed scheme will be show. And a conventional PD control strategy is adopted as a comparison. At the simulation, the initial velocity is 15m/s, that is $s = [\ 15\ 0\ 0\ 0\ 0\ 0\ ]$, and the PD control parameters Kp = 20.4, Kd = 9.0.PPO is coded in python. The PD control and the final comparison is code in Matlab2019.

Set the time-step as 0.005 s. After training with 300000 time steps, the training results are shown in Fig. 2,

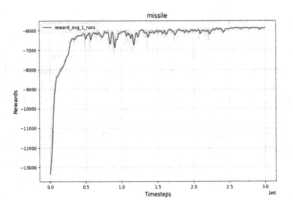

Fig. 2. Change of reward during training

Then test the trained model. The total rewards are −5443.9, The minimum time consumption is computed as 3.18 s.

The comparative performance is shown in Fig. 3 and Fig. 4. Compared with PD, the deep reinforcement learning method can provide the missile with faster vertical separation from water. Pitch can reach 90° faster, and the angular rate at this time is also closer to 0.

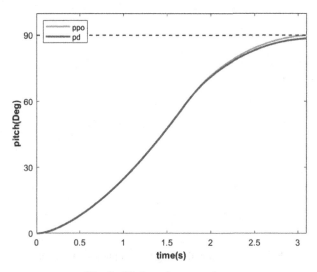

Fig. 3. Pitch angle comparison

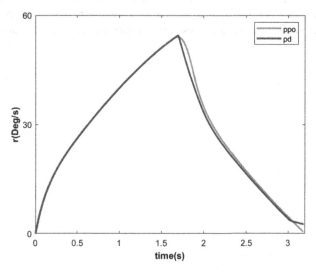

Fig. 4. Angular rate of r comparison.

Although the proposed optimization scheme is a prior and off-line, the process of leaving the water surface is a short time operation that only takes a few seconds. It is known that the depth of the experimental tank is approximately equal to 40 m and the length is 60 m. Therefore this pre-training scheme is feasible. The comparative performance of the change of the missile's position in the longitudinal plane when it exits the water is shown in Fig. 5. The comparative results show that PPO has obvious advantages.

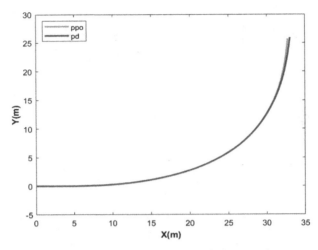

Fig. 5. Position in the longitudinal plane comparison

Finally, the comparison result of control variable is shown in the Fig. 6. The output control variable of PPO is similar to the Bang-Bang control issue which has only two control states, but the switching process from negative to positive is relatively smooth.

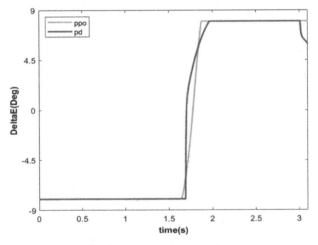

Fig. 6. Thrust angle comparison

5   Conclusion

In this paper, a method based on deep reinforcement learning is used to realize the rapid vertical separation of missiles. By transforming the motion state of the missile into the state quantity in reinforcement learning and the control quantity into the action, the deep reinforcement learning method PPO is used to solve the control problem. To a certain extent, the missile has obtained a faster water velocity. Although this method can not be implemented in real time, the pre-trained model is also robust to new unfamiliar environments. Therefore, it is also expected to be used in actual scenarios in the future.

References

1. Wen, N., Liu, Z., Le, C., et al.: Time optimal control of mini-submarine missile based on control variable parameterization with enhanced time-scaling method. In: 2016 IEEE Chinese Guidance, Navigation and Control Conference (CGNCC). IEEE (2016)
2. Sutton, R.S., Barto, A.G.: Reinforcement Learning: An Introduction. MIT Press, Cambridge (2018)
3. Lecun, Y., Bengio, Y., Hinton, G.: Deep learning. Nature **521**(7553), 436 (2015)
4. Vanvuchelen, N., Gijsbrechts, J., Boute, R.: Use of Proximal policy optimization for the joint replenishment problem. Comput. Ind. **119**, 103239 (2020)
5. Ying, C.S., Chow, A., Wang, Y.H., et al.: Adaptive metro service schedule and train composition with a proximal policy optimization approach based on deep reinforcement learning. IEEE Trans. Intell. Transp. Syst. **PP**(99), 1–12 (2021)

Modeling/Simulation Applications in Education and Training

Research on Interactive Electronic Manual Based on VR Technology

Ma Yongqi[1]([✉]), Cheng Xun[2], and He Qiyun[1]

[1] Institute of Computer Application China Academy of Engineering Physics, Mianyang 62100, Sichuan, China
myq123456a@caep.cn

[2] University of Electronic Science and Technology of China, Chengdu 610000, Sichuan, China

Abstract. Based on traditional IETM and VR technology, the concept of "VR + Manual" is proposed, and the main function design and four-layer architecture design of "VR + Manual" are given. The digital mapping data model is established between the IETM technical data and the digital prototype through the ontology data modeling method, and the seamless connection between the data model and the "digital prototype modeling and simulation engine" is realized. By using the integrating multi-class information visualization method, the visualization of multi-dimensional equipment technical status based on the digital prototype is realized. A data-driven interactive interface generation engine is proposed, which is oriented to five types of IETM data, such as description, procedure, fault, illustrated Parts, and maintenance plan. It realizes the rapid generation of interactive interfaces and the release of the application.

Keywords: IETM · VR · Data model · Visualization · VR + Manual · Framework · Engine

1 Introduction

The interactive electronic technical manual (IETM) uses digital technology to express traditional paper documents, integrates multimedia, database, human-computer interaction, and other technologies, and provides digital display forms such as text, graphics, images, audio, and video, three-dimensional models [1]. It has the functions of visual browsing and interactive operation, and assists in guiding the maintenance, repair, training, and training of weapons and equipment. It has the advantages of diverse content expression forms, efficient retrieval, and data reuse. It can be used on ordinary computers, laptops, tablets, wearable devices, and embedded computers [2]. According to the American military IETM classification, the display style of IETM from level 1 to level 5 is mainly text and graphics, which can not provide users with an immersive, multi-perception, interactive and imaginative experience. In recent years, with the vigorous development of virtual reality (VR) technology, "VR + Games", "VR + Retail", "VR + Education", "VR + Landscape design", "VR + Live broadcast", "VR + Product display" and other products have emerged accordingly [3]. After consulting relevant

© The Author(s), under exclusive license to Springer Nature Singapore Pte Ltd. 2022
W. Fan et al. (Eds.): AsiaSim 2022, CCIS 1713, pp. 583–592, 2022.
https://doi.org/10.1007/978-981-19-9195-0_47

materials, it was found that the concept of "VR + Manual" has not been put forward at present, and few scholars have studied this field. This paper will study how to combine VR and IETM technology to realize "VR + Manual" and put forward personal thoughts from the aspects of data modeling, visualization, and interactive interface, provide some ideas and methods for practitioners and enthusiasts, and contribute a little to the development of IETM technology. As a new attempt, "VR + Manual" completely subverts the traditional manual reading and brings users an all-around visual expression.

2 Related Technologies

2.1 IETM Technology

As a technical means of equipment maintenance support, IETM was produced and developed under the traction of information technology development and military requirements. In terms of standards, there are general international standard systems, such as ASD/AIA/ATA S1000d international specification for technical publications based on a public source database [4]. These standards and specifications regulate IETM data, mainly including data module management, information set design, data module requirements list, data module preparation, illustration, multimedia design, engineering business rules, publication design, and data exchange. In the standard coding system, the standard numbering system (SNS) is divided according to the structure and function of weapons and equipment, which is mainly composed of systems, subsystems, subsystems, and units [5]. When information is generated, the rules of the data module and information set are constructed, the rule of partition and access point is determined, and the creation rules of text, illustration, and multimedia are given.

IETM interactivity refers to the ability of humans and computers to acquire information and knowledge using man-machine dialogue. IETM needs two processes to realize human-computer interaction. First, is the interaction between users and devices [6], that is, the process in which users send operating instructions to computers through the mouse, keyboard, and display screen. The interface size, display layout, color use, font style, and display style of the IETM system are required to meet the requirement of ergonomics. Second, the interaction between the interactive interface and the data in the technical information base [6], that is, the IETM system organizes the data to display to the user according to the functional operation instructions of the interactive interface. The interactive interface functions include access, annotation, transmission and release, diagnosis and prediction, icon function, link, navigation, and tracking. The interactive visual interface display has a navigation panel, text, dialog box, list, steps and procedures, hyperlink, warning and attention, illustration, audio and video, and animation.

The content range of IETM varies according to equipment characteristics, complexity, and business requirements. It usually includes system technical specifications, operation manual, and maintenance manual. The system technical specification mainly describes the operational purpose, composition, tactical and technical performance, use mode, working principle, and technical parameters of the equipment system. The operation manual provides instructions on the user knowledge and operation skills of equipment or subsystem and equipment to help operators master the equipment use methods and operation procedures. The maintenance manual is the guiding technical data for equipment maintenance and repair. It has the maintenance outline, the maintenance procedure, fault report, fault isolation, and illustrated Parts Catalogue according to the information category.

2.2 Virtual Reality Technology

Virtual reality (VR) includes the computer, electronic information, and simulation technology. It is a computer simulation system that can create and experience a virtual world. It uses computers to generate a simulation environment and immerse users in the environment [7]. Virtual reality technology uses the data in real life and the electronic signals generated by computer technology to combine them with various output devices to transform them into phenomena that people can feel. These phenomena can be objects in reality or objects we can't see with our naked eyes [7]. Its core features are immersion, interaction, and imagination. After the development of virtual reality technology 1.0 - concept germination, Virtual Reality Technology 2.0 - technology exploration, virtual reality technology 3.0 - breakthrough development, virtual reality technology 4.0 - industrial application, the key technologies of virtual reality such as 3D modeling technology, 3D reality technology, 3D audio technology, and somatosensory interaction technology have been successfully tackled [8]. This technology has been applied in the fields of national defense and military [9], education and training [10], medical care [11], industrial manufacturing [12], and entertainment culture [13].

VR system usually consists of software and hardware. The software part includes two-dimensional graphics, three-dimensional modeling, VR engine, VR auxiliary development tools, and other software, and the hardware part includes input devices and output devices. It provides a resource environment for software operation and supports the construction of a three-dimensional virtual environment. The input device is the bridge between the user and the computer. The common devices are joystick, data glove, and motion tracker. The output device presents the virtual environment to the user

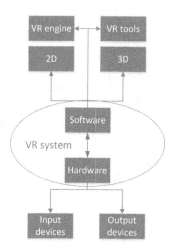

Fig. 1. VR system composition

through vision, hearing, or touch. The common devices are glasses, headphones, and head-mounted displays. The composition of the VR system is shown in Fig. 1.

"VR + Manual" refers to an immersive manual reading experience through processing wearable sensor devices (VR helmets) with the help of VR technology. When reading, the user displays digital weapons and equipment in the virtual world. The manual navigation mode changed from traditional directory navigation or SNS navigation to 3D model product structure navigation. The user cares about where the weapons and equipment are, understands the structure and principle of the equipment by clicking on the hot connection, and carries out equipment disassembly and assembly training through the virtual assembly. Determine the fault location according to the fault code, and the fault tree is generated based on prior knowledge to assist the reader in troubleshooting.

3　"VR + MANUAL" Design and Technology Research

3.1　Functional and Architectural Design

The "VR + Manual" needs to establish an association model between digital equipment and IETM technical data to meet the application scenarios of visual technical status, illustrated Parts, operation training, fault isolation, maintenance, and repair. In the virtual world, the digital equipment is a digital prototype based on the three-dimensional model, which can realize the reading and interactive operation functions of technical instructions, operation manuals, maintenance manuals, fault isolation manuals, illustrated Parts, and components around the digital prototype. The specific functions show in Fig. 2.

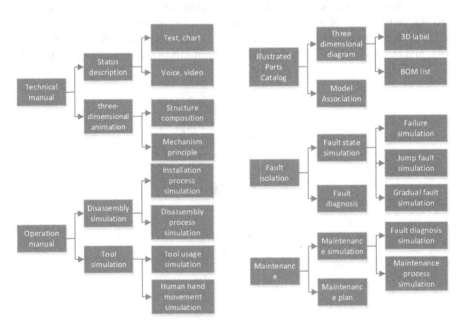

Fig. 2. "VR + Manual" function

The technical manual provides multiple expressions such as 3D models, 3D animation, text, chat, voice, and video. The operation manual provides three-dimensional model disassembly and assembly simulation. The illustrated Parts Catalogue manual supports automatic generation and interactive browsing of illustrations and the product lists, and rapid generation and interaction of product structure explosion diagrams. The fault isolation manual is based on the graphical retrieval and navigation of the fault tree and supports the simulation and treatment of various types of faults. The maintenance manual provides maintenance plan formulation, retrieval management, maintenance parts management, maintenance process guidance, and simulation. From the perspective of domain modeling, the "VR + Manual" function module should include manual production, manual release, and manual reading. The relationship between modules shows in Fig. 3.

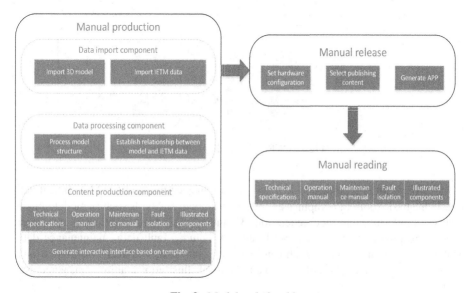

Fig. 3. Module relationship

The manual-making module includes a data import component, a data processing component, and a content-making component. During production, the data import component receives the three-dimensional model of the digital prototype and IETM technical data. The data processing component receives the technical data, processes the model structure, establishes the relationship between the model and IETM technical data, and makes the technical instructions, operation manuals, maintenance manuals, fault isolation manuals, and illustrated Parts catalog manuals based on the template. When the manual is released, set the hardware configuration and select the release content to generate a runnable application. When reading the manual, run the corresponding application program.

According to the relationship between modules and the function of each module in the "VR + Manual", the architecture is designed. The "VR + Manual" architecture includes the foundation, data, business, and application layer, as shown in Fig. 4.

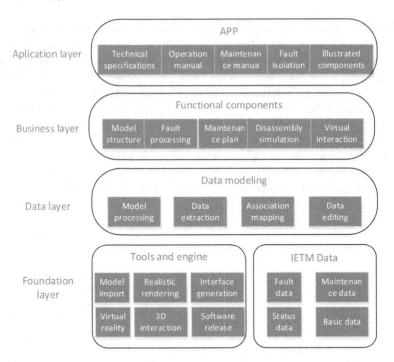

Fig. 4. "VR + Manual" Architecture

The basic layer includes basic tools and graphics engine and IETM technical data set. In the basic tools and graphics engine, environment and tools such as model import, realistic rendering, interface generation, 3D interaction, software release, and virtual reality are provided. IETM technical data set includes fault data, maintenance data, status data, and other data, which can be divided into text, graphics, two-dimensional technical illustrations, three-dimensional technical illustrations, animation, voice, and video.

The data layer centrally processes the digital prototype model data and IETM technical data, establishes the relationship between the digital prototype model data and IETM technical data, extracts the data in the IETM technical data, and superimposes them on the three-dimensional model through data editing.

The business layer encapsulates the underlying business function components, including model structure, fault handling, maintenance plan, disassembly simulation, virtual interaction, and others, to provide services and support for the application layer.

The application layer customizes various application apps according to business needs to meet the needs of different application scenarios. The application includes technical instructions, operation manuals, maintenance manuals, fault isolation manuals, and illustrated Parts Catalogue manual. These manuals can support graphic parts display, operation training, fault detection and handling, and maintenance.

3.2 Data Modeling Based on Ontology

The ontology mapping method [14] establishes the mapping relationship between ontologies and data. It solves the data integration problem of semantic heterogeneity, realizes the digital prototype modeling of the CAD design model of weapons and equipment, realizes the data mapping based on product structure relationship and the technical data mapping based on product life cycle, and generates the association model between the digital prototype and technical data. This association model can be called the "VR + Manual" data model. The data model includes description information, parts catalog, disassembly and assembly steps, fault information, and maintenance plan, as shown in Fig. 5.

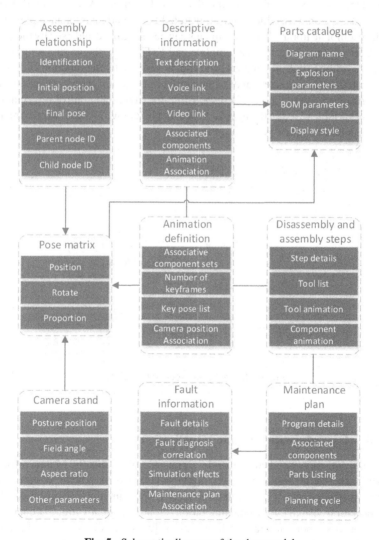

Fig. 5. Schematic diagram of the data model

The description data includes 3D models, 3D interactive animation, text, picture, voice, video, and other data types. The procedure data includes the rapid production of 3D model disassembly and assembly simulation animation, assembly steps, and simulation process data. The fault data includes fault trees, various types of fault simulation, fault handling, and maintenance guidance data. The illustrated Parts and components include diagrams and product lists, the visual configuration of display styles, and exploded view data of product structure. The repair plan category includes data such as repair plans, repair parts, and repair status.

3.3   Data-Driven Interactive Interface Generation Engine

Develop interface modules for five interfaces of Electronic Manuals in IETM [15], including description, procedure, fault, illustrated Parts, and maintenance plan, to form an interface template library. These templates can be recognized and used by the interface engine. After reading the interface template and IETM data, the engine will layout the interface according to this information, and add the event listening function of the control. Each generated interface corresponds to an interface generation engine object. In addition to the interface, the engine object is responsible for saving all control information and some global variables used in the interface for event listening. The data-driven interactive interface generation engine is shown in Fig. 6.

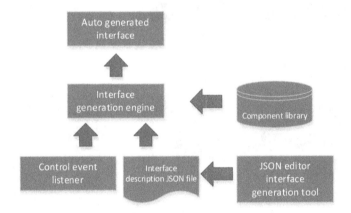

Fig. 6. Interface auto-generation engine

3.4   Multi-class Information Fusion Visualization

Based on the data model and customized interface template, this IETM data information is superimposed on the digital prototype for visual expression. The expression methods include multi-level explosion diagram, interactive 3D animation, 3D technical illustration, information visualization, 3D annotation, cloud-image, texture, 2D chart, image, and hot spot. It supports interactive browsing of five types of electronic manual data, including description, procedure, fault, illustrated Parts, and maintenance plan.

The description data integrate various types of information such as 3D model, 3D interactive animation, text, picture, voice, and video, and provides visual interfaces such as 3D component structure tree, 3D interactive animation, text display, picture editing, and display, voice and interactive display.

The procedure data provides visual interfaces such as rapid production of 3D model disassembly and assembly simulation animation, management of assembly steps and simulation animation, management of disassembly and assembly tools, and simulation and use of disassembly and assembly tools.

The fault data provides a graphical retrieval and navigation interface based on the fault tree, the interactive response between the fault tree and scene events, and effect simulation of common fault types (such as high temperature, fire, fracture, and damage), association mapping management interface between fault tree and fault handler.

Illustrated parts data include automatic generation and interactive browsing of diagrams and product lists, visual configuration of display styles of charts product lists, rapid generation, and interactive interface of product structure explosion diagrams.

The maintenance plan category data includes the plan creation management interface, the visual retrieval of the maintenance plan based on the three-dimensional view, the maintenance spare parts management interface, and the maintenance process simulation interactive operation and management interface.

4 Conclusion

According to the digital needs of weapon equipment integrated support, this paper puts forward the concept of "VR + Manual" based on traditional IETM and VR technology and realizes the electronic manual of equipment technical description, operation training, maintenance, and fault diagnosis through full three-dimensional simulation technology. Put forward the main function design and four-layer architecture design of "VR + Manual", put forward three technical solutions, and use the ontology-based data modeling method to build the "VR + Manual" data model, to realize the seamless connection between the data model and the digital prototype modeling and simulation engine. The visualization method of multi-class information fusion realizes the visualization of multi-dimensional equipment technical status based on the digital prototype. The automatic generation method of interactive interface based on data driving is adopted to realize the rapid generation of interactive interface and the release of the application. No one can refuse something new and interesting. Applying VR virtual reality to IETM can not only improve their boring training, but also attract their attention to promote their interest in learning.

Compared with traditional IETM, the "VR + Manual" enables users to observe and control equipment more vividly and vividly, understand the internal structure and technical mechanism of equipment, improve the maintenance and support efficiency of equipment, and make equipment desktop-level virtual training a reality. The "VR + Manual" can be used for training and training. It can not only shorten the training cycle and improve the training quality but also save money and the loss of real equipment, so that operators and maintenance support personnel are no longer affected by time, site and environment. It also provides a new training method for operators and can effectively solve the problem of low cost-effectiveness of equipment training. Because the "VR + Manual" uses virtual reality technology, there are the following disadvantages in the use process, First, people feel dizzy and nauseous, and their comfort is poor. For example, the glasses are airtight, the lenses are fogged, and the human brain and nervous system are affected to cause dizziness. Second, the wire is bound and easily tripped by the wire. Third, the user experience cost is relatively high and the purchasing power is insufficient.

References

1. Ma, Y.Q., Meng, L.R., Yu, J.: Research on PDF document publishing model in IETM. Microcomput. Its Appl. **36**(24), 87–91 (2017)
2. Ma, Y.Q., Chen, Q.G., Cheng, Z.: Design of IETM tablet reader based on android. Comput. Inf. Technol. **29**(5), 5 (2021)
3. Hu, K., Xiong, W.: Six industry application opportunities for VR. Enterprise Management (9), 2 (2017)
4. Wan, J.Z.: Study of development IETM based on S1000D specification. Appl. Mech. Mater. **190–191**, 249–252 (2012)
5. Jing, H.: Standard IETM development based on S1000D. Manufacturing Automation (A6), 3 (2010)
6. Fei, C.: Research on Key Technologies of IETM Based on S1000D specification. Civil Aviation University of China (2017)
7. Zhou, N.-N., Deng, Y.-L.: Virtual reality: a state-of-the-art survey. International Journal of Automation & Computing (2009)
8. Yang, Q., Zhong, S.: A review of foreign countries on the development and evolution trends of virtual reality technology. J. Dialectics Nat. **43**, 97–106 (2021)
9. Jung, K., Lee, S., Jeong, S., et al.: Virtual tactical map with tangible augmented reality interface. In: International Conference on Computer Science & Software Engineering. IEEE Computer Society (2008)
10. Hodgson, P., Lee, V.W.Y., Chan, J.C.S., et al.: Immersive virtual reality (IVR) in higher education: development and implementation (Revised selected paper) (2020)
11. Wiederhold, B.K., Riva, G.: Virtual reality therapy: emerging topics and future challenges. Cyberpsychol. Behav. Soc. Netw. **22**(1), 3–6 (2019)
12. Liagkou, V., Salmas, D., Stylios, C.: Realizing virtual reality learning environment for industry 4.0 - ScienceDirect. Procedia CIRP **79**, 712–717 (2019)
13. Yoo, S., Parker, C., Kay, J.: Adapting data from physical activity sensors for visualising exertion in virtual reality games. In: Proceedings of the 2018 ACM International Joint Conference and 2018 International Symposium. ACM (2018)
14. Wang, S., Kang, D., Jiang, D.: Survey of ontology mapping. Comput. Sci. **44**(9), 10 (2017)
15. Ma, Y., Meng, L., Cheng, X., et al.: Design of IETM-oriented display style framework. In: 2019 Chinese Automation Congress (CAC) (2019)

Survey on Sharing Technology and Applications of Intelligent Manufacturing Training Equipment Based on Industrial Internet and Man-in-Loop Simulation

Jiaxin Luo[1], Tianhong Lan[1], Tan Li[1(✉)], Song Weining[2], Chen Nanjiang[1], Lin Yanwen[3], Li Runqiang[4], Liu Hairui[5], and Hua Yanhong[6]

[1] Nanchang University, Nanchang, Jiangxi, China
someone8584@sina.com
[2] East China University of Technology, Nanchang, Jiangxi, China
[3] Nanchang Research Institute of Sun Yat Sen University, Nanchang, Jiangxi, China
[4] QuickTech Co., Ltd., Yizhuang, Beijing, China
[5] China Academy of Information and Communications Technology, Haidian, Beijing, China
[6] Jiangxi Vocational College of Mechanical and Electrical Technology, Nanchang, Jiangxi, China

Abstract. High utilization and Precise management of practical training equipment are urgent demands for those assets-holders like vocational skill-training colleges and manufacturing enterprises. New technologies like Industrial Internet and Man-in-Loop Simulation facilitate the connection and virtual training of the industrial equipment, which enable the sharing of Intelligent Manufacturing Training Equipment (IMTE). Comparing with industrial equipment and experimental instruments, some unique key features of IMTE are summarized, such as functional versatility, man-machine security, fault tolerance and easy reproducibility, etc. The technical demands of IMTE sharing are discussed, and the technical architecture of IMTE sharing system is built based on industrial internet and Man-in-Loop Simulation. Key technologies and related applications are reviewed, providing some potential research points and enabling technologies for the future implementation of IMTE sharing system to improve the reusage and management.

Keywords: Training equipment · Training informatization · Industrial internet · Intelligent manufacturing

1 Introduction

Vocational education and skill training play a rising role in recent Chinese economy and society, which generate enormous high-tech human resource for the manufacturing industry and support the great national project of "Made-In-China 2025". In 2019, the

State Council of China released the "National Vocational Education Reform Implementation Plan" [1], which clearly proposes to build a number of high-level vocational education training base that integrates resource sharing services, campus practical teaching services and social technical training services.

Practical training education, especially the training on Intelligent Manufacturing Training Equipment (IMTE) is becoming increasingly important in China and almost every vocational school has made considerable investment to construct training base/laboratory full of IMTEs such as industrial robots, PLC control cabinet, AGVs etc. Nevertheless, problems arise after the great construction on the school side, such as low utilization rate, poor management efficiency of those IMTEs in vocational education colleges and universities, whereas on the industry side those "idle" training equipment are of great usage for the skill improvement of the manufacturing workers. Thus the "sharing economy" of IMTEs from vocational schools to manufacturing workers becomes the promising solution for the problems above.

With the rapid technical development of industrial internet [2] and virtual simulation [3], the online sharing of massive distributed IMTEs become feasible, supporting the connection and data acquisition on the Edge, the equipment status management and analysis in the Cloud, as well as the virtual training and time renting through the Sharing Platform. This paper proposes the new service model of the IMTE sharing based on Industrial Internet and Man-in-Loop Simulation, and surveys the related technologies and applications.

Firstly, we compare IMTE with traditional equipment like industrial equipment and experimental instruments and summarizes some characteristics inherent to the training equipment sharing model. Then we use the characteristics summarized in the previous chapter to construct the architecture of the training equipment sharing system, and the hierarchical relationships and functions of the overall architecture are described. The key technologies and applications of the industrial Internet-based IMTE sharing are reviewed, including edge computing, Man-in-Loop simulation, IoT modeling, data analysis and mining, as well as the man-machine security. Some potential research points and enabling technologies are given at the end for the future implementation of IMTE sharing system to improve the reusage and management.

2   Unique Features and Technical Demands of IMTE

The National Vocational Education Reform Implementation Plan proposes the five-year target to "Create a Batch of High-level Training Bases". And the specific construction index is proposed for the "Construction of 50 high -level vocational schools and 150 backbone majors" [1, 4]. According to the data released by China Education Procurement Network in October 19, 2021, "In September 2021, the number of school training equipment procurement projects reached a new high-period increase of 33%" [5], which shows that the total amount of intelligent manufacturing training equipment in colleges and universities is growing rapidly.

At the same time, the issues such as low equipment utilization, incomplete equipment management, and low levels of informatization are increasingly apparently. Chunqin Xia made effective attempts on the problems of low equipment utilization and low informationization [6]. Hongyan Ke discussed introducing the socialized agencies services to promote the use of scientific instruments and equipment [7]. On the one hand, it can be found that information such as Overall Equipment Effectiveness (OEE) and equipment service time has become a necessary demand for colleges and universities to manage equipment and make a reasonable overall plan for the laboratory [7]. On the other hand, to prevent equipment damage, many companies need to use training equipment before employees using industrial equipment. By using the Intelligent manufacturing training equipment sharing mode based on industrial Internet, it can not only solve the current equipment management and enterprise needs under the purpose of improving the efficiency of equipment, but also promote the high-quality development of vocational and technical education.

2.1 Comparative Analysis of Equipment

Normally equipment is divided into three types, which are the experimental instruments, industrial equipment, and training equipment like IMTEs. Among them, experimental instrument equipment sharing and industrial site equipment on the cloud in the industry are relatively mature. For instance, Xia Zhu analyzed the current status of scientific research instruments and equipment in colleges and universities, and proposed corresponding measures to improve management level [8]; Honghao Gao proposed "the energy consumption monitoring platform based on IPv6 campus" for effective management of colleges and universities instruments and equipment [9]; Wenbin Jiao built a scientific information application platform, which has been fully applied in more than one hundred research institutes of the Chinese Academy of Sciences [10]. However, the research on the sharing of training equipment, especially the IMTEs, is still in its infancy. This chapter will analyze the characteristics of the three types of equipment to study effective technical solutions applicable to the sharing of clouds in intelligent manufacturing training equipment. The characteristics of these three types of equipment are compared and analyzed as shown in Table 1.

2.2 Technical Demands for the Unique Features of IMTE

Integrating Internet and Artificial Intelligence technology, devices sharing on cloud through "Internet +" and "AI+" has become the mainstream choice in the industry [11]. This paper discusses the sharing of IMTE on the cloud based on industrial internet. Through the comparison of the previous section, it can be concluded that high universality, high security, high fault tolerance, easy reproducibility, low utilization rate are the important characteristics of IMTE sharing. Therefore, one should build corresponding functions for its special needs based on the industrial Internet technology system. The followings are the description for the characteristics and functional requirements of IMTE sharing.

High Versatility

An intelligent training equipment needs to provide as many training contents as possible to support students' training, so it is required to cover as many industrial categories as possible. Different hardware modules need to correspond to different Model of Things (MoT). The MoT is a model that abstracts the physical model into a digital model and expresses it on cloud [13]. Intelligent manufacturing training equipment generally includes motor module, electrical module, control module, intelligent robot module, vision module, sensor module, mechanical module, safety protection module and so on. Therefore, the high versatility requires that the intelligent manufacturing training sharing equipment technology framework can support the effective modeling and description of Mot and event instructions of multi-disciplinary modules. In order to meet the characteristics of the universality requirements, this framework can support the effective modeling and description of MoT and its event instructions in variety discipline modules.

High Security

For the users of the equipment are lack of operating experience, the intelligent manufacturing training equipment is highly required in safety for equipment and human. On the one hand, students can safely operate and be familiar with the equipment in the simulation environment, and then carry out practical operation after improving their operation level and experience. Therefore, the intelligent manufacturing training sharing technology framework is needed to realize the establishment of information models and holographic uplink of equipment data to support virtual simulation. On the other hand, before the instruction goes down, the edge computing system security protection system needs to make a security judgment on the instructions from the cloud, and then decide whether to go down to protect the system. For example, it can recognize that the next operation of the operator may lead to action track interference, or sudden power failure during work may cause different dangers. Besides, the misoperation needs to be solved offline, so the system security assessment should cover three layers: the edge layer, the platform layer, and the application layer.

High Fault Tolerance

The typical users are college students, who are basically in a state of complete inexperience in the operation of the actual equipment, so there is a huge risk of misoperation: such as encountering the power supply in an unexpected situation, or mistakenly pressing the reset button or other misoperation during the lease operation. When users misoperate, the system needs to be able to recognize and make right judgments. For example, the system needs to judge whether to accept interrupt or reset instructions when the system is executing key command actions. Because the cloud device is not completely aware of some special situations of the under-cloud device, and to prevent this misoperation reduce the reproducibility of the training equipment and get a better user experience, it is required that when receiving downlink instructions, the edge side device can carry out edge side security identification of the cloud instructions, judge the current device status, mark some dangerous instructions, and decide whether to follow the instructions.

Easy Reproducibility

Easy reproducibility means that students can review their operations during the training, and a common template case makes it easier for students to master the training equipment operation process. During the training period, the practice level of the students varies. For students with poor hands-on ability, the system needs to consider case modules that are more friendly to students, such as one-click automatic operation of module cases, or step-by-step running programs, so that students can quickly get started and use the

Table 1. Equipment comparison (★ refers to the relative level among the three types)

		Experimental Instruments		Industrial Equipment		IMTE	
Precision		Highest precision to complete science experiments	★★★	Lower precision but assure product quality	★★	Lowest precision, just meet the training needs★	★
Specificity		Highly customized functions for specific research	★★★	Certain specificity for large scale production	★★	Various learning contents. Practice more with one IMTE	★
Functional scalability		Reserved interface, hard to expand functions	★	Reserved interface, easy to upgrade and add functions	★★	Module structure, easy to expand new functions	★★★
Stability		High stability over time to ensure the precision	★★★	High stability to ensure consistent production quality	★★★	General functions, providing repeatable training course	★★
Efficiency	Operation efficiency	Precision is much more important than efficiency	★	Rapidest operation response while production	★★★	Smooth operation and easy learning for the trainee	★
	Usage efficiency	5*8, low utilization ratio based on the research task	★	Normally uninterrupted work through 7*24	★★★	5*8, even less on vacation	★
Fault tolerance		Strict experimental procedures, no fault tolerance	★	Strict operation procedures, pre-training required	★★	Highly fault tolerance for misoperation from beginner	★★★

(*continued*)

Table 1. (*continued*)

		Experimental Instruments		Industrial Equipment		IMTE	
Security	Machine Security	Complex operation, experience required for security	★	Specific equipment security protection mechanism	★★	Full protection mechanism to avoid misoperation damage	★★★
	Operator Security	Laboratories may work at high risk in certain tasks	★	Operators may work at risk in certain environment	★★	Trainee Security more important than equipment	★★★
Reproducibility		High uncertainty from materials and environment	★★	High uncertainty from environmental nosie in factory	★★	Follow the course and repeat the tasks to train	★★★

cases to realize the automatic training production line operation. For students who are familiar with equipment operation and can quickly master basic operations, they can quickly skip the beginner stage and enter a deeper level of practical training, such as the practical case of C, C++ and other advanced programming teaching training equipment control provided by the system.

Low Utilization Rate

In colleges and universities, the IMTE is usually arranged and used during the student's class, that is, 5*8 working hours. Except that, the rest of the time is almost completely idle. At present, teachers do not have a detailed grasp of equipment utilization, students' learning rate, students' learning progress, etc. It is difficult for school to carry out detailed management of equipment and scientific course scheduling, and it is difficult to support teachers to effectively analyze the characteristics of students' operating equipment and students' learning progress, so as to achieve personalized teaching arrangements. Therefore, the equipment usage data should be sharing by cloud computing, and providing comprehensive analysis of OEE, training big data analysis and rental transaction management.

3   The Sharing Technical System Architecture of IMTE

Through the analysis of the characteristics of IMTE, it is found that sharing model of IMTE has five key characteristics, such as high versatility, high security, high fault tolerance, easy reproducibility and low utilization rate. Through these characteristics, we can analyze some functional requirements corresponding to this mode. The purpose of sharing intelligent training equipment is to improve the interaction efficiency between people and equipment, not only to meet the requirements for equipment status data collection, but also to meet the needs of people to issue instructions for interactive control of equipment.

The industrial Internet technology system is a technology system for the reliable interaction of people and things. Its essence is based on the network interconnection between machines, control systems, information systems, products and people [12]. Through comprehensive perception of data, real-time transmission, and rapid calculation. Processing and advanced modeling analysis to achieve intelligent control, operational optimization and changes in the way production is organized. The industrial Internet architecture itself has the ability to widely collect multi-source heterogeneous data, that is, it can fully support the high versatility of training equipment, and at the same time, it has higher requirements for network security, which effectively ensures the control security of training equipment. Therefore, this paper constructs a sharing system of IMTE by combining the particularity of industrial interconnection infrastructure and training equipment sharing. The architecture inherits the key features of the industrial Internet architecture, such as high reliability and high data security, which weakens the real-time requirements of industrial equipment to a certain extent, and improves equipment control by adding intelligent security control identification at the edge. It ensures the effective application of the sharing mode of IMTE by using the device data to the cloud. As shown in the Fig. 1, the industrial Internet-based intelligent manufacturing training equipment sharing technology architecture is divided into three layers: the edge,

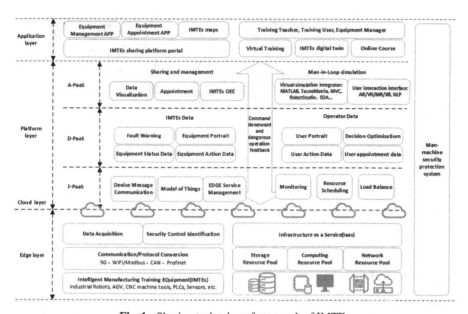

Fig. 1. Sharing technology framework of IMTEs

the platform layer, and the application layer. The following is a detailed introduction to the division methods and functions of each level.

3.1 Edge Layer

The edge layer is close to the data source, which mainly realizes the access, protocol conversion, data collection and security control identification of IMTE. The practical training equipment in this system can communicate with Modbus, CAN, Profinet and other protocols to unify data format and collect data to the edge storage device through 5G, WiFi and other communication methods. High security is a really important characteristics in IMTE sharing system, so by adding security identification of training equipment and control commands in the edge computing device can ensure that the training equipment is not easily damaged by misoperation. As shown in the Fig. 2, the security control identification system mainly realizes the discrimination of dangerous instructions and feedback of equipment failure in the shared mode of smart training equipment. Such as, trainees dangerous operation instructions, equipment physical human interference, equipment power supply sudden power failure and other safety hazards; when trainees operate through the virtual simulation control system, instructions through the platform layer downstream to the edge of the safety control identification system, the system can judge whether the instruction is in safe or in danger through the analysis of the state of the equipment and the position of the robot arm. If it is in a security state, the instruction will be sent to the device and executed, otherwise it will be stopped and alerted that the instruction will lead to danger. If the training equipment fault occurs, the security identification system collects the fault information to alert the user layer of the problem.

Generation and editing of trajectories

There are nine types of trajectory generation methods, hundreds of trajectory parameters and dozens of parameterized trajectory modification methods. There are always one suitable, dozens of parameterized trajectory modification methods, and one is suitable for

Accessible space

The software can not only calculate the reachable space of the robot's flange position, but also calculate the reachable space of the robot's end effector

External tools

There are two common ways to fix the parts: 1. ② Hand held parts and tools fixed

Fig. 2. PQArt Website

3.2 Cloud Platform Layer

The platform layer is the middle layer of the system. It can be divided into three layers according to the type of services: F-PaaS (Foundation-PaaS), D-PaaS (Data-PaaS), and A-PaaS (Application-PaaS). F-PaaS provides platform layer basic services such as

Device Messages Communication, Model of Things, Load Balancing, monitoring and edge management services, and so on. D-PaaS provides big data integration and analysis services, including IMTEs Data and Operator Data. At the same time, the system collects and analyzes operation data such as user behavior data and reservation data. A-PaaS provides artificial intelligence service, virtual simulation platform service and virtual simulation software integration service, and so on. Meanwhile, the data visualization, appointment management and comprehensive efficiency OEE system of equipment built by the big data service of D-PaaS provide corresponding interface service for application layer.

The platform layer uses big data and artificial intelligence technology to analyze the operation reservation records of practical training equipment and students' respectively, and constructs the comprehensive efficiency OEE calculation system of IMTE to analyze and calculate the usage of equipment with different distribution and the ideal usage. Thus, it is available to improve the utilization rate of equipment in terms of deployment of equipment. Big data analysis of students' operations allows for fairer grading and more targeted guidance for students. For the purpose of meeting the generality and easy reproducibility of the training equipment, it can be modularized to build the corresponding object model for each part of the equipment according to the different division of the training purpose. The industrial software operating environment and big data support provided by the platform layer can also provide users of the application layer with a virtual simulation operating platform and a digital twin system to realize the strong interactive features between the equipment and users, so that not only users can conduct virtual training through the virtual simulation platform, but also equipment managers can observe the status of the training equipment through the digital twin system at all times to ensure the security of people or equipment.

3.3 Application Layer

The application layer is the interaction layer between users and the system. The IMTE Sharing Platform portal uses multi-tenant technology to realize virtual partitioning on data and provide customized services for different users. For example, equipment managers and practical trainers can operate the practical training equipment through the system. However, the data between them are independent of each other, and ensure the isolation of data between different users. The map application of IMTE refers to the display of the current status information of practical training equipment in the user's surrounding area through visualization, and the category information displayed to different users is different. In addition, the application of IMTE sharing system also includes applications such as virtual practical training and digital twin, which can also be extended with richer applications through the big data and industrial software integration interface provided by the platform layer.

4 Technologies and Applications

4.1 Model of Things (MoT)

The MoT is an abstracted data model that can unify the data specification of diverse equipment to facilitate the data collection and command downlink, and the MoT is also the basic data model for simulation on the cloud. 'China Mobile IoT Standard White Paper v1.0.0' [13], has defined that 'the thing model is a method of digital semantic description of physical entities, abstracting physical devices as digital models in the cloud'. At present, there are many standard organizations for the thing model, such as ICA, OneDataModel, W3C, Zigbee Alliance, Bluetooth SIG Alliance, OCF, OneM2M, OMA & IPSO and other alliances, which have their own abstract model definition and corresponding description language [13]. Among the mainstream consortia such as 1) Alibaba ICA Consortium defines the thing model as: attributes, services, events, and defines TSL description language; 2) OneDataModel publishes the thing model standard: the ODM abstraction model and SDF description language; 3) CMIC IoT uses JSON as the description language. The MoT is defined as attributes, behaviors, and events. For the behavior explanation is what the device can do; 4) ALL (Industrial Internet Industry Alliance) of the Industrial Internet based on the OPC UA specification information model (3IM), meta-information model as Fig. 3, the object is divided into identification, class and attributes, where the class of service class contains attributes, commands, events, the same as Huawei Cloud's definition of the thing model, respectively, corresponding to the attributes, behaviors, events of the CMIC IoT thing model . And the properties of the objects in 3IM are different from the properties of the thing model, which are divided into static properties and dynamic properties. The device identity is a marker and expression of the device information model, where Sowei Cloud [14] divides the device identity into triads (IPe, address, identifier), which are burned into the gateway or device.

In addition to the definition of the MoT of different union, various MoTs of some Chinese IoT companies are shown in Table 2. Among them, Xiaomi IoT uses four categories, divides services and attributes by function, subdivides and subdivides device functions, and services are device function groups, while attributes are functions. Using such a hierarchical framework is more flexible, and defines a small number of services that can be combined to describe more different products.

Table 2. Some MoT define of Chinese IoT companies

Company	Definition			
Xiaomi IoT	🧍 service	✕ attributes	📑 method	✳ event
Huawei Cloud	✕ attributes	⇆ commands	✳ events	/
Ali cloud	✕ attributes	🧍 services	✳ events	/
CMCC IoT	✕ attributes	🎋 behaviors	✳ events	/
Jingdong Cloud	✕ attributes	📑 methods	✳ events	/
CTWing IoT	⅄ attributes	🧍 services	/	/
Baidu AI Cloud	⅄ attributes	/	/	/

This paper establishes a technical framework based on the industrial Internet, and adopts the vertical structure of ALL-defined information model (3IM) can completely describe all kinds of industrial equipment, which obviously also includes IMTE. Because of the generality requirement, the IMTE object model will generally cover several parts, it involves components such as: motor module, electrical module, control module, intelligent robot module, vision module, sensor module, mechanical module and safety protection module, etc.

4.2 Edge Computing

Edge computing refers to computing the tasks at the end close to the data source. It is aimed at solving the problem of large-scale data processing at the edge, which will cause large computing pressure on the computing center, heavy network bandwidth load, the edge data security, and heavy energy consumption for data transport [15]. The operation object includes the uplink data of the Internet of Everything service and the downlink data from the cloud service [16]. In the IMTE sharing system, the main focus is on the data collection, security control and the convergence of 5G, artificial intelligence, and edge computing.

By establishing a device security model through the edge security system, the edge device state information is collected and analyzed to determine whether the device is in a secure state. Lei W [17] established a security state aware model for assessing the security state of edge devices based on the mapping of data such as storage, communication and computation of edge computing devices to the device state. It is also possible to establish a framework for edge security access and decision making through the edge computing architecture, Junxia Li [17] et al. proposed a software-defined edge computing-based security framework for IoT-enabled healthcare systems. The framework would first require IoT devices to pass a lightweight data validation scheme and then send data from patients to an edge server for processing and analysis, which can be combined with an intelligent software-defined network (SDN) controller for intelligent decision making through edge collaboration.

The IMTE is connected to the edge computing equipment through the 5G network, and it combines artificial intelligence and other technologies to perform edge task scheduling, resource allocation and task offloading, which can provide users with low-latency and high-security IMTE sharing services. Fang Fang [19] et al. summarized some key technologies driving 5G and edge computing. And a win-win model of 5G and edge computing is proposed. Deng Shuiguang [20] et al. divided edge intelligence according to purpose and behavior and discussed related research routes: AI for edge and AI on edge. AI for edge mainly discusses how to use artificial intelligence technology to provide better solutions for the key problems of edge computing. AI on edge discusses how to build artificial intelligence models on edge computing devices. Yang Shu [21] et al. proposed an AI IoT platform based on edge computing with serverless technology. The platform provides a unified service invocation interface and automatically mobilizes resources to meet users' QoE needs. Asim M et al. summarized the application of intelligent computing in key issues of cloud computing and edge computing, such as job scheduling, resource allocation, task offloading, etc. [22].

4.3 Big Data Analysis

Data is part of the core assets of modern information technology enterprises. The common features extracted from a large amount of data can help people to make decisions more accurately. For example, in the IMTE sharing system, the equipment status data, usage data, user reservation data and users behavior data are really important. One can analyze the comprehensive utilization rate (OEE) of the equipment based on the data generated by the equipment. Based on the behavior data of the trainers, one can grade and classify students' operations, and then one can provide corresponding practical training teaching methods for different students' learning situations, so as to achieve the purpose of teaching according to their abilities.

(1) The big data analysis based on equipment status. Li Xinyue analyzed the equipment utilization of 6 usage modes and 4 energy consumption modes of air conditioners in a university in Zhejiang, as well as the usage in 4 years, and studied the relationship between air conditioner usage and energy consumption, which provide support for Energy-saving management and its energy consumption simulation [23], it also provides an analysis method for the equipment analysis mode in the cloud of intelligently-built training equipment. Fengchu Pei [24] provided a feasible solution for online accurate detection of OEE in current equipment-intensive production line clusters for smart shop analysis and optimization. The article uses multilayer perceptron neural network classification for pattern recognition to identify downtime states, and also describes the application scenarios of OEE with its implementation problems in industry. The article also describes the statistical granularity and specific implementation details of OEE, which provides a very suitable reference for the comprehensive utilization analysis of smart manufacturing practical equipment.

(2) The big data analysis of user behavior. Ma Cong [25] evaluated the driving habits of drivers based on OBD technology by establishing a "five-dimensional" evaluation mechanism of driving behavior habits using vehicle driving data. Chen Hui, Bai Jun et al. [26] proposed a MOOC platform user learning prediction framework based on LSTM machine and multi-headed attention mechanism to analyze users' learning behaviors and obtain user and course characteristics, so as to provide a solution to the problem of high withdrawal rate of MOOC platform. To solve the multi-objective flexible job shop scheduling problem with process sequencing, machine selection and employee assignment as sub-problems, Lei Cao [27] et al. proposed the variable neighborhood weed algorithm. The analysis results of the algorithm can provide decision makers to focus on the main problems, assign employees rationally, improve operations and reduce costs. This algorithm, if applied to the smart manufacturing practical training equipment on the cloud, can be useful for teachers to teach students of various learning situations scientifically.

4.4 Man-in-Loop Simulation

Man-in-Loop simulation is a kind of virtual simulation for human-computer interaction, which helps to solve some discipline characteristics such as the low number of physical equipment or some errors examples that cannot be achieved by reality, such as CNC clashing tools. The most important thing is that Man-in-Loop simulation technology can solve the safety problems caused by inexperienced operators of practical training equipment and misoperation. Besides, the Man-in-Loop simulation will contain some strong interaction such as AR, VR, it will enhance the trainee experience and practical operation ability. In general, through the equipment holographic data uplink, the Man-in-Loop simulation system can be established at the application level based on these data and the corresponding equipment modeling. The system mainly covers three scenarios as followings.

The first one is to apply Man-in-Loop simulation to the simulation training before operating real equipment, and simulate certain specific scenarios that are difficult to reproduce. At present, most of the intelligent training robot manufacturers provide corresponding virtual simulation systems, such as the first robot offline programming simulation software in China PQArt (RobotArt) [28], and the core functions are as followings.

The famous foreign robot simulation software robotStudio [29] provides more cutting-edge features: virtual meeting, digital twin, virtual debugging, stop position simulation, augmented reality.

An introduction to RobotStudio features

Virtual Meetings
Is a collaboration feature allowing to share the digital robot solutions in web meetings. The participants are immersed in the virtual room, using a VR headset connected to RobotStudio, where the RobotStudio station can be shared for making design reviews and sales proposals without travelling

Digital Twin
Is a concept to monitor and optimize the automation solution without disturbing the ongoing production. It enables real-time simulation of the production system , like a digital shadow, allowing the users to try changes and do optimization in the virtual world without affecting the production.

Virtual Commissioning
ABBs virtual commissioning solutions speed up commissioning by simulating an exact replica of the production cell in RobotStudio so that all technical issues can be solved in advance. RobotStudio allows to connect to PLCs and other external devices to fully virtually test the complete logic and safety of the cell prior to installing the physical line.

Stop position simulation
Is a feature that visualizes the optimal breaking distance. It simulated the stop position of the robot with millisecond precision for easier use of SafeMove, reduced footprint of cells and for faster and more effective virtual and physical commissioning

Augmented Reality
By using Augmented Reality (AR) technology, you can visualize robot solutions by overlaying the modelled solution over the real-life production environment as a hologram. This is done by visualizing simulations created in RobotStudio through augmented reality glasses or by using our app on a smart phone

Download the RobotStudio® AR viewer app

> iOS
> Android

Fig. 3. RobotStudio Website

The application of Man-in-Loop simulation to student training has had excellent results, such as Gleason [30] et al. trained their basic robotic surgical skills in interns through a virtual reality-based curriculum. The results showed that the usage of a virtual reality-based curriculum improved their robotic skills and also provided objective and automated performance metrics for the trainees across multiple professional standards.

Hardon [31] also designed a robot simulator PoLaRs in the field of robot surgery training. By carrying out 528 experiments on 38 participants, it shows that PoLaRs can also achieve the same simulation training effect. Radi [32] evaluated the feasibility and validity of a new virtual reality robot simulator course. The results showed that the virtual reality course had a high completion rate and excellent feasibility. For the exploration of virtual simulation systems, Zeyang Xia [33] combined the underlying physical models of these two types of simulation platforms so that the rigid-body robot simulation platform like Gazebo and the soft-body robot simulation platform like SOFA, can carry out simulated experiment under the same to architecture. For the impact of virtual reality training on industrial robot operators in manufacturing, Monetti

[34] conducted a drawing and pick-or-place experiment with 24 students in their Unity-based VR model, and the final results showed that 83% of the participants believed that VR helped to familiarize them with real robots, and 75% agreed with the use of virtual tools to train novices.

Secondly, new interactive methods of Man-in-Loop simulation, such as immersive interactive methods, are applied to the intelligent training equipment to enhance the student experience. For example, in the field of robotics, Betancourt [35] proposed an immersive virtual reality method for robotic applications. Bustamante [36] introduced ArmSym, a virtual reality laboratory system, and the results show that ArmSym is able to collect data in prosthetic limbs for immersive experiences. Under the premise of common virtual reality system, Caporasoet proposed a new virtual reality system in paper [37] . With the addition of five surface EMG sensors and a single-axis accelerometer, these inputs give workers awareness of their own physical condition. It is of great significance in the intelligent training equipment for training future workers in the future. In the new interactive mode, the human-computer interaction interface also produces a new mode. Yun [38] et al. proposed a virtual reality-based cyber-physical system (CPS) for creating a 3D operator interface for autonomous human-machine collaboration, which is called I2CPS (immersive and interactive CPS). Huang Sihan [39] et al. summarized and prospected the new generation of operators in the future, and divided the future operation work into strength type, cognitive enhancement type, cooperation type and other types. Equipment sharing is the basis of new operation modes in the future. Virtual simulation of training equipment, digital twins, virtual reality, exoskeleton technology, wearable tracking technology and artificial intelligence technology are the basis of training technology for future operators before operation.

Finally, through Man-in-Loop simulation, the security analysis of the user behavior of intelligent training equipment can be carried out to avoid interference and collision. Among them, ABB's RobotStudio simulation software checks for collisions at programmed points and interpolated paths between them, reducing downtime, enabling remote maintenance and troubleshooting and risk management [40]. And Wei Shen [41] used RobotStudio to simulate the welding trajectory of the ABB1410 robotic arm for virtual simulation design, he performed collision detection on the welding trajectory, detected collisions during movement, and also detected the path reachability. Liqiu [42] et al. used MATLAB and RobotStudio to jointly plan the trajectory of industrial robots to avoid mechanical wear and vibration shocks caused by excessive impulse during motion.

4.5 Security

The security of IMTE can be generally summarized into five categories: device security, control security, network security, application security and data security [43]. As for the sharing of industrial Internet-based IMTE, the security threats mainly exist in the aspects of network communication security, equipment control security and user's operating security. How to ensure user's operating security and device control security is the primary issue to be considered in the security of IMTE.

The security control of the device involves the data security of the command communication process. For this problem, the more popular approach is to introduce blockchain technology into the protection measures of data security. In their literature, Liu Minda [44] et al. summarized the research on the application of blockchain in data confidentiality such as data encryption, authentication, access control, trusted execution, and covert channel. Yuanfei Tu [45] et al. used attribute encryption algorithms to achieve fine-grained non-interactive access control, which effectively controls user privileges and protects data security, dynamically verifies the identification of corrupted data, and ensures the integrity of industrial control system data and communication security. Yankai Sun et al [46] proposed a lightweight private blockchain scheme for industrial control system data integrity protection. Zhu B [47] designed an IoT device monitoring system based on C5.0 decision tree and time series analysis. The system can effectively monitor unknown attacks and further use whitelist matching technique to identify IoT device models and prove the superiority of this method in IoT device monitoring through experiments.

In the IMTE sharing system, it is crucial to ensure the security of the interaction process between users and practical training equipment. By using digital twin technology to simulate the actual environment, the trainers could be more familiar with the operation methods and precautions of the equipment in advance, which can greatly reduce the accident rate. By studying the flaws in the overall interaction system of equipment IoT cloud and applications, observing the details of the interaction between devices, IoT cloud and mobile applications, observing whether the equipment are under attack and the existence of corresponding vulnerabilities, defensive design recommendations can be made to ensure the security of the interaction between the operator and the edge real-world training equipment.

5 Conclusion

This paper starts from the current situation of practical training equipment, summarizes five characteristics of the current situation of practical training equipment by comparing the similarities and differences of college practical training equipment, industrial equipment and chemical equipment, proposes realizing the intelligent management and sharing of practical training equipment through technologies such as Industrial Internet, cloud computing, virtual simulation and big data analysis, and builds the practical training equipment sharing system. A brief overview of key technologies such as MoT, Edge Computing, Man-in-Loop-Simulation, Data Mining and Device Control Security in practical training equipment sharing system is given, and these technologies can effectively support the practical training equipment sharing.

The research and application of training equipment sharing management system based on Industrial Internet is an important issue in the training room construction and the teaching reform in vocational education, and also is a reform and upgrade for the original training room management method utilizing the new technologies such as Industrial Internet, Cloud Computing and so on. Users can enter the user access interface through the application interface provided by the platform layer, and use the applications provided by the system, such as virtual device control, device management and so on. At the same time, device managers can online monitor device information such as the running status, energy consumption information and so on, and then arrange the training courses reasonably. Therefore, the sharing technology of IMTE based on industrial Internet and Man-in-Loop simulation can effectively improve the utilization rate and security, which plays an important role in improving teaching quality, scientific research and social service level.

References

1. Notice of the State Council on printing and distributing the implementation plan of the national vocational education reform. Bulletin of the State Council of the people's Republic of China (06), 9–16 (2019)
2. Khalil, R.A., Saeed, N., Masood, M., Fard, Y.M., Alouini, M.S., Al Naffouri, T.Y.: Deep learning in the industrial internet of things: potentials, challenges, and emerging applications. IEEE Internet of Things J. 8(14), 11016–11040 (2021)
3. Chae, D., Yoo, J.Y., Kim, J., Ryu, J.: Effectiveness of virtual simulation to enhance cultural competence in pre-licensure and licensed health professionals: a systematic review. Clinical Simulation in Nursing 56, 137-154 (2021). (prepublish)
4. The general office of the CPC Central Committee and the general office of the State Council issued the opinions on promoting the high-quality development of modern vocational education. Bulletin of the State Council of the people's Republic of China (30), 41–45 (2021)
5. China education equipment procurement network: in September 2021. the number of school training equipment procurement projects reached a new high, with a ring-on-ring increase of 33% (2021). https://baijiahao.baidu.com/s?id=1714014069930082961&wfr=spider&for=pc
6. Xia, C., Liu, Y.: Construction of university laboratory management system based on information platform. Res. Explor. Lab. 39(11), 246–249 (2020)
7. Ke, H., Jin, R., Liu, Y.: Relying on professional service institutions to innovate socialized service mode of university instruments and equipment. Exp. Technol. Manage. 36(02), 285–288 (2019)
8. Zhu, X., Zhang, G.: Literature review on the management of instruments and equipment used in scientific research in colleges and universities. Res. Explor. Lab. 38(11), 274–277 (2019)
9. Gao, H., Zhang, K., Ni, J., Xu, H.: Experiment equipment energy consumption monitoring platform under IPv6 campus. J. Huazhong Univ. Sci. Technol. (Nat. Sci. Ed.) 44(11), 127–132 (2016)
10. Jiao, W., Shi, G., Liu, L., Yang, E.: Instrument and equipment sharing management system of Chinese Academy of Sciences V3.0. (2016)
11. Yao, X., Jing, X., Zhang, J., Liu, M., Zhou, J.: Towards smart manufacturing for new industrial revolution. Comput. Integr. Manuf. Syst. 26(09), 2299–2320 (2020)
12. Industrial Internet Architecture (Version 1.0) White Paper [EB/OL]. Industrial Internet Industry Alliance (2016)

13. China Mobile Communications Corporation: China Mobile Model of Things Standard White Paper v1.0.0 (2021)
14. sysware cloud. http://www.suoweiyun.com/
15. Shi, W., Sun, H., Cao, J., Zhang, Q., Liu, W.: Edge computing–an emerging computing model for the internet of everything era. J. Comput. Res. Dev. **54**(05), 907–924 (2017)
16. Shi, W., Zhang, X., Wang, Y., Zhang, Q.: Edge computing: state-of-the-art and future directions. J. Comput. Res. Dev. **56**(01), 69–89 (2019)
17. Lei, W., Wen, H., Hou, W., Xu, X.: New security state awareness model for IoT devices with edge intelligence. IEEE Access **9**, 69756–69765 (2021)
18. Li, J., et al.: A secured framework for SDN-based edge computing in IoT-enabled healthcare system. IEEE Access **8**, 135479–135490 (2020). https://doi.org/10.1109/ACCESS.2020.301 1503
19. Fang, F., Wu, X.: A win-win mode: the complementary and coexistence of 5G networks and edge computing. IEEE Internet Things **8**(6), 3983–4003 (2021). https://doi.org/10.1109/JIOT. 2020.3009821
20. Deng, S., Zhao, H., Fang, W., Yin, J., Dustdar, S., Zomaya, A.Y.: Edge intelligence: the confluence of edge computing and artificial intelligence. IEEE Internet Things **7**(8), 7457–7469 (2020). https://doi.org/10.1109/JIOT.2020.2984887
21. Yang, S., Xu, K., Cui, L., Ming, Z., Chen, Z., Ming, Z.: EBI-PAI: toward an efficient edge-based IoT platform for artificial intelligence. IEEE Internet Things **8**(12), 9580–9593 (2021). https://doi.org/10.1109/JIOT.2020.3019008
22. Asim, M., Wang, Y., Wang, K., Huang, P.Q.: A review on computational intelligence techniques in cloud and edge computing. IEEE Trans. Emerg. Top. Comput. Intell. **4**(6), 742–763 (2020)
23. Liu, X., Chen, S., Li, H., Lou, Y., Li, J.: Analysis of air-conditioning usage and energy consumption in campus teaching buildings with data mining. J. Zhejiang Univ. (Eng. Sci.) **54**(09), 1677–1689 (2020)
24. Pei, F., Tong, Y., Yuan, M., Gu, W.: OEE accurate online monitoring for the production line cluster facing with the industrial big data. Computer Integrated Manufacturing Systems, 1–12 (2022)
25. Ma, C.: Study on the Evaluation Method of Driving Behavior Habits Based on OBD Technology. Nanjing University, Master (2016)
26. Chen, H., Bai, J., Yin, C., Rong, W., Xiong, Z.: Behavior based MOOC user dropout predication framework. J. Beijing Univ. Aeronaut. Astronaut., 1–10 (2021). https://doi.org/10.13700/ j.bh.1001-5965.2021.0188
27. Cao, L., Ye, C., Huang, X.: Multi-objective flexible job-shop scheduling based on learning effect. Comput. Integr. Manuf. Syst. **24**(08), 2023–2034 (2018). https://doi.org/10.13196/j. cims.2018.08.014
28. PQArt. https://art.pq1959.com/
29. RobotStudio. https://new.abb.com/products/robotics/robotstudio
30. Gleason, A., et al.: Developing basic robotic skills using virtual reality simulation and automated assessment tools: a multidisciplinary robotic virtual reality-based curriculum using the Da Vinci Skills Simulator and tracking progress with the Intuitive Learning platform. Journal of Robotic Surgery (prepublish) (2022)
31. Hardon, S.F., Kooijmans, A., Horeman, R., van der Elst, M., Bloemendaal, A.L.A., Horeman, T.: Validation of the portable virtual reality training system for robotic surgery (PoLaRS): a randomized controlled trial. Surgical Endoscopy (prepublish) (2021)
32. Radi, I., et al.: Feasibility, effectiveness and transferability of a novel mastery-based virtual reality robotic training platform for general surgery residents. Surgical Endoscopy (prepublish) (2022). https://doi.org/10.1007/s00464-022-09106-z

33. Xia, Z., Chen, J., Gan, Y., Xiong, J.: A coupled model for the simulation of rigid-soft hybrid robot. Robot **43**(01), 29–35 (2021). https://doi.org/10.13973/j.cnki.robot.200050

34. Monetti, F.M., de Giorgio, A., Yu, H., Maffei, A., Romero, M.: An experimental study of the impact of virtual reality training on manufacturing operators on industrial robotic tasks. Procedia CIRP **106**, 33–38 (2022). https://doi.org/10.1016/j.procir.2022.02.151

35. Betancourt, J., Wojtkowski, B., Castillo, P., Thouvenin, I.: Exocentric control scheme for robot applications: An immersive virtual reality approach. IEEE Trans. Vis. Comput. Graph. (2022)

36. Bustamante, S., Peters, J., Schoelkopf, B., Grosse, W.M., Jayaram, V.: ArmSym: a virtual human-robot interaction laboratory for assistive robotics. IEEE Trans. Hum.-Mach. Syst. **51**(6), 568–577 (2021)

37. Caporaso, T., Grazioso, S., Di Gironimo, G.: Development of an integrated virtual reality system with wearable sensors for ergonomic evaluation of human–robot cooperative workplaces. Sensors-Basel **22**(6), 2413 (2022)

38. Yun, H., Jun, M.B.G.: Immersive and interactive cyber-physical system (I2CPS) and virtual reality interface for human involved robotic manufacturing. J. Manuf. Syst. **62**, 234–248 (2022)

39. Huang, S., Wang, B., Zhang, M., Huang, J., Zhu, Q., Yang, G.: Towards human-centric smart manufacturing: framework, enabling technologies and typical scenarios of operator 4.0. J. Mech. Eng., 1–16 (2022)

40. Connolly, C.: Technology and applications of ABB RobotStudio. The Industrial Robot **36**(6) (2009)

41. Shen, W.: Research on virtual simulation design of ABB robot welding operation based on Robotstudio. In: 2020 IEEE International Conference on Artificial Intelligence and Computer Applications (ICAICA), p 894–897. IEEE (2020)

42. Liqiu, Z., Juan, A., Ronghao, Z., Hairong, M.: Trajectory planning and simulation of industrial robot based on MATLAB and RobotStudio. In: 2021 IEEE 4th International Conference on Electronics Technology (ICET) 2021, p 910–914. IEEE (2021)

43. Yu, Y., Chen, Z., Gan, S., Qin, X.: Research on the technologies of security analysis technologies on the embedded device firmware. Chin. J. Comput. **44**(05), 859–881 (2021)

44. Liu, M., Chen, Z., Shi, Y., Tang, L., TCao, D.: Research progress of blockchain in data security. Chin. J. Comput. **44**(01), 1–27 (2021)

45. Yuanfei, T., Yang, G., Zhang, C.: A security scheme for cloud-assisted industrial control system. Acta Automatica Sinica **47**(02), 432–441 (2021)

46. Sun, Y., Zhang, Z.: Industrial control system data integrity protection based on blockchain. Computer Integrated Manufacturing Systems, 1–13 (2022)

47. Zhu, B., et al.: IoT equipment monitoring system based on C5.0 decision tree and time-series analysis. IEEE ACCESS **10**, 36637–36648 (2022). https://doi.org/10.1109/ACCESS.2021.3054044

Modeling/Simulation Applications in Entertainment and Sports

Prediction of Game Result in Chinese Football Super League

Guo Yu[1,2], Jingyong Yang[1(✉)], Xiongda Chen[2], Zhijian Qian[2], Bo Sun[2], and Qingyi Jin[3]

[1] International College of Football, Tongji University, Shanghai, People's Republic of China
yjy777@tongji.edu.cn
[2] School of Mathematical Sciences, Tongji University, Shanghai, People's Republic of China
[3] Shanghai Champion Technology Co, Ltd, Shanghai, People's Republic of China

Abstract. Football is one of the dominant sports in the world and has become a trillion-dollar industry. The top professional football league in China, the Chinese Super League (CSL) with tons of room to grow, is responsible for improving the competitive level of Chinese football and promoting the development of the football industry under the background of big data era. This study tries to predict the result of the game. Among the complete dateset, 1920 team matches of CSL from 2014 to 2017, this study selects 64 variables related to the prediction. The data is fitted with logistic lasso model to analysis and select the variables in the prediction. Then the data set is randomly divided into training set and test set according to the research. With those machine learning classification models trained in the training set and some of them adjusted, this paper finds support-vector machine (SVM) model performing best in the test set, and long-term short-term memory (LSTM) model is applied to predict the outcome of games depending on the data of several previous matches. Then this paper uses the cross-validation method to check and give the validation results. In SVM, the accuracy of prediction reaches 84.54% in the test set, which turns out to be effective. Also, LSTM model gets 63.2% results of the test set and has realistic value for teams. The model in this study is of high credibility for the data from CSL provided by Champion®.

Keywords: Football · CSL · SVM · LSTM · Logistic lasso · Match prediction

1 Introduction

1.1 Significance of Football Data Analysis and Research Needs

Football, which is a technologically complex sport, has become a trillion-dollar industry. The world football industry is known as the 17th largest economy in the world [1]. Football industry accounts for 43 percent of the total output value of the sports industry, far more than other sports industries such as rugby, basketball and volleyball.

The data analysis of CSL football matches has practical significance in assisting coaches in tactical deployment, scientific training of players, selection of players and promoting the development of football lottery industry. Previously, there are some weaknesses in the works on football matches data analysis, varying from imperfect indicators to lacking of data and model construction. By establishing in-depth cooperation with a number of Chinese soccer experts and soccer data company Champion Technology [2], this paper has a deep understanding of the current background of data collection of CSL soccer matches and the urgent need for soccer data analysis.

1.2   Research Status of Football Data Analysis and the Work of This Paper

In traditional football data analysis, due to the difficulty of data collection, a common idea is to select some technical and tactical indicators that are easy to collect for analysis. Moreover, the research can be summarized as follows:

Analysis at the technical and tactical level. The technical and tactical qualities of football players were depicted by directly making factor analysis on some indicators, and factors related to match results were explored by making regression of winning rate factor to them [3]. Generalized linear models were established for some common technical and tactical indicators, and data series inference method was used to find the key indicators for winning the competition [4].

Prediction of competition Results. With the rapid development of statistics and the huge commercial interests of the football betting industry, fans, football workers, the models of the prediction of the outcome of football matches have been constantly explored by experts and scholars.

Based on the requirements put forward by soccer experts, and aiming at the problems of single method and low data utilization rate in current soccer data analysis, this paper mainly does the following work: (1) Clean a large amount of data of Chinese Super League and reduce the dimension of variable classification. (2) Logistic Lasso method [5] was used to explore the factors affecting the outcome of the competition. (3) Establish a long-term short-term memory (LSTM) [6] neural network model capable of processing serial data to predict the winner and loser of Chinese Super Games.

1.3   Data Processing

Data source and basic data information. The original data is from the machine recognition statistics of the CSL match video by Champion Technology, which includes all team data when they have the ball of each CSL match in 2014 to 2017. All the team data contains 1,920 pieces of information and 175 variables. Each team plays 30 games per year, which means each team generates a message for each game. The player data contains 26,288 pieces of information and 101 variables. However, in the original data, there are many variables and some variables are highly correlated, which requires preliminary screening and correction of the original data.

Preliminary data collation. Among the complete dataset, including winning and losing matches in the home and away field, this paper selects 64 variables related to the prediction: pass, set piece, attack, confrontation, defense, fault and foul, goalkeeper specific, organization and so on. After collecting and selecting data, this research excludes some variables leading to the winning and losing of the match directly. All the variables are standardized first, and then divided into training set and test set randomly according to the ratio of 7:3. The training set is fitted with logistic lasso model, and then the paper uses the cross-validation method to check and give the validation results.

Factor analysis. The purpose and use of factor analysis are similar to principal component analysis. Factor analysis can be regarded as the extension of principal component analysis. Factor analysis refers extracts common factors from variable groups, where common factors refer to the hidden factors inherent between different variables.

By means of the FA function of Psych package in R language, the factor rotation method is selected as the maximal rotation of variance [8].

2 Analysis of Factors Affecting the Outcome of Competition

What determines the outcome of a match is very important to the managers and owners of every football club. Understanding these factors can not only increase the training for the team's key weaknesses in the daily training; In addition, it can arrange defense tactics and organize more targeted defense.

The transformation of this problem into a statistical problem is a variable selection problem: on the premise of ensuring the accuracy of the model, features can be screened. Variable selection becomes an important part of statistical modeling for three reasons:

a. In practical problems, usually, in order to reduce the low accuracy of the model due to the lack of important variables, as many variables as possible are chosen. Therefore, the problem of dimension disaster may happen. If some important variables from a large number of variables can be selected to build the model, the dimension disaster problem can be greatly reduced.
b. Good variable selection can remove irrelevant features and reduce the difficulty of modeling. The selected variables can also have a good explanation of problems, that is, the explanatory ability of the model is enhanced.
c. If all variables are put into the model without variable selection, although the accuracy of the model is guaranteed, the model is prone to overfitting: the performance of the model on the test set is far lower than that on the training set, that is, the generalization ability of the model is poor.

In statistical regression or classification problems, Forward Selection, Backward Selection and Stepwise Selection are three different methods of selecting variables depending on indicators. Methods of variable selection based on penalty term include elastic network (ridge regression and Lasso regression are special cases), SCAD and MCP.

In this paper, the Logistic Lasso model is adopted and implemented with the help of R language's 'Glmnet' package based on Coordinate Decent [9].

2.1 Logistic Lasso Model Steps

Data set processing steps:

1. Based on the original data not processed in Sect. 3, 64 variables are selected according to the following dimensions: passing, set piece, attack, confrontation, defense, error and foul, goalie characteristic, organization, home and away. For the 63 variables except home and away, the average data of each team in this season was calculated and added into the independent variable as a new variable, so the total number of independent variables was 127. There are no variables directly related to the outcome of the match, such as the number of shots, assists, key passes, on target, off target and so on. Because these variables are obviously related to the results, the variables selected by this model are of little significance for practical guidance. Therefore, these variables are not chosen into the model.
2. The matches with the winning or losing result are selected;
3. Standardization was carried out for continuous variables, classification variables such as home and away ('HOME_AWAY') and match results were converted into factor type, and the data set was randomly divided into training set and test set in a 7:3 ratio (Figs. 1 and 2).

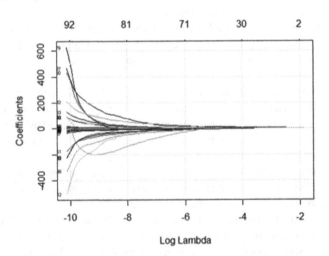

Fig. 1. Path diagram of coefficient solution of Logistic Lasso method.

With the coefficient solution path diagram and cross validation, the value of lambda is selected. The order in which each variable enters the model can be seen. In the coefficient solution path diagram, on the far right of the horizontal axis is the first variable to enter the model, which can also be regarded as the most important variable in the diagram of cross validation, the best value of lambda is at the lowest point of the red curve. At this point, the number of variables should be 27. The dotted line to the right of the minimum represents a concise model within one standard deviation. The corresponding number

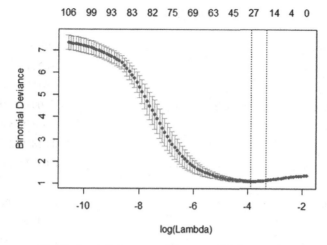

Fig. 2. Logistic Lasso model cross validation results.

of variables is 19. In order to reduce the error of the model, the value of lambda at the lowest point is selected, and the result of variable selection is (Table 1):

Table 1. Logistic Lasso model variable selection results.

The variable name	Coefficient
The success rate of forward passes	7.425791097
The passing rate in midfield	5.842123565
Successful long pass rate	2.07735183
Home and away, H at home, G away	1.081262211
Backcourt possession rate _ season	0.788538398
Successful free kick	0.720138043
Successful corner kick rate _ season	0.277156788
The goal ball goes over half court	0.168423815
Number of yellow cards	0.121007645
1/3 zone offensive count _ season	0.103950278
The pass success rate in the 30-m zone	0.072354657
30 m indirect free kick _ season	0.072143579
Backcourt foul	0.039971124
Number of clearance	0.035884594
Tactical corner kick, first place outside the penalty area	0.034780365

(continued)

Table 1. (*continued*)

The variable name	Coefficient
Faults _ season	0.02442884
foul	0.02138589
Number of successful passes in the backcourt	−0.000453545
Blocking pass	−0.004583549
Cross	−0.018550273
Free kick up front	−0.040730134
Blocking shot	−0.098871118
Blocking shot	−0.126598985
Goalkeeper diving	−0.147002229
Number of red cards	−0.772137098
Backcourt pass success rate	−0.817451517
An own goal	−1.413176222
Intercept	−16.25163566

To sum up, the above table has a positive impact on the outcome of the game: the success rate of passing in the front, the success rate of passing in the middle, the successful long pass rate, home and away variables; Negative effects on the outcome of the game are: own goal, backcourt pass success rate, red card number, goalkeeper save and other variables. The negative effect of the success rate of passing in the back court on the victory of the game can be explained as follows: when the attacking team have the possession, it cannot advance to the front court to organize the attack quickly due to the lack of attacking organization ability; the team is forced to frequently pass the ball in our backcourt to avoid mistakes when under pressure from opponents; This leads to a generally high pass-rate in the backcourt for the weaker teams (Table 2).

Forecast the test set:

The prediction of the test set can obtain the confusion matrix as follows:

Table 2. Confounding matrix of Logistic Lasso model prediction results.

Classification	Actual Value		
Prediction Outcome		+1	−1
	+1	40	21
	−1	13	41

ROC curve:

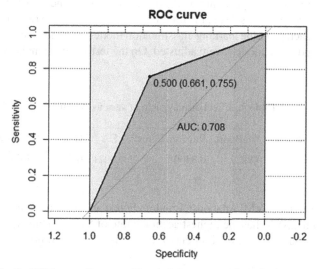

Fig. 3. ROC curve diagram of Logistic Lasso model prediction results.

At this time, the AUC is 0.708, indicating that the variables screened by the model are effective for classification (Fig. 3).

2.2 Similarity Measure of Data

In general, variables can be divided into the following three categories according to the measurement scale: continuous variable, categorical variable and Ordinal.

For different data types, different measures are used to measure the similarity between the two variables. Distance is often used as a measure of similarity between samples, while similarity coefficient is often used as a measure of similarity between variables.

The commonly used distance measures are the Minkowski Distance, the Canberra distance and the Mahalanobis Distance [10].

3 Predict the Outcome of the Chinese Super League Competition

With the continuous development of machine learning and deep learning, there are more and more application scenarios. For the prediction of football matches, most scholars and professionals in the industry use Logistic regression, BP neural network and other classification models for the prediction, but the effect is not ideal, and it is difficult to guarantee the authenticity of data and results. Football is a very complex sport and it is difficult to make a very accurate prediction of the outcome of a match, especially when comprehensive data are difficult to obtain. In this chapter, the data collected by Champion Technology will be used to predict the outcome of the competition by using popular machine learning methods and deep learning methods.

3.1   Triple Classification Models [14]

After processing the statistical data of 1920 games of the Chinese Super League from the 14th season to the 17th season, the training set and test set were randomly divided in a ratio of 7:3. Each machine learning classification model was trained on the training set respectively, and some models were adjusted. On the test set, each model is represented as (Table 3):

Table 3. Effect display table of various classifiers.

SVM	Decision Tree	AdaBoost	Random Forest	LDA	QDA	Naïve Bayes	KNN
0.8454	0.7743	0.8229	0.8300	0.8211	0.5023	0.7621	0.6093

It can be seen that SVM, Adaboost, Random Forest and LDA have high classification accuracy, and the SVM has the highest accuracy rate of 84.54% (Fig. 4).

The confusion matrix of the four classifiers is visualized as follows:

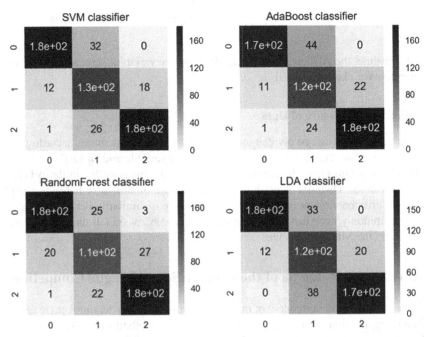

Fig. 4. Confusion matrix visualization of SVM, Adaboost, RandomForest and LDA.

As can be seen from the figure above, SVM has the best performance in predicting the winner and the winner. The prediction performance of random forest is similar to THAT of SVM, but slightly lower than that of SVM. In general, Adaboost, random forest and LDA have similar performance.

It can be seen from the above that, if the data performance of the two sides of the match is known (variables related to the result have been removed), the final result of the match can be judged better through the classifier.

Since the technical statistics of the two sides can't be obtained before the game, we cannot predict the winner with above model. Therefore, in this paper, 480 matches in the 14th season according to the three keywords are ranked by seasons, round number and team ID, and the following attempts are taken according to the suggestions of football experts:

Method 1. The average of the first three or five games of the two teams are taken as a prediction of the next performance, and this prediction is used as an input to the classifier that has been trained to predict the outcome of the match.

Method 2. The first three or five games of the two teams are used as input, training classifier; and results are predicted with this classifier.

Method 3. The team's performance is regarded as sequence data, training sequence model LSTM to predict the team's data performance, and the prediction is used as input in the above trained classifier to predict the result of the match.

Method 4. Same as method 3, the team performance was treated as sequence data to train the sequence model LSTM, but at this time, the last layer of the sequence network was changed to Softmax classification layer, and the result of the next match was directly judged.

3.2 LSTM Prediction Model

LSTM. Long-term short-term Memory (LSTM) model is a special type of RNN network (Recurrent neural network) [11]. RNN is the generic term for a series of neural networks capable of processing sequential data. In many cases, the LSTM has been a great success and has been widely used.

LSTM networks are easier to learn long-term dependencies than simple circular architectures. It can be used to test artificial data sets with long-term learning ability [12] and has achieved the most advanced performance in challenging sequential processing tasks [13].

With the help of football experts, this paper assumes that the technical and tactical performance of a team is a series of data, that is, the next data performance of a team has a strong correlation with the data of its previous matches. Based on this assumption, LSTM model is established and obtains good results.

LSTM prediction of the outcome. In the test set for the 2014 season, there were 125 games, 48 games for 0(minus), 36 games for 1(draw), and 41 games for 2(win). Among the results of the above four methods:

Method 1 was a complete failure. No matter taking the average data of the first three games or the first five games as input, the output results of the classifier were all 2(wins). In this case, the accuracy rate was 32.8% of the test set for the matches with the winning results.

In method 2, with the first three matches as input, the best classifier achieved 48% classification accuracy in the test set.

In method 3, the last layer of LSTM uses mean square error as the error measure, and the Keras framework command is: model. Compile (loss = 'mean_squared_error', Optimizer = 'Adam'). The output of this model is taken as the input of the classifier mentioned above. The result is the same as that of method 1, and the match result is judged as the winner.

In method 4, the last layer of LSTM is Softmax layer. Cross entropy is used as the measurement of error. The command of Keras framework is: Model.add ('Softmax'). Model. Compile (loss = 'categorical_crossentropy', optimizer = 'adam', metrics = ['accuracy']). By adjusting the parameters, the accuracy of the model in the test set is 63.2%.

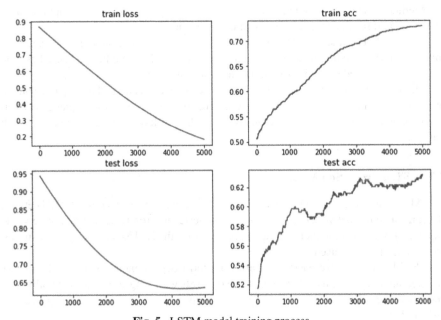

Fig. 5. LSTM model training process.

In Fig. 5, the 'loss' shows MSE of the results of the model, while 'acc' means the accuracy of the prediction, which both change as the number of iterations increases.

4 Results and Conclusions

In this paper, the data of Chinese Super League provided by Champion Technology Company is cleaned and sorted. Screening the variables by the Logistic Lasso model and determining the parameters through cross verification, the key factors to win the game and unfavorable factors are found. Four classifiers, including SVM, Adaboost and Random Forest, performed well with the highest accuracy of 84.54%, which indicates that there is a strong correlation between the team performance data collected at present and the result of the match. In LSTM model, the data of the previous three matches of the were input and the results of the matches were output. The accuracy rate reached 63.2% in the 2014 season test set.

Different from previous about predictions of match, this paper introduces Logistic Lasso model to select variables, also adds LSTM, a time series model to describe the continuous performance of a team. After training the complete model, this paper finds the combination of Logistic Lasso and SVM has the highest accuracy. What's more, the researchers hope to predict the outcome of a match before its beginning, which means those data haven't existed. So, this paper uses data of several previous matches with LSTM to accomplish this goal, which shows good results.

In the future, faced with a huge number of variables in one football match, researchers, coaches, and players will have more confidence to analysis and forecast the results based on this model. Alternatively, they'll be able to adjust the tactics in game referring to the coefficients of variables. Even before a game, the outcome of it can be predicted. Though the accuracy of this prediction LSTM isn't as high as the Lasso-SVM model, LSTM model gives the advice and opportunity for the team to prepare in advance.

References

1. Zhang, J.: On the Chinese football industries. Sport Sci. **01**, 1–4 (2001)
2. Shanghai Champion Technology Co., Ltd., Shanghai. http://www.champdas.com
3. Wang, K., Lv, X., He, J.: Regression analysis of soccer players' technical and tactical quality level and winning rate factor. J. Capital Inst. Phys. Edu. **24**(2), 146–150 (2012)
4. Liu, H., Peng, Z.: Big data analysis of soccer skills and tactics performance – based on generalized linear model and data series inference method. J. Sport **24**(2), 109–114 (2017)
5. Ao, X., Gong, Y., Li, J.: Football match result prediction based on disc data. J. Chongqing Technol. Bus. Univ. **33**(6), 85–89 (2016)
6. Alcaraz, J.C., Moghaddamnia, S., Peissig, J.: Efficiency of deep neural networks for joint angle modeling in digital gait assessment. EURASIP J. Adv. Signal Process. **2021**(1), 1–20 (2021). https://doi.org/10.1186/s13634-020-00715-1
7. Jolliffe, I.T.: Principal Component Analysis and Factor Analysis, pp. 129–135. MIT Press, Cambridge (2004)
8. Mulaik, S., Hirsch, J., Schonemann, P.: Multivariate Behavior Research, pp. 159–171 (1992)
9. Hastie T., Tibshirani R., Friedman J.: The Element of Statistical Learning, 2nd edn., vol. 192. Springer, Cham (2009)
10. Nielsen, F.: The statistical Minkowski distances: closed-form formula for Gaussian mixture models. In: Nielsen, F., Barbaresco, F. (eds.) GSI 2019. LNCS, vol. 11712, pp. 359–367. Springer, Cham (2019). https://doi.org/10.1007/978-3-030-26980-7_37
11. Aydin, O., Guldamlasioglu, S.: Using LSTM networks to predict engine condition on large scale data processing framework. In: International Conference on Electrical and Electronic Engineering, pp. 281–285. IEEE (2017)
12. Gers, F., Schmidhuber, J., Cummins, F.: Learning to forget: continual prediction with LSTM. Neural Comput. **12**(10), 2451–2471 (2000)
13. Hochreiter, S., Schmidhuber, J.: Long short-term memory. Neural Comput. **9**(8), 1735–1780 (1997)

14. Chand, N., Krishna, C.: A comparative analysis of SVM and its stacking with other classification algorithm for intrusion detection. In: 2016 International Conference on Advances in
Computing, Communication and automation (ICACCA 2016), pp. 40–45 (2016)
15. Chen, X., Jin, Q.: Factors' analysis and result prediction of Chinese super league. In: 25th
Annual Congress of the European College of Sport Science (2020)

Author Index

Printed in the United States
by Baker & Taylor Publisher Services